Digestive Physiology and Nutrition of Ruminants

Volume 3—Practical Nutrition
Second Edition

by
D. C. Church, Ph.D.
Ruminant Nutritionist
Senior Author & Editor

ISBN 0-9601586-7-7

Published and Distributed by

O & B Books, Inc.
1215 NW Kline Place
Corvallis, Oregon 97330
United States of America

Telephone 503-752-2178

TABLE OF CONTENTS

Table of Contents of Other Volumes of

DIGESTIVE PHYSIOLOGY AND NUTRITION OF RUMINANTS

Volume 1 – Digestive Physiology, 2nd edition, 1975

Volume 2 – Nutrition, 2nd edition, 1979

Chapter 1—Introduction

by D.C. Church

For the reader who has read Volumes 1 and 2 of DIGESTIVE PHYSIOLOGY AND NUTRITION OF RUMINANTS or any other works on ruminant animals, it is apparent that a tremendous amount of research effort has been expended on these species in recent years. While most of the subject matter covered in these volumes has dealt with the more basic aspects of digestive or nutritional physiology, it should be pointed out that there has also been a considerable amount of effort directed toward the more applied aspects of feeding and nutrition of ruminant animals. The reasons for this, obviously, are that ruminants supply things that man needs, wants, or likes. Meat, milk and cheese provide a very substantial amount of the human diet in many countries. Also, various by-products such as wool, leather, animal feed, fertilizers, tallow, pharmaceuticals, etc., find their way into commercial channels at competitive prices with similar products from other sources. In addition to the uses just mentioned, various species of bovines, buffalo, camels, and camiloids still serve as draft animals in some underdeveloped countries; however, this usage is rapidly diminishing. Game and domestic species managed for recreational purposes make up a small but increasing percentage of the total number.

Although it may be true that animal products are not essential dietary components for humans other than the young, meat, milk and cheese are very satisfying to the palate. In addition they provide high quality proteins, vitamins, minerals and energy. No doubt these nutrients could be supplied more economically from plant sources if one thinks only of the economic aspects of buying food in the market today. However, the satiety value of animal products may also satisfy some psychological urge, therefore adding to man's appreciation of and desire for meat.

Ruminant animals, at least some species, have the ability to exist and thrive on a tremendous variety of diets. For domestic cattle their diets may range from very low quality roughage which may not allow maintenance of body weight to a feedlot ration that may be 90+% grain. The rumen allows the animal to eat such a diverse diet and survive. Thus, domestic ruminants typically are fed relatively large amounts of diet ingredients not suitable (in any quantity) for man, poultry or swine. Secondly, ruminants harvest forage from millions of acres of rough and arid land areas that cannot be harvested economically except by herbivorous animals. If these lands were not grazed by ruminant species, very little food of value to man would be produced on these lands.

Lay persons unfamiliar with livestock production are often prone to point out that ruminants, particularly fattening beef, consume vast amounts of grains or other feedstuffs which could be used directly by humans for food. The suggestions are also put forward that swine or poultry, which show better feed conversion, should be grown instead of beef or sheep. The efficiencies of different forms of livestock production are shown in Table 1-1. Note that these values include the cost of rearing and maintaining breeding stock as well as the final production phase. Data in this table illustrate that beef production is inefficient as compared to the others shown. However, milk production is at least as efficient as most other forms of animal protein production.

One thing to remember is that as much as 35% of the total beef going to the market goes directly off of grass, either as grass fat young cattle or as cull breeding animals. Most of these would have received only very small amounts of concentrate feed. The same comment applies to sheep as a much higher percentage of lambs are marketed directly off the ewe and never see the feedlot. Table 1-2 shows the usage of grains as livestock feed as of 1975. The data in this table point out that all grain usage only accounts for about 26% of total feed usage.

Even though feedlot cattle or dairy cows may consume relatively large amounts of grains or other concentrates, a substantial amount of the total feed consumption is not considered to be edible for humans. The effect of this on efficiency of production of

Table 1-1. Overall efficiency with which farm animals produce protein.[a]

Product	Level and/or rate of output	Grams of protein produced/Mcal of DE input[b]
Eggs	200 eggs/year	10.1
	*236	12.6
	250	13.7
Broiler	*3.5 lb/12 wk; 3.0 lb feed/lb gain	11.9
	3.5 lb/10 wk; 2.5 lb feed/lb gain	13.7
	3.5 lb/8 wk; 2.1 lb feed/lb gain	15.9
Pork	200 lb; 6 lb feed/lb gain	5.0
	*200 lb/6 mo; 4 lb feed/lb gain	6.1
	200 lb; 2.5 lb feed/lb gain	8.7
	Biol. limit (?); 2 lb feed/lb gain; no losses	12.1
Milk	7,936 lb/yr; no concentrates	10.5
	*10,900 lb/yr; 22% of energy as concentrates	12.4
	11,905 lb/yr; 25% of energy as concentrates	12.8
	20,000 lb/yr; 50% of energy as concentrates	16.3
	30,000 lb/yr; 65% of energy as concentrates	20.5
Beef	*1,100 lb/15 mo; 8 lb feed/lb gain	2.3
	1,100 lb/12 mo; 5 lb feed/lb gain	3.2
	Highly intensive system; no losses	4.1

*Data represent average USA situation under good management conditions.
[a]From Reid et al (1978)
[b]These data include the dietary energy cost of rearing and maintaining the breeding stock, and of reproduction and mortality, as well as that of production itself.

Table 1-2. US grain production and animal feed usage in 1975 (millions of tons).[a]

	Feed grains			Food grains			
Item	Corn	Sorghum	Oats & barley	Wheat	Rice	Rye	Total
Produced	161.3	20.8	19.5	57.2	4.8	0.6	264.2
Used as feed	99.6	14.3	13.2	1.6	0	0.2	128.9
Percent as feed	61.7	68.8	67.7	2.3	0	33.3	48.8

Total livestock feed

Item	Grains	By-products	Harvested roughage	Pasture	Total
Million tons	128.9	36.6	89	243	497.5
% total feed	25.9	7.3	17.9	48.8	100.0

[a]Based on USDA statistics for 1975.

Table 1-3. Inputs and returns from animal production.[a,b]

Product	Total energy & protein				Human edible energy & protein			
	Energy		Protein		Energy		Protein	
	Input, Mcal	Return, %	Input, kg	Return, %	Input, Mcal	Return, %	Input, kg	Return, %
Milk	19960	23.1	702	28.8	4555	101.1	111.5	181.4
Beef	20560	5.2	823	5.3	1869	57.1	39.9	108.8
Swine	1471	23.2	66	37.8	588	58.0	29.0	86.0
Poultry	23.2	15.0	1.2	30.0	11.2	31.0	.48	75.0

[a] Data from Bywater and Baldwin (1979)
[b] Inputs are calculated as DE and digestible protein and include cost of maintaining breeding herds and flocks.

edible energy and protein is illustrated in Table 1-3. Note that milk production results in a net increase in both edible energy and protein and that beef production results in a net increase in edible protein. This means of expressing efficiency is more appropriate to illustrate the competitive nature between animals and humans for edible materials. There are, of course, a number of other considerations that go along with potential shifts in feedgrain to foodgrain production. These have been discussed in more detail by Bywater and Baldwin (1979).

Thus, we see that ruminant species occupy an important niche in modern agriculture. Time will determine how ruminants will fit into the total picture as the world becomes more complex. If the human population continues to proliferate at its current rate, it seems most likely that the feeding of large amounts of cereal grains in feedlots—such as is the current practice in the USA—will decrease also. However, there are many areas in the world where much larger numbers of animals could be produced if adequate forage were available or if diseases and pests could be controlled. It seems likely, therefore, that ruminants will continue to be vitally important to man for many decades to come.

IMPORTANCE OF ADEQUATE NUTRITION

There are a variety of factors operating in today's agricultural businesses which make it imperative that the nutrition and feeding of animals be at least reasonably adequate. In the USA, as well as in many other countries, the over-riding factor is the cost-price squeeze on producers. As a general rule, the price of animals and by-products from them, if corrected for inflation, may have increased slightly over the past couple of decades. Some products—such as wool and leather—may have even decreased in price due to pressure from synthetics. On the other hand, the cost of labor, land and capital has gone up at a much more rapid rate with the result that many small or inefficient operators have been forced out of the business or suffer from a reduced income.

The typical producer response to low prices in an agricultural market is to increase production or to improve the efficiency of production. Thus, the livestock producer is forced to take a reduced income, to accept a reduced income/unit marketed, or to become more efficient if he is to remain in business. Gross efficiency may usually be improved by achieving a higher turnover rate. In this connection, it should be pointed out that finishing cattle in well-run modern feedlots may be expected to gain over 3 lb/day (1.5 kg) at a feed conversion of about 7 units of feed/unit of gain as compared to about 2 lb/day (ca. 1 kg) at a conversion of about 8-9:1 a couple of decades ago. Comparable changes have occurred in the dairy industry. Some of the factors bearing on the need for adequate nutrition in situations demanding high production and efficiency are discussed in succeeding paragraphs.

Development of Large, Specialized Operations

With respect to beef cattle, there has been a tremendous growth of large feedlots for finishing cattle over the past 30 years. These lots supply an increasingly larger and larger share of the total market and are somewhat reminiscent of the poultry industry a few years ago. Feedlots are currently in operation that have a capacity of around 100,000 head of cattle, and there are a great many in operation that have a capacity of 35,000 to 50,000 head. With the small producer, particularly a farmer-feeder, a slight reduction in feed conversion or a reduced rate of gain or a slightly higher cost of feed may not seem important. However, for the large feedlot that has 50,000 hungry mouths to feed each day, a small change can mean a tremendous difference in operating expense and profit in a year's time. Thus, these large operations find it economical to take advantage of many things that the smaller operator cannot use or may not consider feasible. Many of the large lots use highly specialized nutrition and management consultants to advise on feeding practices, nutrition, and overall management of the animals they feed. In a well-run operation the result is a rapid turnover at relatively low profit per animal.

In the dairy industry there is also the same trend for development of larger and larger operations with the tendency to maintain large herds of cows relatively close to metropolitan areas. Some of the operators buy freshening heifers or cows and milk them only for a short time—perhaps for only one lactation. The animals may be maintained exclusively in drylots and are culled quickly if they don't produce up to expectations.

With respect to veal calves, there is a trend for specialization here, also. Management practices and nutritional knowledge have developed to the point that young calves can be raised with relatively few disease problems and so that they will grow rapidly and produce a high quality of veal for specialized markets.

In sheep feeding, the trend toward a few big lamb feeding operations has been evident for many years. Generally, lambs that are not finished sufficiently by the end of summer grazing are often purchased by feeders who finish them in big lots as required for the market. In the eastern part of the USA, there is a trend toward more frequent lambing and weaning of younger lambs which may be finished over slotted floors, a procedure which helps to control internal parasites.

Each of the situations mentioned—where large numbers of animals are confined and pushed to gain rapidly—may be expected to make the nutrition of the animal a more critical factor than if production were maintained at a lower level in less crowded conditions.

Greater Stresses on Animals

Several factors may contribute to greater stressing effects on animals in today's agriculture. Of these, the movement of animals throughout the world by ships, air, and ground transportation may result in very rapid dissemination of disease organisms and their vectors.

Close confinement of large numbers of animals is more apt to result in disease outbreaks. In addition, the close confinement must, in many cases, result in psychological and sociological stresses that have unknown effects on animal performance and nutrient requirements.

The process of assemblying large numbers of animals from diverse backgrounds must result in a variety of stresses. Animals may be transported over long distances in adverse weather. Their previous exposure to disease and parasites will differ, and sudden confinement with strangers are all adverse factors. Many range cattle or sheep have never consumed any feed but forage, and suddenly they are confronted with strange surroundings, strange feed in bunks, water in odd containers, and are subjected to loud and totally unfamiiar noises. The stress of weaning may also be imposed on top of these other stresses.

In many of the large beef feedlots it is the practice to dehorn and castrate where required, brand, inoculate for a variety of diseases, administer vitamins of various kinds and, perhaps, treat for stomach worms or other parasites in one quick trip through the squeeze chute—all this being done to an animal that probably has not recovered from stresses to which it has previously been exposed. If nothing else, the ability of cattle to withstand such a combination of stresses convinces one that they are extremely hardy animals. Furthermore, most of them recover and do surprisingly well in subsequent months in the feedlot. Death losses, as a whole, are remarkably low in well-run feedlots.

Economic Pressure to Increase Plant Production

Agricultural production of various plant products faces the same type of cost-price squeeze that applies to the animal industries. As a result, one solution to this problem is the application of increasing amounts of fertilizers and of cultural or management practices that serve to increase yields. One possible result of the increased yields is the production of plant tissues which may be deficient in one or more of the mineral elements, particularly the trace elements. In the western parts of the USA, long-time irrigation of sandy soils may result in the same thing, resulting sooner or later in malnutrition of animals which consume the plant tissues.

Fertilization under some conditions may result in toxicities, as well. One example of this may occur from the heavy use of N fertilizers, particularly on soils borderline or deficient in S. In this situation the plant tends to accumulate much higher levels than normal of non-protein amino acid N as well as nitrate-N. Plants grown under these conditions can easily accumulate enough nitrate-N to be toxic. A second illustration applies to acid soils where either Mo or Ca will result in an increased production of legumes. Mo applications may be used without markedly increasing plant tissue levels or at least not to an objectional level. However, when both Ca and Mo are applied, the plant then may accumulate toxic quantities of Mo. This would, of course, reduce performance of animals consuming such forage in quantity.

These examples illustrate some of the problems that may occur as a result of nutrient imbalance in the soil, although the frequency of occurrence is poorly documented. Undoubtedly, many other analogous but unknown situations occur from time to time but are not recognized at this time. The incidence of such situations will increase with time unless remedial measures are taken.

Increasing Use of Waste and By-product Feedstuffs

In an effort to hold down the cost of production, most animal producers are increasingly turning more and more to the use of any apparently useful product which is competitive price-wise that the animal will consume without apparent harm. Materials having potential value for feed include garbage, sewage, secondary bacterial sludge from many different sources, poultry litters, manures, papers, wood and bark products, etc. Currently, several experiment stations are vigorously working on the utilization of poultry and ruminant manures which are beginning to present disposal problems in some heavily populated areas. Numerous papers have appeared in recent years dealing with use of wood products, paper, garbage, sewage sludge, etc. A considerable effort on a world wide basis has been expended to develop means of improving usage of low quality fibrous material such as straw, corn stalks, etc.

Many feeders now use field and cannery residues of crops raised for human consumption as well as residues of crops grown for seed or other uses. In the area where the author lives—the Willamette Valley of western Oregon—large acreages of land are used for the production of grass seed of various kinds. The general practice is to burn off the grass fields as a means of disposing of excess straw and for disease and pest control. This, of course, produces a lot of smoke and the non-farm inhabitants have prevailed upon the legislature to force a deadline when fields may no longer be burned. Consequently, there is pressure to provide some means of using this low quality straw for animal feed or other alternate uses. Other by-product feeds used in this area include mint silage (residue left after extracting mint oil), pea-vine silage and hay, spent hop residue, and grass seed screenings. A similar list with different products could be given for any area in the world where animals are produced.

The various examples just mentioned of different kinds of materials that are or may be used as feed illustrate the diversity of potential feedstuffs available to feeders in some areas. The nutrient content of these feeds-stuffs is also extremely diverse and will require considerable care in developing rations and feeding programs to take optimum advantage of these different feed sources. As an example, the mint silage mentioned previously has a chemical composition similar to that of alfalfa hay or silage. However, digestion trials indicate that the digestibility of the protein is only about 25%, probably due to the heating needed to extract the mint oil. This clearly reinforces the thought that competent nutritional advice is often required to take advantage of many of these by-product feedstuffs.

Increasing Use of Synthetic and Purified Products

There has been a tremendous increase in the usage of urea over the past 30 years in ruminant rations. While this has often resulted in a reduced cost of crude protein in the formula, it has not always resulted in efficient utilization of the added N due to limiting factors, such as readily available carbohydrates, or to any of several required mineral nutrients. Inappropriate use has often resulted in toxicity, as well. However, there is no reason to believe that usage will not increase, but this will require, again, competent nutritional wisdom to take advantage of the lower cost of N from such products. Other examples of this type may be expected to come up from time to time in the future.

DEVELOPMENTS IN THE FIELD OF RUMINANT NUTRITION

It might be of interest for some of the readers of this book to discuss briefly some of the developments in ruminant nutrition that have taken place in relatively recent times. It is not the intent to discuss this subject point by point, but rather to cover it in broad perspective. Detailed documentation of this type is available in many technical journals. One article that covers a 50-year period is presented in the Journal of Animal Science, Vol. 17, Number 4.

Importance of Basic Research

The art of feeding animals is gradually being replaced by the sciences of nutrition, management, and disease control, although the feeder who is observant and practices applied animal behavioral methods still has much to recommend him. It is true that some practical livestock people without much knowledge of nutrition have, on occasion, come up with new applications of nutrition or management that have met with very good success. These have usually been innovative men who are observant and resourceful. An example here that might be used is the widespread use of vitamin A under feedlot conditions. Where many scientific people believed that the vitamin should not be required in the amounts being used, practical usage showed a good response.

Nevertheless, this writer is convinced that the bulk of the progress in ruminant nutrition in recent years and most of whatever may come in the future has been or will be largely due to the development of a better understanding of animal metabolism and nutritional knowledge of the type that has been presented in Vol. 1 and 2. The importance of basic research to point the way for improvements in practical nutrition or to find the answers to practical problems can be documented many times over. Recent examples that come to mind include the story of Se, vitamin E and nutritional muscular dystrophy (see Ch. 5 and 9, Vol. 2). Other examples might include the development and application of knowledge on rumen fermentation, etc., that permit the practical usage of urea and other NPN compounds under very diverse conditions. Of current interest is the basic research background that has shown that solubility of proteins is very important to their usage, information having been built up from rumen studies, abomasal infusions, digestibility and N retention trials, etc. This information provides the basis for treating proteins so that they are not completely hydrolyzed in the rumen but will be digested in the small intestine, a situation which may result in improvement in N utilization. These as well as many other examples could be cited which show the eventual value of basic research to practical nutrition.

Areas Where Improvements Have Been Made

The actual changes in feeding practices that have taken place over the past couple of decades are relatively minor in many cases but the overall result has been higher and more efficient production. If an uninformed person were to visit a feedlot or dairy as it was in 1950 and a modern one today, the thing he probably would be most impressed with would be the increase in size of the operation and the number of animals. Greater mechanization would be apparent, of course, but the animals are still eating silage and/or hay and some type of concentrate mix; part of the ration may be cubed or pelleted and some of the grains may be flaked, but otherwise it would not be drastically different. These changes are very undramatic as compared to changes that have occurred in medicine, instrumentation, space, transportation, etc. Some of the developments that have taken place over the past 20 years or so are discussed briefly in subsequent paragraphs.

Protein and N. There has been a tremendous development of information on rumen metabolism of nitrogenous compounds with the result that there is a partial understanding of factors that affect rumen metabolism, N recycling, excretion, and requirements. Protein requirements can be specified with a reasonable accuracy in terms of N required/unit of metabolic size, but there is less certainty when expressing crude protein requirements as a percentage of a complete ration or in cases where a variety of factors affect N metabolism. Knowledge is adequate to permit use of relatively high levels of non-protein-nitrogen for most classes of animals.

Energy. Much information has been gained on energy metabolism of the volatile fatty acids and their importance to the ruminant animal. Fermentation and digestion of fibrous and readily available carbohydrates are now well understood. Procedures based on a slaughter technique or a factorialization method have been developed that allow calculation of the amount of energy required for a given performance of most classes of animals (Ch. 8, Vol. 2). These methods suffer from the fact that it may be expensive to obtain necessary data and that the results are probably not always applicable to field conditions that differ from the experimental conditions.

Minerals. An appreciable amount of information has been developed which provides the basis for estimating absorption, excretion and need by the rumen microorganisms and the overall need of the host. Important advances have been made with respect to mineral metabolism and milk fever, Mg tetany and urinary calculi (see Ch. 12, 13, Vol. 2). The needs for some of the trace minerals have been fairly well quantified and some important nutrient interactions have been elucidated, namely Se-vitamin E, Cu-Mo-S, and further information on Ca-P. A practical need for Co, Cu, Mn, Se and Zn have been demonstrated in ruminants in many parts of the world in addition to the needs for I and Fe that have been known for a longer period of time. Further information has been obtained on the toxicity of a variety of mineral elements.

Vitamins. Information on rumen metabolism of vitamins A, D and E has been published leading to a better understanding of their needs. With respect to vitamin A, a practical response has been shown in many situations, particularly where low-fiber rations are commonly used. Se-vitamin E interrelationships are much better understood, and the role of vitamin D in animal metabolism is more clearly understood. With respect to the B-complex vitamins, the role of cobalamin and Co has been clarified. The need for B-complex vitamins in practical rations is not clear, although most evidence indicates little need except for the young animal before rumen function develops.

Taste, Appetite and Feed Intake. A considerable amount of recent information has been developed on this subject (see Ch. 11, Vol. 2), information indicating that rumen concentration of the volatile fatty acids as well as other chemicals may have an inhibitory effect on appetite. Other information shows that feed intake of fibrous feeds is restricted because the gastrointestinal tract limitations are more restrictive than other factors.

Rumen Dysfunction. Information that has been obtained on bloat has characterized what happens with respect to rumen fermentation, rumen motility, etc., but the relative importance of several proposed causative factors—plant pectins, plant proteins, or mucinolytic microorganisms—remains to be seen. Acute rumen digestion has also been fairly well characterized and various methods have been devised to prevent it.

Metabolic Disorders. Information at hand appears to be adequate to allow most dairymen to prevent the majority of the milk fever that might otherwise occur but control of ketosis is less certain. Progress has been made on grass tetany and other problems; urinary calculi can be controlled by appropriate methods in the feedlot. Further work is needed on problems such as diarrhea and problems caused by various toxins.

Feed Additives. The ready access to cheap sources of antibiotics and sulfa drugs was widely accepted by feeders and they have been used extensively. Management practices have been developed for routine feeding as well as for stressed animals. Medically, antibiotics, sulfas or other drugs such as various worm medicines have been a great help to the animal industries. Other additives tend to come and go. Some in this class might include: enzymes, oyster shell, sand, dried rumen organisms, tranquilizers, etc.

Hormones. The use of hormones or synthetic hormones is probably one of the more widespread practices that have been accepted by those finishing cattle and lambs for the market. A variety of different products are on the market that may be given orally or subcutaneously and which will produce a favorable response in daily gain and feed conversion with minor side effects. The most advantageous use of these products requires optimal nutritional conditions, however.

Feed Preparation and Processing. The increased usage of cereal grains and their by-products for finishing cattle and lambs and for feeding of dairy cows has resulted in a marked increase in interest in improving animal usage of these feedstuffs. A variety of methods for processing grain–rolling, moist heat treatments, extruding, pelleting, popping, ensiling, et.–have been used to improve various cereals. Likewise, research has shown that pelleting or cubing of low quality forage will result in increased feed intakes and more rapid gains although milk fat percentages may be reduced when fed to dairy cows. The result has been a tremendous increase in use of processed grains for finishing cattle and in the use of cubes made of coarsely chopped forage for dairy cows. However, increasing fuel costs cast some doubt on the feasibility of continued use of some processing methods.

THE NEED FOR FURTHER NUTRITION RESEARCH

It might be appropriate to point out some of the areas which the writer feels are in need of further research. Even though a tremendous amount of information is currently available on ruminant animals, further information is needed on almost any topic; if nothing else, to refine information on requirements under more specific conditions or for high producing animals. Some areas of deficient knowledge are probably more limiting than others. It is also probable that few nutritionists would agree on the order in which any priority might be applicable. Some promising areas are discussed in subsequent paragraphs.

Soil, Plant and Animal Interrelationships

This is an area, particularly as applied to the mineral elements, that needs some concerted effort among soil, plant and animal scientists to study some of the likely limiting factors. One or two examples were mentioned previously, the effect of high N fertilization on nitrate build-up and its toxicity to animals and the effect of Ca fertilization on Mo uptake in plants. Undoubtedly, many other relationships remain to be clarified. In addition, it might be noted that we have almost no idea of how ingestion of soil or water contribute to the overall nutrient supply needed by ruminants.

Subclinical Nutrient Deficiencies

This area is somewhat similar to the one mentioned previously and applies, particularly, to the mineral elements. Recent surveys in several areas indicate that the trace element content of plant tissue may often be below levels considered to be required by ruminant species. The increasing use of fertilizers and other management and cultural tools to increase plant production makes it more likely that subclinical deficiencies will become more and more common in the future. Information on seasonal composition of plants, the effect of soil types, weather conditions, fertilization practices, plant species, etc., are needed here as well as studies on the effect on the animal; means of clearly identifying subclinical deficiencies are also needed.

Effect of Stress on Nutrient Needs

Ruminant animals are subjected to a wide variety of stresses–infectious diseases, parasites, sociological and psychological stresses, noise, environmental contaminants, toxins of various types, etc.–which affect their health and well-being. However, a survey of the literature (see Ch. 16, 17, Vol. 2) shows there is insufficient quantitative data on which to evaluate nutritional needs for alleviating stress or for preventing undue stress.

Nutrient Interactions

This would seem to be another fertile field for intensive research. Some of the well-known interactions include: protein-energy, Ca-P, Ca-P vitamin D, Ca-Zn, Cu-Mo-S, Se-vitamin E. Others are known to occur, but information is inadequate and incomplete.

Nutrient-Genetic Interactions

This is also a topic on which relatively little information is available on ruminant animals although a limited amount of information is available on other species. It would seem to be a profitable field for concentration.

Appetite and Feed Intake

One reason for the current interest in this area is the hope that ruminants can be stimulated to over eat to a greater degree in order to improve their rate of production and efficiency of production—something on the order of the European custom of force-feeding geese might be envisioned, if feasible. If consumption of nutrients could be increased without detriment to the animal, this might be a means of reducing production costs.

Rumen Function

One of the big losses in the normal ruminant is that of methane which is synthesized in the fermentative processes. This synthesis, which at maintenance accounts for 8-10% of gross energy intake, appears to serve no useful purpose. In vitro studies show that methane production can be inhibited. There may also be other means, such as the antibiotic rumensin, of improving efficiency of rumen fermentation which at best is an inefficient system. Other areas related to rumen function needing more work include the perpetual bloat problem, acute indigestion and studies related to adaptation of animals to sudden changes in the diet.

Metabolic Disorders

Ketosis is still a considerable problem in high-producing dairy cows and some solution, either prevention or treatment, is needed which is more effective than current practices. Other problems that might be mentioned are displaced abomasum and pulmonary emphysema.

Nutrition of Wild Species

It is likely that wild species may become of more importance as managed resources; thus, more information is needed on nutritional needs, likely deficiencies, metabolic problems, etc. Fortunately, more information is being developed, but at a rather slow rate.

CONCLUSIONS

The discussion presented in this chapter is intended to point out some of the major nutritional or nutritionally related problems facing our modern, highly specialized animal industries, to mention some of the developments that have taken place in relatively recent times, and to suggest some areas that would seem to be critical with respect to overall improvement in the nutritional health and wellbeing of ruminant animals.

This brief discussion is intended to set the stage for more technical and specialized discussions of subject matter that follows in subsequent chapters. If nothing else, perhaps it will point out to practical livestock people or others involved in the industry the eventual benefits of spending money for research on the more basic areas as well as on more applied subjects.

References Cited

Bywater, A.D. and R.L. Baldwin. 1979. Alternative strageties in food animal production. Unpublished, Univ. of California, Davis.

Reid, J.T., O.D. White, R. Anrique and A. Fortin. 1978. Research needs in the energetics of livestock production. Paper presented at the Amer. Dairy Sci. meetings, East Lansing, MI.

Chapter 2—Feedstuffs Used by Ruminants

by D.C. Church

INTRODUCTION

This topic is one which could, if desired, fill an entire book rather than a short chapter. There is a tremendous amount of information on many different aspects of feeds and their use, thus it is primarily a decision of what should be included. Since some information is presented in the individual chapters on suitable feedstuffs for specific uses, the intent here is to present a general overview of the nutritive properties and limitations of roughages and concentrates as they are related to the needs of ruminant animals.

A moderate amount of information on nutrient content of different feeds or plants (herbage) is given in Appendix Table 1. For readers who wish additional information of a tabular nature on composition and digestibility, the writer would suggest NAS (1971) for data on North American feeds; this publication is currently under revision. Similar information is available on feeds from Latin America (McDowell et al, 1974) and on tropical feeds (Gohl, 1975). Other sources include the old standby, Morrison's Feeds and Feeding (1958), which is, unfortunately, out of print, and a book by Crampton and Harris (1969) as well as other lists which are published from time to time in various trade magazines. Other sources are available for European readers.

Tabular examples of research data have not been included in this chapter to substantiate statements that have been made. If the reader wants more detailed information than has been given by writers on specific chapters, then you are referred to Ch. 7-10 of Church (1977) where considerably more detail is given on chemical composition of feedstuffs and on animal performance.

ROUGHAGE

First, a few simple definitions are in order. **Herbage** is sometimes defined as the total vegetative material available to a grazing animal while **forage** has been defined as vegetation which can be consumed by an animal. Both terms are used with unharvested vegetation. **Roughage**, on the other hand, is more of a nutritional term which generally includes plant stem and leaf material and sometimes flowers or seeds either green, dry or harvested. Roughage may also include high fiber feeds such as cottonseed hulls, corn cobs, sawdust, etc. There are numerous exceptions, but roughage is usually restricted to feeds that have low physical densities and high (>18%) crude fiber contents.

Herbage is the natural food for all ruminants, domestic or wild. Ruminant species appear to have a competitive advantage over other non-ruminant herbivores, at least with respect to digestibility of herbage, in particular poor quality vegetation. Of course, different ruminant species have adapted to different environments and to the plants which grow there. Thus, we don't expect a reindeer to consume the same type of diet as does an African buffalo.

Contributions from Herbage

Granted that ruminants are adapted to and require a herbivorous diet, what does herbage supply that the animal needs or wants? The two most important things are a source of required nutrients and the bulk required for the ruminant gastro-intestinal tract. Other minor plant biochemicals may have some effects on nutrient utilization or on animal health.

Nutrients. Ruminants must consume from one source or the other the necessary protein (or N source), energy, fatty acids, minerals and vitamins which are necessary for the rumen microorganisms and/or the host's tissues. For any particular animal species there are normally many different plants which can be utilized and, if the opportunity presents itself, which will be utilized throughout a normal growing season. When given a choice, selectivity of diet components changes as different plants grow, mature and become dormant on different time schedules throughout a growing season.

Grazing animals, particularly selective eaters, are capable of and normally do select out the higher quality plant parts which contain more protein and less fiber than is present in the herbage available to them (Ch. 6). Of course, when vegetation is in short supply, selectivity is more difficult.

It should be pointed out that some of the mineral nutrients may be obtained in substantial amounts from water supplies (see Ch. 2 of Vol. 2) or by ingestion of soil (Healy, 1973). However, quantitative estimates from either source are not well documented.

Bulk. The ruminant stomach and gut are adapted to and require, over long periods of time, a bulky, fibrous diet. Of course, if we consider the many species of domestic and wild ruminants, there is a considerable latitude in dietary requirements which are related to variations in stomach anatomy and relative stomach size in the various species (see Ch. 2, Vol. 1, Ch. 17).

Although cattle and sheep can be maintained on diets which are quite dense (<10% crude fiber) for extended periods of time, a much higher level of feedbunk management and more careful ration formulation are required to maintain adequate performance. A moderate amount of indigestible material (fiber) promotes salivary production and normal stomach and intestinal motility and helps to maintain a uniform flow of digesta through the gastro-intestinal tract (see Ch. 6, 7, Vol. 1). Herbage or roughage available from harvested crops is the only source of adequate levels of fiber to satisfy these needs, with the possible exception of some wood products.

When domestic ruminants are given a choice of roughage of some type and concentrates, they will always consume appreciable amounts of roughage. The relative proportion of roughage and concentrate that will be taken depends on a number of factors such as relative palatability of both roughage and concentrate and metabolic needs of the animal. Domestic species generally will over consume on concentrates during periods when energy requirements are relatively low. For example, dairy cows during late lactation or the dry period, and ewes after lambs have been weaned. Wild species generally do not get excessively fat, so it is a logical conclusion that genetic selection has encouraged animals to eat more concentrates. Nevertheless, ruminants have a craving for roughage and we can conclude that their system demands it.

NUTRITIONAL CHARACTERISTICS OF HERBAGE

Proteins

Proteins of plants (usually classed as leaf or seed proteins) may be extremely complex. The leaf proteins are usually metabolic (i.e., concerned with growth or biochemical functions of cells), whereas seed proteins are stored reserves. True protein will usually average 75-85% of the total crude protein (CP), although it may be less than this at times. Solubility and digestibility of protein generally decreases with advancing maturity. A variety of non-protein-nitrogen (NPN) compounds are usually present which may range from simple compounds such as nitrate to complex peptides. Some non-protein amino acids are believed to contribute to taste and palatability of plants.

The protein content is generally considerably higher in legumes than grasses, especially as plants become more mature. Protein content is greatly affected by leaf:stem ratios, since stems are invariably lower in protein than leaves.

Carbohydrates

Carbohydrates are usually divided into non-structural and structural. The non-structural compounds (sugars, starches, fructosans) usually comprise <20% of dry tissue, although the levels may be much higher than this in very young plants. These compounds are highly digestible and are utilized efficiently.

Structural carbohydrate (hemicellulose, cellulose, pectin) is a term that is essentially synonymous with fiber. The amount increases in plant tissue, especially stems, with advancing maturity. Solubility and digestibility also decrease with maturity.

Lipids

The lipids of nutritional significance are those found primarily in the chloroplasts and include a variety of complex lipids, many of which contain large percentages of C16, C18:2 and C18:3 fatty acids. Lipid content generally declines with maturity.

Minerals

The mineral content of herbage is variable

but for some minerals it is markedly affected by plant species. Note in Table 2-1 that herbs and legumes are appreciably higher in Ca and Mg and they are also generally higher in P than grasses.

Table 2-1. Mineral content of herbs, legumes and grasses.[a]

Species	Ca	P	K	Mg
Herbs				
Yarrow	1.3	0.39	3.7	1.0
Burnet	1.3	0.26	1.8	1.1
Plantain	1.6	0.32	2.7	0.6
Chicory	1.4	0.48	4.6	0.6
Legumes				
Trefoil	1.6	0.34	2.3	0.76
Alsike Cl.	2.0	0.33	2.1	0.62
Alfalfa	2.1	0.39	2.6	0.55
Sainfoin	1.0	0.38	2.5	0.79
Grasses				
Peren. ryegrass	0.46	0.25	2.0	0.21
Cocksfoot	0.42	0.25	2.3	0.22
Timothy	0.41	0.23	2.2	0.25
Meadow fescue	0.44	0.25	2.1	0.26
Tall fescue	0.30	0.24	2.0	0.32
Chewings fescue	0.23	0.19	1.6	0.21

[a]From Thomas et al (1952). Mean of 5 cuts.

With regard to the trace elements, there are marked differences in concentrations found in different plant species and it is impossible to make a blanket statement that applies to the trace elements since soil conditions affect uptake of some elements more than others.

The usual levels and ranges of mineral elements expected in mixed herbage are shown in Table 2-2 as compared to British recommended requirements. Note that the "normal" mineral content usually brackets requirements. Thus, it is well to remember that any of the mineral elements may be borderline or deficient in forage, depending on some of the many factors which affect mineral uptake by plants.

Non-Nutritive Compounds

Plants contain large numbers of biochemical compounds which are not usually considered as nutrients and which are normally present in rather small amounts. Some of these compounds may have appreciable effects on either nutrient utilization or on animal metabolism and health.

Lignin. A considerable amount of research information is available on lignin because of its negative association with herbage digestibility. Lignin, itself, is considered to be indigestible by animals but it also inhibits herbage digestibility. This occurs partially because of physical effects which inhibit swelling of plant fibers (which inhibits access by microorganisms or digestive juices). Grinding may

Table 2-2. Mineral content in pastures in relation to requirement.[a]

	Herbage content		Desirable content	
Element	Range	Normal	Fattening sheep	Lactating cow
Macro elements, % of DM				
Ca	0.04-6.0	0.2-1.0	0.5	0.52
P	0.03-0.68	0.2-0.5	0.25	0.42
S	0.002-2.12	0.05-1.0	0.07	0.15
Mg	0.03-0.75	0.1-0.4	0.06	0.15
Trace elements, ppm of DM				
I	0.07-5.0	0.2-0.8	0.12	0.80
Fe	21-1,000	50-300	30	30
Co	0.02-4.7	0.05-0.3	0.1	0.1
Cu	1.1-29	2-15	5	10
Mn	9-2,400	25-1,000	40	40
Zn	1-112	15-60	50	50
Se	0.01-4,000	0.03-0.15	>0.03	>0.03

[a]From Butler and Jones (1973)

overcome some of the physical inhibition. Perhaps even more important are cross-links between lignin and carbohydrate polymers (hemicellulose, cellulose). These cross-linkages inhibit enzyme activity as the enzymes studied will not hydrolyze a carbohydrate chain closer than 2-3 sugars away from the cross links (Wong, 1973). Consequently, a substantial amount of the carbohydrate residue cannot be hydrolyzed and used by rumen microorganisms. Lignin is also bound to proteins and carbohydrates in cell walls, resulting in low digestibility of cell walls.

The lignin content increases with maturity in nearly all herbage. It is not uniformly distributed, but is found in higher concentrations in stems than in leaves. While the content in leaves of most plants doesn't increase greatly, the proportions of leaf:stem decrease with maturity, resulting in an increase of lignin in the whole plant. Generally, the lower the quality of roughage, the higher the lignin content.

As mentioned, grinding will usually increase digestibility somewhat. Treating highly lignified forage with strong alkali or acid improves digestibility since these chemicals cause some swelling of the plant fibers and cleave some of the lignin-carbohydrate bonds. Ensiling, however, has little effect on lignin content of herbage.

Plant Phenolics. This group of compounds is large and complex with many different functions and properties ranging from relatively simple compounds responsible for some colors to others which have estrogenic properties to polymers such as lignin and tanin (Wong, 1973). A variety of factors affect concentration of the phenolics, including genetic, light, age, stage of growth and disease.

When ingested by animals, most phenolics are detoxified and excreted as conjugates with sulfuric or glucuronic acids. Some degradation of flavonoids or alteration of isoflavones is known to occur in the rumen, and animal tissues are capable of methylation and dehydroxylation of some phenolics. Tannins, among other things, are capable of precipitating proteins (making them less soluble) and are believed to be one of the factors which inhibit froth formation and bloat in the rumen (see Ch. 17, Vol. 1).

Organic Acids. A variety of organic acids may be found in most plants, primarily in the leaves. Acids found in relatively large amounts include malic, malonic, quinic, citric, shikimic, *trans*-aconitic and oxalic. These acids, which are usually biochemical intermediates, may account for 3-10% of the dry weight of the whole plant. Accumulation in grasses is high during cool, cloudy weather and may be increased by N fertilization. When present in high concentrations, this fraction could account for a substantial portion of the digestible tissue since we would assume that they would either be metabolized in the rumen or completely absorbed.

It is recognized that oxalic interferes with Ca utilization and often causes toxicity if animals are not adapted to high levels; *trans*-aconitic and citric appear to be involved in grass tetany; and some others (malic, quinic, shikimic) are known to produce taste responses in ruminants. However, there is no information of consequence relating these organic acids to nutritional needs of ruminants or to utilization of nutrients from herbage.

Other Plant Biochemicals. Numerous other groups of compounds are found in many plants. Alkaloids comprise a rather large group which, apparently, has no function in the plant other than to discourage consumption by animals, including insects (Culvenor, 1973). Alkaloids are more commonly found in toxic concentrations in forbs and browse than in grasses, but toxicity does occur with some grasses such as phalaris, ryegrass and fescue species. Legumes also contain alkaloids, but problems of toxicity are largely restricted to large shrubs and tree species. Pyrrolizidine alkaloids, which produce liver damage in cattle although much less in sheep and goats, are found in a variety of species (forbs, browse) and they may cause severe problems in some areas.

Other plant biochemicals that may have detrimental effects on animals include the cyanogenic glycosides, certain sterols and saponins.

IMPORTANT FACTORS AFFECTING HERBAGE NUTRITIONAL VALUES

The nutritional characteristics of herbage depend on many different factors. A number of these variables have been discussed in Ch. 6. Others will be discussed briefly here. Of the various factors which alter quantity or quality of herbage, probably the most important are: species, stage of maturity, climate and geography (including moisture availability, latitude, temperature, humidity), and soil fertility. Brief comments will be made on these factors.

Stage of Maturity

It has been documented many times that stage of maturity has a pronounced effect on nutrient content of forage (see Ch. 7 of Church, 1977), and this is true whether we are dealing with herbage from temperate or tropical zones (see Ch. 15) or dryland range areas (Ch. 6). Lush, young plant tissues, whether grass, forbs or browse, are characterized by high moisture content. Nutrient content (dry basis) is always high in CP and relatively high (the highest it will be) in readily available carbohydrates (RAC) such as various sugars and starch. The fibrous components—cellulose, hemicellulose, pectins, lignin—are low at this stage. As the plant matures the protein and RAC content decrease sharply accompanied by concomitant increases in the fibrous components. Lipid content usually declines as does mineral content. Furthermore, the NPN content generally decreases, particularly with advanced stages of maturity and the proportion of soluble proteins (metabolic proteins) decreases as the plant's need for growth is reduced.

With increasing maturity the fibrous portions of the plant increase and become less soluble and digestible. Digestibility by animals is highly positively correlated with protein content and negatively with fiber. Thus, digestibility decreases more or less continuously with advancing maturity. Cool season grasses may decrease in energy digestibility at the rate of 0.3-0.5%/day for each day that harvest is delayed for the first cutting. The decrease is less drastic for subsequent cuttings.

Browse herbage may not show the marked changes observed in grasses and forbs, partly because a fair amount of the vegetation is perennial, woody growth and, in species that are not deciduous, leaf drop occurs on and off throughout the year. These two characteristics result in less marked changes, but the same trends would be apparent in leaf tissues of browse species.

Species

There are pronounced differences in nutrient content of vegetation (forage) from different species and, of course, between major groups of plants. Data have been shown on mineral content and differences have been mentioned on protein content. Other examples are given in Ch. 6 and 15.

Latitude and Altitude

Latitude may have pronounced effects on herbage because of the changes in temperature and light intensity. Individuals living out of the tropics tend to think of the tropics as a hot, humid area. While this may be, mean temperatures are probably not much warmer than in many subtropical areas, and maximum temperatures are usually lower. Of course, the reason for this is that day length at the equator is quite uniform throughout the year. Mean daylight hours (sunrise to sunset) increase progressively at midsummer as we go from the equator to the poles. We see from this that daylength is an important factor, allowing plants to compensate for cooler nighttime temperatures or shorter growing seasons, etc.

Also of importance (on latitude) is the angle of the light rays (and other invisible radiation). Maximum absorbance of radiation occurs when the target is at a direct right angle (90°) to the radiation source. In the tropics this occurs essentially all year long. In the extreme northern or southern latitudes daylength may be quite long in midsummer, but light absorption is reduced considerably because of this reduced angle.

Tropical areas, generally, are characterized by wet and dry seasons rather than hot and cold seasons. Consequently, species of plants which tolerate tropic conditions are usually different than those which can tolerate temperate or arctic conditions. For example, tropical grasses generally have much higher cellulose contents than temperate grasses.

The effect of altitude can best be illustrated by vegetative changes occurring on a mountain or in deep canyons (i.e., Grand Canyon) where conditions can be observed ranging from arctic to tropical (minus the seasonal changes related to sunlight and latitude). The effects on the plant are primarily a combination of temperature, humidity, and wind plus the effect of more ultraviolet penetration at higher altitudes.

Fertilization

With regard to fertilization, the effect depends, in some cases, on whether the forage is pastured or harvested and removed. The effect also depends on soil nutrient supply. With P and K, addition of N will increase P and K content in herbage if soil content is high but will decrease it if soil level is low. Phosphate fertilizers generally cause less changes than N in other elements. High levels of K tend to depress forage Mg.

NUTRITIVE VALUE OF DIFFERENT ROUGHAGES

Pasture, Rangeland

This topic is covered in detail in Ch. 6, but a few comments are in order here. Overall, the nutritive value depends on the plant species available and the stage of maturity when consumed. As a general rule we can say that pastures will supply adequate or excess protein as long as they are actively growing, but not enough for young or high producing animals when plants are mature or dormant. Energy consumption is not adequate to allow maximal milk production by high producing cows or fattening cattle, but probably is close to maximum for growing lambs. Macromineral supplementation may be required, particularly NaCl and phosphate and trace minerals may be required in specific areas.

Greenchop (Soilage)

The use of greenchop is not widespread although it is used in some instances for dairy operations or in a few finishing programs for cattle and it is a fairly common practice in many undeveloped countries where green feed is hand cut for a few head of animals. The general nutritional characteristics are similar to pasture, although overall quality is probably better than the grazing season average for pasture from the same source. The reason for this statement is that greenchopping is not normally done unless there is a good supply of vegetation. However, greenchopping does reduce selectivity and could result in an overall reduction in nutrient quality during short periods of time. Greenchopping allows maximal utilization of forage because of more complete recovery (no trampling losses or contamination with manure) but close attention to animal needs and fuel costs may not always make it a feasible operation.

Silages

Many different crops, crop residues or cannery residues from food crops have been used to make silage. However, the majority of the silage made in North America is made from corn or sorghum plants or from forage, primarily grass, legumes or grass-legume mixtures. The nutritive properties of these are discussed briefly.

Grass-Legume Silage. Grass or legume silages, if harvested at an optimum stage, are high in CP (20% ±) and carotene but only moderate in digestible energy. Low-moisture silage (haylage) is a very palatable feed and has less of the N present as NPN than high-moisture silage. The low-moisture silage will usually stimulate consumption of dry matter equal to hay while high-moisture silage will normally be consumed in lesser amounts than for hay. The disadvantages of low-moisture silage are that rather complete elimination of oxygen must be accomplished, otherwise the silage is subject to excessive heating with resultant low digestibility, particularly of protein. High-moisture silages, on the other hand, are subject to greater losses by seepage or by excessive metabolism of proteins which may result in poor quality feed.

Corn and Sorghum Silages. Either of these crops make very good silage which may have 70% or more TDN because of the high grain content. Grain content may be on the order of 47-50% from corn, although somewhat lower for sorghum silage.

Both silages are relatively low in CP (8% ±). This can be corrected with excellent results by adding urea or ammonia to increase the level to 10-11% on a dry basis (see Ch. 9). Limestone has also been used as an additive. This helps to correct a natural Ca deficiency and buffers acid production allowing somewhat greater production of lactic acid. Usually, optimal results may be expected if corn is harvested at the dent stage when starch deposition is complete and with dry matter contents of 33 to 37%.

Silages are very satisfactory feeds for domestic ruminants, particularly older animals. Young animals may have difficulty in consuming enough for maximal production. However, maximum performance cannot be obtained with fattening lambs or cattle or with dairy cows without feeding additional grain.

Hays

Hays, of course, have the same general nutritive properties as the forages they were made from. Since hays are generally made from more mature vegetation than would be typical for a pasture stage (pre-boot for grasses and prebloom for most legumes), the nutritive value would normally be lower than for optimal pasture stage. We would expect some reduction in CP and energy content and some increase in fiber and lignin in a typical hay vs pasture from the same herbage. On the other hand, good quality hay will have a greater nutritive content than mature or post-ripe pasture.

Drying, if done carefully, does not greatly alter nutritive value although digestibility of protein may be reduced slightly. This statement applies to forage dried artificially or under ideal field conditions. Probably a high percentage of hay suffers some losses in the field related to excessive drying or damage from dew or rain. Leaf loss and respiration losses always occur (see Ch. 7 in Church, 1977). There generally is some reduction in protein and some increase in NPN during drying. Carotene is lost during drying, but high levels can be maintained if the forage does not heat during drying or if the hay is not bleached excessively by sunlight. Vitamin D content increases with exposure to sunlight during drying.

The stage of maturity, as with pasture herbage, is extremely important in making high quality hays. This simply cannot be over emphasized if forage is to make up a high proportion of the diet of high producing animals.

High quality legume hays will support moderate growth rates of cattle and sheep and moderate milk production. They provide excess protein but adequate energy for lactating beef cows. Grass hays are usually lower in CP (8-12%) and energy than legumes, but they make very satisfactory partial diets for growing animals and are excellent during maintenance and pregnancy of beef cattle. Sheep and goats will eat grass hays, but maximum consumption will not approach that of legume hays.

Crop Residues

Crop residues such as straws, corn and sorghum stover, corn cobs, cottonseed hulls and similar high fiber feeds are normally available in large amounts in most areas. These materials are all characterized by their low protein content (3-6%), high fiber and lignin, and low digestible and net energy. Carotene content and vitamin contents are nil, and mineral content is apt to be low.

When properly supplemented with limiting nutrients mentioned above, low quality roughages are satisfactory as the primary or only roughage for beef cows except during late pregnancy and lactation when some better quality feeds are required. Limited amounts (perhaps ½-¾ of the diet) are satisfactory for younger animals growing at moderate rates. They have only limited use for dairy cows; some could be used in the dry period. For fattening cattle these roughages can be used in moderate amounts (perhaps up to 20-25% of the ration) without reducing gain although feed conversion is apt to be reduced.

Sheep and goats don't like most of these low quality roughages. They will eat them if forced to, so such feeds can be used to some extent for dry ewes or in pelleted finishing rations for lambs.

HIGH ENERGY FEEDS

The majority of the high energy feedstuffs is provided by the cereal grains in North America. Of these, corn grain is by far the most widely used followed by grain sorghums, oats and barley. Other common sources include mill feeds (primarily from wheat), molasses, beet pulp and fat. Other ingredients are available, but quantities available are quite limited.

Cereal Grains

The principal component of cereal grains is starch which ranges from ca. 41% in oats to 72% in corn. The starch is highly digestible (90+%) and, thus, accounts for the high energy values. Fiber content generally is inversely related to starch content and to overall feeding value. Protein content usually ranges from 10-14% but may vary depending on starch content, soil fertility and genetic differences. Fat content is low (1.7-5%) as is the carotene content, although grains are moderately good sources of vitamin E. All cereal grains are extremely low in Ca (0.02-0.09%) but are moderate sources of P (0.3-0.5%) and Mg (0.11-0.22%). The NRC rating of the cereal grains for ruminants is shown in Table 2-3.

Table 2-3. Relative NRC rating of cereal grains for ruminant animals.

Grain	Dig. CP	TDN	NEg
Corn	100	100	100
Barley	131	91	95
Milo	95	88	83
Oats	132	84	77
Wheat			
Hard red	152	97	96
Soft white	115	97	96

The known chemical differences in starch composition or protein content and solubility have not been sufficient to explain the differences shown in Table 2-3. Suffice it to say, all of the grains are excellent feed, although some require more processing than others to achieve maximal performance (see Ch. 3). Wheat is more prone to produce digestive disturbances, particularly in unadapted animals.

Miscellaneous cereals such as triticale (wheat x rye cross), rye, rice, millets and other sorghums are not important feed sources on a national basis, although they may be important in some local areas.

Milling By-Products

Milling of wheat provides substantial amounts of by-product feeds (primarily bran, middlings, shorts, wheat mill run) which find some use in ruminant feeds, primarily in rations for lactating cows or in some commercial feeds. These ingredients are moderate in protein and energy and high in P. Small particle size may create problems if the ingredient is used in rations in large amounts.

Corn mill feeds are more limited and hominy or hominy feed are the principal products. Hominy is an excellent feed equal to or superior to corn grain. Small quantities of mill feeds are available from other grains, but the amount is exceedingly small and not of much significance for ruminants.

Molasses

Most molasses is derived from sugar production from sugar cane with lesser amounts from beets and other sources. The principal nutrient in molasses is sugar (50-60%) along with some of the mineral elements (especially K and Ca). The sweet taste and sticky nature (which reduces dust) are of interest as well as its function as a binder for pelleting.

Fats and Oils

Fats from various sources have been used with good success in feedlot rations for cattle and in milk replacers. Only limited use is made of fats in dairy rations or for sheep. In recent years the price has not been attractive.

Miscellaneous

Many other by-product or waste materials are used for feeding. These include dried pulp from beets or citrus fruits, whey, cull potatoes, residue from many cannery or freezing operations, dried bakery product, and miscellaneous root crops. Only a very limited number of research reports are available on many of the more unusual materials.

NITROGEN SOURCES

Proteins

There is a relatively long list of ingredients that are satisfactory protein sources for ruminants (see Appendix Table 1), but most of these are not available in large quantities. The principal supplementary sources are soybean and cottonseed meals followed by corn gluten meal and lesser amounts of other oilseed meals (linseed, peanut, rape, safflower, sunflower, etc.), brewery and distillery by-products and other miscellaneous sources.

For ruminant animals there is relatively little difference in performance when animals are fed these different sources which range from 25-50% CP. Some are excellent sources of P and most have at least moderate P levels. Fiber and energy vary as does protein content and digestibility. Provided no palatability problems occur, all can be used in moderation and the most common souces mentioned in the first paragraph can be used at high levels with no problems. Only in a few cases are toxins present which restrict usage.

Amino acids are, of course, critical for preruminants (in milk replacers). There is some evidence that high producing animals such as superior dairy cows, may benefit from protein sources with good concentrations of methionine or lysine. Active research is under way to improve protein utilization by selecting dietary sources high in limiting amino acids which will escape metabolism in the rumen but will be digested in the gut (see Ch. 3, Vol. 2). It appears that protein solubility may be a critical factor for maximal efficiency, but a considerable amount of uncertainty still exists on this topic.

NPN

Urea is the primary source of NPN although small amounts of mono- and diammonium phosphate are used and some anhydrous ammonia may be used in silages or liquid feeds.

Urea is a very satisfactory N source for feedlot cattle or in other situations where it is fed along with high starch rations. It is used less efficiently as a component of a liquid supplement containing mostly molasses. In addition, there is a considerable controversy on its use for high producing dairy cows (see Ch. 9). Utilization can often be improved and chances of toxicity reduced when urea is combined with other feeds, such as alfalfa, and then pelleted (Dehy-100) or with grain and extruded (Starea).

Urea use is encouraged because it costs considerably less than other sources of N. In many situations some urea is quite beneficial when fed with protein sources. It can, of course, be toxic (and lethal) if misused (see Ch. 17, Vol. 1).

Other Miscellaneous N Sources

Wet or dried manures or poultry litter are finding increased uses as N sources. They may be mixed in feed directly or ensiled which improves the palatability. Bacterial sources (derived from water cleanup operations from food processing plants, paper pulp mills, etc.) appear to be a very promising potential source. In the future we should expect to see more and more of the N supply coming from these sources.

References Cited

Butler, G.W. and D.I.H. Jones. 1973. In: Chemistry and Biochemistry of Herbage, Vol. 2. Academic Press, New York.

Church, D.C. (ed.). 1977. Livestock Feeds and Feeding. O & B Books, Inc., Corvallis, OR.

Crampton, E.W. and L.E. Harris. 1969. Applied Animal Nutrition, 2nd ed. W.H. Freeman and Co., San Francisco.

Fleming, G.A. 1973. In: Chemistry and Biochemistry of Herbage, Vol. 1. Academic Press, New York.

Gohl, B. 1975. Tropical Feeds. FAO Feeds Information Centre, Rome.

Healy, W.B. 1973. In: Chemistry and Biochemistry of Herbage. Vol. 2. Academic Press, New York.

McDowell, L.R. et al. 1974. Latin American Tables of Feed Composition. Dept. of Animal Science, Univ. of Florida, Gainesville, FL.

Morrison, F.B. 1958. Feeds and Feeding, 22nd ed. Morrison Pub. Co., Ithaca, NY.

NAS. 1971. Atlas of Nutritional Data on United States and Canadian Feeds. Nat. Acad. Sci., Washington, D.C.

Thomas, B. et al. 1952. Emp. J. Exp. Agr. 20:10.

Chapter 3—Feed Preparation and Processing

by W.H. Hale

INTRODUCTION

In the last 25 years there have been marked changes in the beef and dairy industries in the USA. Feedlots with over 1,000 head on feed now account for more than 60% of the fed cattle. As a rule, the commercial beef feedlots must be highly mechanized and, as a result, dietary roughage levels have declined.

There has also been a corresponding increase in the size of dairy herds, and at the same time milk production/cow has increased markedly. A portion of the increase can be attributed to genetic improvement but, no doubt, the major improvement has been due to the increased intake of energy by the lactating cow. Although the ME intake from roughages has increased due to changes in types of roughages and improved harvesting and storage techniques, the increased ME intake of the dairy cow is due primarily to increased grain intake.

The development of high grain feeding systems with both finishing beef cattle and lactating dairy cows coincided with abundant supplies of grain due to low energy input for producing the grain and the high ME production/acre from grain when compared to forage. If the average yield/acre of alfalfa and corn for 1976 is considered, corn produced 7,200 Mcal of ME/acre as compared to 5,100 for alfalfa hay if used for feedlot beef. The equivalent NEg for finishing cattle was 3,100 Mcal for corn compared to 1,600 for alfalfa. It is, then, no surprise that grains are the desired energy source for finishing beef cattle and for high producing dairy cows.

This period of increased grain intake for ruminants was also a period of low fossil fuel prices. This period also saw the development of grain processing systems which require a high energy input. During the last few years fossil fuel energy values have more than doubled and further increases may be expected relative to the prices of feeds. Thus, it is apparent that energy use must be used efficiently. Consequently, more emphasis is being placed on processing methods for grain which require less energy. The discussion will relate to the benefits of grain and roughage processing methods with special emphasis on energy requirements.

GRAIN PROCESSING

Need for Processing

The level of grain needed in ruminant diets depends upon the rate of production desired from the animals. With low rates of production (such as with mature beef cattle and growing feeder cattle), roughages may be used entirely. However, when high rates of production are desired, the energy intake of the diet must be increased by adding grain or other high energy supplements.

Blaxter (1962) showed that fat synthesis in the ruminant on good hay was 780 Kcal/kg of feed as compared to 1,950 Kcal of fat synthesized/kg of corn. It is apparent from these values that it would be difficult to fatten cattle on good hay alone (in a short period of time) as total feed intake and the rate of fat synthesis would limit production. Modern day steer fattening rations may contain as much as 85% grain, and the grain may supply up to 90% of the useable energy of the ration. Consequently, any improvement in efficiency of utilization of the grain will be reflected in improved gains and/or improved efficiency.

Prior to 1960 little attention was given to the method of grain processing other than grinding, cracking or steam rolling. Steam rolling was used to reduce the dustiness of grain preparation and kill the viable weed seed. In his report "50 Years of Progress in Beef Cattle Nutrition", Riggs (1958) stated that rolling or crushing had little advantage over grinding.

Following the reports of improved utilization of steam processing and flaking of sorghum grain and corn (Matsushima et al, 1965; Hale et al, 1966) many new methods of grain processing were developed. Many of the methods developed were experimental and received very little use in the field (Hale and Theurer, 1972). Three of the newer sophisticated methods (popping, exploding and micronizing) are still in limited use. The extruding, roasting, and pressure cooking processes are no longer used to any extent.

Whole Grain

Corn is the only feed grain that can be fed satisfactorily in whole form to cattle. The other feed grains need to be processed by some method. Corn grain is highly flexible in cattle feeding systems in that it may be fed whole, cracked, ground or flaked. When eating, adult cattle apparently masticate feed only for sufficient salivation to permit swallowing and, as a result, whole feed grains appear to be swallowed with very little mastication. It is generally accepted that during the process of rumination very little whole grain is regurgitated to be maticated, although recent documentation of this cannot be found. Finishing steers fed 90% whole shelled corn diets are rarely observed to ruminante. The probable reason for lack of regurgitation of whole grain is the high specific gravity of the grain.

Grain processing for sheep is not as important as with cattle since sheep masticate grains thoroughly. Sheep digest non-protein organic matter and protein of sorghum grain to a 10-20% greater extent than do cattle (Bucanhan-Smith et al, 1968). Furthermore, method of processing and level of grain in the diet of sheep appears to have little effect on digestibility (Wilson et al, 1973). This suggests that caution should be taken in extrapolating digestion data on processing methods from sheep to cattle.

The farmer-feeder of the midwest for many years has fed finishing steers whole shelled corn ad libitum plus supplement with limited roughage. With corn silage, cracked corn was fed in addition to the corn silage in order to raise the concentrate level of the total diet. In the last several years there has been a renewed interest in whole shelled corn feeding due to the energy cost of processing as well as equipment and installation costs for the newer processing units.

Unfortunately, very little information is available on the feeding of shelled corn as generally recommended by commercial companies. Current recommendations are that cattle fed whole shelled corn receive a diet consisting of 90% whole corn plus 5% roughage and 5% supplement.

Vance et al (1972) fed corn either as whole shelled or crimped with 6 levels of 40% DM corn silage. The concentrate portion of the ration contained 88.5% corn together with protein and mineral supplementation. Considering all silage levels, the data showed that dry, whole, shelled corn need not be cracked or ground for optimum performance in high energy rations for growing and finishing steers (Table 3-1). If it is assumed that corn silage contained approximately 50% roughage, then the roughage level in the two comparisons was 9.6%. Daily gain, feed intake and feed efficiency were similar between the two treatments, indicating that the whole shelled corn was an effective feeding system with corn silage. With a ration containing 90% shelled corn, 5% supplement and 5% cottonseed hulls, Hale et al (1979) reported equivalent gains and feed requirements with a 82% cracked corn ration containing 10% alfalfa hay.

In a digestion study, Vance et al (1972) compared whole shelled and crimped corn with corn silage fed at 0 or 20 lb/head/day. As might be expected, DM digestibility was depressed due to the corn silage feeding level. However, there was no effect on starch digestibility related to silage level or corn preparation. In the case of the high corn silage ration, a portion of the starch digestion was from the corn silage. These data indicate that whole shelled corn, even when fed at high levels, is utilized efficiently by finishing steers. Observations by the author indicate that much of the whole shelled corn which appears in the feces has been digested to some extent; however, no estimate of the extent of digestion is available at the present time.

It is probable that steers receiving whole shelled corn in high concentrate rations are susceptible to dramatic changes in weather conditions and poor feeding management. Founder frequently becomes a serious problem with whole shelled corn rations, particularly with Zebu type cattle, probably due to the eating habits of those cattle.

At the present time it would appear that wheat and barley should be processed by some method if it is included in the steer finishing or dairy ration. It is considered absolutely essential to process sorghum grain prior to feeding to any type of cattle.

Grinding or Dry Rolling

No recent information is available on the feeding value of ground or dry-rolled grain

Table 3-1. Effect of processing methods on performance of feedlot cattle.

Comparison	Daily gain, lb	Daily feed consumption, lb	Feed conversion, lb	Improvement in conversion, %	Source
Whole shelled corn	2.73	18.5	6.78	2	Vance et al (1972)
vs crimped corn	2.66	18.4	6.92		
Whole shelled corn	2.25	14.8	6.58	5	Gerken et al (1971)
vs ground corn	2.19	15.2	6.94		
Cracked corn	2.32	19.1	8.23	-8	Clanton and Woods (1966)
vs pelleted corn	1.96	17.4	8.88		
Dry-rolled sorghum	2.90	21.8	7.52	6	Hale et al (1966)
vs steam flaked	3.04	21.4	7.04		
Ground corn	2.65	18.9	6.88		Matsushima & Montgomery (1967)
vs 1/32" flake	2.82	17.3	6.14	11	
1/12" flake	2.70	18.0	6.66	3	
High moisture corn*	2.40	15.7	6.54	13	Tonroy et al (1974)
vs dry corn*	2.44	18.3	7.50		
High moisture corn*	2.73	20.6	7.63	9	Utley & McCormick (1975)
vs dry corn*	2.40	20.1	8.38		
High moisture sorghum*	2.02	24.3	12.03	12	Riggs and McGinty (1970)
vs dry sorghum*	2.00	27.4	13.70		

*90% DM basis

other than the discussions in relationship to some of the new processing methods. It appears that roughage level may affect grain utilization; these aspects are discussed elsewhere. Nevertheless, grinding of grain is an extremely useful processing method for small farmer-feeder operations. It is economical as the same piece of equipment may be used to process hay and grain. It has been shown that the digestibility of ground sorghum grain is not increased when compared to dry-rolled grain in rations containing 75-85% sorghum grain (Buchanan-Smith et al, 1968). With rations containing 50% roughage, the digestibility of dry-rolled sorghum grain was higher than on rations containing 98% grain (Saba et al, 1964; Keating et al, 1965). The digestibility of ground corn in dairy rations is discussed in a later section.

Gerken et al (1971) fed all-concentrate rations containing 90% corn with the corn either ground or whole. Daily gain and feed intake were similar between the two treatments, however feed requirements were lower for the whole shelled corn (Table 3-1). In a digestion study with an all-concentrate ration containing 75% corn, White and Hembry (1971) found no difference in the digestibility of the energy of the ration whether the corn was ground or whole.

Pelleted Concentrates

Prior to the development of the new grain processing systems in the 1960's, grinding and pelleting of the concentrate portion of beef cattle rations received considerable attention in an effort to reduce the dustiness of the feed. Baker et al (1955) reported that pelleting of high concentrate rations would lower feed requirements but frequently reduce daily gain. Clanton and Woods (1966) fed corn as cracked or pelleted corn in rations with 75% corn. When the corn was pelleted, daily gains were reduced by 16% and feed requirements increased by 9% (Table 3-1). This was due in part to a lowered feed intake of the pelleted grain. The authors also found that method of roughage preparation (pelleted or ground) did not affect performance.

The supplement portion of a finishing ration which contains the protein, vitamins, minerals or other ingredients may be pelleted in order to prevent these from settling out. For the farmer-feeder, pelleting of the supplement is common. Complete pelleted feeds would have the advantage of improved mechanical handling, but there appears to be little interest in pelleted feeds due to performance of the animals and the high energy cost of pelleting the grain portion or the complete feed.

Steam-Processed-Flaked

In the commercial feeding areas of the USA, the most common method of grain processing is steam-processed and flaked which was developed in the mid 1960's. The flaking system is merely an extension of the steam rolling process. The moisture content of the grain is raised to approximately 18% while in the steam chamber. It is then run through the rollers to produce a quality control flake. The weight per volume of excellent flaked grain will be about 45% of the volumn weight of the whole grain. A desirable flake will weigh between 25 and 27 lb/bushel. To accomplish this the rolls must be set extremely close together and the capacity of the roller mill will be reduced by 45-50% of that for dry rolling.

The moisture content of the flake from the roller will be 16-18% and, if the material is to be stored for more than 24 h, a drying system must be included in the process. In commercial operations the major portion of grain is fed the same day it is processed and thus requires no drying.

A comparison of dry-rolled and steam-flaked sorghum grain is given in Table 3-1. There was a 6% improvement in feed efficiency, however the ration contained only 58% grain. Later departmental reports from various institutions indicated that response to flaking was much greater than observed in this trial and a good portion of that was apparently due to improved quality control and increased grain levels in the ration.

Steam processing and flaking of corn results in improvement of utilization of the corn. Table 3-1 gives a comparison of 2 degrees of flaking compared with ground corn. Daily gain was improved by flaking, feed intake was slightly reduced and, as a result, feed requirements were improved by 11% with the desirable flake. This trial points out the importance of flake quality. Various in vitro evaluations also indicate the significance of quality control in grain flaking systems (Hale, 1973). Husted et al (1968) noted that digestibility of milo was not improved if the steamed grain was cracked rather than flaked when compared to dry-rolled grain.

Flaked grain should be handled in such a method so that the integrity of the flake is retained as much as possible. Garrett et al (1971) ground sorghum grain flakes which had been processed by pressure cooking. When the flakes were ground, daily gain was reduced and feed requirements increased.

Based on observations with the scanning electron microscope, Harbers (1975) found that the starch granules of flaked sorghum grain had a very irregular mass which made the granules more vulnerable to enzyme attack. The irregular formation due to flaking resulted in the loss of bifringence. This is commonly referred to as gelatinization, and Pfost (1971) showed this to be equivalent to approximately 40%.

Gelatinization has been used as a measure of quality control with flaked grains. While 40% to 50% gelatinization appears to be desirable, DeBie and Woods (1964) showed that completely gelatinized corn could markedly reduce performance of finishing steers. McNeill et al (1975) indicated that gelatinization is only one of the factors affecting improved utilization of flaked grains and that a secondary factor is rupture or release of the starch granules from the protein matrix which surrounds the granule. For example, starch digestibility in reconstituted sorghum grain was high, yet there was no gelatinization. However, in reconstitution there is a complete disruption of the protein matrix in the endosperm. Flaking either corn or sorghum grain results in increased digestion of the starch in the rumen when compared to dry-rolled grain (Waldo, 1973).

There is some question as to how much wheat or barley should be processed prior to inclusion in high concentrate rations of cattle. In general, the information available indicates that steam processing and flaking or reconstitution of wheat and barley will not improve feed efficiency. Calculations by Waldo (1973) show that barley starch digestibility in the rumen is extremely high and not affected by processing. Christensen et al (1969) fed rations containing 97.5% barley to Holstein steers; the grain was dry-rolled, dry-rolled and pelleted, steam-rolled and pelleted and steam-rolled, ground and pelleted. All processes increased gain over the dry-rolled barley without any effect on feed conversion. With diets containing 84% barley,

Garrett et al (1971) did not observe any improvement in digestibility from processing. It appears that the primary advantage of some type of barley processing other than dry rolling or grinding relates to the feed intake and not to improvement in utilization. Similar observations have been reported with wheat. Arnett (1971) fed a ration containing 42% flaked sorghum grain plus 42% wheat prepared by dry rolling, early harvested, flaked, extruded, and whole wheat. Regardless of the method used there was no marked effect upon daily gain or feed requirements with exception of whole wheat. When whole wheat was compared to dry-rolled wheat, feed requirements were increased by 16%. It must be remembered that the wheat only supplied half of the grain in the total ration. Aimone and Wagner (1977) observed that micronizing wheat improved feed intake and average daily gain when compared to dry-rolled wheat, however, there was no effect upon feed requirements. The general observation has been that extensive processing of wheat or barley will not improve utilization but may result in an improved rate of gain due to increased feed intake. It is generally recommended for high concentrate rations that wheat or barley be processed by some method such as minimum steam rolling in order to prevent dustiness. With sheep, Cornett et al (1971) noted lower N digestibility with steam-flaked wheat when compared to dry-rolled wheat and attributed it to a more rapid rate of passage related to the many fine particles of the fragil flake. Many years ago Morrison (1956) recommended that grains not be finely ground. The recommendation appears to be as valid today as it was then.

Other High Energy Processing Methods

Popping, micronizing and exploding were developed during the cheap fossil fuel prices in the 60's. These methods expand the grain and completely disrupt the endosperm structure of the kernel. At the present time there appears to be little commercial use of these methods. The reader is referred to Hale and Theurer (1972) for a complete description of these processes. If quality control is practiced with these methods, the effect on gain and feed requirements are similar to those of steam processing and flaking (Table 3-7).

High Moisture Grain

Early-harvested high moisture grain may be defined as grain harvested from the field at 25-30% moisture and stored in a suitable structure. At this moisture level the grain can be combined from the field and this method of grain processing has come into a considerable use since the development of modern grain harvesting equipment.

One of the purposes of hearvesting grain in this manner is to permit feeding of much higher grain levels than can be accomplished with corn silage. Grain will combine satisfactory at 25-30% moisture, but at higher levels harvesting is difficult. High moisture grain is highly adaptable in respect to storage in that it can be stored whole, ground or rolled. Grinding or rolling into storage permits horizontal storage. If corn is stored whole, it should be rolled or ground prior to feeding. If sorghum grain is not rolled or ground into storage, then it should be rolled or ground prior to feeding.

Beeson and Perry (1958) reported that corn harvested at 32% moisture and ground into storage was 10-15% more efficient in promoting gains than ground ear corn at a 15% moisture level (equivalent moisture basis). In a trial with early harvested corn grain (26% moisture), steers fed for 184 days on a 60% corn ration with corn silage as the roughage showed a 13% improvement in feed requirements when compared to dry-ground grain (Table 31-). Feed intake was reduced with the high moisture grain as compared to dry-rolled grain. In this trial the high moisture grain was rolled prior to feeding. Tonroy and Perry (1974) reported improved in vitro DM digestibility with early-harvested corn grain as compared to dry-rolled grain. Utley and McCormick (1975) harvested corn grain at 18% moisture and added sufficient water to raise the level to 26% moisture prior to storage. A 95% grain ration was fed to steers, but the method of corn preparation prior to feeding was not given. When the early-harvested high moisture grain was compared to dry grain, gain was increased by 0.33 lb/day and feed requirements improved by 13% (Table 3-1).

Sorghum grain may also be early-harvested at moisture levels of 25-30%. This is particularly attractive if birds are a problem with ripened grain in the field. The same general principles apply to early harvesting milo. If the sorghum grain is to be stored whole, it probably can best be done in upright structures; for horizontal structures it should be ground. If either the early-harvested corn or sorghum is stored in horizontal pits, they must be well covered to prevent undue spoilage.

Table 3-1 gives the average of 4 trials in which early-harvested high moisture sorghum grain was fed to finishing steers. Average daily gain was not affected, however feed requirements were reduced by 12%. The high feed requirements were due to the fact that the diets contained relatively low levels of grain and the primary roughage was cottonseed hulls. As with corn, the feed efficiency was improved with early-harvested sorghum grain as compared to dry-rolled grain. Very little sorghum grain is early-harvested in the USA. This is apparently due to the fact the feedlots in the milo producing areas are commercial lots and early harvesting as a system has largely been with farmer-feeder operations. If a commercial feeding operation is to utilize early-harvested high moisture grain as a source of grain for the major portion of the year, an extremely large storage capacity would be required. For reconstitution or steam flaking, much smaller storage facilities are necessary. However, there are a few commercial feeding operations which take advantage of early-harvested high moisture grain.

The improvement in efficiency is believed to be due to the fact that in early-harvested grain the starch and protein components are not in the crystalline state as with dry grain. Certainly, the early-harvested grain would require considerably less wetting in the rumen for microbial digestion.

The use of high moisture grain, particularly corn, is a viable system of grain processing which has a low energy requirement. Frequently, high moisture grain can be purchased on an equivalent dry matter basis for a lower price than dry corn. The reason for a lower price is the cost of artificial drying which is required if the grain is to be stored without ensiling or preservatives.

Reconstitution

Reconstitution is defined as a method by which grain at normal moisture storage levels is reconstituted to 26-30% moisture and stored in an oxygen limiting structure. This system was first proposed by Hale et al (1963) based on laboratory evaluations with sorghum grain. Moisture level, storage temperature, storage time and physical properties of the grain are important for proper reconstitution as illustrated in Table 3-2. It appears that the moisture level must be at least 25% for proper reconstitution.

Sorghum grain must be reconstituted whole if the reconstitution process is to occur. Table 3-3 shows the effect of moisture level and storage time on milo reconstituted in the whole or ground form. There was little effect on in vitro DM digestibility of milo reconstituted whole at the 21% moisture level. The data suggest that at 28% moisture there was some reconstitution by 2 days which was

Table 3-2. In vitro digestibility (%) of reconstituted sorghum grain as affected by storate time, temperature and moisture level.[a]

	Moisture %		
Temperature, F°	18	26	30
10 days			
41	36	39	36
75	33	44	46
20 days			
41	38	42	43
75	37	47	49
30 days			
41	37	42	44
75	35	44	51

[a]Neuhaus and Totusek (1971)

Table 3-3. Effect of grinding sorghum grain prior to reconstitution on in vitro digestion.[a]

	Days of storage		
Moisture %	2	8	16
Grain reconstituted whole			
21	35	34	34
28	39	40	44
Grain reconstituted ground			
21	31	31	32
28	33	32	34

[a]From Neuhaus and Totusek (1971)

Table 3-4. Effect of reconstitution on performance of feedlot cattle.

Comparison	Daily gain, lb	Feed intake, lb	Feed conversion, lb	Improvement in feed conversion, %	Source
Sorghum					
Ground or dry-rolled*	2.44	19.0	7.99		
vs reconstituted*	2.44	16.9	6.93	11	Riggs & McGinty (1970)
Corn					
Dry-rolled	2.44	18.3	7.50		
Reconstituted	2.44	17.0	6.97	11	Tonroy et al (1974)

*90% DM basis

markedly increased with storage time of 16 days. With grinding prior to reconstitution there was no effect of storage time or moisture level, indicating that it is necessary to store sorghum grain in the whole form if adequate reconstitution is to occur.

A summary is given in Table 3-4 of three cattle feeding trials in which sorghum grain was reconstituted to 28% moisture. The daily gains were the same between the 2 treatments, however feed intake was reduced due to reconstitution and feed efficiency was improved by 11%. The improvement in feed efficiency is similar to improvement shown by steam processing and flaking.

Corn may also be reconstituted, however data with corn reconstitution are limited. It is much simpler to use early-harvested high moisture corn grain for feeding. If the corn were to be reconstituted, it would need to be harvested dried and then reconstituted and this would add considerable cost to the processing. The effect of reconstituted corn as compared to dry-rolled corn is given in Table 3-4. The improvement in utilization is similar to that noted for steam-flaking corn.

In practice the problems associated with reconstitution of grain are related to the type of structure and moisture level necessary for proper reconstitution. Since whole grain cannot be packed satisfactorily in horizontal storage, it is necessary to store the material in some type of upright structure. Currently, oxygen limiting structures are favored to prevent excessive loss of grain DM. It is difficult to raise the moisture level of stored grain (10-14% moisture) to 30% in a short time. This can be accomplished with water heated to 130° and application of the water at the time the grain is elevated into the storage unit. Using hot water defeats one of the ultimate objectives of this method, that is to reduce energy requirements. To date, wetting agents have not been successful. Experimental data are urgently needed on methods by which the moisture level may be easily increased if reconstitution is to become a viable processing method.

The complete mechanism of action by which the process of reconstitution improves utilization of grain by ruminant animals is not fully understood. Histological studies of sorghum grain which has been reconstituted with the proper moisture level for 21 days reveal a complete disruption of the endosperm (Sullins et al, 1971). The protein matrix becomes disrupted and frees the starch granules and protein bodies. The initial stages of reconstitution appear to be closely related to the early stages of germination. In germination the embryo of the seed secretes gibberellines which then migrate to the aleurone layer of the seed and result in production of amylases and proteases (Luchsinger, 1966). The amylases stimulate starch solubilization and the proteases increase protein hydrolysis. The peripheral endosperm, which is adjacent to the aleurone, is affected markedly by enzyme activity. It is this portion of the endosperm which is thought to be relatively indigestible to rumen bacteria and intestinal enzymes (Rooney and Clark, 1968). This aspect of enzyme activity in reconstituted milo is oxygen dependent and the oxygen in the storage unit is probably depleted in 48 h;

if not, the grain would sprout in the storage unit. Based on in vitro studies, this aspect of grain modification accounts for approximately 40% of the total improvement of the utilization of reconstituted grain noted in feeding trials (Newhaus and Totusek, 1971).

No documentation is available, but it is believed that after the first 48 h of reconstitution bacterial fermentation results in the remainder of the improvement noted. The pH of reconstituted sorghum grain after proper storage is approximately 4.4. Loynachan (1970) added high levels of antibiotics to reconstitution flasks and observed only a slight reduction in nylon bag DM disappearance.

This method of grain processing holds considerable promise, particularly if the moisture elevation necessary for reconstitution can be accomplished successfully in a short period of time. The major advantage of the method is the low fossil fuel energy input as the processing takes advantage of the production of endogenous enzymes within the grain. However, a more thorough understanding of the basic aspects of reconstitution is necessary if this method is to be widely accepted.

Processed Grains for Dairy Cattle

Due to the effect of grain on milk fat %, lactating dairy cows are usually fed low levels as compared to the high level used in finishing rations for feedlot cattle. The actual grain intake for lactating cows is rarely over 50-55% of the total diet.

Bade et al (1973) fed 52% sorghum grain either as ground or reconstituted to dairy cows yielding 50 lb of milk/cow/day. Concentrate intake was reduced due to reconstitution; however, there was no effect on FCM production and the efficiency of milk production was improved. The reconstituted sorghum grain resulted in a slight reduction in milk fat %. Netemeyer et al (1977) fed cows a 50:50 concentrate roughage ratio with either ground (dry) or reconstituted sorghum grain. The concentrate mixture contained 80% grain. Total milk yields, FCM and feed intake were not affected by grain processing, however the fat % in the milk was slightly higher for the cattle fed the reconstituted grain. There was no effect upon digestibility of ration components or efficiency of milk production due to grain treatment.

Brown et al (1970) fed a 45% concentrate diet with the concentrate either pelleted or steam processed and flaked. The actual grain level in the concentrate was 33% and consisted either of 55% sorghum grain and 18% barley or 18% sorghum grain and 55% barley. There was no dry-rolled grain control. Grain processing had no effect upon daily milk yield, however fat % were reduced when the concentrate was pelleted. There were no significant effects on rumen volatile fatty acid production due to concentrate processing. Flaking of sorghum grain markedly improved its utilization in starter rations for day-old dairy calves when compared to dry-rolled grain (Schuh et al, 1970).

Moe and Tyrrell (1977) found ground corn to be more digestible than whole or cracked corn in a 40% roughage ration containing 45% corn grain. Daily milk yield was significantly higher on the corn meal diet, but milk fat % was reduced when compared to the whole corn diet. The effect was apparently due to low DM digestibility of the whole corn diet as compared to the corn meal diet (Table 3-5). In this respect, Saba et al (1964) noted that digestibility of dry-rolled milo (50% hay diet) was higher than on a 15% roughage diet. Roughage levels appear to have an affect upon the digestibility of grains and, in the case of lactating dairy cows, the roughage levels are sufficiently high to permit a high digestibility of ground grains.

Clark et al (1973) compared whole or rolled dry corn and whole or rolled early-harvested corn. Roughage was fed at libitum with dry grain equivalent offered at the ratio of 1 kg/2.75 kg of milk produced. There was no difference in milk production between dry corn and early-harvested corn, however rolling of either of the corn preparations increased milk yield and FCM production. Milk fat %

Table 3-5. Digestibility of whole and ground corn diets for lactating dairy cows.[a]

Item	Whole	Ground
Dry matter, %	61	67
Organic matter, %	62	68
Ether extract, %	60	76
Crude protein, %	51	51

[a]From Moe and Tyrell (1977); diets contained 45% corn.

Table 3-6. Comparison of corn processing systems.[a]

Item	Processing method			
	Dry-rolled	Flaked	Whole shelled	Early harvested & reconstituted
Daily gain, lb	2.75	2.75	2.75	2.84
Daily feed, lb[b]	18.97	17.44	18.92	18.31
Reduction, %	---	8.1	---	3.5
Feed/lb gain, lb[b]	6.90	6.34	6.88	6.45
Grain level, %	74	74	78	80
Improvement in ration efficiency, %	---	8.1	---	6.5
Improvement in grain efficiency, %	---	10.9	---	8.1

[a]Average 150 days of feed, initial weight 550 lb.
[b]90% DM basis

was not affected by treatment. Rolling corn prior to feeding significantly improved TDN content of the corn. Chandler et al (1975) obtained similar findings with dry ground and early-harvested corn.

It appears that early harvesting of corn is a satisfactory grain source for feeding to lactating dairy cows. There appears to be no undesirable effect on milk fat production and milk yield, but there may be an improvement in efficiency with which the grain is utilized.

Summary of Grain Processing Systems

For comparison of the effect of various grain processing systems the author reviewed approximately 200 trials and 50 were used in the final evaluations. The major portion of the trials were from department publications and Feeders' Day Reports with a limited number from scientific journals. No attempt has been made to cite all references as many are not readily available. Some adjustments of the data were necessary since all grain processing systems were not compared in all trials. Only finishing trials were included because the significance of grain processing in growing rations has not been determined.

The summary of the corn processing trials (Table 3-6) indicate little effect on daily gain, with the possible exception of early-harvested and reconstituted grain which were somewhat higher than with the other three methods. Flaking of corn resulted in a 8.1% reduction in feed intake and ration efficiency was improved by 8.1%. Early-harvested or reconstituted grain showed only a small reduction in 90% DM intake and feed requirements were improved by 6.5%, which was somewhat less than that observed with flaked grain. With whole shelled corn, gains and feed intake were very similar to dry-rolled grain and as a result there was no improvement in ration efficiency. The advantage of feeding whole shelled corn is the small amount of equipment required. It probably requires the lowest energy input of any of the systems. If it is assumed there are no associative effects due to processing, then the improvement in corn grain utilization due to processing was 10.1%.

The effect of several processing methods on efficiency of utilization of finishing rations formulated with sorghum grain is given in Table 3-7. For the comparison, popped, exploded and micronized grain have been included under one heading since the experimental data with these grains indicate that similar results between the three can be expected, provided adequate quality control is practiced.

All three processing methods improved average daily gain by approximately 8% when compared to dry-rolled grain. Feed intake was lower on all of the processed grain comparisons compared to dry-rolled grain. Feed requirements were reduced by 11% due to steam processing and flaking when compared to dry-rolled grain. A similar improvement was noted when the popped, exploded and micronized grains were included under one heading. Reconstitution of milo grain resulted in the largest effect on feed requirements. This, in part, may be due to the fact that

Table 3-7. Comparison of sorghum grain processing systems.[a]

Item	Processing method			
	Dry-rolled	Flaked	Reconstituted	Popped, exploded and micronized
Daily gain , lb	2.56	2.76	2.76	2.76
Daily feed, lb[b]	18.66	17.85	17.42	17.85
Reduction, %	---	4.3	6.6	4.3
Feed/lb gain, lb[b]	7.29	6.47	6.31	6.47
Grain level, %	74	74	78	74
Improvement in ration efficiency, %	---	11.2	13.4	11.2
Improvement in grain efficiency, %	---	15.1	17.2	15.1

[a]Average 140 days on feed, initial weight 540 lb.
[b]90% DM basis

grain levels in the reconstituted trials were somewhat higher than with the other comparisons. If it is assumed that there is no associative effect on ration utilization due to grain processing and that the entire improvement in ration utilization is due to the processed grain, then the improvement in sorghum grain utilization is on the order of 15%. It is apparent that utilization of sorghum grain can be improved markedly by proper processing.

If Tables 3-6 and 3-7 are compared, it is apparent that average daily gains of cattle are similar when receiving either sorghum grain or corn processed by some system other than dry rolling. In addition, feed requirements are very similar between corn and milo with the newer processing methods. The data also show that the improvement in utilization in sorghum grain due to proper processing is much larger than can be expected with similar methods used for processing corn.

ROUGHAGE PROCESSING

Harvested roughages are fed to ruminant animals when grazing is not available or under intensive systems of production such as finishing cattle and lactating cows maintained in dry lots. Methods of harvesting and storage of roughage probably relate more to the climatic conditions during harvesting than any other factor. In areas of heavy rainfall during the forage harvesting season, silage making is more common. However, in the arid areas, baling is the most common method of harvest. Due to the many types of roughages, stage of maturity and harvesting conditions, nutritional values vary widely. In the case of the low quality roughages (which are very useful in maintenance and growing rations for ruminants), the cost/ton may approach that of good quality roughage. The reason for this (in the case of cereal straws) is the low production/acre which increases the harvesting costs.

Independent of maintenance use, roughages are added to rations not only to supply energy but to impart certain physical properties which make the rations acceptable to the animals. On a long-term basis some roughage is necessary in rations in order to maintain good rumen function and prevent digestive disorders. In the case of dairy cows, roughage is necessary in order to maintain satisfactory butterfat levels in the milk (see Ch. 9).

In either all-roughage or finishing rations, preparation of the roughage becomes important not only from the standpoint of efficiency of utilization but also from the standpoint of economics because of roughage loss during feeding. Research on methods of harvesting and preparation for feeding are limited as compared with grains. This is probably due in part to the fact that such research applies to restricted geographical areas.

Many methods of harvesting and storage have been used and abandoned for various reasons. At one time it was common to stack hay in the field and feed from the stack. This requires a high labor input even though mechanical equipment is available now for hauling and stacking. This system of hay storage is still useful for the small operator, however, even for the small farmer it is often accomplished with custom operators due to the high cost of equipment. In the past barn dryers were used to some extent for storage of long hay. With barn dryers the hay could be harvested with 30% moisture and dried by forced air. This method requires a high labor and energy input, and consequently, has nearly disappeared. More details on equipment and processing methods will be found in the book by Church (1977).

Baled Roughage

Baling is probably the most common method used in developed countries to harvest roughage. Forage is cut and permitted to dry in the field; in areas of high humidity, a hay conditioner may be used at the time the hay is cut. For proper baling the moisture level must be sufficiently low (15-20%) at the time of baling to prevent spoilage of the forage after baling. In areas of high rainfall baled hay must be stored under cover while in dry, arid areas (such as the southwestern USA), the bales may be stacked in the open. In order to increase efficiency of the baling process, bales are usually tightly packed and, as a result, are very heavy. This has an advantage for transportation, but presents problems in handling. For example, in the southwestern areas the baling operation has become almost entirely mechanical and the bales are rarely handled by man. Mechanical stackers are in use and for transportation the bales are loaded 25-30 at a time with grapling hooks and unloaded and restacked by the same system.

In certain sections of the USA, large round bales have become popular. These are produced by rolling the hay in the windrow between the ground and the bale or forming a bale off the ground by rolling the hay on belts, rollers or chains. The bales may weigh between 1,000 and 1,800 lb, depending upon the equipment used. The outside of the bale is tight compared to the inside core which permits higher moisture content at the time of baling. The bales cannot be tightly packed. The common method of handling the large round bales for feeding is with a bale forklift mounted on a tractor. Special racks have been developed to minimize loss, but, if the unrolled hay is fed on the ground, only one bale should be offered to minimize waste. If round bales are used in mixed rations, special grinders are available which will grind one bale at a time (Hillman and Loghan, 1978; Max, 1978).

Field-Chopped Roughage

Hay may be chopped out of the windrow into a wagon and blown into a barn or into a fenced type of feeder to permit animals to feed directly from storage. Field-chopped hay has a low bulk density and requires large storage areas. This method of handling roughage has limited use if large quantities of roughages are to be handled.

Ground Roughage

It is usually recommended that roughages be coarsely ground or chopped prior to feeding even if roughage is the sole feed. If it is to be incorporated into mixed diets, then some chopping or grinding is required. When roughage is the sole feed, grinding will reduce waste and refusal of the unpalatable portions. With lambs, feed wastage was 2.4% for ground hay as compared to 16.2% for hay fed in the long form (Reynolds and Lindhal, 1960). The author made measurements on wastage with baled alfalfa for beef cattle and found it to be approximately 10% of the total amount fed, but the wastage did not include many of the fine particles which could not be recovered from the ground. At a maintenance feeding level, waste will be much lower than at a production level. It is generally recommended that long hay should be ground into 1-2" particles for either all-roughage diets or for incorporation into mixed diets and that stovers be chopped or coarsely ground prior to feeding to prevent excessive wastage.

Addition of 1% fat at the time of grinding is a good dust control practice which also increases the available energy content of the mixture. Due to recent pollution control regulations, dust control in many areas is essential at the time of hay grinding. Water may also be added at the time of grinding for dust control, but the author's observation shows it to be less effective than fat. Unless water addition is carefully controlled, it increases the difficulty of grinding.

With fattening rations containing low levels of roughages, the usual recommendation is that the roughage not be finely ground. With modern beef finishing rations containing as

Table 3-8. Effect of mechanical processing roughage on animal performance.

Animals, Treatment	Daily gain, lb	Feed consumption lb	Feed conversion lb/lb gain	Author(s)
Steers, growing				
Ground alfalfa hay	2.05	15.7	7.66	Hale et al (1969)
+ 9% molasses	2.30	17.4	7.57	
Steers, growing				
Suncured alfalfa hay				
Baled	2.73[a]	21.2[ab]	7.77	Lofgreen (1969)
Cubed	2.90[b]	22.3[b]	7.69	
Dehydrated alfalfa hay				
¼" pellet	2.37[c]	18.8[c]	7.93	
¾" pellet	2.77	21.0[a]	7.58	
Steers, finishing				
Long alfalfa hay	1.80	17.6	9.78	Weir et al (1959)
Pelleted alfalfa hay	2.17	20.5	9.45	
Steers, finishing				
Pelleted alfalfa hay	2.31	23.3	10.1*	Oltjen et al (1971)
All-concentrate ration	2.79	16.0	5.7*	
Steers, finishing				
Pelleted bermudagrass hay	2.33**	---	---	Utley et al (1975)
73% shelled corn ration	2.84**	---	---	

*Dressing % were 55.4 and 59.9% for alfalfa and concentrate, respectively; **Dressing % were 54.3 and 59.8% for roughage and corn diets, respectively.

[a,b,c]Statistically different ($P < .05$)

little as 10% roughage, it is extremely difficult to determine if roughage processing has an effect on feed intake. With properly processed grain, molasses and tallow additions, the importance in the degree of grinding of the roughage is probably minimized. Addition of molasses to an all-roughage ground diet increased feed intake. A 9% molasses addition to ground alfalfa increased feed intake by 1.7 lb/day and increased gain by 0.25 lb with no effect on requirements (Table 3-8). The feeding value of long and ground roughages is discussed in the following sections in relation to other roughage processing methods.

Cubing

Cubing is a process by which coarsely ground roughage is pressed into a cube about 1.3" square which may be of variable lengths. Cubes from good quality alfalfa hay will weigh 25-32 lb/ft^3 as compared to 40 lb for hay pellets. Alfalfa hay is cut and left in windrow until it contains ca. 10% moisture. At the time of cubing, water is sprayed back on the hay to raise the moisture level to ca. 14%. This is done because hay in the windrow containing 14% moisture will not cube satisfactorily. The advantage of hay cubes occurs because of the ease of transportation and mechanical handling. However, the cost of cubing is such that cubes are used primarily in special situations where other forms of hay packaging present problems.

With animals weighing 300 lb and up, cubes may be fed without further processing. If the cube is to be used in a mixed ration, it must be crushed by some method prior to incorporation into the diet, but grinding through a hammer mill is not recommended because of the fine particles produced.

Attempts to cube grass hay and low quality roughages have not been as successful as with legume hays. Cubing of low quality roughages is usually uneconomical due to the high cost of cubing and the low energy value of the roughage. In the author's area of the USA, cotton gin trash has been cubed for feeding of range cows during periods of range forage shortage. The advantage of cubes under these conditions is that they can be spread on the ground and no feed bunks are required.

Lofgreen (1969) compared alfalfa hay processed as baled, cubed and dehydrated pellets harvested from one field on the same day. Table 3-8 shows that the cubes produced more rapid gain as a result of a greater feed intake. Steers fed the ¼" dehydrated pellets ate less and gained less than any other treatment. This study suggests that there is no particular advantage in feeding a pellet of a high quality roughage when compared to baled or cubed hay.

Cubed roughages, particularly alfalfa, are attractive to dairymen as a forage source. Anderson et al (1975) reported higher roughage intake with cubed alfalfa as compared to baled alfalfa when concentrate feeding was held constant (Table 3-9). In this study total milk production was increased and milk % decreased somewhat. There was no significant difference due to treatment. Reports available are conflicting in respect to the effect of alfalfa hay cubes on milk fat %. It is the author's opinion, based on observations of cubing, that the stage of maturity and the length of cut may be quite variable. In most reports, cube quality is not discussed and it is probable, if the alfalfa at the time of cubing was fairly immature and the particle size of the cube was small, that milk fat percent may be depressed. This would be particularly true with high producing cows which are fed restricted amounts of roughage. However, there is no doubt that feed intake for both beef cattle and dairy cattle is increased by cubing when compared to baled or long alfalfa hay (Lofgreen 1969; Wallenius and Bryant, 1972; Anderson et al, 1975).

Table 3-9. A comparison of baled and cubed alfalfa for lactating cows.[a]

Item	Baled	Cubed
Milk production, lb/day	59.6	64.5
Milk fat, %	3.5	3.2
Roughage/day, lb	20.2	25.7
Concentrate/day, lb	22.7	22.4
Roughage refusal/day, lb	0.6	0.9

[a]Anderson et al (1975)

Dehydration

Green forage may be preserved by dehydrating the forage at high temperatures. The only two forage products on which national price quotations are given are dehydrated and sun-cured alfalfa pellets. Dehydrated alfalfa hay pellets are usually guaranteed to contain 17% protein with a high carotene content. The primary use of dehydrated and sun-cured alfalfa pellets is as a supplement rather than the primary source of roughage. The data presented in Table 3-8 indicate that the feed intake of animals receiving only dehydrated alfalfa pellets is probably lower than on equivalent roughage which had been baled or cubed. In cattle finishing areas in which no dry roughages are fed and corn silage is the sole source of roughage, it is common to add 1-2 lb of either dehydrated or sun-cured pellets/steer/day.

Pelleted Roughage

During the 1950's and 60's, considerable attention was given to the use of pelleted roughages for beef cattle. Pelleted roughage will weigh about 40 lb/ft^3 as compared to 5-6 lb for long hay. It can be transported more economically for longer distances than long hay. In recent years the energy cost of pelleting roughages has risen dramatically and the interest in pelleting has declined. The effects of pelleting roughage are greatest with low quality roughages and, as the forage quality improves, benefits are reduced (Beardsley, 1964).

With 2/3 timothy-1/3 alfalfa hay, Johnson (1964) found that voluntary consumption increased when the hay was ground or pelleted as compared to long or chopped hay. Digestibilities of organic matter and energy were the highest with the long hay and lowest with the pelleted hay (Table 3-10). Weir et al (1959) fed long or pelleted hay to finishing steers. Pelleted hay increased consumption and gain, however feed requirements were not affected materially, although they were lower on the pelleted hay than with the long hay (Table 3-8). Pelleting roughages usually reduces digestibility of the roughage since the feed intake is usually higher on the pelleted roughages (Campling and Milne, 1972). Even though there is a depressed digestibility of DE due to pelleting, the utilization of the NE may be improved to a point where the NE per unit of weight may actually be higher than with the unprocessed roughage (Greenhalgh and Wainman, 1972).

Table 3-10. Intake and digestion by lambs of hay processed by four methods.[a]

Item	Long	Chopped	Ground	Pelleted
Voluntary consumption[b]	52	50	59	73
Digestibility, %				
Organic matter	62	55	50	51
Energy	58	52	47	47

[a]Johnson et al (1964). Hay was alfalfa-timothy.

[b]$BW_{kg}^{0.75}$

In recent research with pelleted roughages for beef cattle, the comparisons have been between pelleted all-roughage rations and high concentrate rations to determine the feasibility of finishing cattle on all-roughage rations. Oltjen et al (1971) compared an all-pelleted alfalfa diet with added minerals to an all-concentrate diet containing approximately 91% corn. Steers were fed ad libitum for 160 days. Steers on the all-concentrate ration gained 0.48 lb more than steers on the pelleted alfalfa ration and feed requirements were reduced approximately 50% (Table 3-8). Dressing % was reduced markedly on the pelleted alfalfa ration when compared to the all-concentrate ration. However, when corrected to a constant dressing %, the concentrate-fed steers gained 0.7 lb/day more than the forage-fed steers although there was little difference in carcass grade. In a second series of studies (Dinius et al, 1975) with heifers, results were similar to those in the first trial. Dressing % was 5.4 percentage points lower on the pelleted alfalfa than on the all-concentrate feed and carcass grade a full grade lower. These studies also indicated that a pelleted mixture of clover and timothy hay produced similar performance to the pelleted alfalfa diet. These studies indicate that cattle may be finished on pelleted alfalfa diets to a desirable finish, but ultimately economics of the system must be taken into consideration. One of the problems with all-forage finishing diets as compared to high concentrate finishing diets is the difference in the amount of manure produced from the two feeding systems and the cost of the additional manure removal with all-forage diets.

Utley and McCormick (1975) compared a 73% ground, shelled corn diet with bermuda grass hay pellets as the entire diet. Steers on the all-forage diet were permitted some grazing during the period on feed. Daily gain and dressing % were significantly lower on the all-forage diet (Table 3-8). Carcass grade on the all-forage diet steers was good– as compared to good+ for the high concentrate diet.

A novel approach has been dehydrating and pelleting the entire corn plant (or sorghum) and comparing this to corn silage (Karn et al, 1974). Comparison of the two corn harvesting systems indicated no significant difference in gain, DM intake or feed requirements, however the feed requirement on the dehydrated corn plant was 4% greater than on the corn silage. Dehydration of the corn plant will probably result in less total DM loss than when harvested as corn silage.

Nelson and Sims (1974) substituted 10.5% dehydrated alfalfa pellets for an equivalent amount of corn silage DM with lactating cows and observed no difference in performance between the two groups as related to gain and FCM production. Brown et al (1977) replaced 10, 30 and 50% of the alfalfa cubes in a 50% concentrate ration with pelleted or non-pelleted cottonseed hulls. They noted no undesirable effect due to the pelleted cottonseed hulls on total milk production (53 lb/day) or milk fat %. Milk fat % tended to increase with increasing levels of pelleted cottonseed hulls.

Alkali Treatment of Low Quality Roughages

Approximately half the total weight of grain producing plants is represented by the roughage portion of the plant. Over 200 million tons of these roughages are available yearly in the USA. Due to low digestibility (below 50%), such roughages find little use in ruminant feeds other than on maintenance rations and for growing cattle (Klopfenstein, 1979).

Garrett et al (1979) fed a 72% rice straw diet when the rice straw was treated with either 4% NaOH or 5% anhydrous NH_3. The diet contained 17% concentrate. Table 3-11 gives the results of a feeding trial in which pelleted alfalfa hay served as the control. Treatment of the rice straw with either NaOH or NH_3 markedly increased feed intake and gain when compared to the non-treated rice straw control. Daily gain for the rice straw treated

Table 3-11. Effect of treating roughage with hydroxide on animal performance.

Animals, Treatments	Daily gain	Feed consumption	Feed conversion	Author(s)
Cattle				
Alfalfa hay	1.56 lb	20.3 lb	13.0	Garrett et al (1979)
Rice straw				
Control	0.50	17.7	35.4	
NaOH-treated	1.56	25.1	16.1	
NH_3-treated	1.17	20.4	17.4	
Steers				
Corn cob control	0.66 lb	9.0 lb	13.6	Koers et al (1970)
NaOH-treated	1.61	11.9	7.4	
Lambs				
Wheat straw control*	36 g	908 g	25	Hasimoglu et al (1969)
NaOH-treated*	160	1226	8	

*Organic matter digestibility was 49 and 60%, respectively, for the control and NaOH treatments.

with NaOH was very similar to the alfalfa control. Feed required/unit of gain was reduced by approximately half due to alkali treatment. Garrett et al (1979) observed little response to alkali treatment of rice straw if the diets contained only 36% rice straw. With lambs, a 13% improvement in the digestibility of organic matter was observed due to treatment with NaOH but a significant depression in N digestibility was observed.

Koers et al (1970) fed steers a ration containing 80% corn cobs and 20% soybean meal. Treatment of the corn cobs with 4% NaOH improved both gain and feed requirements (Table 3-11). Lambs fed a 80% wheat straw plus a 20% soybean meal showed a marked improvement in gain when the straw was treated with 4% NaOH (Table 3-11), and organic matter digestibility of the straw was increased by 22%. The increase in digestibility of low quality roughage due to alkali treatment is due to increased cell wall digestibility (Mowat and Ololade, 1970). N digestibility also appears to be increased when NH_3 has been used but decreased with NaOH. Klopfenstein (1979) estimates the final ration should not contain more than 2% of alkali residue if optimum performance and digestibility are desired.

Alkali treatment of low quality roughages holds some promise, but there are many obstacles to overcome such as the polluting effect of the alkali and cost of alkali treatment. Klopfenstein (1979) estimated that the total cost of alkali treatment (Na, Ca hydroxides) would be in the range of $25-$30/ton. This cost would certainly detract from its use in many areas. However, in many countries where feed supply for ruminants is extremely short, alkali treatment of low quality roughages may be a realistic approach. This statement particularly applies to use of NH_3 for treating low quality roughages.

Literature Cited on Grain Processing

Aimone, J.C. and D.G. Wagner. 1977. J. Animal Sci. 44:1088.

Arnett, G.W. 1971. J. Animal Sci. 33:275.

Bade, D.H. et al. 1973. J. Dairy Sci. 56:124.

Baker, F.H. et al. 1955. Kansas Agr. Expt. Cir. 320.

Beeson, W.M. and T.W. Perry. 1958. J. Animal Sci. 17:368.

Blaxter, K.W. 1962. The Energy Metabolism of Ruminants. Charles Thomas Pub. Co., London.

Brown, W.H. et al. 1970. J. Dairy Sci. 53:1448.

Buchanan-Smith, J.G. et al. 1968. J. Animal Sci. 27:525.

Chandler, P.T. 1975. J. Dairy Sci. 58:682.
Christensen, D.A. et al. 1968. J. Animal Sci. 48:263.
Clanton, D.C. and W. Woods. 1966. J. Animal Sci. 25:102.
Clark, J.H. et al. 1973. J. Dairy Sci. 56:1531.
Cornett, C.D. et al. 1971. J. Animal Sci. 32:716.
DiBie, W.H. and W. Woods. 1964. J. Animal Sci. 23:872 (abstr.).
Garrett, W.N. et al. 1971. Hilgardia 41:123.
Gerken, H.J. et al. 1971. J. Animal Sci. 52:379 (abstr.).
Hale, W.H. 1973. J. Animal Sci. 37:1075.
Hale, W.H. and C.B. Theurer. 1972. In: Digestive Physiology and Nutrition of Ruminants, Vol. 3. 1st ed. O & B Books Inc., Corvallis, OR.
Hale, W.H. et al. 1963. Proc. West. Sec. Amer. Soc. Animal Sci. 14:30.
Hale, W.H. et al. 1966. J. Animal Sci. 25:392.
Hale, W.H. et al. 1979. Arizona Cattle Feeders' Day report, May.
Harbers, L.H. 1975. J. Animal Sci. 41:1496.
Husted, W.T. et al. 1968. J. Animal Sci. 27:531.
Keating, E.K. et al. 1965. J. Animal Sci. 24:1080.
Lynachan, T. et al. 1971. J. Animal Sci. 31:248 (abstr.).
Luchsinger, W.W. 1966. Cereal Sci. Today 11:69.
McNeill, J.W. et al. 1975. J. Animal Sci. 40:335.
Matsushima, J.K. et al. 1965. Proc. West. Sec. Amer. Soc. Animal Sci. 16:73.
Matsushima, J.K. and K.L. Montgomery. 1967. Colorado Agr. Expt. Sta. Farm and Home Res.
Moe, P.W. and H.F. Tyrrell. 1977. J. Dairy Sci. 60:752.
Morrison, F.B. 1956. Feeds and Feeding, 22nd ed. The Morrison Publishing Co., Ithaca, N.Y.
Netemeyer, D.T. et al. 1977. J. Dairy Sci. 60:748.
Neuhaus, V. and R. Totusek. 1971. J. Animal Sci. 33:1321.
Pfost, H.B. 1971. Feedstuffs, Feb. 27, p. 24.
Riggs, J.K. 1958. J. Animal Sci. 17:981.
Riggs, J.K. and D.D. McGinty. 1970. J. Animal Sci. 31:991.
Rooney, L.W. and L.E. Clark. 1968. Cereal Sci. Today 13:259.
Saba, W.J. et al. 1964. J. Animal Sci. 25:533.
Schuh, J.D. et al. 1970. J. Dairy Sci. 53:475.
Sullins, R.D. et al. 1971. Cereal Chem. 48:567.
Tonroy, B.R. et al. 1971. J. Animal Sci. 39:931.
Tonroy, B.R. and T.W. Perry. 1974. J. Animal Sci. 41:495.
Utley, P.R. and W.C. McCormick. 1975. J. Animal Sci. 38:676.
Vance, R.D. et al. 1972. J. Animal Sci. 35:598.
Waldo, D.R. 1973. J. Animal Sci. 37:1062.
White, T.W. and F.G. Hembry. 1971. J. Animal Sci. 33:305 (abstr.).
Wilson, G.F. et al. 1973. J. Agr. Sci. 80:259.

Literature Cited on Roughage

Anderson, M.J. 1975. J. Dairy Sci. 58:72.
Beardsley, D.W. 1964. J. Animal Sci. 23:239.
Brown, W.H. et al. 1977. J. Dairy Sci. 60:919.
Campling, R.C. and J.A. Milne. 1972. Proc. B. Soc. Animal Prod. p. 53.
Church, D.C. 1977. Livestock Feeds and Feeding. O & B Books, Inc., Corvallis, OR.
Davis, C.L. 1979. In: Regulation of Acid-Base Balance. Church & Dwight Co. Inc., Piscataway, N.J.
Dinius, D.A. et al. 1975. J. Animal Sci. 41:868.
Garrett, W.N. et al. 1979. J. Animal Sci. 48:92.
Greenhalgh, J.F.D. and F.W. Wainman. 1972. Proc. B. Soc. Animal Prod. p. 61.
Hale, W.H. et al. 1969. J. Animal Sci. 29:773.
Hasimoglu, S.T. et al. 1969. J. Animal Sci. 29:160 (abstr.).
Hillman, D. and T. Logan. 1978. Beef Cattle Science Handbook, Agriservices Foundation, Clovis, CA.

Karn, J.F. et al. 1974. J. Animal Sci. 38:1974.
Klopfenstein, T. 1979. Fed. Proc. 38:1939.
Koers, W. et al. 1970. J. Animal Sci. 31:1030 (abstr.).
Lofgreen, G.P. 1969. California Cattle Feeders' Day Rpt.
Max, G.D. 1978. Beef Cattle Science Handbook, Agriservices Foundation, Clovis, CA.
Mowat, D.N. and B.G. Ololade. 1970. Can. Soc. Animal Prod. Proc. 35 (abstr.).
Nelson, D.K. and J.A. Sims. 1974. J. Dairy Sci. 57:630 (abstr.).
Oltjen, R.R. et al. 1971. J. Animal Sci. 32:327.
Reynolds, P.J. and I.L. Lindahl. 1960. J. Animal Sci. 19:873.
Utley, P.R. and W.C. McCormick. 1975. J. Animal Sci. 41:495.
Wallenius, R.W. and J.M. Byrant. 1972. J. Dairy Sci. 55:692 (abstr.).
Weir, W.C. et al. 1959. J. Animal Sci. 18:805.

Chapter 4—Feed Additives and Growth Promotants

by D.C. Church

A feed additive is generally defined as a feed ingredient of a non-nutritive nature which will stimulate growth or other types of performance, improve the efficiency of feed utilization or which may be beneficial in some manner to the health or metabolism of the animal. Most of these compounds would be classified as drugs (a substance used as a medicine). Many of the commonly used feed additives are antimicrobial agents which may include antibiotics or antibacterial agents (synthetic compounds which kill or inhibit the growth of bacteria) and antifungal agents (natural or synthetic compounds which inhibit the growth of fungi). Hormone-like substances may also be included.

Depending on the nature of the effect on animals, feed additives in this group have been described as growth permittants, growth effectors, growth promotants or rumen additives. If the compound results in improved animal performance, it is termed a production improver, regardless of its particular effect on the animal (see Ch. 4 of Church, 1977).

Feed additives have been used extensively in the USA and in many other countries during the last 30 years. In very recent years, restrictions enforced by regulating agencies have curtailed or threatened to curtail large-scale usage because of suggested but unproven development of resistant strains of microorganisms that might be more pathogenic to man or which might be resistant to antimicrobial agents used in treating diseases in man. While it is undoubtedly true that feed additives are misused from time to time, it also is true that they have been extremely beneficial to livestock producers under our modern methods of production. It is to be hoped that regulations do not become so strict as to obviate the use of proven and efficient production improvers. For the benefit of readers interested in this particular topic, recent reports dealing with possible hazards to animals and man include those of Kiser (1976), Solomons (1978), Van Houweling and Gainer (1978) and Visek (1978).

Information on approval and usage of feed additives is published in the Federal Register. However, the Feed Additive Compendium (Anon., 1979) is a commercial publication that is quite easy to use and one which is updated monthly.

ANTIMICROBIAL AGENTS

Antimicrobial agents currently approved for use in ruminant rations in the USA are shown in Table 4-1. Information is given on the species and class of animal, dose range, claims by the manufacturer, and withdrawal time required for animals going to slaughter. Generally, the antimicrobial agents are believed to act by controlling or modifying microbial populations in the rumen, gut or body tissues. Some of the effect may be due to control of subclinical diseases while others may help to control diseases such as bacterial enteritis, respiratory disease or liver abscesses. In nearly all cases the antimicrobials are administered in very small amounts, ranging from <1 mg/hd/d to <1 g/hd/d, the dosage level depending on the drug, animal species and class, and the purpose for which the drug is used. An older series of articles dealing with the mode of action of antimicrobials include papers by Cravens and Holch (1970), Kemp and Kiser (1970), Wallace (1970) and Weston (1970).

Antibiotics, in general, are more effective in young, growing animals. Older data show that low level use of antibiotics (15-80 mg/day) tends to result in an increased feed intake, an improvement in feed conversion and usually some increase in growth rate. Weekly intramuscular injections of larger doses (400 mg) may work just as well as daily oral intake. Generally, the most pronounced responses with young animals occur in a situation where *E. coli* infections are high and Roy (1970) has suggested prophylactic use of the tetracycline antibiotics at a level of 125 mg/day during the first 3 weeks of life in such cases, even when calves have had adequate colostrum. Antibiotics, as a rule, will help to reduce the incidence or severity of several different types of diarrhea.

Table 4-1. Antimicrobial agents and hormone production improvers approved as feed additives for ruminants in the USA by the Food and Drug Administration in late 1979.[a]

Compound	Animal	Dose range*	Claims**	Withdrawal time before slaughter
Antibiotics				
Bacitracin	Feedlot cattle	35 mg/hd/d	ADG	none
Bacitracin methylene	Feedlot cattle	70 mg/hd/d	LA	
disalicylate	Feedlot cattle	250 mg/hd/d for 5 d, none for 25 d	LA	
Bacitracin zinc	Feedlot cattle	35-70 mg/hd/d	ADG, FE	none
Chlortetracycline	Calves	0.1-0.5 mg/lb BW	ADG, FE, DR	
	Beef & non-lactating dairy cows	70-750 mg/hd/d***	ADG, FE, LA, DR, FR RD, AP	48 h at 350 mg/hd/d or higher
	Beef	5.0 mg/lb BW for 60 d	AP	10 d
	Dairy cows	0.1 mg/lb BW/d	DR, FR, RD	
	Sheep	20-50 g/ton feed	ADG, FE	
Erythromycin	Feedlot cattle	37 mg/hd/d	ADG, FE	
Monensin	Feedlot cattle	5-30 g/ton feed	FE	
Neomycin sulfate	Cattle, sheep, goats	70-140 g/ton complete feed	DR	cattle, 30 d sheep, lambs, 20 d
	Calves, lambs, kids	200-400 mg/gal RMR	DR	
Oxytetracycline	Calves (0-12 wk old)	0.05-0.1 mg/lb BW	ADG, FE	
	Calves, starter feeds or milk replacers	0.5-5 mg/lb BW/d; or 25-75 mg/hd/d; or 50-100 g/ton dry feed		5 d if fed in RMR or 2 g/hd in dry feed
	Feedlot cattle	75 mg/hd/d or 0.5-5 mg/lb BW/d	ADG, FE, DR, BT, LA DR, RD	
	Dairy cattle	75-100 mg/hd/d	MP, BT, DR	
	Sheep	10-20 g/ton	ADG, FE	
		20-100 g/ton	DR, ET	
Tylosin	Feedlot cattle	8-10 g/ton feed	LA	
Antimicrobial Combinations				
Chlortetracycline/ Sulfamethazine	Beef cattle	350/350 mg/hd/d for 28 days	ADG, RD	7 d
Monensin/Tylosin	Beef cattle	20-30/10 g/ton	FE, LA	
Oxytetracycline/ Chlortetracycline	Calves	50 g/ton feed	DR	
Oxytetracycline/ Neomycin	Calves	50/35-140 g/ton to 100/70-140 g/ton	DR	
	Calves	8-100/100-200 mg/gal RMR	DR	
		or		
		40-200/200-400 mg/gal RMR	DR	
Hormone Production Improvers				
Melengestrol acetate	Beef heifers	0.25-0.50 mg/hd/d	ADG, FE, ES	48 h
Thyroprotein (iodinated casein)	Dairy cows	0.5-1.5 g/100 lb BW	MP	

[a]From Anon. (1979).

*Abbreviations used: d, day; BW, body weight; hd, head; RMR, reconstituted milk replacer.

**Abbreviations used: ADG, average daily gain; FE, feed efficiency; MP, milk production; DR, diarrhea; FR, foot rot; RD, respiratory diseases; AP, anaplasmosis; LJ, lumpy jaw; ET, enterotoxemia; LA, liver abscesses; BT, bloat; ES, estrus suppression.

***See reference cited above for more details.

A summary of some of the older data on cattle is shown in Table 4-2. These data indicate that young, growing-wintering cattle responded somewhat more than finishing cattle. This might be expected as cattle of this age are usually subjected to more drastic ration changes, perhaps more inclement weather, and are more prone to sickness than older cattle or feedlot cattle. There are numerous reports from experiments with feedlot cattle which do not indicate any statistical improvement in performance due to

Table 4-2. Effect of feeding low levels of antibiotics to beef cattle.[a]

Item	Growing-wintering		Finishing	
	Controls	Antibiotic	Controls	Antibiotic
Daily gain, kg	0.56	0.61	1.16	1.24
Feed/unit gain	12.70	12.02	9.85	9.37
Number of animals	5353		2354	
Number of days on feed	112		117	
Improvement in gain, %	8.9		6.7	
Improvement in efficiency, %	5.4		4.9	

[a]From Wallace (1970); Antibiotics were fed continuously, mostly at 70 mg/hd/d. Antibiotics involved were chlortetracycline, oxytetracycline and bacitracin.

the feeding of antibiotics. However, there is usually a decrease in abscessed livers, particularly with an antibiotic such as tylosin, on rations that are prone to produce this condition (Brown et al, 1973). The usual level approved for antibiotics is about 70-80 mg/hd/d, or when complete rations are used, the antibiotics may be included at a level of 5 mg/lb of feed (11 mg/kg). However, this varies with the particular situation (Table 4-1).

When used at higher levels (i.e. up to 700 mg/hd/d for cattle), antibiotics such as the tetracyclines or some of the combinations of antimicrobials have often been helpful in counteracting stresses associated with transportation and adjustment to new conditions. However, these higher levels are not approved for long-term feeding periods.

With lactating dairy cows only two antibiotics, chlortetracycline and oxytetracycline, are approved for long-term use in the USA. The older literature does not indicate any improvement in milk production or feed efficiency on either a short-term or long-term usage (Foreman et al, 1961). Consequently, there has been little experimental interest in the subject.

With regard to sheep, note in Table 4-1 that only chlortetracycline, oxytetracycline and neomycin are approved for use with sheep. Research evidence, such as that shown in Table 5-3, indicates that antibiotics or antibiotics plus some of the sulfas are equally as effective with sheep as with cattle. However, because of a much smaller marketing potential and because of the very high costs in getting FDA approval, most of the manufacturers have not tried to get approval for use with sheep.

Table 4-3. Effect of chlortetracycline and sulfamethazine on performance of feeder lambs.[a]

Item	Treatment			
	Control	CTC	S	CTC + S
Daily gain, g	184	210	214	239
Feed consumed, kg/d	1.28	1.36	1.34	1.41
Feed conversion, kg feed/kg gain	6.96	6.48	6.26	5.90

[a]From Calhoun and Shelton (1973)

In recent years penicillin has been withdrawn as an approved feed additive for ruminants. Only one antibiotic, monensin, has been approved recently and, to date, in combination only with tylosin (see Table 4-1). A good many research reports on monensin have been generated in the past few years, only a few of which will be cited here. In studies with cattle on growing-finish rations, results in general suggest some slight increase in rate of gain (0-10%) but a reduction in feed consumption accompanied by greater rumen production of propionic acid, an altered acetate: propionate ratio, and an increase in feed efficiency (Raun et al, 1976; Mowat et al, 1977; Prange et al, 1978; Steen et al, 1978). Although rumen studies do not indicate any marked shift in microbial populations or other parameters other than an increased propionic acid production (Dinius et al, 1976), it seems likely that some shift in populations must occur to account for this change. It is assumed that the increased propionic acid accounts for the reduction in appetite (see Ch. 11, Vol. 2). An example of some data from one trial is shown in Table 4-4.

Table 4-4. Effect of monensin on growing-finishing cattle.[a]

Item	Control	Monensin
Growing phase		
Daily gain, kg	0.63	0.70
DM consumed, kg	4.67	4.62
DM intake/gain	7.47	6.63*
Finishing phase		
Daily gain, kg	0.86	0.88
DM consumed, kg	5.40	4.91
DM intake/gain	6.32	5.57*

[a]Excerpted from Mowat et al (1977)
*Statistically different

In studies with monensin fed to grazing cattle or cattle fed low quality roughage, the results indicate that the antibiotic increased gain and improved feed efficiency or reduced losses in mature cows during the winter. There is information showing that grazing times are reduced but with no effect on milk yield or composition (Potter et al, 1976; Moseley et al, 1977; Turner et al, 1977; Lemenager et al, 1978; Owens et al, 1978).

HORMONE PRODUCTION IMPROVERS

These products which are hormone or hormone-like in nature, are on the approved list at this time. Melengestrol acetate has been quite effective with heifers. It results in a suppression of estrus and improved gain and feed efficiency (Young et al, 1969; Zimbelman et al, 1970; Hawkins et al, 1972).

Until 1979 diethylstilbestrol (DES) was approved as an oral additive or a subcutaneous implant for feedlot cattle. It has now been withdrawn by the FDA. In the writer's view it is quite unfortunate that the FDA has bowed to political pressure and pressure from hysterical consumer's groups. DES was withdrawn because the drug can, if given in massive doses, cause cancer in laboratory rats. The FDA has ignored scientific evidence on cattle which shows that only minute traces (parts/billion) of DES or some degradation product may be found in liver or kidney tissues; none has been identified in muscular tissues. DES is quite effective in stimulating deposition of nitrogen resulting in more rapid gain, improved feed efficiency and less fat deposition in animals fed to the same weight.

Iodinated casein (thyroprotein) is the only other hormone or hormone-like compound approved as a feed additive for lactating dairy cows. Results on growing-finishing lambs or cattle have not shown favorable responses in experimental studies. However, with lactating cows data indicate an increase in milk production. An example of the effect of short-term use is shown in Table 4-5. In this experiment cows in a warm climate were given thyroprotein after having been in production for ca. 2 mo. and were fed for 6 weeks at a rate of 2.5 g/100 kg BW. In more recent work Schmidt et al (1970) studied the effect of feeding thyroprotein for most or all of the lactation period. When thyroprotein feeding was started 50 days postpartum, maximum increase in milk production occurred during the 1st 2 weeks and averaged 7% increase in production (Fig. 4-1); these cows generally continued to produce more milk until about 18 weeks postpartum, although the response lasted longer in first lactation cows. However, total lactation response was

Table 4-5. Effect of feeding thyroprotein on daily milk production.[a]

Treatment	Total milk, kg	4% FCM, kg	Milk components, %			
			Fat	Solids	SNF	Protein
Control	17.2	17.3	4.14	13.29	9.17	3.65*
Thyroprotein	22.8**	22.4**	3.98	13.16	9.09	3.39

[a]From Stanley and Morita (1967); * ** Statistically different.

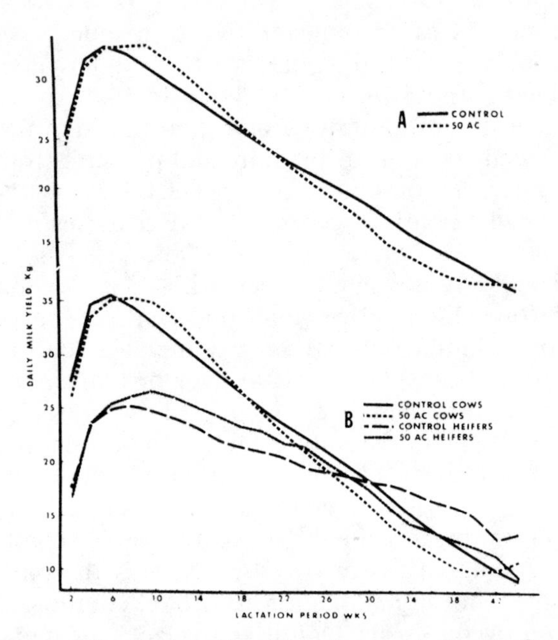

Figure 4-1. Daily milk yield of cows during the 3-yr experiment. A. Cows were fed no thyroprotein (control) or 10 g of thyroprotein daily beginning 50 days postpartum (50 AC). B. Treatment as in A but divided as to cows calving the first time (50 AC heifers) or cows calving for at least the 2nd time (50 AC cows). From Schmidt et al (1970).

not different from control and thyroprotein-fed cows. Feeding of thyroprotein resulted in marked reductions in body weight and condition and these cows required more services/conception and had longer calving intervals than control cows. These data indicate no response at the peak of milk flow in high-producing cows. Feed consumption will usually increase or a relatively high weight loss will occur, or both. Thus, the research information indicates that thyroprotein feeding might best be utilized, if at all, with some cows that are past their peak of production, particularly cows that might be overly fat.

SPECIAL PURPOSE ADDITIVES

A number of other special purpose additives have been approved for use in ruminant diets by the FDA in the USA (Table 4-6). Compounds in this group have been developed for rather specific purposes such as control of some specific disease, metabolic disorder, or pest. In contrast to the antimicrobials which are usually fed for relatively long periods of time, this latter group (except for EDDI and poloxalene) are usually fed for relatively short periods of time.

Table 4-6. Special purpose feed additives under control of the Food and Drug Administration in the USA.[a]

Compound	Animal	Dose range	Claims*	Withdrawal time
Ammonium chloride	Range cattle	0.75-1.25 oz/hd/d	UC	
	Fattening cattle	1-1.5 oz/hd/d	UC	
Amprolium	Calves	5 mg/kg BW for 21 d or 10 mg/kg for 5 d	CC	24 h
Coumaphos	Cattle	0.091 g/100 lb BW for 6 d	GW	
Decoquinate	Cattle	0.5 mg/kg BW/d	CC	
Ethylenediamine dihydriodide (EDDI)	Cattle (except lactating dairy cows)	400-500 mg/hd/d for 2-3 wk or 50 mg/hd/d in feed or salt	FR, LJ	
	Sheep	12 mg/hd/d	LJ	
Methoprene	Cattle	46-180 g/ton	FP	
Phenothiazine	Sheep, goats	12.5-25 g/d or 1 g/hd/d in feed or salt	GW	
	Cattle (except lactating dairy cows)	10 g/100 lb BW or not more than 70 g; or 2 g/hd/d in feed or salt	GW	
Poloxalene	Cattle	1-2 g/100 lb BW/d	BT	
Propylene glycol	Dairy cattle	0.25-0.5 lb/hd/d prior to and after calving	KT	
Rabon	Cattle	0.07 g/100 lb/d	FP	
Ronnel	Beef, non-lactating dairy cows	5.5% in mineral block**	FP, GP	10 d
Thiabendazole	Cattle	3-5 g/100 lb BW	GW	3 d
	Sheep, goats	2-3 g/100 lb BW	GW	30 d

[a] From Anon. (1979)
*Abbreviations are: UC, urinary calculi; CC, coccidiosis; GW, gastrointestinal worms; FP, fly prevention; GP, grub prevention; KT, ketosis; FR, foot rot; LJ, lumpy jaw.
**For more detailed information see reference above.

Except for ammonium chloride, poloxalene and propylene glycol, the other compounds listed in Table 4-6 are more appropriately discussed in a book dealing with general livestock management rather than nutrition. Information on poloxalene has been discussed in Ch. 17 of Vol. 1 regarding prevention of pasture bloat and information on propylene glycol as related to prevention or treatment of ketosis has been discussed in Ch. 12 of Vol. 2.

Many other special purpose additives are used in ruminant rations from time to time. Included are chemicals for flavoring and seasoning, anticaking agents, surfactants, dust control agents, emulsifiers, buffers, pellet binders, etc. Many but not all are on the GRAS list (generally recognized as safe). For further details on this topic consult Smith (1979).

References Cited

Anon. 1979. Feed Additive Compendium. The Miller Pub. Co., Minneapolis, MN.
Brown, H. et al. 1973. J. Animal Sci. 37:1085.
Calhoun, M.C. and M. Shelton. 1973. J. Animal Sci. 37:1433.
Church, D.C. (ed.). 1977. Livestock Feeds and Feeding. O & B Books, Inc. Corvallis, OR.
Dinius, D.A., M.E. Simpson and P.B. Marsh. 1976. J. Animal Sci. 42:229.
Foreman, C.F., N.L. Jacobson and A.E. Freeman. 1961. J. Dairy Sci. 44:141.
Hawkins, D.R., H.E. Henderson and H.W. Newland. 1972. J. Animal Sci. 35:1257.
Kiser, J.S. 1976. J. Animal Sci. 42:1058.
Lememager, R.P., F.N. Owens and R. Totusek. 1978. J. Animal Sci. 47:262.
Moseley, W.M., M.M. McCartor and R.D. Randel. 1977. J. Animal Sci. 45:961.
Mowat, D.N., J.W. Wilton and J.G. Buchanan-Smith. 1977. Can. J. Animal Sci. 57:769.
Owens, F.N., K.S. Lusby and R. Totusek. 1978. J. Animal Sci. 47:247.
Potter, E.L. et al. 1976. J. Animal Sci. 43:665.
Prange, R.W., C.L. Davis and J.H. Clark. 1978. J. Animal Sci. 46:1120.
Raun, A.P. et al. 1976. J. Animal Sci. 43:670.
Roy, J.H.B. 1970. The Calf–Nutrition and Health. 3rd ed. Penn. State University Press, University Park, PA.
Schmidt, G.H., R.G. Warner, H.F. Tyrrell and W. Hansel. 1970. J. Dairy Sci. 54:481.
Smith, B.W. (ed.). 1979. Feed Industry Red Book. Communications Marketing Inc., Edina, MN.
Solomons, I.A. 1978. J. Animal Sci. 46:1360.
Stanley, R.W. and K. Morita. 1967. J. Dairy Sci. 50:1097.
Steen, W.W. et al. 1978. J. Animal Sci. 46:350.
Turner, H.A., R.J. Raleigh and D.C. Young. 1977. J. Animal Sci. 44:338.
Van Houweling, C.D. and J.H. Gainer. 1978. J. Animal Sci. 46:1413.
Visek, W.J. 1978. J. Animal Sci. 46:1447.
Young, A.W., L.V. Cundiff and N.W. Bradley. 1969. J. Animal Sci. 28:224.
Zimbelman, R.G. et al. 1970. J. Amer. Vet. Med. Assoc. 157:1528.

Chapter 5—Ration Formulation

by D.C. Church

INTRODUCTION

Ration formulation is a topic which deserves some space in any book dealing with applied animal nutrition. The actual mechanics of formulation and the use of such methods as algebraic solutions, Pearson's square, double Pearson's square, slack space, reserved feedstuffs, etc., are relatively simple and straightforward. These have been discussed in detail elsewhere (see Church, 1977). If the reader is not familiar with these techniques, some time should be spent in review of the topic. On the other hand, formulation with the use of computers (linear programming) is more complex. Some discussion on this topic will be found later in the chapter.

The variability in ruminant diets is immense. They may range from milk replacers for preruminants to diets essentially of only poor quality roughage to those that are practically 100% concentrate. This tremendous variability greatly complicates formulation for optimal performance (high, yet efficient production) because we do not, at this time, always have enough information to relate simple chemical analytical data information on digestibility, etc., to animal performance and because we do not have enough data to always know what factors affect animal requirements. Some of the problems which may result in inefficient formulas are discussed in the following pages.

NUTRIENT REQUIREMENTS

In order to develop a satisfactory diet, we must have some knowledge of the limiting nutrients needed by an animal in a specific situation and of the amounts of nutrients required in the diet. In the USA, the National Research Council (NRC) publications on beef cattle, dairy cattle, sheep, and a forthcoming publication on goats are the generally accepted standards used for domestic ruminants. These publications represent the opinions of different committees which have organized the publications. Generally, the standards have been quite acceptable, although some obvious discrepancies may be noted from time to time. At any rate the NRC publications provide a generally satisfactory starting point and tables delineating these requirements are given in the Appendix. In many instances experienced nutritionists may modify specified requirements in a given situation on the basis of experience under similar requirements in a given situation on the basis of experience under similar conditions with a given class and species of animal. This is to be expected since ruminant animals are produced under widely different conditions, many of which may have some effect on nutrient requirements. In contrast, poultry are produced in rather standard conditions regardless of the area and, in addition, basic rations are rather similar regardless of where the birds are grown. Consequently, it is much more difficult to refine the dietary requirements of ruminants to the degree that can be accomplished with poultry or swine.

Nutrient requirements of healthy animals may be altered by a considerable number of genetic and environmental factors. Some of these which have been identified include:

Genetic: species, breed, strain

Production related: age, sex, pregnancy, lactation and level of lactation, growth and rate of growth, desired carcass fatness, and quality and quantity of milk produced.

Other environmental: disease, nutrient deficiencies, climatic factors such as temperature and humidity, muddy lots, wind, rain and other miscellaneous stresses.

Although they may not always alter the requirements for absorbed nutrients, factors such as level of feed consumption, energetic and physical density of the diet, feed additives, growth stimulators, and feed processing methods may greatly affect efficiency or completeness of digestion and metabolism of absorbed nutrients. Thus, if we are basing animal needs on chemical composition or digestibility, these factors usually alter efficiency and have the same final effect as if they altered nutrient requirements. For example, if we feed lambs a pelleted hay-grain diet, consumption will be increased greatly, as compared to a non-pelleted diet, and the lambs will gain more/day yet on a diet with a lower concentration of digestible protein and energy.

The NRC publications generally are developed to make allowances for some differences related to species, occasionally for breeds and for body size, sex, age, pregnancy, lactation and milk fat percentage, and rate of growth. Information is not, at this time, sufficient to include the marked effect of the environmental factors into the nutrient requirement tables. Illustrations of proposals to utilize additional information are discussed later in the chapter.

ANIMAL VARIATION

Animals of similar breeding do not always have the same nutrient needs, even though it is economically feasible to treat them as if they do. Previous exposure to various stresses (disease, injury, etc.) may have altered their capabilities. Hormone stimulus (thyroid, growth and sex hormones) can easily alter metabolism as well as physical activity. In addition, some individuals are more susceptible to a given stress than others. In animal research there are many documented instances which show marked differences in nutrient requirements. The author happens to be an example of an individual with an inordinately high requirement for riboflavin. Thus, it is well to remember that animals differ in nutrient requirements just as they differ in taste preferences, hair color or temperment. The consequence (for ration formulation) is that our objectives should be to satisfy the needs of most of the animals. A few will receive far more than needed; most will get just what they need; and a few may be underfed. Only in the case of outstanding animals (such as superior dairy cows) is it feasible to feed on an individual basis so that they are not underfed.

FEEDSTUFF VARIATION

The nutrient content of any given dietary ingredient (plant material) is dependent upon the species and variety, weather and water supply, soil fertility and fertilization, insect or disease damage and losses or damage occurring during harvesting and storage. Consequently, it is not always easy to look at a table giving feedstuff composition and arrive at a decision which clearly describes the ingredient which you may wish to use in a ration. These comments apply, in general, to all farm- or ranch-raised roughages and grains. Grain by-products, protein supplements and other manufacturing by-products are usually standardized for a major nutrient such as protein. Thus, these particular feedstuffs are apt to be less variable. The end result of the variation in feedstuffs and in animal requirements is that ration formulation is not likely to be as exact as one might think it is or is necessary to be.

EXPECTED CONSUMPTION

If we can agree that nutrient requirements are reasonably well documented, then one of the other important factors which must be partially resolved is that of total feed consumption. It is well known that feed processing, physical density and overall feed quality, among other things, will affect feed consumption.

Thus, if we can be sure how much feed will be consumed when animals are fed ad libitum, we can, then, devise a satisfactory diet, at least in terms of nutrient content. However, when we add other ingredients or provide supplements for roughage, consumption frequently changes, sometimes as a result of increased (or decreased) rumen metabolism. For example, if we feed cattle a poor quality roughage such as cereal straw, consumption is likely to be less than 1% of body weight/day. However, if we provide a satisfactory protein-mineral supplement, straw consumption may almost double. On the other hand, if considerable grain is fed to cattle or sheep on pasture, the inevitable result is that less pasture forage is consumed. Other factors which may modify consumption include feed processing methods which alter density of either concentrates or roughages. Likewise, use of feed additives or growth stimulators may alter feed consumption. Consequently, it is not always possible to predict with much precision how much feed will be consumed when a particular change is made in a supplement or complete ration. At this time we cannot go to a

Table 5-1. Effect of reducing protein consumption of cattle in cold or hot environments.[a]

Trial	Mean temp., °C	ADG, kg/day[b] Control	ADG, kg/day[b] Adjusted	PER[c] Control	PER[c] Adjusted	SBM replaced, g/head/day
1	1.6	1.07	1.15	0.98	1.25	110
2	1.6	0.91	0.82	1.03	1.06	110
3	2.3	1.06	1.10	1.19	1.43	150
4	25.9	1.11	1.14	0.79	0.91	130
5	25.7	1.27	1.15	1.11	1.16	160

[a]From Ames et al (1980); [b]differences were not significant; [c]PER = protein efficiency ratio and = units of weight gain/unit of protein consumed.

table and find the answer for anything other than a given set of feed ingredients upon which research data have been obtained.

Some information on expected consumption is given in the appendix tables which specify nutrient requirements. Other suggestions are given in the various chapters in this book dealing with the different species and classes of animals.

ADJUSTING ANIMAL REQUIREMENTS

Temperature and Protein Requirements

One possibility for improving efficiency of protein utilization is to reduce the protein content of rations during periods of heat or cold stress. The practical application of this has been demonstrated with lambs (Ames and Brink, 1977) and cattle (Ames et al, 1980). Note in Table 5-1 that approximately ¼ lb of soybean meal could be replaced/day in finishing rations for cattle without affecting daily gain and with an improved PER. The explanation for this is that energy requirements for maintenance increase with either hot or cold stress (see Vol. 2, Ch. 8, 16), while there is little if any change in protein requirement. Thus, protein consumption can be reduced without reducing performance.

In order to calculate the amount of protein to feed, it is first necessary to estimate the effect of thermal stress on gain. Formulas used by Ames et al (1980) are:

Cattle

Cold ADG (kg) $= 1.4 + 0.013C$

Hot $= 1.42 + 0.12C - 0.003C^2$

Sheep

Cold ADG (g) $= 112.12 + 6.99C$

Hot $= 213.16 + 3.75C - 0.24C^2$

where C = ambient temperature in °C.

These formulas give the reduction in daily gain. The protein requirement for maintenance is subtracted from total consumption and that allowed for growth is reduced proportionately to the reduction in gain.

Based on consulting experience in commercial lots, Rohwer (undated) has developed some formulas for estimating the effect of temperature on consumption of cattle. Where effective temperature (wet bulb thermometer) is less than –7°C,

$$\text{Temp. effect} = 1.0 + [(°C+7.0) \times 0.015] - [(°C+7.0 \times 0.0175)^2].$$

When effective temperature is greater than 23°C, then

$$\text{Temp. effect} = 1.0 - [(°C-23.0) \times 0.01].$$

Intake = normal intake x temp. effect. He also suggests a formula for cold stress adjustment for NEm for cattle which is:

$$BW_{lb}^{0.75} \times [42.6+(17.93-0.05425 \times BW^{0.75}-0.55535 \times °F)].$$

Based on personal experience the writer cannot vouch for either of these proposals. However, this approach for modifying ration specifications for environmental stress is logical and deserves close scrutiny. With added information and experience, this approach certainly should result in some refinement in formulation and in more efficient use of expensive nutrients.

Other Adjustments

It has been known for many years that different breeds mature at different body weights. Thus, at a given body weight, a large sized animal, such as a Simmental steer, would be expected to be less mature and have less body fat than a small breed such as Angus. Since the energy (and protein) requirements are inversely related to body fat (because the remaining tissue has less muscle and protein) or directly to lean body mass, it is logical to feed animals according to their physiological maturity. Fox et al (1977) have developed some proposals based on carcass evaluation of animals with different sizes (called frame size). An abbreviated version of their relative size rankings is shown in Table 5-2. For example, we see that a small steer weighing 720 lb should have the same body composition as a large framed steer weighing 1080 lb.

Table 5-2. Estimated body weights at which body fat and protein are similar.[a]

Sex, frame size				
	Shrunk body weight, lb			
Steers				
Small	400	560	720	880
Average	500	700	900	1100
Large	600	840	1080	1320
Heifers				
Small	330	460	590	720
Average	400	560	720	880
Large	470	660	840	1030
	Chemical composition, % of BW			
Fat	14.9	19.5	24.2	28.8
Protein	19.5	18.6	17.6	16.5

[a]Data from Fox et al (1977)

Suggestions for modifying NEg requirements of animals of different frame sizes are shown in Table 5-3. Note that the multipliers suggested apply to animals of the same weight but with different frame sizes. The same approach is applicable to thin, average and fat animals since thin animals gain more rapidly because of compensatory responses and because the weight gain in the early stages is primarily water and protein which requires only 12-15% as much energy an an equivalent

Table 5-3. Proposed modifications in animal requirements or feed energy values.[a]

Item	Multipliers	
	NEm	NEg
Animal requirements		
Frame size		
Steers		
Small	---	1.18
Average	---	1.00
Large	---	0.87
Heifers		
Small	---	1.15
Average	---	1.00
Large	---	0.89
Body condition[b]		
Very thin	0.95	0.90
Average	1.00	1.00
Very fleshy	1.05	1.10
Breed		
Other than Holstein	1.00	1.00
Holstein	1.12	1.12
Environment		
1. Shade, no mud, no chill stress	1.00	---
2. Outside lot, dry climate or barn and lot with hard surface	1.05	---
All others are outside lots		
3. Well mounded, bedded	1.10	---
4. Min. mud, wind protection	1.15	---
5. Some mud, wind protection	1.20	---
6. Moderate mud, no wind protection	1.25	---
7. Frequent deep mud, no wind protec.	1.30	---
Ration Energy Values		
No antibiotics or hormones	1.00	1.00
Monensin	1.10	1.10
Other antibiotics	1.04	1.04
Growth stimulants	1.12	1.12
Monensin & growth stimulants	1.20[c]	1.20[c]

[a]Modified from Fox et al (1977); [b]The writer has changed this adjustment group from ration adjustment to one for animal requirements as this seems more logical. [c]Fox et al give a multiplier for Monensin and growth stimulators of 1.10, a value which does not seem to be reasonable.

gain in body fat.

Fox et al also suggest a correction for Holsteins. Most likely, other breed differences should be adjusted for as there is a fair amount of evidence showing differences in maintenance requirements of different breeds.

Suggested alterations in NEm for environmental stress are also shown. It would be difficult to be very precise on these factors, but the approach is logical.

It is also logical to adjust either animal requirements or feed energy values for the effect of additives and growth stimulants (primarily hormones) since those listed do,

Table 5-4. Use of adjustment factors for animal requirements or energy values of feeds.

Item	NEm	NEg
NRC requirement for steer gaining 1.2 kg/day[a]	6.89	6.55
If very thin (0.95 NEm, 0.9 NEg)	-0.34	-0.65
If Holstein (1.12 NEm, NEg)	+0.83	+0.79
If outside lot (#3) (1.10 NEm)	+0.69	---
Total	8.07	6.69
Increase	17.1%	2.1%
If NE of ration is[b]	2.00	1.35
With monensin and growth stimulator (1.20 NEm, NEg)	+0.40	+0.27
Total	2.40	1.62
Increase	20.0%	20.0%

Feed requirement for	Maintenance	Gain	Total
Average steer, kg feed	6.89/2.00 = 3.445	6.55/1.35 = 4.85	8.29
Modified requirements	8.07/2.40 = 3.36	6.69/1.62 = 4.13	7.49
Reduction in requirements			9.6%

[a]Values as Mcal; [b]Values as Mcal/kg feed.

usually, result in more rapid gain and/or more efficient gain. These types of adjustments have not been programmed into the NRC tables, so this approach is a logical means of improving the precision of estimating changes in ration requirements.

The use of some of these adjustment factors is illustrated in Table 5-4. With the particular adjustment factors used, it turns out that feed requirements would be reduced 9.6%. If these adjustments work out in practice, then they will, indeed, allow more precise ration formulation, but it will take some time to see how they work in practice.

RATION SPECIFICATIONS

In formulating rations our objectives are to provide nutrient needs in an economical physical package that will be consumed in desired amounts. By themselves, tables giving nutrient content of feedstuffs are not sufficient. The tables don't tell us if the feed causes diarrhea if consumed in large amounts, if it has rumen stimulatory properties, if the physical density may be limiting or if the physical nature of the feed may limit its use in milling equipment. Consequently, we need to use specifications other than nutrients when designing rations. Whether we are formulating by hand or with a computer, the limitations mentioned above apply.

Restrictions on Feed Ingredients

There are many times (probably a majority of the time) when we will want to use some specific limitations on individual feed ingredients, and these may be done for different reasons. Some examples are given.

Minimum Ingredient Requirement. For cattle finishing rations two examples immediately come to mind. The author likes to use some minimal amount of alfalfa hay in complete rations because (1) it is a high quality feed much liked by cattle and (2) because of his opinion that alfalfa has rumen stimulatory properties not reflected in the usual analytical data. The second example is for molasses. This ingredient is quite useful as a palatability factor (sweetness) and as a means of reducing dustiness in dry rations or as a carrier for micro ingredients. Another reason for including minimal amounts might be strictly to use up a feed on hand that otherwise might not be included in the ration. Feeds may be included because of their fiber content (bulkiness), laxative properties, or for some specific nutritional reason that may or may not be defined in tables of composition. Cottonseed, usually ground, has become a favorite ingredient for dairy rations because of the general response of an increase in butterfat percentage when fed to lactating cows.

Maximum Amount Limited. As with minimal requirements, there may be several reasons to restrict quantities of specific feedstuffs. Urea is a good example here. Since it usually costs less than other N sources, urea must be restricted when using computer formulation, otherwise more may be used than animals will tolerate without some reduction in performance. Also, we might wish to put some limit on total NPN in a ration including that in all feedstuffs, not just urea. Likewise, fats might be included in excess, although their price usually prohibits excessive use.

When molasses (or other liquids) is cheap, restrictions are often necessary because of limitations on the amount that can be handled by milling equipment or because the amount that might be used may not give optimal animal performance. Other feeds such as cull potatoes (or other high-moisture feeds) may need to be restricted because of previous experience demonstrating that consumption of dry matter and performance will be reduced if they are not limited. However, this may usually be accomplished indirectly by requiring minimal amounts of fiber. One other reason for a restriction might be to stretch out the supply of a given ingredient so that it will be available over a longer period of time.

Other Ration Limitations

With major limiting nutrients, complete rations or supplements are usually based on minimal amounts of protein, energy, Ca, P and, possibly, other macro or micro mineral elements. Since the ratio of Ca:P is sometimes critical, formulation may be more appropriate on this basis than on minimums. For example, if a legume hay such as alfalfa makes up a majority of the diet, the Ca content will usually exceed the requirement, P may or may not reach the requirement, and we might end up with a Ca:P ratio exceeding the optimal upper level of about 2.5:1. We might, then, deem it desirable to add P above the required level so that the Ca:P ratio is within the desired range.

Protein:energy ratios have found extensive use for monogastric species. They have not, as yet, come into widespread use for ruminants. However, we might expect increased use in the future when utilization of bypass protein, soluble and insoluble N are sorted out to a workable degree, but this is not the case at this time. In some instances N:S ratios might be feasible to use.

Limitations on fiber are often used, particularly for lactating dairy cows. This is done because of the well known effect of low fiber rations on milk fat depression (see Ch. 9). In the opinion of the writer, this is not a very effective and versatile means of solving this problem. For example, if we have hay available from the same source as long (baled or stacked), chopped, ground, cubed or pelleted, then the fiber concentration should be the same in each product, provided no losses occurred in processing. Obviously, pelleted hay does not produce the same production response as long hay because of much greater consumption and more rapid passage through the rumen. Furthermore, the fiber restrictions do not always give the same type of response when used with dry roughages as opposed to wet roughages (i.e., silage). Thus, the fiber limitation is largely limited to a restricted usage rather than as a versatile means of defining ration characteristics.

Dairy nutritionists also generally put some upper and lower limitations on the amount of concentrate which should be fed to lactating cows. In the opinion of the author, nutritionists who use fiber (crude or ADF) and concentrate limitations to help define a ration are really trying to come up with a measure of physical density. It is well known (see Vol. 1) that the ruminant gastro-intestinal tract evolved and is adapted to consumption of herbage, although there are great differences in the capability of different species to utilize herbage of different qualities. Consequently, it seems logical that utilization of some measure of physical density (sometimes called bulk density) in conjunction with caloric density (units of available energy/unit of DM) should allow a more complete characterization of a diet.

Physical density data have been included for some of the feedstuffs listed in Appendix Table 1. They are not readily available on many roughages, partly because such measurements are more tedious to collect than for concentrates.

Mertens (1980) has presented data which suggest that NDF (neutral detergent fiber) is a reasonably good measure of density when the roughage in the ration was ground. In contrast to the other fiber values, NDF contains all of the normal fibrous components (lignin, cellulose, hemicellulose). NDF is also highly correlated to digestibility (negatively), rumination (positively), and intake (negatively). A graph showing response of dairy cows to different NDF levels is shown (Fig. 5-1). This may not, however, resolve the problem of changes in density which occur after feed processing.

Table 5-5. Readily available carbohydrate (RAC) content of some feedstuffs.[a]

Feedstuff	RAC, %[b]	NFE, %[c]
Alfalfa hay, med. qual.	23	34-40
Grass hay, med. qual.	17	40-42
Barley, Pac. Coast	65	65-67
Beet pulp, mol. dried	23	54-56
Corn, dent	71	72-73
Cottonseed meal, sol.	21	25-27
Molasses, cane	54	46-47
Wheat, Pac. Coast white	76	74-75
Wheat mill run	39	52-54

[a]From Church (1972); [b]Expressed at % reducing sugars, as fed basis; [c]Expressed on as fed basis or 90% DM.

Readily Available Carbohydrates (RAC)

RAC (sugars, starches) has had only extremely limited use in formulating rations. One reason might be that analytical data on feedstuffs are not commonly available. For the feedgrains, NFE values are comparable to RAC (Table 5-5), but values for other feedstuffs are generally lower when NFE is computed from the proximate analysis. This is, no doubt, because some of the hemicellulose of fibrous materials is soluble in reagents used for the crude fiber analysis, resulting in higher NFE values than would be the case if they were derived only from starch and sugar.

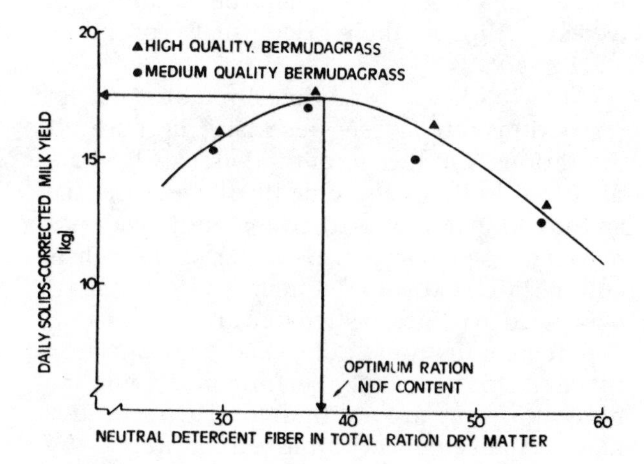

Figure 5-1. Determination of the optimal neutral detergent fiber content of the total ration resulting in maximum solids-corrected milk production using the quadratic regression equation. Courtesy of D.R. Mertens (1980).

Table 5-6. Performance of cattle fed diets formulated using RAC values.[a]

Item	ADG, lb/day	Feed conversion	DE conversion
Experiment 1 (1.43 Mcal DE/lb)			
Low RAC (43%), no urea	2.97	6.34	9.00
Low RAC, 30 lb urea/T	2.82	6.08	8.59
High RAC (51%), no urea	2.89	6.42	9.11
High RAC, 30 lb urea/T	3.07	6.28	8.91
Experiment 2 (1.46 Mcal DE/lb)			
45% RAC	3.08	6.40	9.18
50% RAC	3.06	6.12	8.85
55% RAC	3.14	6.22	8.92

[a]From Church (1972)

Some of the author's data on use of RAC in formulating fattening rations for cattle are shown in Table 5-6. It would appear that use of RAC could result in some improvement, particularly in rations utilizing relatively large amounts of urea. Further research evidence is needed to confirm or refute use of RAC in formulation.

Data on Processed Grains and Roughages

Although considerable research effort has been expended in evaluating processed feeds, appropriate information (energetic utilization, particularly) has not found its way into the NRC publications. As a result, it limits the use of processed feeds on the basis of tabular data. It is to be hoped that this deficiency will be remedied in the future.

EXAMPLES OF HAND FORMULATION

For the benefit of the less experienced readers we will go through two examples of formulation. Thus, a few simple thumb rules are in order. After listing requirements (or modifying them), we must have a list of available feed ingredients. Remember that simple formulas (or requirements) can be met with simple mixtures, although it may or may not be desirable to have more complex formulas than needed. More complicated specifications require use of more ingredients if formulation is to be exact, and adding unnecessary specifications will invariably increase formula cost.

The question frequently comes up whether to formulate on a dry basis or as fed basis. Rations done on a dry basis are easier to compare to requirements, particularly when wet ingredients such as silages are included in the ration. On the other hand, formulation on a dry basis can result in odd weights and percentages in the as fed formula. From a practical point of view, if weighing is not done on an easily controlled scale (and sacks must be handled by hand), it is best to avoid fractional weights, using parts of sacks, etc. One example will be done on as fed and the other on a dry basis.

First Example

For purposes of illustration we will put together a simple protein supplement which is to be pelleted. We will specify that it is to have 30% CP with 1/3 of the CP to be from urea, 0.6% P and a Ca:P ratio between 1.0 and 1.5 and with 10,000 units of vitamin A added/lb. In addition, we will specify that it is to have 5% molasses (pelleting aid) and some alfalfa hay. This formula will be done on an as fed basis since its guarantees are based on as fed weights. This means that the nutrient contents of the ingredients must be multiplied by the DM percentage since the tabular data in Appendix Table 1 are on a dry basis. The feed ingredients that will be used in both examples are shown below. As fed percentages are shown in ( ) and have been rounded off for convenience.

Ingredients to be used in examples

	DM %	CP, %	TDN, %	Ca, %	P, %
Alfalfa hay	90	18.4(16.6)	57(51)	1.25(1.12)	0.23(0.21)
Corn silage	28	8.4(2.4)	70(20)	0.28(0.08)	0.21(0.06)
Corn grain	89	10.0(8.9)	89(79)	0.02(0.02)	0.35(0.31)
Cottonseed meal	92	44.8(41)	75(69)	0.17(0.16)	1.31(1.20)
Molasses	75	4.3(3)	91(68)	1.19(0.88)	0.11(0.08)
Brewers grains	92	28.1(26)	66(61)	0.29(0.27)	0.54(0.50)
Urea	98	287 (281)	--	---	---
Dical. phosphate	96	---	--	21.0(20.2)	18.6(17.9)
Limestone	98	---	--	33.8(33.1)	---

Step 1. Calculate the urea required and the CP and TDN from molasses. Urea required = 1/3 of 30/2.81 = 10/2.81 = 3.56 lb/100 or %

	%	CP	TDN
Urea	3.56	10.0	--
Molasses	5.0	0.15	3.4
Total		10.15	3.4

Step 2. Calculate the nutrient concentration needed in the remaining part of the supplement. We will reserve 1% slack space for adjusting Ca and P, so the CP still needed = 30 - 10.15/.9044 = 21.95%, and the TDN needed = 65 - 3.4/.9044 = 68.11% [The .9044 comes from 100 - (3.56 + 5 + 1)].

Step 3. Use double Pearson's square and solve for CP and TDN.

Mix 1, exact CP, low TDN (<68.11%) — Calculate TDN

Alf. hay 16.6 ⟶ 21.95 ⟶ 19.05 = 78.07% x 51 = 39.82

CSM 41 ⟶ 21.95 ⟶ 5.35 = 21.93% x 69 = 15.13

Total 24.40 — 54.95% TDN

Mix 2, exact CP, high TDN (>68.11%)

Corn grain 8.9 ⟶ 21.95 ⟶ 19.05 = 59.34% x 79 = 46.88

CSM 41 ⟶ 21.95 ⟶ 13.05 = 40.66% x 69 = 28.05

Total 32.10 — 74.93

Combine mixes and solve for TDN

Mix 1 54.95 ⟶ 68.11 ⟶ 6.82 - 34.13%

Mix 2 74.93 ⟶ 68.11 ⟶ 13.16 = 65.87%

Total 19.98

Step 4. Calculate ration %.

	%	Ca	P
Urea (required)	3.56	--	--
Molasses (required)	5.0	0.044	0.006
Alfalfa hay 100 x .9044(.7807 x .3413)	24.10	0.270	0.051
CSM, 100 x .9044(.2193 x .3413) = 6.77 } 100 x .9044(.4066 x .6587) = 24.22 }	30.99	0.050	0.372
Corn grain, 100 x .9044(.5934 x .6587)	35.35	0.007	0.110
Slack	1		
	100.00	0.371	0.539

Step 5. Sum up Ca and P and calculate amount needed.

The amount of Ca and P in the supplement is shown on the right above. We need only a small amount of P (0.6 required - 0.539 = 0.061%). This can be provided by adding a small amount of dicalcium phosphate (0.061/.179 = 0.34 lb/100). This ingredient also supplies some Ca (0.34 x .202 = 0.069 lb), thus we now have 0.44% Ca (0.069 + 0.371). For a minimum ratio of 1:1 Ca:P, we need more Ca. The needed Ca (0.6 - 0.44 = 0.16 lb) can be provided with 0.48 lb of limestone (0.16/0.331). We now have used up most of the slack (0.34 + 0.48 = 0.82). The remainder could be used for a vitamin A premix,but it would probably be more feasible to mix it with some of the CSM. Thus we will add 0.18% TM salt and the supplement is now complete as shown:

Urea	3.56
Molasses	5.0
CSM	30.99
Corn	35.35
Alfalfa	24.10
Dical	0.34
Limestone	0.48
TM salt	0.18
Vit. A	+

With the procedures used we have now met specifications exactly for CP, TDN, Ca, P, minimum vitamin A, and for limits put on urea and molasses.

Second Example

For this example we will use NE values for computation. As explained by Church (1977), there are problems when trying to use NEm and NEg values since the ratio of NEm:NEg varies from feed to feed. Preston (1979) states that "the most accurate way to use NE values to formulate rations would be to use NEm value plus a multiplier x the NEg value all divided by 1 + the multiplier; the multiplier is the level of feed intake above maintenance relative to maintenance". The value we might derive from the modified requirements (Table 5-4) would be:

$$NE = \frac{NEm + (4.13/3.36)(NEg)}{1 + (4.13/3.36)} = \frac{2.40 + (4.13/3.36)(1.35)}{1 + (4.13/3.36)} = \frac{4.059}{2.229} = 1.821$$

Preston suggests that this method would be feasible to use in computer formulation. When used with hand formulation, it is necessary to compute a new value for each change in the level of feeding. For example, if the NEg in the above example was 2.26 rather than 3.36, the answer would be 1.722. We will use this procedure in this example. However, the writer suspects that most nutritionists would probably not bother to use it for hand formulation.

Step 1. Establish requirements.

For this example we will formulate a ration on a dry basis for a 350 kg steer gaining 1.2 kg/day. The tabular values are 6.23 Mcal NEm and 5.92 Mcal of NEg. This animal should consume about 8.1 kg (18 lb) of dry matter, resulting in the need for NEm of about 1.89 Mcal/kg and NEg of about 1.25 Mcal/kg (NRC values are higher, but response to additives and growth stimulators aren't included).

Step 2. Compute adjusted NE values.

If we use Preston's procedure, it will require 3.29 kg of feed (6.23/1.89) for maintenance and 4.74 kg of feed for gain (5.92/1.25), so the formula would be:

$$NE = \frac{1.89 + (4.74/3.29)(1.25)}{1 + (4.74/3.29)} = \frac{3.69}{2.44} = 1.51$$

The mean of the NE values (1.82 and 1.25) is 1.57, thus we need to reduce the mean values of individual feed ingredients by about 4% (1.57 – 1.51/1.51 = 3.97%). Presumably, this will improve our precision in formulation when using NE values. For this particular example we will only need to use three ingredients to solve for NE and CP. The computations on the NE values are as shown:

Ingredient	NEm	NEg	Mean	Mean x .96
Corn silage	1.56	0.99	1.275	1.22
Corn grain	2.18	1.43	1.805	1.73
Brewers grains	1.42	0.83	1.125	1.08

Step 3. Formulate using double Pearson's square.

Mix 1, exact NE, low protein (<11%) — Calculate CP

Ingredient	Target NE	Parts	%	CP
Corn silage 1.22	1.51	0.22	= 43.14% x 8.4	= 3.62
Corn grain 1.73		0.29	= 56.86% x 10.0	= 5.69
		0.51		9.31

Mix 2, exact NE, high protein (>11%)

Ingredient	Target NE	Parts	%	CP
Corn grain 1.73	1.51	0.43	= 66.15% x 10.0	= 6.62
Brewers grains 1.08		0.22	= 33.85% x 28.1	= 9.51
		0.65		16.13

Combine and solve for CP

Mix	Target CP	Parts	%
Mix 1 9.31	11	5.13	= 75.23%
Mix 2 16.13		1.69	= 24.77%
		6.82	

Step 4. Calculate ration % on dry and as fed basis.

	% dry	DM, %	Amount as fed*	% as fed
Corn silage, 100(.4314 x .7523)	32.45	28	115.89	60.52
Corn grain, 100(.5686 x .7523) = 42.78 } 100(.6615 x .2477) = 16.39	59.17	89	66.48	34.72
Brewers grains, 100(.3385 x .2477)	8.38	92	9.11	4.76
	100.00		191.48	100.00

* Calculated by (% in dry formula/% DM)100

In this example we have fulfilled exactly the specifications for NE and CP but have not done anything for other nutrients or additives. Generally, for complete rations with wet ingredients, it is preferable to formulate on a dry basis as shown here, particularly for inexperienced persons.

COMPUTER FORMULATION

A number of computer programs are now available which do not require any expertise in programming by the user. These programs may be available through some of the state extension services or through many feed mills or nutrition consultants. There are two main types of programs. One is designed to produce a least cost formula. It considers only specifications on rations and nutrient content of feed ingredients along with cost of the ingredients. The other type which is intended to formulate for maximum profit, considers other factors such as price of milk or meat and certain other production costs in addition.

With computer formulation many more specifications and/or restrictions are feasible as compared to hand calculation. However, it must be remembered that added specifications and restrictions require more complex formulas and normally result in some increase in costs, although it may only be a very slight increase.

Least cost formulations of rations for animals are feasible because they have specific nutritional requirements which must be provided through the ration for the animal to grow rapidly and efficiently. Cattle (or other animals) do not require any particular feed ingredient so long as the ration provides sufficient amounts of each of the required nutrients. Therefore, the most economical ration is one which combines feed ingredients to fulfill these nutrient requirements at the least possible cost.

Nutritionists have strived for many years to balance between nutrient requirements, palatability and cost of feed ingredients to reduce animal production costs. The electronic computer can rapidly consider a very large number of potential feed ingredients and nutritional requirements which must be "matched." Accuracy and speed of calculation are the major advantages of computer formulation. Conventional and least cost formulation follow different types of logic. By conventional formulation, logic proceeds as follows: The nutritionist first judges how much concentrate and roughage should be fed. If a 15% corn silage (percent of air dry ration) is selected, then the remaining 85% becomes corn plus supplement. Next the nutritionist checks how much protein, Ca and P is provided by the mixture of corn and corn silage. Then sufficient proteins, minerals and possibly vitamins are added to balance the ration. The nutritionist made several judgements (level and source of roughage and grain, and source of supplemental vitamins and minerals) without considering economics.

A second type of logic is used for computer formulation. First, the nutrients and the minimum or maximum amounts of each nutrient is entered. Then for each feed to be considered, the current price, nutrient composition, and percentage limitations are considered. Using this information, the computer picks the mixture of feed ingredients which will satisfy the nutrient levels specified. This procedure is the reverse of the conventional formulation procedure where the feeder or nutritionist first decided what he wanted to feed and then adjusted the mixture to meet the animal's nutritional requirements.

The success obtainable from either type of ration formulation depends upon the nutritional specifications set for each commodity. These must describe accurately how these feeds match the animal's nutritional requirements. As new information on nutrient requirements appears, the nutrient limits must be changed. Conventional widsom and thumb rules aid one in conventional formulation to cover certain nutrient needs without special consideration or thought. Many of the early least cost computer formulations were disappointing because the computer lacks this wisdom and all appropriate constraints were not included by the computer program.

The California Net Energy System for maintenance and gain gives excellent results in linear programming for feedlot cattle. Neither TDN or ENE [(NEm+NEg)/2] will rank feeds in the order at which they supply energy for gain by feedlot cattle. In least cost feed mixing for feedlot cattle, minimum requirements must be set for net energy for gain (NEg), protein, Ca, P, K, and either roughage or crude fiber. In addition, other nutrients or nutrient ratios such as metabolizable protein, urea fermentation potential, soluble protein or energy:protein ratios may be included. Any nutrient or ratio in the series may be set as a minimum, a maximum, or an exact (equality) requirement. Some programs also include a rounding function. When this is used, quantities of feed can be included to the nearest 50, 20, 1 lb, etc. This is a realistic approach which should result in less errors in the feed mill.

Limits may also be included on individual feed ingredients. For example, salt can be locked into the formula at a fixed amount. Certain ingredients like molasses may need to be restricted for either mechanical or nutritional reasons to a level below some upper limit. Through these ingredient maximums or minimums, the nutritionist applies his skill and knowledge to the formulation beyond that controlled by nutritional specification. For example, in some types of rations, dehydrated alfalfa meal may have a value higher than would be indicated by its known nutrient composition. Including this product at some lower limit in the ration can improve the final ration. Where ingredient limits are truly necessary, they should be set; however, indiscriminate limits can result in a fixed rather than a least cost mixture. As nutrient composition or availability from feedstuffs change, the input for that feed must be updated.

One limitation of conventional linear programming is nutrient density within the mix. If nutrient levels specified for the mix are low, the computer will add a low cost ingredient as a filler to reduce the cost of the ration. While the ration price may be very low, the filler may be of no benefit to cattle. In this manner, a least cost mixture may not produce least cost production of cattle. Conversely, if the nutrient density of the mix is specified too high, the computer is forced to use feeds with concentrated nutrients which will raise the cost of production even though the mix is least cost at the specifications set.

A second limitation is that "associative effects" of feeds are not considered. Especially with unprocessed grains, a moderate level of roughage addition will drpress digestibility and feed efficiency. Since these effects appear to vary with several factors (roughage type and level, grain processing and feed intake) in a poorly understood manner, more animal research is necessary before these restrictions and interactions can be considered in least cost formulation.

When the ration is completed, most programs will, if desired, print out a complete analysis on whatever nutrients have been included on different feed ingredients, even though a particular nutrient may not have been used in the formulation. This may be helpful, at times, in identifying potential deficiencies.

One example is given of a beef finishing program which is available through the California Extension Service. In this program the operator lists the feeds that are to be considered along with their prices as shown in Table 5-7. This particular program has some built in constraints (limitations) for some ingredients as shown on the right side of the table.

Input is also required on sex, breed, if the previous ration was high roughage, if rumensin is used, if the ration is a growing or finishing ration, a value for shrinkage of cattle into and out of the feedlot, and a cost for daily yardage. You can pick any one or several feeding phases and a begining and ending body weight are required. The ration can be obtained on an as fed and/or dry basis. It does not ask for input on nutrient requirements, although the built in constraints can be modified for individual situations.

For the example used (English breed steers, fed high roughage background ration, no rumensin, with a beginning weight of 500 and a final weight of 700 lb, and with 3% shrinkage), the program came up with the ration shown in Table 5-8. The nutrient requirements built into the program were (dry basis) for NEm (0.87 Mcal/lb), NEg (0.55 Mcal/lb), crude protein (12.44%), Ca (0.53%), P (0.45%), and K (0.66%). The Ca:P ratio was restricted to a range of 1.2:1 to 2.5:1. Table 5-8 shows the ration on as fed and dry percentages, estimated daily and total consumption, cost, and the range of prices at which these particular feed ingredients would be selected. For these ingredients which were not selected, a price is shown in Table 5-9 at which a specific ingredient would come into the formula. This type of information serves as a useful buying guide, although the prices

Table 5-7. Feed cost and constraints used.

Feed name	As fed, $/cwt	Min	Max, % DM
Corn silage, 30% DM	1.25		100.00
Alfalfa hay, 25% CF	5.00		100.00
Barley, 46-48#	6.00		100.00
Corn, dent #2	5.60		100.00
Beet pulp, mol dr.	5.90		45.00
Cottonseed meal, 41%	9.00		30.00
Brewers grain, Cal	5.50		30.00
Dical. phos.	18.00		1.50
Limestone, gr	2.30		2.00
Molasses, cane	5.50		10.00
Urea, 46% N	12.00		1.00
Feather meal	15.00		15.00
Bakery waste	7.00		15.00

Table 5-8. Ration analysis.

	DM basis		As fed basis						
Feeds used	DM %	% in ration	% in ration	lb/ Hd/Dy	lb total	Total cost	Price $/cwt	Lower range	Upper range
Corn silage, 30% DM	30	41.79	68.38	21.17	1717.61	21.47	1.25	1.17	1.29
Corn, dent # 2	90	52.90	28.85	8.95	724.75	40.59	5.60	5.20	5.90
Cottonseed meal 41	90	3.01	1.64	0.51	41.20	3.71	9.00	6.38	10.51
Urea, 46% N	100	1.00	0.49	0.15	12.33	1.48	12.00	-65.87	33.27
Dical. phos	100	0.83	0.41	0.13	10.19	1.83	18.00	-0.61	74.71
Limestone, gr	100	0.48	0.24	0.07	5.93	0.14	2.30	-0.97	19.74
				30.97	2512.00	69.21			

Table 5-9. Feeds not used.

Feed	% DM	As fed price At formulation	Incoming
Alfalfa hay	90	5.00	4.35
Barley	90	6.00	5.55
Beet pulp	90	5.90	5.65
Brewers grain	90	5.50	4.78
Molasses, cane	75	5.50	4.29
Feather meal	90	15.00	11.79
Bakery waste	90	7.00	5.87

(called shadow prices) are subject to change with different constraints on ingredients or nutrient requirements. For this particular situation the program gives an estimated daily performance of 3.01 lb gain with a feed conversion (dry basis) of 5.05 at a cost of 34¢/lb, which includes a 20¢ daily yardage charge.

A number of LP's are available on a public basis and an unknown number on a private basis through feed milling companies or nutrition consultants. The public programs do not, generally, allow formulation for maximum profit. For example, with the illustration used here, more information would be required. The cost of cattle coming into the lot and the selling price of those leaving would be an essential requirement. In addition, some imput would be required on death losses, veterinary expenses, cost of capital, interest, depreciation, etc. in order to cover most of the expected expenses which are associated with an enterprise such as feeding cattle.

References Cited

Ames, D.R. and D.R. Brink. 1977. J. Animal Sci. 44:136.
Ames, D.R., D.R. Brink and C.L. Willms. 1980. J. Animal Sci. 50:1.
Church, D.C. 1972. Feedstuffs 44(37):37.
Church, D.C. (ed.). 1977. Livestock Feeds and Feeding. O & B Books, Inc., Corvallis, Oregon.
Fox, D.G. and J.R. Black. 1977. Michigan Agr. Expt. Sta. Res. Rpt. 328, p. 141.
Mertens, D.R. 1980. Distillers Feed Conference Proc. 35:35.
Preston, R.L. 1979. Feedstuffs 51(35):3A.
Rohwer, G. Undated. Bar Diamond Co., Parma, Idaho.

Chapter 6—Part A. Nutrition of Livestock Grazing on Range and Pasture Lands

by M.M. Kothmann

Classical animal nutrition is based on the concept of determining the nutreint requirements of an animal and the nutritive value of feeds, then formulating a ration which meets the requirements of the animal. Nutrition research is generally conducted in confined areas such as pens or metabolism stalls and the diet is controlled. Nutrition of grazing animals has unique characteristics. The nutrient requirements (energy) of the animal may be increased by activity associated with grazing and travel or by increased environmental stress such as high or low temperatures. Also, the nutritive value of forage consumed is difficult to determine because the animal selects various combinations of plant species and plant parts. Nutrition of the animal is influenced primarily by management of the land resource and animals and not by formulation of rations.

Nutrition for grazing animals is concerned with management factors such as stocking rates, season of grazing, grazing systems, plant species available, kinds and classes of animals best suited for the forage resource and formulation of supplements to alleviate specific nutrient deficiencies. In this chapter I will consider first the forage resource, followed by an examination of how animals interact with vegetation in selecting diets and the use of supplementation to correct nutrient deficiencies in the diet. Part B addresses the effects of grazing systems on range livestock production.

FORAGE RESOURCES

Principles relating to nutrition of grazing livestock are the same for range and pasture lands. However, applications may vary for several reasons. Pasture lands generally consist of monocultures or combinations of two plant species. The species are selected and planted by the manager. Management is usually applied to control soil fertility (fertilization), competition from other species (weed control), and sometimes moisture availability (irrigation). Rangelands usually consist of complex mixtures of many plant species representing grasses, forbs, and woody plants. Management of rangelands is usually less intensive than on pasture lands. Fertilization and irrigation are very seldom applied and weed control, if utilized at all, is applied at infrequent intervals.

Because of differences in the characteristics and management of the forage resources on range and pasture lands, different management and cultural practices are emphasized to influence nutrition of grazing animals. The most important factors on pasture lands have been selection and breeding of improved plant materials, establishment of new plantings, fertilization technology, methods of harvesting and storing forage during periods of excess production, weed control, and grazing management. On rangelands grazing management has been the most important management practice with brush control second and reseeding of ranges third. Few other cultural practices are applied to rangelands.

Forage resources can generally be described by two components, availability and nutritive value. The nutritive value of a forage depends upon its nutrient content, the digestibility of the nutrients, and the amount the animal will consume (intake). Factors which affect availability and nutritive value of forage will be discussed under the general topics of forage classes, maturity and environment.

Forage Classes

Forage species may be grouped into three classes—grasses, forbs and browse. Grasses are members of the family *Poaceae,* forbs are herbaceous annual and perennial dicots and monocots other than grasses, and browse consists of perennial woody shrubs and vines utilized as forage by grazing animals. Characteristic differences exist among classes and within classes. Variation in nutritive value exists among species within a forage class, among individual plants within a species, and between leaf and stem fractions within a plant.

For example, leaves of heath aster (*Aster ericoides* L.), a perennial forb extensively distributed in North America, were significantly higher in percent N than stems (Kothmann and Kallah, 1978; Fig. 6-1). Nitrogen in leaves ranged from 1.9-2.3% and in stems from 1.1-0.3% between March 10 and April 1. In vitro dry matter digestibility (IVDMD) of leaves was high throughout the year ranging from 81% on March 10 to 71% on November 5 (Fig. 6-2). IVDMD of stems declined rapidly with maturity from a high of 60% in March to a low of 25% in November. Concurrent with the decline in nutrient content of stems was a significant increase in the relative proportion of stem to leaf material within the plants.

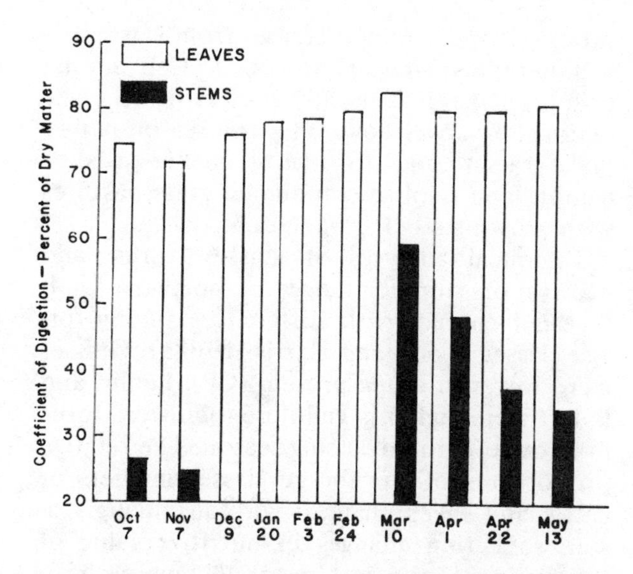

Figure 6-2. In vitro coefficients of digestion of heath aster aerial parts sampled at different stages of development from October 7 to May 13 near College Station, Texas. Taken from Kothmann and Kallah (1978).

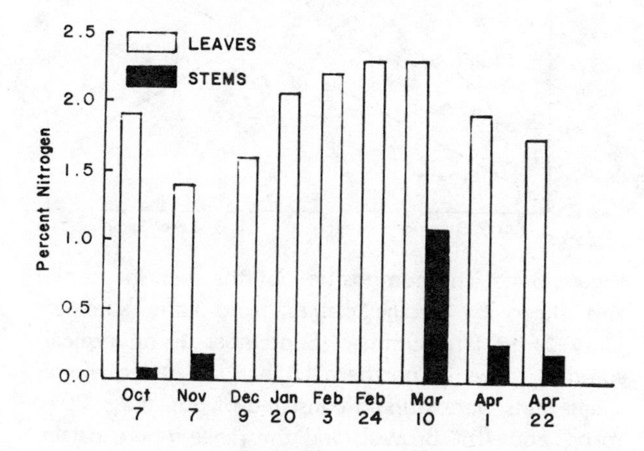

Figure 6-1. Total N (%) in heath aster stems and leaves sampled at different stages of development from October 7 to April 22 near College Station, Texas. No stem material was available from winter rosettes. From Kothmann and Kallah (1978).

Table 6-1. Mean composition and nylon bag dry matter digestibility (NBDMD) values of browse samples. Component values other than dry matter are percent oven-dried weight.[a]

	Spring		Summer	
	Twigs	Leaves	Twigs	Leaves
Dry matter	21.5	28.2	46.9	42.6
Cell wall contents	35.8	26.1	67.4	35.0
Acid det. fiber	28.1	20.7	54.0	26.6
Acid det. lignin	11.3	9.0	16.5	10.4
Silica	0.22	0.37	0.12	0.62
NBDMD	74.9	78.6	35.9	67.4

[a]From Short et al (1973)

Analysis of composite twig and leaf samples from 23 browse species in the southeastern USA indicated high digestibility of both fractions during spring (Table 6-1). Twig samples, containing only 5 cm terminal segments of current-year's growth, declined 39 units of digestibility from spring to summer; whereas, leaves only declined 11 units. Thus, with browse as with forbs, early growth of both leaves and stems is highly digestible, but the digestibility of the stems declines rapidly to very low levels as plants mature. Digestibility of leaves remains relatively high as long as they are alive.

Characteristic differences in digestible dry matter content of different groups of grasses and legumes were illustrated by Riewe (1976) (Fig. 6-3). Warm season perennial grasses tend to have the lowest digestibility and legumes the highest with cool season perennial grasses and annual grasses being intermediate. Total dry matter production of these forage groups tends to follow an inverse trend with warm season perennial grasses being the most productive.

Grasses generally can not provide adequate digestible energy to meet the requirements for lactating dairy cows, whereas a high quality legume pasture may. Growing animals require higher levels of energy in their diets to achieve optimum growth rates than warm season grasses can provide. The cost of forage production generally increases from warm season perennial grasses to cool season annual grasses. Thus, it is usually preferable to graze mature, dry beef cows on warm season perennial grasses and to utilize higher quality legume and cool season annual grass pastures with growing stock or lactating cows.

Chemical analyses of grasses, forbs, and browse on summer ranges of northern Utah revealed characteristic differences among forage classes (Cook and Harris, 1968a). Grasses were lower in crude protein (CP), lignin, and P, but were higher in cellulose than were forbs or browse with advancing season. The relative proportions of the forage classes in diets of cattle and sheep changed with advancing season, reflecting changes in nutritive value of the forage (Fig. 6-4). Grasses declined in both

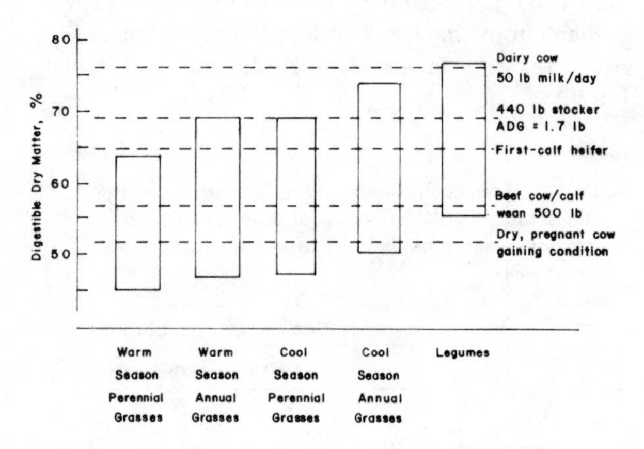

Figure 6-3. The relationship between digestible dry matter content of forage classes and the nutrient requirements of cattle. From Riewe (1976).

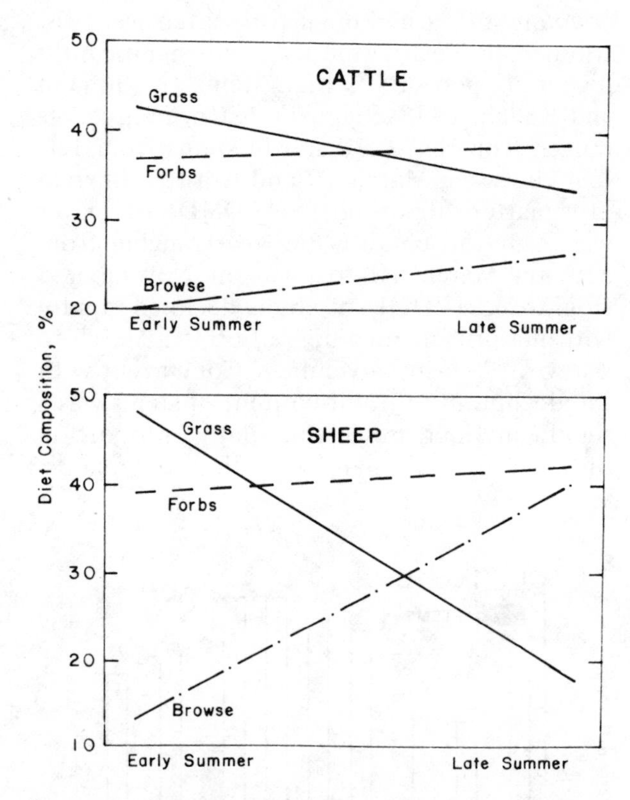

Figure 6-4. The composition of the diet for cattle and sheep by forage classes from early summer (July 1) to late summer (September 1) on typical summer range in northern Utah. The range where sheep diets were studied consisted of 65% grass, 20% forbs, and 15% browse, and the range where cattle were studied consisted of 25% grass, 40% forbs and 35% browse. From Cook (1956).

Table 6-2. Average chemical content of grass, forbs and browse[a] collected from typical mountain range during early summer (7/1 to 7/15) and late summer (8/15 to 9/1).[b]

	Grass		Forbs		Browse	
Item, %	Early	Late	Early	Late	Early	Late
Ether extract	2.3	2.4	4.3	3.1	4.2	6.3
Protein	8.3	4.2	10.6	8.8	12.3	10.8
Lignin	9.7	12.3	9.7	11.6	15.6	16.1
Cellulose	38.7	44.5	26.0	29.1	20.5	23.7
Other carbohy.	35.6	31.4	38.7	38.6	41.0	37.2
Ash	5.4	5.2	10.7	8.8	6.4	5.9
P	0.27	0.21	0.42	0.32	0.31	0.33

[a]Averages include 11 grasses, 25 forbs and 7 browse spp, all of which are common on summer ranges of northern Utah.
[b]From Cook and Harris (1968a)

cattle and sheep diets, forbs increased slightly, and browse increased significantly from early to late summer. These shifts in diet selection maintained higher levels of CP and P in the diets.

Comparison of the percentage of grasses, forbs, and browse in the diets with their percentages in the available forage (Fig. 6-4) indicates that sheep selected most strongly for forbs. The percent grass in sheep diets was below the percent in the available forage. During early summer, cattle selected strongly for grasses, but as grasses matured they were replaced in the diets with browse.

Average daily gains of sheep and cattle decline as vegetation on spring and summer ranges matures; however, gains of lambs and calves decline less than gains of ewes and cows (Cook and Harris, 1968a; Tables 6-3, 4). This is a common occurrence in the production of lambs and calves from rangelands and represents an important principle. It is not always possible to meet the total energy requirement for lactating cows or ewes from grazed forage, nor will the forage meet the nutrient requirements of young, growing animals. However, the cow and/or ewe meets the nutrient requirements of the young animal through conversion of low quality forage

Table 6-3. Average daily gain for sheep and cattle on mountain ranges during the summer grazing season.[a,b]

	Pounds per day gain			
	Sheep		Cattle	
Period	Ewes	Lambs	Cows	Calves
June 8 to July 16	0.19	0.64	1.56	1.67
July 17 to August 5	0.15	0.60	1.00	1.42
August 6 to Sept. 15	0.07	0.48	0.58	1.18

[a]Lambs weighed an average of 71.1 and calves weighed an average of 351.2 lb.
[b]From Cook and Harris (1968a)

Table 6-4. Nutrient content and average daily gain for sheep and cattle for different stages of growth of introduced wheatgrasses compared to native foothill grasses and the recommended standard for females on the range during the first 8 weeks of lactation.[a,b]

					Pounds per day gain			
	Dig. protein,	Dig. energy,	TDN,	P,	Sheep		Cattle	
Species and season	%	Kcal/lb	%	%	Ewes	Lambs	Cows	Calves
Crested wheatgrass								
early	10.6	1,578	69.2	.16	0.37	0.56	1.5	2.3
late	3.9	991	50.4	.12	-0.25	0.39	0.3	1.6
Pubescent wheatgrass								
early	11.8	1,401	68.6	.18	0.36	0.54	1.9	2.1
late	3.1	1,078	55.2	.15	0.06	0.40	0.2	1.3
Tall wheatgrass								
early	11.7	1,325	62.8	.18	0.26	0.53	1.2	2.3
late	7.0	1,109	56.2	.16	-0.43	0.44	0.8	1.5
Intermediate wheatgrass								
early	10.0	1,208	59.4	.19	0.28	0.66	1.6	2.2
late	5.4	1,169	59.6	.17	0.22	0.52	0.5	1.7
Russian wildrye								
early	8.1	1,172	59.6	.16	---	---	1.7	2.3
late	7.4	1,142	58.6	.15	---	---	1.1	2.0
Native foothill grasses								
early	7.4	1,396	65.9	.23	0.10	0.58	1.0	1.5
late	4.6	1,142	56.7	.18	-0.12	0.53	0.1	1.1
Recommended	5.4	1,120	57.0	.22	---	---	---	---

[a]Early and late was May 1 to May 15 and June 10 to June 20 for all species except Russian wildrye which was June 15 to June 20 and July 8 to July 15.
[b]From Cook and Harris (1968a)

to milk. When energy intake of the cow or ewe is below their requirement, energy stored as fat can be mobilized to support lactation. It is usually much more economical to store excess energy as fat in the cow or ewe during periods when high quality forage is available, than it is to feed it as a supplement during periods of deficiency.

Spring and fall present critical problems to range livestock producers in the western USA because of limited forage availability on foothill ranges. Several species of introduced wheatgrasses are widely used to provide forage during these periods. When five introduced species were compared with native wheatgrasses (Table 6-4), all species met the nutrient requirements of lactating animals for digestible protein and energy during early spring, but only three species met the requirements in late spring.

Crested wheatgrass *(Agropyron cristatum)* is well suited for early spring grazing, since it starts growth earlier than most other species. With proper grazing it can be utilized by sheep from as early as April 15 until about June 8 during most years, and cattle can graze it about 2 weeks longer. To maintain vigorous production, two pastures should be used so that early and late spring grazing can be alternated. Intermediate wheatgrass *(Agropyron intermedium)* and Russian wildrye *(Elymus junceus)* are more suitable for late spring and early summer grazing than crested wheatgrass. Cattle do well on these species until as late as August 1 in most years; however, sheep do not do well on foothill ranges after about the third week of June. Intermediate wheatgrass stands are not maintained on lower foothill ranges receiving less than 13 inches annual precipitation, but Russian wildrye will grow in lower rainfall areas (Cook, 1966).

Desert ranges in the intermountain region of the western USA support primarily grass-shrub communities with few forbs. They are grazed primarily by dry ewes and cows during the winter when forage quality is relatively low. The average composition of animal diets from three major plant communities is shown in Table 6-5. As on summer ranges, grasses are lower than browse in CP and P, but do contain higher levels of ME. Diets consisting of browse would be deficient in energy, but would contain adequate levels of CP, P and carotene to meet the maintenance requirements of cows and ewes (Cook et al, 1954).

Table 6-5. Average vegetation composition of diets of grazing sheep and content of the critical nutrients used in appraising nutrient value of desert ranges for winter grazing in the Great Basin for 3 major vegetation types.[a]

Item	Browse	Grass	Average
Predominately grass range			
Veg. comp. of diet, %	24	76	100
Dig. CP, %	4.9	0.8	2.1
P, %	0.12	0.06	0.08
ME, Kcal/lb	616	821	737
Predominately saltbrush range			
Veg. comp. of diet, %	61	39	100
Dig. CP, %	4.6	0.9	3.1
P, %	0.12	0.07	0.10
ME, Kcal/lb	619	757	628
Predominately sagebrush range			
Veg. comp. of diet, %	70	30	100
Dig. CP, %	4.8	0.7	3.7
P, %	0.12	0.06	0.12
ME, Kcal/lb	575	803	619

[a]From Cook and Harris (1968a)

Diets composed of all grasses would contain adequate energy but would be deficient in CP, P and carotene. Where animals can select from both grasses and browse, the need for supplements can be reduced.

The value of plant communities as a forage resource varies not only with the quality and quantity of forage produced, but is also determined by the kinds of animals utilizing the resource. A study of cattle and deer utilization of aspen, pine, and mixed aspen-pine communities in the Black Hills National Forest of South Dakota revealed that understory production was inversely related to overstory density (Kranz and Linder, 1973). Aspen communities appeared to represent better feeding areas for both deer and cattle than mixed aspen-pine. However, use by white-tailed deer *(Odocoileus virginianus)*, estimated by fecal pellet group density, was greatest in mixed aspen-pine. Cattle use, estimated by fecal chip density, was greatest in aspen and least in pine.

Introduced warm season, perennial grasses commonly used in tame pastures of the southeastern USA are rather highly digestible at initiation of growth in the spring (McCartor and Roquette, 1976; Fig. 6-5). The rate of decline in forage quality from spring to summer varies widely among species and varieties. Common lovegrass *(Eragrostis curvula)* declines most rapidly and reaches its lowest digestibility in August. Digestibility of Coastal

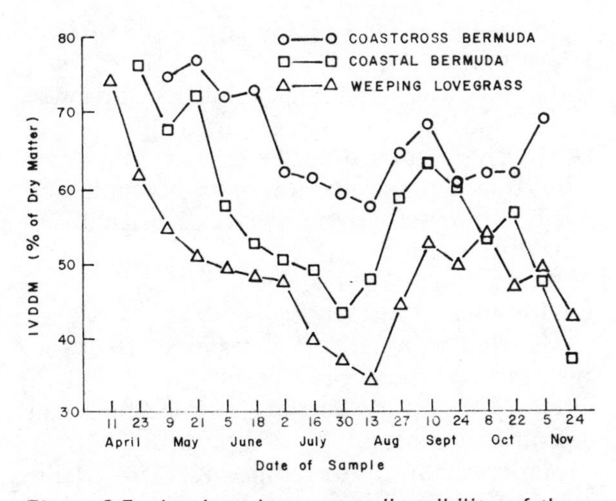

Figure 6-5. In vitro dry matter digestibility of three grasses throughout the growing season at Overton, TX. Taken from McCartor and Roquette (1976).

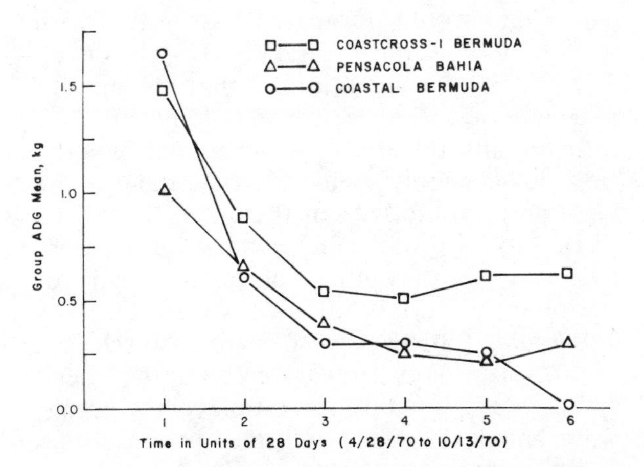

Figure 6-6. Average daily gain (ADG) time trends on Pensacola bahiagrass, Coastal bermudagrass and Coastcross-1 bermudagrass pastures in Louisiana. From Chapman et al (1972).

bermudagrass *(Cynodon dactylon)* declines more rapidly than Coastcross bermudagrass, a variety selected for higher digestibility, but both reach a low point in mid-summer. Late summer rains during most years, combined with reduced light intensities and temperatures, result in fall growth which increases digestibility of the forage, but not to the levels of early spring. Digestibility of Common lovegrass and Coastal bermudagrass declines again after mid-September, but Coastcross bermudagrass remains relatively high through November. The suitability of the three forages for different classes of cattle will vary seasonally with Coastcross bermudagrass providing significantly higher quality forage than the other two grasses during summer and late fall.

Average daily gains (ADG) of steers grazing on Coastcross-1 bermudagrass, Coastal bermudagrass, and Pensacola bahiagrass *(Paspalum notatum)* were compared between April 28 and October 13 (Fig. 6-6). When the stocking rate was 6 steers/0.81 ha, Pensacola bahiagrass and Coastal bermudagrass produced comparable ADG, but both were inferior to Coastcross-1 bermudagrass. The superiority of the Coastcross was manifested during summer and fall as would be indicated from the digestibility measurements at Overton, Texas (Fig. 6-5).

Legumes are the only forbs which have received significant attention as forage species for seeded pastures. Martin and Anderson (1975) evaluated the nutrient content of 12 weed species. Redroot pigweed *(Amaranthus retroflexas)*, common lambsquarters *(Chenopodium album)* and common ragweed *(Ambrosia artemisiifolia)* had nutrient composition and digestibility equivalent to that of high quality alfalfa *(Medicago sativa)*. They concluded that some of the common weed species found in new stands of perennial forages growing in fertile soils of Minnesota do not decrease the nutritive value of hay or pasture if utilized at relatively early stages of maturation. Generalizations are not possible in comparisons of forage quality among "weed" and crop species. Comparisons must be based on specific examples where species and environmental conditions are described.

Maturity

Maturity is a major factor affecting forage quality. Emphasis is generally placed on the decline in nutritive value associated with maturation of forage. However, Wilman and Wright (1978) evaluated the early stages of regrowth in three grasses after defoliation with and without applied N. S.22 Italian ryegrass, S.24 perennial ryegrass and S.37 Cocksfoot were studied for 21 days immediately following cutting or grazing. Digestibility remained essentially constant, but the percent cell contents and the percent of CP in cell contents increased from day 3 to 21, whereas water soluble carbohydrates decreased. The proportion of the digestible material which is very readily available to the animal evidently increased appreciably during the first 3 weeks.

This trend would normally be associated with increased voluntary intake (Osborn et al, 1974). These data imply that grazing of regrowth less than 3 weeks old may not be nutritionally beneficial to the animals in addition to generally being detrimental to the vigor and productivity of the plant.

Distribution of N and water soluble carbohydrates within individual plant organs of reproductive tillers of Cocksfoot *(Dactylis glomerata)* changed with maturity (Davies, 1976). Ten days after the mean date of head emergence, the most digestible plant organs were the leaf blades and the inflorescences and the least digestible were the lower internodes. At the mature hay stage the ranking was leaf blades>upper sheaths>inflorescence >lower internodes>upper internodes. At all stages in the development of the crop from 10 days after head emergence onward the upper organs of the head-bearing tillers were richest in N; whereas, the basal portion of the stem was highest in water-soluble carbohydrates. At 10 days after head emergence, the organs in the upper part of the profile were more digestible than the basal organs, but this distinction became progressively less marked, and at the most mature stage the upper internodes were in all instances less digestible than the lower.

The increasingly polarized distribution of N and water soluble carbohydrates that accompanies plant maturity has many implications for forage management and even plant selection and breeding. Based on his research, Davies (1976) recommended cutting cocksfoot near the ground (4 cm) to improve yields of hay or silage without reducing digestibility. This is related to the growth form of the plants and would vary among species and varieties within a species.

Seasonal changes in digestibility of three browse species, two grasses, and mixed forb samples were reported by Short et al (1974) (Table 6-6). Honeysuckle (*Lonicera* spp.) and greenbrier (*Smila* spp.) are evergreen and elm (*Ulmus* spp.) is a deciduous tree. Honeysuckle grows most of the year and digestibility of its leaves was relatively constant at a very high level. Its twigs were also highly digestible except during the summer. Early spring growth of leaves and twigs of greenbrier and elm were highly digestible, but digestibility declined by summer. The digestibility of mature greenbrier twigs and elm leaves picked up from the ground in January was very low. Digestibilities of panic grasses and mixed forbs were lower than digestibilities of browse leaves and the cool season grass, Elbon rye. Obviously, a good mix of species would allow a grazing animal maximum opportunity to select a nutritionally adequate diet at all seasons of the year.

Environmental Influences

Many environmental factors may affect forage quality but only temperature, light intensity, moisture, fertility and leaching will be considered here.

Plant response to temperature and light intensity varies between temperate (C_3) and tropical (C_4) grasses with species having different optimum levels. Dry matter production is increased by higher light intensities and temperatures but water soluble carbohydrate content is reduced by high temperatures. High light intensity increases water soluble carbohydrates. Dry conditions reduce dry matter production and water soluble carbohydrates with a resulting increase in percent CP and ash. Low light intensities and reduced water availability can result in accumulation of NO_3 in plants (Deinum, 1966).

The effects of temperature, light intensity and moisture give rise to predictable seasonal changes in forage quality in temperate regions. In tropical regions forage quality is affected primarily by variations of moisture, stage of maturity, species and fertility (see Ch. 15). In temperate regions spring grass has high CP and low fiber contents; summer grass has low CP and high fiber contents; autumn grass has high CP and average fiber contents.

Plant response to fertility is primarily reflected in dry matter production although some changes in chemical composition of forage have been reported. Changes in chemical composition can be confounded with differences in the relative proportion of plant parts (leaf, stem and inflorescence) and in stage of maturity. High levels of soil N may alter leaf to stem ratios and retard plant maturity. Ellis and Lippke (1976) reported the response of Kleingrass *(Panicum colaratum)* and bermudagrasses to levels of N fertilization at 3 locations (Table 6-7). CP increased at high levels of N but digestibility was not affected.

Leaching of forages, especially mature forages, is one of the most important factors affecting nutrition of range animals. Young,

Table 6-6. Percentages of cell wall constituents and nylon bag *in vitro* dry matter digestibility (IVDMD) of forage collected on four dates from grass, forb and browse species in the southeastern United States.[a]

Species	Date	Cell walls		IVDMD	
		Leaves	Twigs	Leaves	Twigs
Honeysuckle	Apr.	20.7	41.2	86.5	70.6
	Aug.	30.2	66.4	84.3	35.0
	Oct.	28.6	38.5	84.7	66.3
	Jan.	26.0	40.9	84.9	66.2
Greenbriers	Apr.	22.8	30.7	91.2	87.5
	Aug.	43.9	77.5	65.7	18.0
	Oct.	40.6	78.7	66.5	18.2
	Jan.	38.5	76.0	73.3	20.9
Elm	Apr.	21.7	38.9	89.7	78.3
	Aug.	37.9	64.5	54.4	44.9
	Oct.	38.0	62.3	51.4	42.0
	Jan.	61.4	62.4	26.6	47.0
Panic	Mar.	76.5[b]		61.6[b]	
	May	75.2		68.2	
	Aug.	62.4		61.0	
	Nov.	66.8		50.0	
Elbon rye	Nov., vegetative	36.5		93.8	
	Mar., boot	60.1		89.2	
	May, seeds present	70.1		59.6	
Mixed forbs	Apr., immature	43.8		68.5	
	July	54.2		51.5	
	Oct.	61.6		42.1	
	Jan., mature	74.8		26.7	

[a]Adopted from Short et al (1974)
[b]Samples of panic, Elbon rye and mixed forbs were whole plant composits of leaves and stems present on those dates.

Table 6-7. Effect of nitrogen fertilization on crude protein (CP) content and its lack of effect on dry matter digestibility (DMD).[a]

Texas[1]						Homer, Louisiana[2]			Oklahoma[3]		
Kleingrass			Coastal bermuda			Coastal bermuda			Coastal bermuda		
N	CP	DMD	N	CP	DMD	N	CP	DMD	N	CP	DMD
lb/acre	%	%	lb/acre	%	%	lb/acre	%	%	lb/acre	%	%
25	5.9	56.1	25	6.7	53.8	0	10.4	51.4	0	11.5	64.9
100	6.6	57.1	100	7.6	54.8	600	15.8	53.1	400	18.4	66.0
200	9.8	55.8	200	10.6	53.4	1200	16.9	52.1	1400	18.9	65.0
300	10.8	57.4	300	10.8	55.1	--	--	--	--	--	--

[a]From Ellis and Lippke (1976)
[1]Data of Buentello and Ellis (1969); one application 6 wk. prior to cutting.
[2]Data of Rainwater (1974); 6 applications 4 wk. prior to cutting.
[3]*In vitro* data of Webster et al (1965); 4 applications.

actively growing plant tissue is relatively immune to loss of mineral nutrients and carbohydrates; whereas, more mature tissue approaching senescence and fully mature tissues are very susceptible to leaching. Leaves from healthy vigorous plants are much less susceptible to leaching than are leaves which are injured, whether injury be induced by microorganisms, insects and other pests, adverse climate, nutritional and physiological disorders, or by mechanical means (Tukey, 1970).

K, Ca, Mg and Mn are the inorganic nutrients usually leached from live plants in largest quantities. Those minerals most resistant to leaching are Fe, Zn, P and Cl. Of the major organic constituents, carbohydrates are most likely to be leached from live plant material. Quantitive losses of proteins, amino acids and organic acids from live plant material are slight (Tukey, 1970).

The intensity and volume of rain affect the extent of leaching. Rain which falls as a light drizzle, continuously bathing the foliage, will remove considerably more nutrients than will a greater quantity of water which falls in a short period of time. Leaves which are wetted and dried are subsequently wetted more easily, indicating that intermittent rain tends to overcome the hydrophobic characteristics of some leaves. Dew is very important as a leaching agent, especially in seasons and climates where rainfall is low. Fog which accumulates on leaves and then drips to the soil may be a very effective leaching agent. Losses of carbohydrates by leaching increase as both temperature and light intensity increase; whereas, mineral losses are not affected. Live plants with adequate nutrient supplies are able to replace nutrients leached by brief infrequent rains. However, during prolonged periods of rain, as during the wet season in the tropics, the growth and yield of the plant may be severely limited by its inability to absorb nutrients by the roots and replace them in sufficient quantities to oversome leaching losses (Tukey, 1970).

Another major loss of nutrients from foliage as plants mature is translocation. The major elements, N, K and P, and nonstructural carbohydrates are readily mobilized and translocated from senescent plant tissues (Larcher, 1975). Thus, with maturity, forage quality declines as a result of the combined effects of reduced nutrient content from leaching and translocation and reduced digestibility of the fiber fraction.

DIET SELECTION

It is a well documented fact that under almost all circumstances livestock graze selectively on range and pasture lands. Both animal and forage attributes affect diet selection. Animal attributes include species, class of animal, productive function, prior conditioning and experience, climatic variables and perhaps other factors, the sum of which determine *preference.* Factors affecting diet selection which are attributed to the vegetation are those affecting *palatability* of forage and include chemical composition and physical characteristics such as texture, pubescence or presence of spines. Preference and palatability cannot be applied separately in practice, since preference for a given animal has a unique set of forage characteristics which define palatability (see Ch. 11, Vol. 2). For example, because of different forage preferences between a goat and a cow, characteristics of palatable forage would differ in certain respects for the two species of animals.

Certain expressions of forage preference are similar among all kinds of livestock. Leaf is preferred over stem and green tissue is preferred to mature or dead forage. These preferences generally result in the selection of diets having nutritive value higher than the average of the forage available. This was illustrated by Arnold (1960) using sheep continuously grazing *Phalaris tuberosa* pasture at a stocking rate such that consumption exceeded growth. Standing forage was partitioned into stem, green leaf, and mature leaf daily for a 6-day grazing trial (Table 6-8). The first day's grazing did not alter the proportions of the 3 fractions; however, N content of all 3 fractions was lower on day two. On days two and three, leaf was selected in preference to stem resulting in an increase of percent stem in the standing forage from 33.7 to 63.6%. N content of stem and mature leaf fractions continued to decline to day six.

Cattle exhibited similar preference for leaf over stem when grazing Coastal bermudagrass pasture (Table 6-9). During both December and June there was more stem than leaf in the available forage, but the diets contained only a small percentage of stem. Digestibility of the leaf and stem fractions in the diets was

Table 6-8. Changes from day 1 through to day 6 in the proportions of three fractions of *Phalaris tuberosa* and their nitrogen content due to continued selective grazing.[a]

	Stem		Green leaf		Mature leaf	
Day	Avail. DM, %[b]	N in OM, %[c]	Avail. DM, %	N in OM, %	Avail. DM, %	N in OM, %
0	34.1	1.21	19.4	3.56	46.6	2.72
1	33.7	1.17	18.7	3.42	47.5	2.55
2	58.2	1.07	10.5	2.64	31.1	2.31
3	63.6	1.04	9.2	2.42	27.2	1.84
4	57.6	1.00	14.1	3.76	28.3	1.72
5	63.8	0.93	6.3	4.21	29.9	1.73
6	69.9	0.89	11.3	3.97	18.9	1.49

[a]From Arnold (1960)
[b]Available dry matter
[c]Nitrogen in organic matter

Table 6-9. Digestibility and proportion of plant parts in available vs. consumed Coastal bermudagrass pasture.[a]

		Available			Consumed		
Date		Whole	Leaf[b]	Stem[c]	Whole	Leaf	Stem
12-73	Wt %[d]	100	34	66	100	82	18
	Kg/100 kg W[e]	1200	408	792	1.60	1.31	.29
	IVDOM[f]	35	40	36	42	45	39
6-74	Wt %[d]	100	41	59	100	90	10
	Kg/100 kg W[e]	800	328	472	3.11	2.80	.31
	IVDOM[f]	55	63	48	65	65	57

[a]From Ellis (1978)
[b]Leaf lamine, not including leaf sheath.
[c]Stem and leaf sheath.
[d]Weight of individual fraction as % of total fraction.
[e]Weight of available forage in plot per 100 kg body weight of grazer.
[f]*In vitro* digestibility of organic matter standardized with a forage sample of known *in vivo* digestibility of organic matter.

always higher than the mean for leaf and stem fractions in the available forage.

Similar results were obtained for selection of leaf and stem by sheep grazing alfalfa (Arnold, 1960). Selection of leaf over stem was accentuated as the plants matured and quality of stems declined relative to leaves. As with *Phalaris,* sheep initially consumed some stems but then switched almost totally to leaves until leaves were no longer available and they were forced to eat the remaining stems (Fig. 6-7). These data indicate clearly the effects of increasing utilization on the quality of diets the animal can select. Heavy grazing pressures will force increased utilization of low quality forage, reducing nutrient intake of the animal.

The effects of chemical composition of the forage on diet selection is not simply expressed since many chemicals interact with the animal's senses during the grazing process. Plant characteristics such as CP, P, carotene and moisture content and digestibility of forage are generally positively associated with palatability. Lignin, cellulose, cutin, silica, cell walls, alkaloids and various terpenes and essential oils are negatively associated with palatability (Marten, 1978). Simple classifications of plant palatability based on limited chemical characteristics such as CP, crude fiber, or digestibility are inadequate when applied to the diverse kinds of forages found on rangelands.

Broad generalizations can be made about dietary preferences of cattle, sheep, and goats for forage classes; however, these are subject to influence of the composition of the available forage. Cattle generally select diets based on grasses with limited amounts of either forbs or browse, depending upon their respective availabilities (Fig. 6-8).

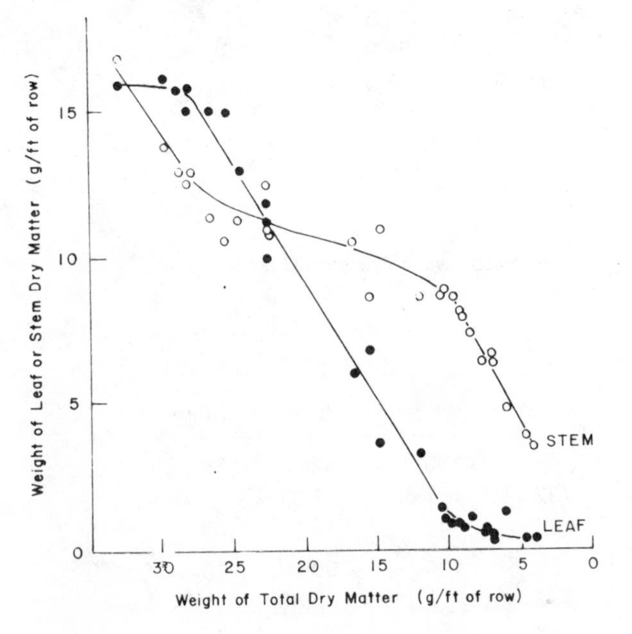

Figure 6-7. Changes in relationship between stem and leaf weight to total available dry matter of alfalfa as a result of selective grazing. From Arnold (1960).

Figure 6-8. Cattle select grasses as the primary component of their diet but palatable forbs, such as heath aster, may contribute significantly to their nutrient intake.

Anderson (1977) found that Hereford heifers in the Rolling Plains of Texas selected a relatively high percentage of forbs during May, but utilized few forbs during the fall (Fig. 6-9). Such seasonal shifts in diet are common to all range livestock. Bryant and Kothmann (1979) found that sheep (Fig. 6-10) and Angora goats (Fig. 6-11) also selected more forbs during spring than during fall and winter. The highest percentages of browse were selected during fall and winter when few forbs were available. Sheep consumed more grasses and forbs but less browse than did Angora goats. Forage classes were generally selected in proportion to their ability to provide green foliage with grasses utilized most heavily by cattle, forbs by sheep, and browse by goats.

Selective grazing has certain advantages to the animal but presents a variety of problems to the manager. Animal performance can be enhanced by allowing them to graze selectively. In fact, the average forage quality on many ranges is not adequate to support livestock production and only by allowing the animals to graze selectively can livestock production by sustained economically. On pastures where the average forage quality is adequate, selective grazing can "stratify" the

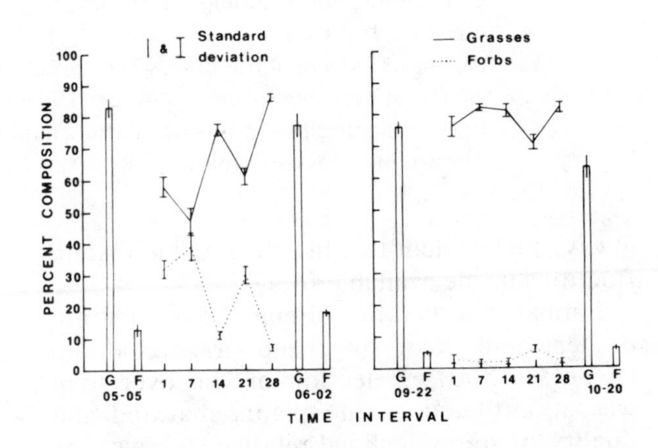

Figure 6-9. Standing crop and diet composition (%) of grasses and forbs on SDG paddock C. Dates on which the standing crop of grassed (G) and forbs (F) were evaluated are below the histograms. Diets were sampled at 7-day intervals during two 28-day grazing periods. From Anderson (1977).

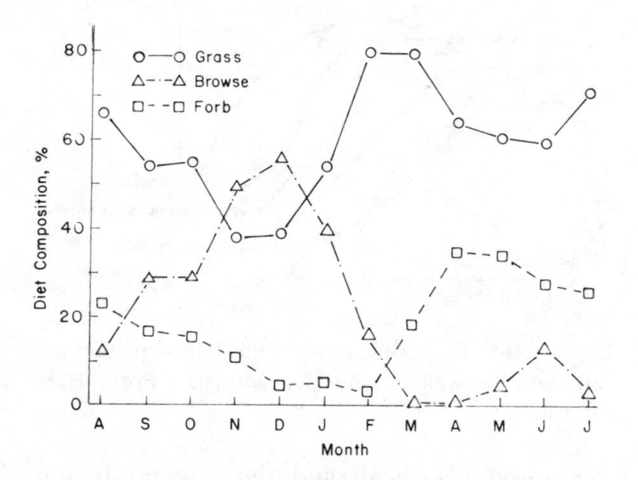

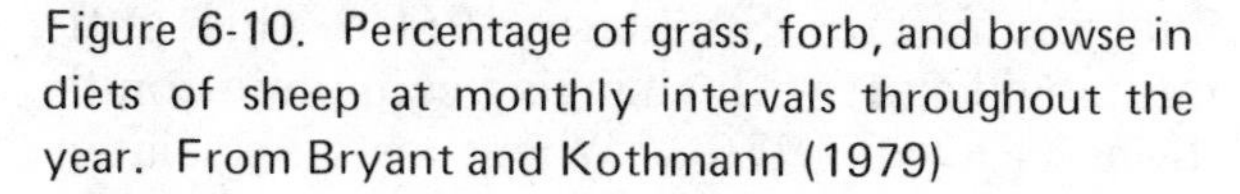
Figure 6-10. Percentage of grass, forb, and browse in diets of sheep at monthly intervals throughout the year. From Bryant and Kothmann (1979)

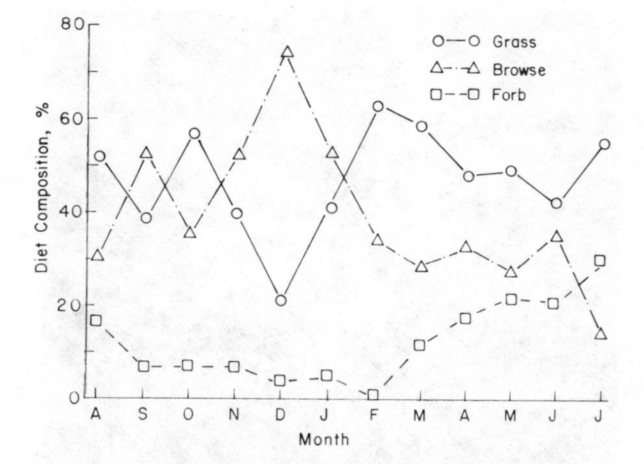

Figure 6-11. Percentage of grass, forb, and browse in diets of Angora goats at monthly intervals. From Bryant and Kothmann (1979).

forage so that a portion is not of suitable quality (Table 6-8). The removal of green leaves, leaving stems and mature leaves causes a shift in diet quality with increasing utilization (Fig. 6-8). To avoid this problem on tame pastures, forage can be utilized at an immature stage or can be strip grazed, forcing utilization of all of the plant during one or two days. This maintains the average nutrient intake at an acceptable level and prevents several days of luxury consumption followed by several days of deficient intake which may occur with longer grazing periods in rotation grazing.

On rangelands, the major problems caused by selective grazing are related to species and area selectivity. Most rangelands support diverse floras with species varying in palatability (Fig. 6-12). Also, topography may be steep and rough with large paddocks encompassing a variety of soil types. Distance from water to forage may be in excess of two miles. All of these factors accentuate selective grazing and much of range management is directed toward alleviating the problems caused by selective grazing. Adjustments of stocking rates and season of grazing and development of specialized grazing systems are methods commonly used to alleviate problems of *species selective* grazing. Development of additional water and fencing and strategic location of mineral supplements and feed grounds are primary methods directed to problems of *area selective* grazing.

Figure 6-12. Selective grazing is a major problem in range utilization. Animals graze some plants too closely while leaving other plants undergrazed. This leads to deterioration of the range.

NUTRIENT INTAKE FROM GRAZED FORAGE

Nutrient intake is a function of the amount of feed an animal consumes and the content and availability of the nutrients in the feed. The amount of forage a grazing animal will consume is a function of the availability and quality of the forage, the kind of grazing management applied, and the kind, class and physiological condition of the animal. We have already considered some of the factors affecting nutrient content of forages and the process of diet selection by the animal. Understanding the interrelationship of the plant and animal components in range and

Figure 6-13. Esophageally fistulated animals (upper) may be used to determine the botanical and chemical composition of diets selected by grazing animals. Sheep equipped with fecal collection bage (lower) provide estimates of total fecal output. Analysis of the feces allows estimates to be made of total nutrient intake and digestibility.

pasture systems is vital to providing proper nutrition for grazing animals (Fig. 6-13).

Grazing Management

Grazing management should always be one of the first considerations of the livestock manager since it is the major factor affecting animal production/head and/acre. Consideration should first be given to properly adjusting the stocking rate. Hart (1978) reviewed research relating animal production and stocking rates (Fig. 6-14). As stocking rate increases, gains/animal decrease, but gains/acre increase to a point beyond which further increases in stocking rate reduce both gains/animal and/acre (Petersen et al, 1965). This general relationship applies to all grazing lands and the optimum stocking rate will be in the range between the rate producing

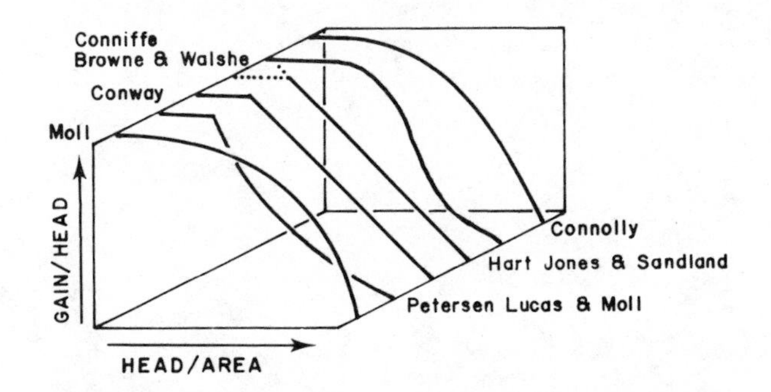

Figure 6-14. Proposed relationships between stocking rate and liveweight gain per animal. From Hart (1978).

maximum gain/head and the rate producing maximum/acre (Stoddart, 1960). This optimum will shift with changing economic conditions (Riewe, 1976). When costs associated with an individual animal are high relative to cost/acre of land, stocking rate should be adjusted to improve gain/animal. Stocking rates should be shifted to increase gain/acre when costs associated with land are high relative to costs/animal.

In most developed countries, costs of land are high relative to costs of animals. Thus, land owners generally seek to maximize production/unit of land. On tame pastures intensive cultural practices are applied to maintain the productivity of the vegetation, and the land is capable of sustaining intensive agriculture. However, many rangelands are not capable of sustaining intensive agriculture and heavy stocking rates designed to maximize animal production/unit of land cause deterioration of both vegetation (Kothmann et al, 1978) and soil (Knight et al, 1980) conditions.

Rangelands are frequently grazed yearlong or throughout the season when grazing is possible. Selective grazing results in the desirable forage species and the most accessible parts of the pasture receiving the most frequent and intensive grazing. The result is reduced vigor and productivity of these components, and over a period of years less desirable species invade. Thus, the effect of continuous grazing at heavy stocking rates is a long-term decline in productivity of the range.

Grazing systems have been designed to offset the effects of selective grazing and to enhance range improvement through plant succession (Merrill, 1954; Kothmann et al, 1971). The components of a grazing system are the number of pastures and herds of livestock and the length of grazing and rest periods. Livestock are concentrated on a portion of the land to allow the vegetation on the remaining land in the system to grow for a period of time without grazing. Livestock are rotated among pastures to provide scheduled periods of grazing and rest for each pasture.

To design a grazing system for a given unit of land, the manager must consider three basic aspects of grazing management. (1) *stocking rate* is determined by the number of animals allocated to the system; (2) *stocking density* is determined by the number of pastures and herds for any given stocking rate; and (3) *stocking pressure* is determined by the length of the grazing periods for a given stocking rate and stocking density.

Both stocking density and stocking pressure can be changed in a grazing system without changing stocking rate, but changes in stocking rate will always affect stocking density and pressure. Increasing stocking density will result in more uniform distribution of grazing, thus, grazing systems which concentrate animals on smaller portions of the land, i.e. more pastures and fewer herds, will promote better grazing distribution. Increasing stocking pressure reduces diet quality (Anderson, 1977; Allison, 1978; Taylor et al, in press), thus, reducing the length of grazing periods should improve animal performance. Long rest periods which allow significant amounts of forage to mature between grazing periods will reduce animal production unless stocking rates are light. With all of these factors to consider, the design of a successful grazing system, which meets both plant and animal requirements, is not a simple exercise (Fig. 6-15). More detail on grazing systems is presented in Part B of this chapter.

Figure 6-15. Successful range livestock production is dependent on a combination of good management of both animals and rangelands.

Animal Units

Rangelands are grazed by different kinds and classes of animals each requiring different amounts of forage. The "animal-unit" (AU) is a concept designed to allow expression of the forage demand by an aggregate of grazing animals, since adding numbers of sheep and steers and cows does not provide a useful figure. The forage requirement of an animal-unit is defined as a constant 12 kg per day. The "animal-unit-equivalent" (AUE) for any animal may be calculated by dividing that animal's forage requirement by 12 kg/AU day. The number of animals multiplied by this animal-unit-equivalent will give the number of animal-units.

To calculate stocking rates the time factor must be incorporated. This may be done by multiplying the number of AU by the number of days they grazed on an area giving animal-unit-days (AUD). The total number of AUD grazed on an area during a year is the stocking rate. AUD may be converted to animal-unit-months (AUM) or animal-unit-years (AUY) by dividing by 30 or 365, respectively.

The AU and AUE should not be confused with substitution ratios or competition indices used for different kinds of animals on ranges. Different kinds of animals have different dietary preferences and grazing behavior. For example, goats and deer will utilize more browse and steeper, rougher topography than cattle. However, cattle will utilize relatively level grasslands more efficiently than goats or deer. Thus, the suitability of rangelands for different animal species varies with vegetation, topography and other factors. The purpose of the AU is simply to express the total forage demand in common units. The suitability of any given range for different kinds of animals must be determined and substitution ratios worked out independently of the AUE.

Establishing Stocking Rates

The best procedure for arriving at the correct stocking rate for an area is experience combined with records of animal production and vegetation trends. If estimates of annual forage production are available, a moderate stocking rate for yearlong grazing can be set by allowing 4X as much forage/animal as its predicted intake for the year. Allowing 2-4X the predicted intake will result in heavy stocking. Field experience of the author indicates that under yearlong grazing, approximately one-fourth of the total usable forage produced on rangeland can be consumed by grazing animals under proper stocking (Kothmann and Mathis, 1971). This relationship may change with grazing systems employing high stocking densities (see Part B).

Forage disappearance on a mid-grass range in north Texas equaled forage consumption at a stocking pressure of 10 kg/au/da, but at 50 kg/au/da, forage disappearance was twice as great as intake (Allison and Kothmann, 1979) (Table 6-10). Stuth (personal communication) found a similar relationship between stocking pressure and forage disappearance for Kleingrass and Coastal bermudagrass. These data indicate that the efficiency of forage harvest by grazing is a function of stocking pressure. Additional research is needed to determine how this relationship may be used to increase the efficiency of conversion of forage to red meat.

Table 6-10. Average forage disappearance and organic matter intake for four levels of stocking pressure averaged across three grazing trials and two periods within each trial.[a]

Stocking Pressure	Forage disappearance, kg/au/da	Intake, kg/au/da
10 kg/au/da	8.5	8.4
20 kg/au/da	12.0	9.4
40 kg/au/da	12.7	8.6
50 kg/au/da	16.3	8.6

[a]From Allison and Kothmann (1979)

Forage Availability and Quality

Average daily gain of grazing animals is related to both quality and availability of forage. Duble et al (1971) examined these relationships using 6 warm season perennial grasses (Fig. 6-16). When DM digestibility of forage was >60%, gains of yearling heifers were not affected until forage availability fell below 500 kg/ha. As the digestibility of the forage declined, the relation between gains and availability changed. The reason for this is the variability of forage quality within the standing crop. Standing crop is relatively homogenous when forage is young and digestibility is high. As forage matures and average digestibility of standing crop declines, there is more variation. The obvious conclusion is that stands of uniformly high quality forage can be grazed closely with little effect on animal performance. However, where the quality of forage in the standing crop varies significantly, animal performance will be much more sensitive to grazing intensity.

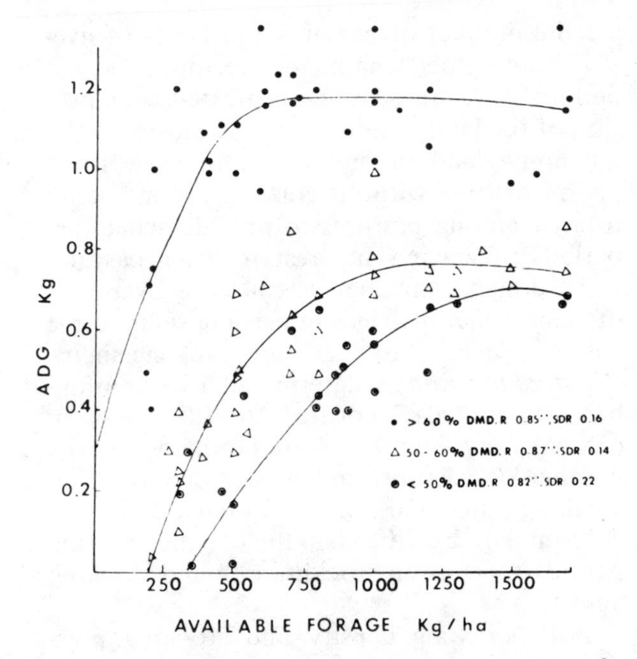

Figure 6-16. Polynomial regressions, $Y = a + bX + cX^2 + dX^3$, between animal performancy (Y) and available forage (X) on warm-season grasses at three ranges in dry matter digestibility (SDR = standard deviation from regression). From Duble et al (1971).

There is an asymptotic relationship between animal gain and mean pasture availability. Willoughby (1959) found the asymptote occurred at approximately 1,400 lb green forage (air dry basis)/acre. Johnston-Wallace and Kennedy (1944) reported that 1,000 lb (dry weight) of green forage was required to obtain maximum daily intake. Their research also indicated that a mean forage height of 4-6 inches was required to obtain maximum intake. From these studies it can be seen that increasing production of forage during peak growth periods will have little effect on gains/animal; however, even small increases in green forage when availability is below 500 lb/acre will result in significant increases in gain/animal (Willoughby, 1959).

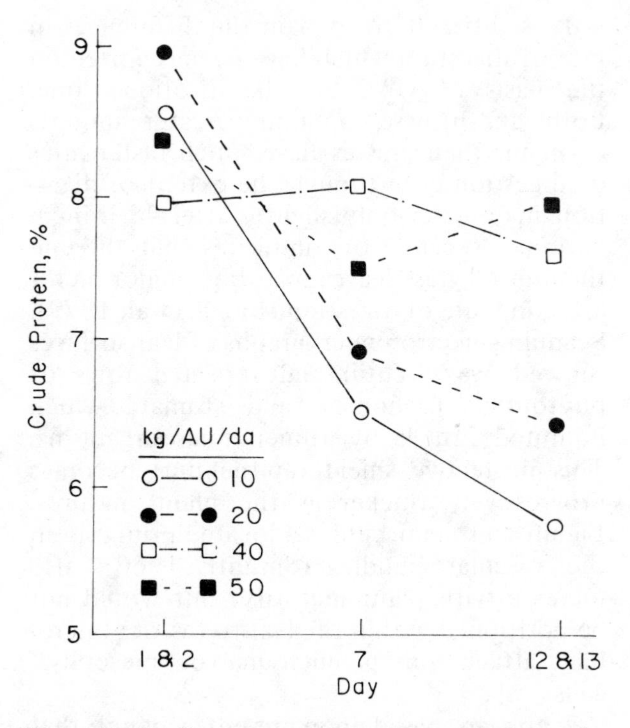

Figure 6-17. Percentages of crude protein in cattle diets for four levels of stocking pressure during a 14-day grazing trial on north Texas rangeland during July. From Allison (1978).

Allison (1978) investigated the effects of forage availability on CP content of cattle diets on rangeland in north Texas. He created 4 different stocking pressures by allowing different sized areas on a uniform stand of vegetation for a 14-day grazing period. CP content of diets declined progressively during the grazing trial on the two most intensive stocking pressures (10 and 20 kg/au/da), but did not change appreciably at the less intensive stocking pressures (40 and 50 kg/au/da) (Fig. 6-17). Intake averaged 7 kg/da/au during the trial.

Taylor (1966) investigated the relation between animal performance and forage availability and height on pastures of S.24 perennial ryegrass. Stubble height after grazing (in grazed areas only) was the best indicator of grazing intensity and animal performance. Animal performance fell as the mean height of herbage in grazed areas fell below 9.7 cm. At a mean height of 3.8-5.6 cm in grazed areas, animal performance reached maintenance. While this relation can be expected to vary for plants having different growth forms, the management implications would be similar.

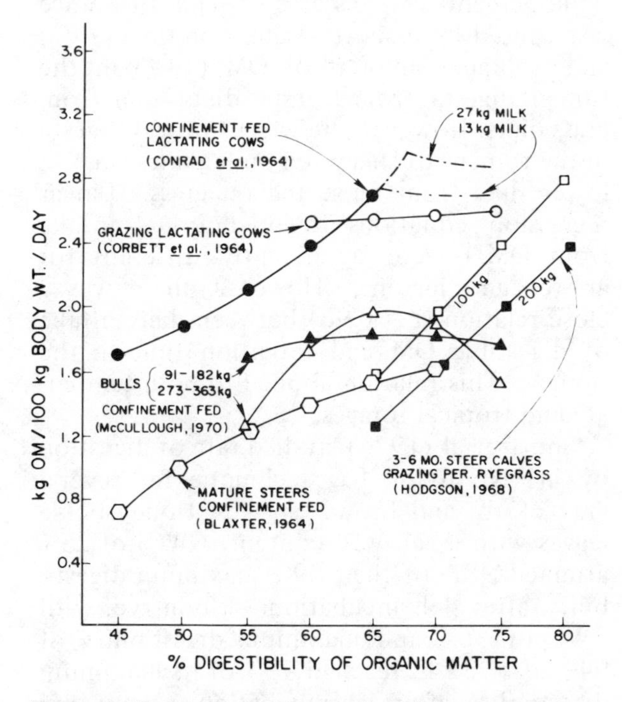

Figure 6-18. Forage intake vs. digestibility as reported by various investigators. From Ellis (1978).

The relation between forage intake and digestibility of forage has been investigated by many workers (Fig. 6-18). Intake is affected by age, size and physiological condition of animals but certain principles apply generally. Metabolic control of intake functions above the point where the animals' nutritive requirement for energy is met. However, on most ranges and pastures, digestibility of the forage limits intake. Thus, digestibility of forages has a dual effect on animal performance as it reflects the availability of energy in the feed and also affects the amount of feed that will be consumed. As digestibility of forage declines, nutrient intake will decline even more rapidly.

The relation between digestibility and intake is affected by the class of forage. Thornton and Minson (1973), working with tropical and temperate grasses and legumes, concluded that digestibility was a good basis for comparing nutritive value of forages within a forage class, but may not be valid when comparing grasses to legumes because of differences in voluntary intake (see Ch. 15). Voluntary intake of sheep was 13.7% higher for legumes averaging 53.2% OMD than for grasses averaging 63.0% OMD. Voluntary intake was 28% greater for legumes having the same percent OMD as grasses. This difference was caused by a shorter retention time (17%) and a higher amount of OM (14%) in the rumen digesta from legume diets than from grass diets. However, the weight of wet digesta in the rumens of sheep fed on legumes was 7% lower than from those fed on grass. Linear regression equations to predict OM intake from OMD were significantly different for grasses and legumes. However, there was a close relation (r = 0.96) between daily intake of digestible OM and retention time in the rumen. This relation applied to both temperate and tropical forages.

Short et al (1974) studied rate of digestion by in vivo nylon bag technique for several grass, forb, and browse species. Honeysuckle leaves with a cell wall content (CWC) of 25% attained 98% of their 89% maximum digestibility after 4 h incubation. Elbon rye, with CWC of 36% and maximum digestibility of 94% after 32 h, reached 75% of its maximum digestibility after 4 h incubation. A mixture of mature forbs having 56% CWC, attained 80% of its maximum nylon bag digestibility of 49% within 4 h. DM digestibility of mature woody twigs of 70% CWC reached 72% of the 33% maximum digestibility in 4 h. In sharp contrast to these rapid rates of digestion, a composite sample of mature grasses with 76% CWC reached only 27% of its maximum digestibility (57% after 168 h incubation) in 4 h. These sharply contrasting rates of digestion would be expected to have a significant effect on DM intake and nutrition of grazing animals. Comparisons of forage quality among different classes of forage based upon digestibility or CWC are probably not valid unless an adjustment is made for different rates of digestion and retention time in the rumen.

The causes for differing rates and extents of digestion have been attributed to many different factors with CWC and extent of lignification being considered the most important (Moore and Mott, 1973). There is generally a good correlation between lignin content and forage quality if comparisons are made within a plant species (Hart et al, 1976). Although not as good, the correlation between CWC or percent lignin and forage quality holds for grasses as a class. Distinct differences in forage quality exist between temperate and tropical grass species with temperate grasses generally having higher energy values (Laksevela and Said, 1978).

It is difficult to explain the differences in rate of digestion which have been reported on the basis of CWC and lignification alone. Forbs and browse containing greater amounts of lignin than grasses have much faster rates of digestion even though the extent of digestion may differ only slightly after 48 h incubation. Recent work indicates that the epithelium of grass leaves may be a major factor affecting rate of digestion (Brazle et al, 1979). Scanning electron micrographs of leaf surfaces showed waxy cutin and repeated rows of phytoliths, trichomes, and stomata which inhibited attack by rumen microorganisms. This protective shield (epithelium) becomes progressively thicker as the plant matures. Lignin and hemicellulose, located primarily in the vascular bundles (Sinnott, 1960), also increase with plant maturity, but would not be positioned within the leaf to restrict microbial attack on parenchyma or mesophyll cells.

It appears, based upon present evidence, that the potential extent of digestion is related to the degree of lignification, but the rate of digestion is affected more by morphological characteristics of leaf structure. Specifically, development of cutin in the epithelium and the arrangement of vascular bundles in the leaf. Forb and browse leaves with a branching vascular system break down more rapidly than do grass leaves with parallel vascular bundles. This subject needs to be thoroughly researched using the new tools available today.

Evaluating Forage Quality by Chemical Analysis

Energy and protein are the major nutrients which need to be evaluated in range forages. The levels of these nutrients are usually positively associated, but the correlation is not extremely high (Cook et al, 1977). Therefore, to identify forage quality adequately, both protein and energy should be considered.

The gross energy content of forages does not vary greatly among plant species or with plant maturity when expressed on an OM basis. However, digestible energy (DE) varies significantly with plant species and maturity.

Crude fiber, lignin, cellulose, CP, and cell walls (neutral detergent fiber) are chemical constituents which have been used in regression equations to predict DE (Short et al, 1974; Hart et al, 1976; Cook et al, 1977).

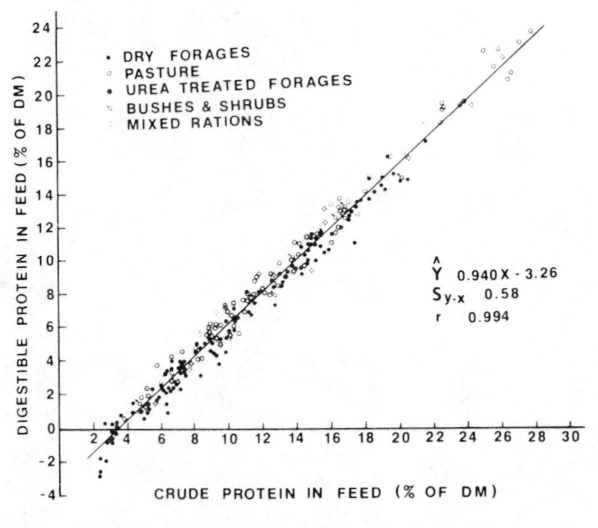

Figure 6-19. Relationship between the crude protein and apparently digestible crude protein concentration of feeds. From van Nierkerk et al (1967).

Best results are obtained when the regressions are used with closely related plant species. When the equations are extended from warm to cool season grasses, or from grasses to forbs and browse, lack of fit increases greatly.

In vitro digestibility techniques have also been used to estimate forage digestibility. Such techniques require more extensive laboratory equipment and the presence of donor animals for rumen innoculum, but they provide a more direct estimate of the DE content of a forage. Considerable variation may occur between different "batches" or analyses with in vitro techniques. Therefore, it is advisable to include a standard forage of known digestibility with each batch to correct for this variation. A close relationship (r^2 = 0.956) exists between digestible OM and DE (Rittenhouse et al, 1971; Jeffery, 1971; NRC, 1970). In vitro techniques may estimate apparent digestibility (Tilley and Terry, 1963) or true digestibility (Van Soest et al, 1966) of forage. To convert digestible OM to DE, Jeffery (1971) developed the equation DE = -0.218 + 4.92 DOM (r = 0.939), where DOM is the coefficient of digestion for OM and DE is in units of Kcal/g. This is very similar to the equation used to convert TDN to DE (NRC, 1969).

Protein may be evaluated as CP or digestible CP. The relation between the percentage of CP in the diet and digestible CP in the diet was determined to be linear (Fig. 6-19). The estimated true digestibility of dietary CP is

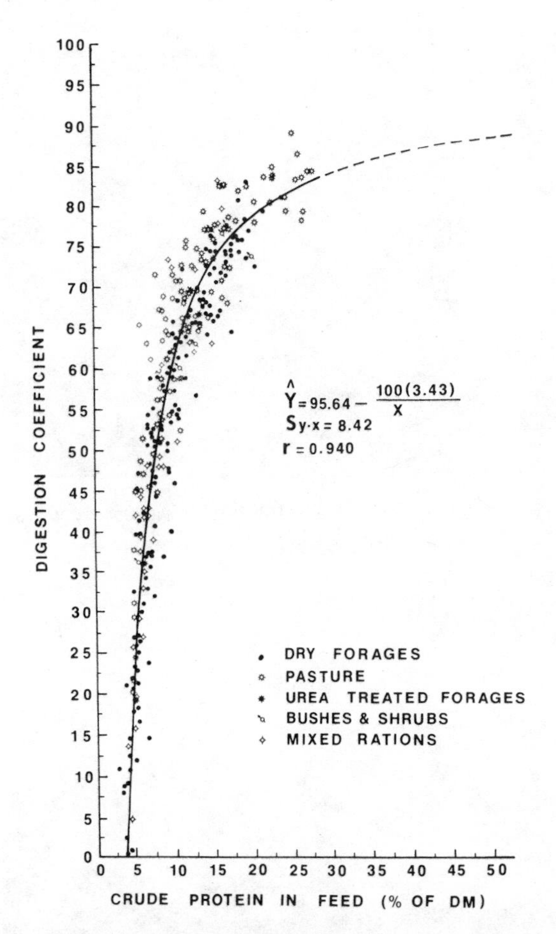

Figure 6-20. Relationship between the crude protein concentration of feeds and its apparent digestibility. From van Niekerk et al (1967).

94% and metabolic fecal N is 3.26 g/100 g DM consumed. The relationship between CP in the diet (x) and its digestion coefficient (y) was determined to fit the general form $y = a + bx^{-1}$. Fitting 361 observations from South Africa by least squares techniques, Van Niekerk (1967) obtained a realistic equation (Fig. 6-20) which agreed well with the linear equation. From these data and similar findings by other researchers, it appears that CP is a suitable measure of the protein content of the diet of grazing animals.

SUPPLEMENTATION OF GRAZING ANIMALS

Livestock grazing on green, actively growing forage generally obtain adequate levels of most nutrients. When they are forced to utilize mature or dormant forage, nutrient deficiencies may be expected.

Figure 6-21. Emergency feeding is necessary when forage is temporarily unavailable, for example if covered by snow.

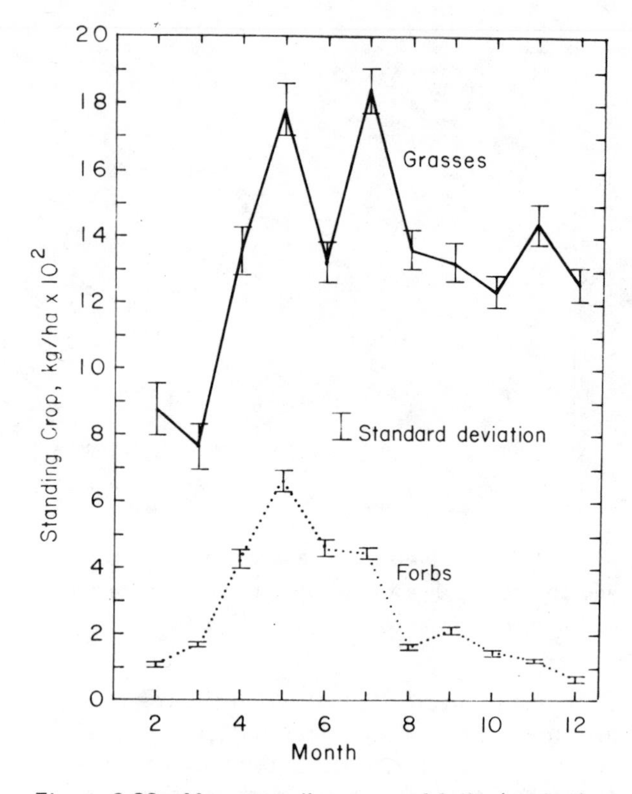

Figure 6-23. Mean standing crop of forbs (excluding Texas broomweed) and grasses within a 20-ha, continuously-grazed paddock for eleven sampling dates during 1975. From Anderson (1977).

Figure 6-22. Maintenance feeding is necessary when pastures are grazed so closely that dry matter availability limits forage intake of animals.

Supplemental feeding is the practice of providing limited amounts of feed containing concentrated levels of the deficient nutrients. Sometimes supplemental feeding is confused with emergency or maintenance feeding. Emergency feeding should be practiced to meet the total requirement of animals on range or pasture when forage is temporarily unavailable for grazing, such as if covered by snow (Fig. 6-21). Maintenance feeding is required when pastures are overstocked and forage availability drops to levels which restrict intake, resulting in a DM deficiency (Fig. 6-22). In emergency and maintenance feeding situations, the objective is generally to feed a balanced diet to carry animals through a stress period. This section will examine the role of supplemental feeding for range and pasture livestock.

Significant body reserves of some nutrients, such as vitamin A and energy, may be accumulated by grazing animals. However, there are small body reserves of most minerals, protein and water and the animal's dietary requirements must be met on a regular basis.

Supplementation of minerals and vitamins will not be considered in detail in this chapter. Vitamin A is the only vitamin commonly found deficient in diets of range livestock. Body reserves may be adequate for 90 to 120 days. Most commercial range supplements contain synthetic vitamin A and it may also be supplied by injection. Na and P are the minerals which are most widely deficient and adequate provisions should be made to supply them. Na should be provided throughout the year. P is required yearlong in some regions which have P-deficient soils. P should be provided whenever livestock are grazing mature

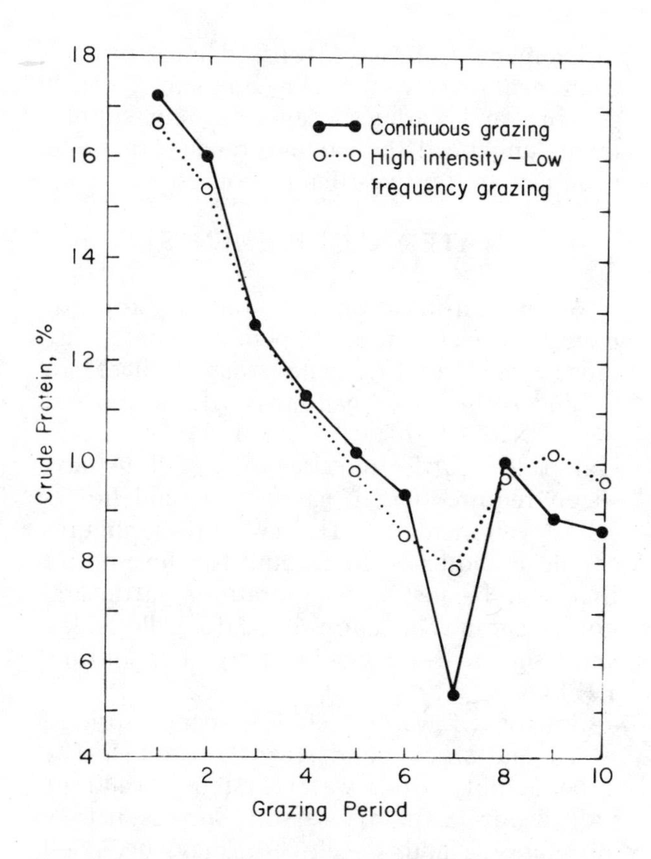

Figure 6-24. Trends in the crude protein content of diets of Hereford heifers for ten 28-day grazing periods beginning March 13, 1975, for 2 types of grazing management. Each sampling point consists of 5 diet collections made on 7-day intervals. From Anderson (1977).

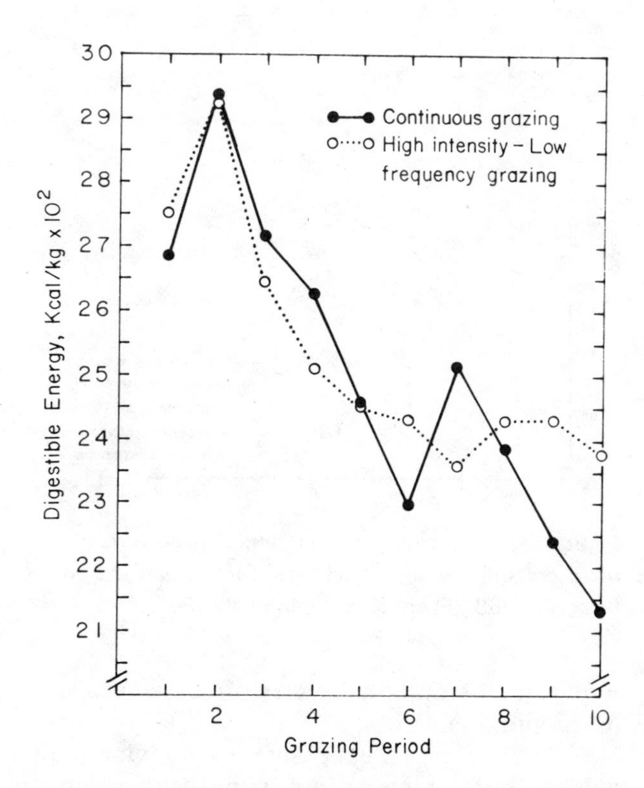

Figure 6-25. Trends in the digestible energy content of diets of Hereford heifers for ten 28-day grazing periods beginning March 13, 1975, for 2 types of grazing management. Each sampling point consists of 5 diet collections made on 7-day intervals. From Anderson (1977).

or dormant vegetation. Recent research indicates that K may be deficient in diets of cattle grazing on dormant forage (Kothmann and Hinnant, 1976; Karn and Clanton, 1977).

Protein and energy are both commonly deficient during certain periods of the year on rangelands. Whenever protein is deficient, animal production will be greatly reduced and utilization of the energy in the forage diet will be hampered. Growth will be reduced in stocker animals and reproduction and lactation will be reduced in breeding stock. Provided protein is adequate, cows and ewes can, to a limited extent, draw on body reserves of energy (fat) during periods of dietary deficiency without any losses in production.

Both amount and quality of range forage varies greatly during the year. Anderson (1977) found that the maximum standing crop of forage was reached in June in the Rolling Plains region of North Texas (Fig. 6-23). The highest quality diets were obtained in March and April at the beginning of the growing season with quality declining until late summer (Fig. 6-24, 6-25). Late fall growth increased CP and energy content slightly, but diet quality was still much lower than during spring. The mean weights of cows calving on these ranges during December to February were least in early spring and highest during June (Fig. 6-26) (Kothmann et al, 1971). The high levels of energy in the spring diets were stored as body fat and used the following winter when energy intake was at submaintenance levels.

This principle of utilizing periods of high quality forage to store energy reserves for later periods of deficiency has great economic importance. If animals are stocked too heavily or otherwise managed during the growing season to prevent accumulation of energy reserves, the cost of supplemental feeding will increase markedly. It is important to schedule breeding and weaning to try to obtain an optimum combination of the animal's nutrient

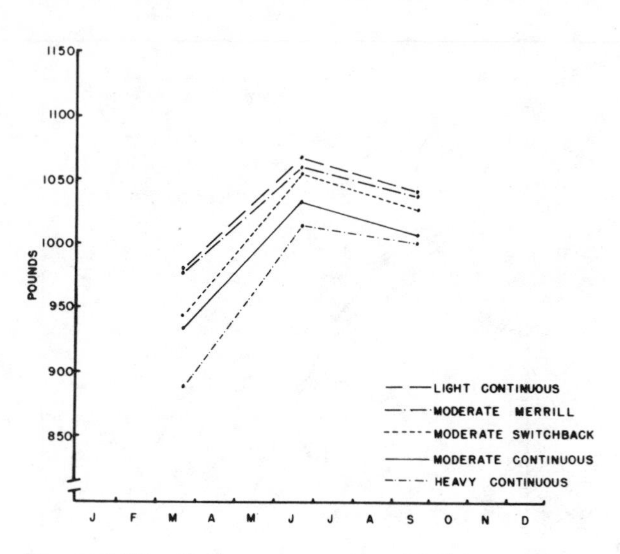

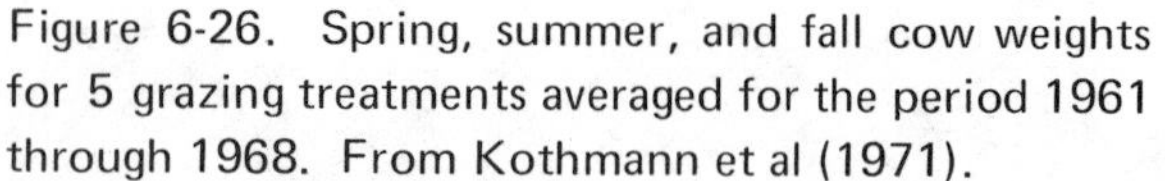
Figure 6-26. Spring, summer, and fall cow weights for 5 grazing treatments averaged for the period 1961 through 1968. From Kothmann et al (1971).

requirements and the cycles of forage availability and quality.

Not only is it expensive to supplement energy, but intake and digestibility of forage are affected differently by protein and energy supplements. Both intake and digestibility are enhanced by the addition of protein-rich supplements to roughage diets low in N (Ch. 3, Vol. 2). Supplemental protein may increase intake of low quality mature forage by 50 to 100%. The addition of supplements containing large amounts of starch generally depresses the digestibility of cellulose and with low quality forages also depresses the digestibility of N.

Studies evaluating protein vs energy supplements for sheep on desert winter range in Utah showed that cottonseed meal increased daily intake of range forage, but grain supplements had no effect or tended to decrease daily intake of forage (Cook and Harris, 1968b). The quantity of digestible protein consumed daily in the supplement and the range forage was actually decreased by feeding corn or barley in some trials. High protein supplements substantially increased the levels of ME consumed by the sheep. Corn supplements generally increased energy in the daily ration, but barley reduced daily intake of energy because of the reduced daily intake and depressed digestibility of cellulose and other carbohydrages in the range forage. It was concluded that range livestock producers could benefit more from feeding a protein supplement such as cottonseed meal or soybean meal than from energy supplements such as corn and barley because protein supplements enhanced the value of range forage and produced better livestock responses.

WATER REQUIREMENTS

Water consumption by adult livestock of medium weight in a temperate climate may range from 7 to 17.5 gallons/day for beef cattle and from 1 to 4 gallons/day for sheep and goats (NRC, 1974; Ch. 2, Vol. 2). As a general rule, cattle require 8-10 gallons/day, sheep require 0.75-1 gallon/day, and horses 10-12 gallons/day. However, these figures should be adjusted to account for the amount that will be lost by evaporation. Cattle and horses should be allowed 12-15 gallons/day and sheep 1-1.5 gallons/day (Vallentine, 1971).

Sources of water include streams, springs, wells, and stock ponds (earthen tanks). The dependability of a water system should be judged during the dry season. Springs, if they are fenced and developed, may be good sources. Wells powered by windmills and pumps may be a major source. However, their development depends to a large degree on geology, the depth to water, and the risk involved in hitting water. Earthen tanks have become economically competitive with wells. However, if they are not properly constructed, much water can be lost by evaporation and seepage. For this reason it is better to build deep tanks with a small surface area to volume ratio rather than large shallow ones. The low maintenance cost of tanks is one of their primary advantages. A herd of 100 head of cattle requires about 1,000 gallons/day. This means a tank capacity for a year of 365,000 gallons or 1.1 acre-feet, plus extra for losses and reserves. If the purpose is to provide water for livestock, there seems little justification to build tanks larger than 10 acre-feet (Peterson and Heath, 1963).

Water systems utilizing wells require adequate storage and drinking space at troughs. The best design for troughs seems to be rectangular with sides sloping in towards the bottom to prevent cracking due to frost. A pole should be placed across the top to prevent livestock from getting into it. A flat board allowed to float in the trough prevents birds and rodents from drowning and decomposing in the water. Also, a piece of canvas tied to the inside and allowed to dip into the water provides a means of escape for rodents that might fall in. A good rule of thumb is to allow one linear foot of open water/10 head of cattle (Hubbard, 1975).

Many different problems can arise from deficiencies in the livestock watering systems. These deficiencies include not having enough waterings, not having enough water at each place, poor distribution of watering places, and poor water quality (Vallentine, 1971). Some problems related to livestock include loss of weight, decreased calf crop and weaning weights and, in extreme cases, death. Overgrazing of forage near water is frequently a problem as is under-utilization of forage a long way from water. The overgrazed area near water is usually termed a "sacrifice area".

Several benefits to be derived from increased water developments include more uniform distribution of grazing, possible reduction of need for supplemental feed, and reduced herd size to allow better management. Reducing water intake reduces DM intake. For this reason cattle should never go longer than 48 h without water. The water to DM ratio in the rumen remains relatively constant when water intake is restricted. The animal adjusts through a reduction in moisture excreted in urine and feces and by reducing DM intake. Reduction of water consumption has also been reported to cause a reduction in CP digestibility (Balch et al, 1953; French, 1956; Thornton and Yates, 1968; Asplund and Pfander, 1972).

Livestock water sources can vary greatly in quality. About 60% of the livestock water used nationwide is from ground water with the remainder coming from surface sources (USDI, 1968). Usually, surface sources are lower in mineral content than ground water (see Ch. 2, Vol. 2).

All mineral elements essential as dietary nutrients are present to some extent in water (Shirley, 1970), and it is generally believed that elements in water solution are available to the animal as much as if they were consumed in dry feed or mineral blocks (Shirley et al, 1957). Water may contribute part of the animal requirement of S, I, Ca, Cu, Co, Fe, Mn, Zn and Se (EPA, 1973), but the amount contributed is subject to a great deal of variation depending upon the kind and class of livestock, amount of water consumed and the concentrations of these minerals in the water. In addition, the difference between the minimum requirement and the level at which toxic effects may occur is very small in some cases. For these reasons the NRC (1974) suggests that water not normally be relied upon as a source of essential minerals. Of greater importance are the physical, chemical and biological properties of the water and their effect on water palatability and animal health.

Physical properties of water affecting livestock are primarily due to temperature variation of the water, turbidity, intoxication and deprivation. Although cattle will thrive on water that is near freezing temperature during cold weather, they will more readily consume water warmed to near their own body temperature. Conversely, cool water is more readily consumed during warm weather (Cunningham et al, 1964). Water deprivation will affect feed efficiency and produce digestive disturbances, while intoxication, although very rare, may aid in producing such disturbances as laminitis in horses and cattle (Herrick, 1971).

There are a variety of substances suspended or dissolved in livestock water that may influence palatability or be potentially harmful. These include inorganic elements and their salts, biologically produced toxins, parasitic or disease carrying organisms and man-made pollutants, particularly fertilizers and pesticides. Concentrations at which these substances render water undesirable are subject to many variables. Turbidity caused by suspended particles of clay and organic matter does not appear to be an important factor in itself with regard to palatability or animal health. In addition, both short and long term effects and interactions with other substances must be considered (EPA, 1973). Substances found in quantities not toxic to the consuming animal may accumulate to levels undesirable for those who consume livestock products. These factors make it difficult to determine safe levels for toxic substances.

Salinity problems are more universal in scope and are more likely to be encountered by stockmen. Total salt or mineral content (also expressed as total dissolved solids), while not indicating a single contaminating substance, is a common measurement that carries significance in determining livestock water quality. Highly mineralized water can cause physiological disturbances (and even death) in animals. Even small salt concentrations can result in a decrease in palatability (Weeth and Lesperance, 1965; Dollahite and Armstrong, 1974).

From time to time watering places may become infested severly by "blooms" of blue-green algae that produce toxins seriously affecting livestock. Toxicity of these blooms is extremely variable depending upon which species and strains of algae present, types and numbers of associated bacteria, growing conditions, animal health and amount of toxin consumed (Gorham, 1964). Livestock poisonings have also been observed following rapid decomposition of algal blooms. Shilo (1967) suggests that this may be due in part to botulism poisoning resulting from anaerobic conditions accompanying decomposition. Algae can be controlled by use of 1 ppm of Cu sulfate added to the water (Hubbard, 1975).

The purity of water consumed by livestock has far reaching implications and there are many ways in which livestock water can become contaminated. Certain contaminates may hinder livestock directly through losses by death, losses in production, or interference with reproductive processes. They may also contaminate milk and meat to the point that human consumption is undesirable. It is necessary to understand the possible hazards in order to ensure an adequate supply of good quality drinking water for livestock.

Part B. Evaluation of Livestock Needs on Designing Grazing Systems for Rangeland

INTRODUCTION

The need for controlling livestock grazing to prevent excessive use of plants was recognized about the turn of the last century by range managers and researchers. Deferment was recommended, and various schemes of rotating the deferment were developed to allow all range units to benefit. The usual approach to controlling defoliation effects of animals has been to stock conservatively and periodically remove all animals from the range.

The predominant considerations in the design of grazing systems have been plant related factors such as levels of non-structural carbohydrates and times of shoot growth, root growth, flowering, seed maturity, germination and seedling establishment. Since plants are the source of all primary production in grazed ecosystems, their continued survival in a vigorous condition is essential.

Although the plant is the sole source of primary production, animals, either domestic or wild, are generally the only means a rancher has of deriving income from this plant production. Unfortunately, the needs of livestock have not been adequately considered during the design and development of most grazing systems.

The key to successful animal production on most ranges is selective grazing by the animals. The average nutrient content and digestibility of forage on most ranges will not support high levels of animal production. Mature forage provides only enough nutrients for maintenance. Animals need to be able to consume immature green forage in order to obtain high enough levels of nutrient intake to support production. Allowing animals to select the most nutritious parts of the total available forage increases nutrient intake and improves animal production. However, it is this selective grazing which creates the need for grazing systems.

ANALYSIS OF GRAZING SYSTEMS

The objectives of various grazing systems were summarized by the Arizona Interagency Range Committee (1973) as follows: distribute utilization; restore vegetation on sacrifice areas; maintain forage density; maintain forage composition; meet nutritional needs of livestock; avoid stress on animals; reduce supplemental feeding; and minimize labor costs.

Four of these objectives deal primarily with vegetation, three with livestock needs and one with economic considerations. Most grazing systems have primary emphasis on improving or maintaining species composition of range vegetation with minor emphasis on meeting the nutritional needs of livestock.

Decisions Required

Before attempting to show how the needs of grazing animals can be integrated into grazing systems, we will examine the management components of grazing systems. Under continuous grazing there are 5 basic decisions which must be made with respect to grazing management. These are: stocking rate; kind and class of animals; pasture size; water location; and supplement locations. Each of these decisions is important, but they are decisions that are made infrequently. When a grazing system is implemented, there are several additional decisions which the manager is required to make. These are: land area per system; number of pastures per system; number of herds per system; and grazing cycle (length of rest periods and length of grazing periods). Depending upon the kind of grazing system, these decisions may be fixed when the system is initiated with little change thereafter, or they may be re-evaluated within seasons every year.

Stocking rates are usually set under continuous grazing and minor adjustments are made annually or semi-annually; however, the number of animals generally does not vary greatly except during prolonged droughts. Under continuous grazing, stocking rate is the only variable the manager can adjust, thus allowing little flexibility in responding to drought seasons (Vaughan-Evans, 1978). Grazing systems provide a hierarchy of decisions relating to stocking rate, stocking density and grazing pressure. However, with deferred rotation and rest rotation systems there still is limited flexibility. High intensity-low frequency systems offer more flexibility in management, but the greatest degree of flexibility is possible under short duration grazing. It may be illustrated by the following comments.

Stocking rate. This is the first and most basic decision and is determined by the number of animal-unit-days of grazing that are allocated to a given area of land during a year.

Stocking density. This is generally the second level of decision making and is a function of the stocking rate and the number of pastures per herd. For example, if 4 pastures are available, grazing systems can be designed using 1, 2 or 3 herds, thus changing stocking density while stocking rate (the total number of animals used) may remain constant.

Grazing pressure. Stocking rate and stocking density both affect grazing pressure which is the ratio of forage demand by the animals to forage available on the pasture. It is also affected by the length of the grazing period and grazing cycle. Thus, the manager of a grazing system can manipulate grazing pressure by changing the length of grazing and rest periods without changing stocking rate or stocking density.

From this brief analysis of these three components of grazing systems, it should be apparent that decision making related to grazing management is much more complex under grazing systems than under continuous grazing. However, there is a large difference between the complexity of a simple deferred rotation or rest rotation grazing system and a highly developed short duration grazing system.

Types of Grazing Systems

Many reviewers have summarized literature comparing continuous grazing to rotational grazing systems with no regard to the kind of grazing system being considered. Since the variation in grazing systems is almost infinite, it is desirable to classify and describe them according to common characteristics. This has been done here by describing deferred-rotation, rest rotation, high intensity-low frequency, and short duration grazing as four different types of grazing systems (Fig. 6-27). Continuous and seasonal grazing are generally not considered grazing "systems". The four types of grazing systems are characterized as follows.

Deferred rotation (DR). These systems are based on the concept of providing seasonal deferment which is rotated among pastures. The deferment is rotated so that each pasture is deferred during each season (Wambolt, 1973). Systems are generally designed using a fixed number of pastures/herd of livestock. Carrying capacity is calculated on the total land area in the system and should be set conservatively. Deferment periods generally vary

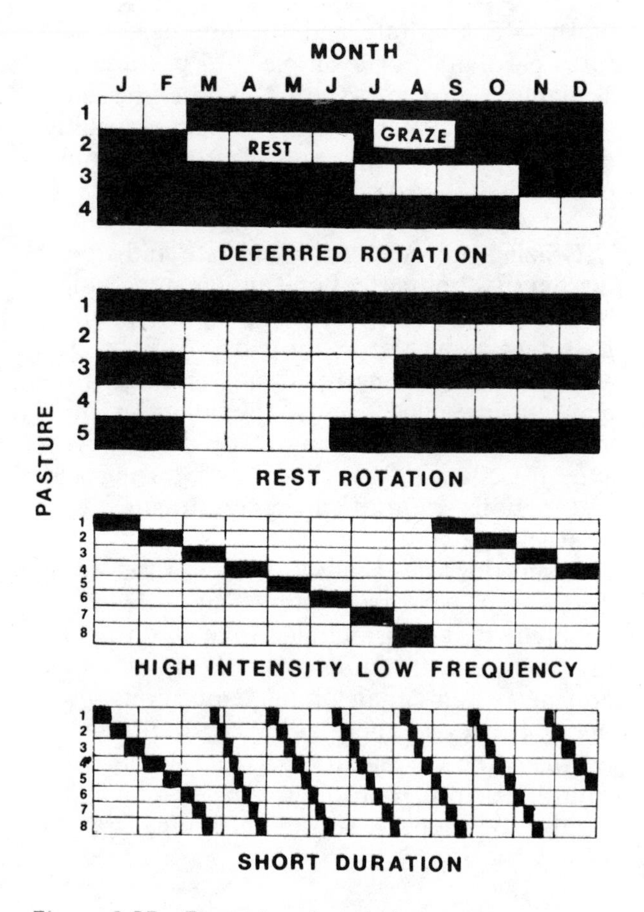

Figure 6-27. Examples of typical systems representing four different types of grazing systems, deferred rotation, rest rotation, high intensity-low frequency, and short duration.

from 3 to 6 mo., but may be as long as 12 mo.

Rest rotation grazing (RR). Pastures are rested from grazing for a full year with the rest rotating among pastures. Deferments are provided for seed production and for seedling establishment, but these deferments always occur during the same season. Two to 5 pastures may be used in the systems (Hormay, 1970). Rest rotation systems differ from deferred rotation systems in that deferments are not rotated seasonally, carrying capacity is calculated based only on that portion of the range that is open for use each year, and generally a much smaller percentage of the total area is available to grazing during the period of active plant growth.

High intensity-low frequency (HILF). These systems are based on the use of intensive grazing periods with relatively long rests. They require at least 3 pastures/herd. It is difficult to separate HILF systems from short duration grazing since there is a continuous gradient from one type to the other. The examples given in Fig. 6-27 illustrate typical systems for each type. Criteria for separation are that HILF systems generally have grazing periods greater than 2 weeks, rest periods longer than 60 days and grazing cycles greater than 90 days. Carrying capacity is based on the total land area in the system. A light to moderate degree of use must be maintained to prevent excessive declines in animal production.

Short duration grazing (SDG). These systems also have 3 or more pastures/herd, but are characterized by relatively short grazing and rest periods. Grazing periods should be less than 14 days (preferably 7 days or less). Rest periods vary from 30 to 60 days, but do not exceed 60 days. Grazing cycles are generally short enough to allow 6 or more full rotations/year (Savory, 1979). Carrying capacity is based on the total land area in the system. The degree of utilization desired at the end of a forage production-utilization cycle is greater under this system than under the others. Because of the high degree of control over both frequency and intensity of defoliation, a higher degree of use is possible without detriment to plant or animal production.

These four types of grazing systems do not include all possible types, but they represent the major approaches that have been used. Therefore, they will be used as the basis for consideration of animal needs in grazing systems.

Animal needs as related to grazing systems may be separated into two broad categories, nutritional and handling stress. Nutritional needs will be related to the quantity and quality of forage available to and consumed by grazing animals. Handling stress may result from moving animals from pasture to pasture and in some systems from mixing different herds during the year. Since grazing systems concentrate animals on only a portion of the total land area, herd size is generally larger than under continuous grazing and this may stress animals during watering and feeding. Large herds may also cause problems during calving and lambing.

NUTRITIONAL NEEDS OF ANIMALS

The nutritional needs of animals may be met by feeding supplements or conserved forages such as hay and silage. Complementary pastures may be developed to provide for deficient periods. However, this section will address only the effects of different types of grazing systems on the nutrient intake from range forage by grazing livestock.

Pastures and ranges which are grazed only during the active growing season present fewer nutritional problems than those that are grazed yearlong or primarily during the dormant season. Forage quality is highly correlated with the age (maturity) of the forage and with plant part. Forage quality declines with maturity and the relative proportion of leaf:stem also declines.

Mature forage generally is only adequate to meet maintenance requirements of animals and may require supplementation of protein, P and vitamin A. To achieve reasonable levels of animal production, animals must consume primarily immature leaves. Grazing systems which allow much of the forage to mature prior to grazing will never achieve good animal production. Also, any grazing system which restricts the animal's opportunity to graze selectively from young, live leaves will reduce animal production.

The degree of sensitivity with which nutrient intake of grazing animals responds to changes in grazing pressure varies seasonally (Taylor and Deriaz, 1963; Hart, 1978). During seasons when almost all of the available forage is green leaves, grazing pressure has little effect on nutrient intake until availability limits DM intake. As the proportion of dead forage and stems increases in the standing crop, the sensitivity to grazing pressure increases. However, when almost all forage is dormant, the sensitivity to grazing pressure is again reduced. Thus, the sensitivity of nutrient intake by grazing animals to grazing pressure is a function of the degree of variation in forage quality within the standing crop. By determining the proportion of green and dead herbage in the available forage, it is possible to predict periods when animal performance will be most sensitive to grazing pressure (Fig. 6-28).

Deferred rotation grazing systems maintain low grazing pressure at all times by spreading animals over half or more of the land area in the system and using conservative stocking rates. This, combined with the beneficial effects of periodic deferments, is the primary reason for generally good animal performance under DR. However, these systems are very sensitive to stocking rate. For example, under the Merrill system (4 pastures-3 herds-12 mo. graze-4 mo. rest), if proper stocking is 16 ac/AUY, then each pasture will be grazed 12 mo. at 12 ac/AUY and rested for 4 mo. If the overall stocking rate were set at 12 ac/AUY, then each pasture would be grazed at the rate of 9 ac/AUY for 12 mo. and rested 4 mo. Because of the concentration of animals on pastures for relatively long periods of time under DR systems, it is very important that the stocking rate be set conservatively (Bishop and Birrell, 1975; Wambolt, 1973).

Rest rotation grazing was initially developed for use on high elevation summer range (Hormay and Talbot, 1961). However, it has also been applied to other types of ranges including range grazed yearlong. Hormay (1970) recommends that 20-40% of the range receive a full year's rest and in special situations more may be rested. In addition to this, additional land is deferred during the growing season to allow for seed production and seedling establishment. This results in livestock being restricted to a relatively small portion of the range during the major growth period. Because of this restriction of grazing area, stocking rates must be reduced significantly

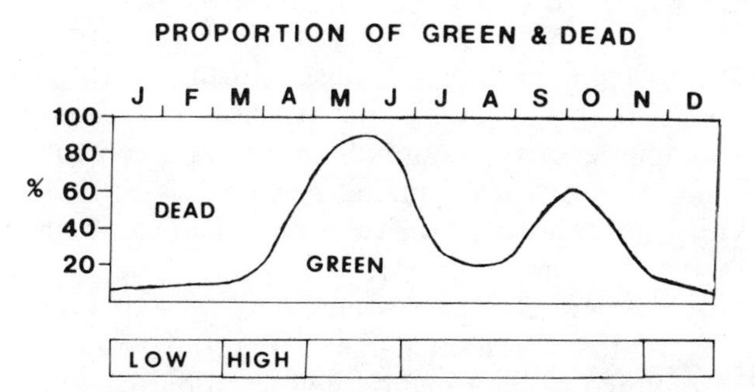

Figure 6-28. The relative proportion of green and dead material changes seasonally on ranges. The sensitivity of nutrient intake by grazing animals to changes in grazing pressure is a function of the homogeneity of forage quality in the standing crop.

to prevent reduced animal performance and damage to the spring grazed range.

Another effect of RR grazing is accumulation of much mature forage. During yearlong rests mature forage accumulates and may hamper efforts of animals to obtain green forage in subsequent grazing periods. Deferment until after seed maturity also presents animals with primarily mature forage. Thus, by design, RR grazing systems may be expected to reduce individual animal production if stocking rates are maintained or to reduce animal production/unit of land area if stocking rates are cut.

HILF grazing systems have some of the same disadvantages noted for RR. When stocking densities are increased significantly by concentrating animals from several pastures onto one, grazing periods in excess of 2 weeks result in intense grazing pressure. The result is reduced nutrient intake during the latter part of the grazing periods (Anderson, 1977; Allison, 1978; Taylor et al, 1980).

HILF grazing systems have a grazing cycle that exceeds 3 mo. which results in some pastures not being grazed during periods of active plant growth. Thus, the opportunity for harvest of this high quality forage is lost. HILF systems also result in accumulation of much mature forage because of long rest periods. Therefore, these systems have two primary negative impacts on animal nutrition. First, the rest periods allow much of the forage to mature prior to grazing and, second, the relatively long grazing periods restrict the opportunity for selective grazing, forcing animals to consume lower quality forage.

In all three types of grazing systems (DR, RR, and HILF), the ability to control defoliation of individual plants is lacking, thus rest periods are long. This allows time for plants which were grazed too frequently and intensively to recover vigor.

Development of Short Duration Grazing

Short duration grazing provides an opportunity to control both frequency and intensity of defoliation on essentially all plants. Because overgrazing can be prevented, full use of the range vegetation is possible and a high degree of management control can be exercised over both the quantity and quality of forage available to the grazing animals. Because SDG represents such a significant departure from traditional range grazing systems, its history and development will be traced briefly.

Non-selective grazing was first reported by Acocks (1966). It was developed as an applied conservation program on ranches needing veld improvement in South Africa. During the decade of the 1960's, it evolved to short duration grazing (Goodloe, 1969; Howell, 1978; Vaughan-Evans, 1978; Savory, 1979).

The original concept was to graze pastures intensively during a 2-week period, forcing livestock to utilize much of the unpalatable but grazable forage. It was noted that this was "hard on the animals", and suggestions were made to evaluate the needs for supplements and to use stock having low nutrient requirements. The primary purpose of the system was range reclamation. After reclamation was achieved, it was recommended that selective grazing schemes be used for improved livestock performance. A 12-mo. rest was considered an essential preliminary to initiation of this intensive stocking system.

Acocks (1966) set 2 weeks as the practical minimum grazing period for ranchers beginning non-selective grazing and 6 weeks as the minimum rest period. He noted that longer rest periods may be necessary for vigorous seeding and seedling establishment on depleted ranges, but the need for them will become less as reclamation is achieved.

Goodloe (1969) noted that there was a shift in emphasis from resting for seed production towards the positive effects of the grazing periods. These included "chipping" of soil crust by hoof action and putting standing dead grass down to litter. There was also considerable emphasis placed on improved uniformity of utilization resulting from high stocking densities applied during short grazing periods (Savory and Parsons, 1980). Similar shifts in emphasis from long rest periods to short grazing periods and relatively short rest periods have been reported by Corbett (1978) and Merrill (personal communication). Both of these ranchers began using HILF grazing systems during the early 1970's. Both grazing and rest periods have been shortened to improve animal performance until the systems now reflect a low intensity of SDG.

Grazing Distribution

Wambolt (1973) stated that grazing distribution is the major problem that grazing systems seek to solve. The most difficult kind of grazing distribution problem is unequal use of individual plants, even of the same species. This is frequently not even recognized as most attention is given to area and species selective grazing problems. Grazing with a high stock density and at a heavy stocking rate promotes uniform utilization of forage. However, when forage is too short, animals cannot obtain adequate nutrient intake (Allden and Whittaker, 1970).

Under continuous grazing or DR and RR systems, it is not possible to control the frequency of defoliation. This leads to deterioration of the vegetation. HILF grazing allows better control of frequency of defoliation; however, grazing periods greater than 2 weeks allow animals to regraze plants that have produced regrowth and some of the more palatable plants which have not produced significant regrowth. This results in some plants being grazed at a high frequency and intensity, interspersed with long rest periods which are required for these plants to recover.

An examination of grazing behavior will assist in understanding the defoliation effects under the different grazing systems. Hodgson (1966) investigated heavy and moderate rates of set-stocking on S23 ryegrass. Under heavy stocking there was a relatively large number of frequently-grazed tillers, whereas under moderate stocking there was a large number of infrequently grazed tillers. Tillers were defoliated every 7-8 days with heavy stocking, and every 11-14 days with moderate stocking. Sheep expressed a strong tendency to graze those plants having the greatest green leaf length. The green leaf length removed at each defoliation averaged 40% under heavy and 27% under moderate stocking. Although sheep selected tillers having the greatest green leaf length, there was a tendency for tillers that were defoliated early during the study to be defoliated more frequently later on than those which were not grazed initially (Hodgson and Ollerenshaw, 1969). One of the major effects of defoliation on plant growth was rapid development of secondary tillers. This produced a plant with more leaf and less stem than an ungrazed plant, although total biomass produced was similar or even less on the grazed plant.

Gammon and Roberts (1978), studying native vegetation in Rhodesia, found that from 5-15% of the grass tillers were grazed during any given 4-week period on pastures continuously grazed at the recommended stocking rate. The recommended rate was conservative. They noted an interaction between severity and frequency of defoliation. Plants grazed most frequently were also grazed most intensively. Tillers grazed leniently or not at all at the first part of the season were less likely to be grazed later in the season or at high intensities.

Frequency of defoliation fell into a bimodal pattern with the most palatable species having about 35-40% of tillers defoliated at intervals exceeding 63 days or not at all, while about 10% were defoliated at least every 14 days (Fig. 6-29). The less palatable species had a higher percentage of tillers defoliated at intervals ≤14 days than the palatable species. This could be interpreted to indicate that very young forage of these species was readily acceptable to grazing animals but palatability declined more rapidly with maturity. About 50% of the tillers of less palatable species were defoliated at intervals exceeding 63 days (Gammon and Roberts, 1978). The high proportion of tillers which are defoliated infrequently or not at all appears to be a major factor contributing to the low efficiency of forage utilization on most rangelands. Obtaining proper use of

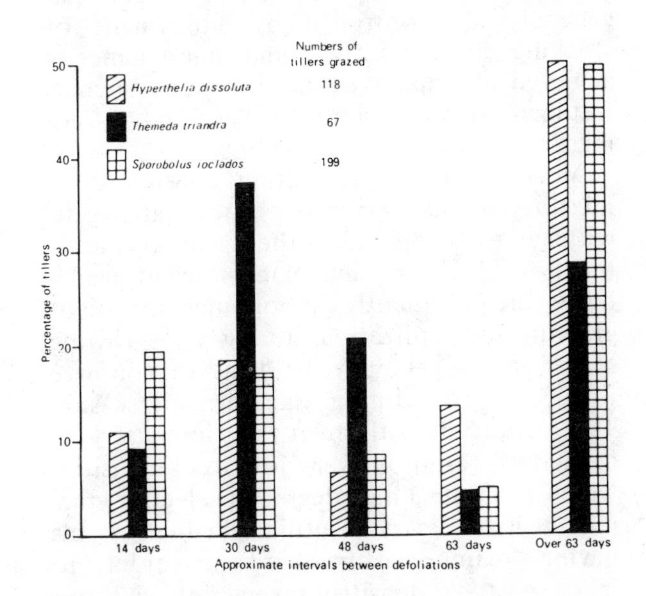

Figure 6-29. Percentages of tillers regrazed at various intervals after first being grazed between 17 December 1971 and 15 February 1972, under continuous grazing by steers on rangeland in Rhodesia. From Gammon and Roberts (1978).

these tillers could increase the carrying capacity of most rangelands substantially.

When comparing the study of Gammon and Roberts with Hodgson and Ollerenshaw, certain parallels are evident. As stocking density declines, the frequency of defoliation declines on an increasingly large proportion of the tillers. However, even at relatively low stocking densities, some of the tillers are still utilized at a high frequency and intensity. Thus, high stock densities are required to achieve uniform utilization of forage. However, rest periods are required to prevent a portion of the plant population from being defoliated too frequently and intensively.

Grazing Intensity and Pasture Type

The variable responses of intensive rotation grazing schemes reported in the literature are probably a function of pasture characteristics. Tame pastures, consisting of species selected for high resistance to grazing and growing on good soils with favorable moisture and nutrient availability, generally respond well to continuous grazing. High carrying capacities result in high stocking densities. A large proportion of the current forage production is harvested within 4-6 weeks of the time it is produced, and conservation of standing crop for grazing during the dormant season is seldom practiced. Selective grazing effects can generally be controlled by adjustment of stocking rates. Lastly, stand maintenance is achieved principally by cultural inputs such as fertilization, weed control, tillage and reseeding.

Most rangeland vegetation consists of a mixture of species, varying in their ability to withstand grazing and in their growth characteristics. Moisture and plant nutrient availability are frequently suboptimal for plant growth and utilization during the growing season must generally be light to conserve forage for use during the dormant season. This, coupled with relatively low carrying capacities, results in very low stock densities which promote a high degree of selective grazing. It is necessary to utilize grazing systems having many pastures/herd in order to increase stock densities several fold without corresponding increases in stocking rate.

Optimum Number of Pastures

The optimum number of pastures/herd, from the stock density-utilization aspect, should be a function of the heterogeneity of the vegetation and the length of the growing season in relation to the grazing season. The less variability in vegetation that would cause selective grazing problems and the longer the growing season in relation to the grazing season, the fewer the number of pastures required/herd. With relatively short growing seasons and long dormant seasons, more pastures will be required to achieve adequate stock densities to promote uniform untilization.

Evaluation of the factors relating to defoliation of individual tillers indicates that defoliation effects can be manipulated to advantage where selective grazing problems exist. On rangelands where approximately half of the tillers contribute little to the grazed forage, it is essential that utilization patterns be changed to bring these tillers into production (Gammon and Roberts, 1978). However, attempts to do this simply by increasing the stocking rate under continuous, DR, RR or HILF grazing results in deterioration of either plant and/or animal production.

The objective of SDG is to utilize high stock densities for short periods to obtain a *single* defoliation of *all* tillers during a grazing period. Since less than half of the leaf area of the tiller is generally removed during a single defoliation, injury to the plant is minimal and beneficial effects may result from stimulation of tillering. Uniform utilization will produce a stand of vegetation in which differences in plant maturity within and among species are minimal. This will aid in reducing selective grazing.

The art of achieving this uniform utilization under SDG, while maintaining acceptable animal performance, requires careful consideration of the balance between stocking rate, stocking density and grazing pressure. Very little quantitative research is available to provide guidance in this task, and no set formula can be applied that will work under all conditions. At the present time the best advice appears to be for operators attempting to utilize SDG to become thoroughly acquainted with the principles and objectives of SDG and to develop grazing plans which function acceptably.

An example may help illustrate the interrelations of stocking rate, stocking density and grazing pressure. Assume that a pasture of 1,000 acres has a carrying capacity of

20 ac/AUY under continuous grazing and is stocked at that rate. It would be stocked with 50 cows. During the growing season stock density and grazing pressure are both low, allowing animals to graze very selectively. This results in many plants not being grazed and thus becoming mature and low in nutritive value and palatability. Doubling the stocking rate still has little effect on stock density, changing it from 20 to 10 ac/cow.

Now let us assume that the 1,000-ac pasture is subdivided into 10 pastures of 100 ac each. Using the stocking rate of 50 cows, the stocking density is 2 ac/cow, or if stocking rate is doubled, stocking density becomes 1 ac/cow. Thus stocking density may be increased by a factor of 10X without changing stocking rate.

Assume that a SDG grazing system is initiated using the 10 pastures and 50 cows and is designed to start with a 50-day grazing cycle. At the end of 5 days in each pasture only about half of the plants have been utilized and the remainder are ungrazed. The manager is faced with the questions: should the number of animals be increased, should the grazing period be extended, or should the pastures be subdivided further? The first question relates to stocking rate, the second to grazing pressure and the last to stocking density.

The number of animals (stocking rate) must be based on the total grazable forage available during the year and not on the degree of use during any one grazing period in SDG. The forage production will be harvested during 5-7 grazing periods within a year. Changing the length of the grazing period affects the length of the grazing cycle which should be based on the growth rate and maturity of the plants. Therefore, extending the grazing period to increase utilization may create problems of over-mature forage later in the season. Increasing the number of pastures increases the stocking density and it also affects the graze:rest ratio. The more pastures/herd the shorter will be the time of grazing in relation to the rest period. The situation will need to be evaluated in light of these considerations to determine which course of action is correct. Sometimes a combination of changing animal numbers, grazing cycle and numbers of pastures may be required.

It is obvious that for any given level of forage availability as stocking density is increased by creating more pastures/herd, the length of the grazing period becomes more critical. The more pastures/herd that a grazing system has, the shorter the grazing period must be. Hodgson (1966) and Hodgson and Ollerenshaw (1969) demonstrated that at high stock densities most tillers will be defoliated twice within a 2-week period. Gammon and Roberts (1978) found that even under relatively low stock densities there was a significant percentage of plants being defoliated at intervals of 2 weeks or less. Thus, to achieve the objective of limiting the frequency of defoliation to one time per grazing period, the grazing periods should be limited to 1 week or less.

If SDG is attempted with as few as 3-5 pastures, the grazing periods will be too long or the rest periods will be too short to obtain the desired effects. Also, with only a few pastures stock density is not increased sufficiently to give uniform utilization unless stocking rates are relatively high. To maintain grazing periods of <7 days and rest periods of 42 days, a minimum of 7 pastures will be required. However, with only 7 pastures there will be very limited flexibility with respect to changes in the sequence of rotation or in length of the grazing periods and cycles. Increasing the number of pastures in the system allows more flexibility in management. With many pastures, livestock may be combined or split into several herds and the speed of rotation may be varied significantly.

Vaughan-Evans (1978) listed 5 important principles which have emerged from 12 years experience with SDG.

1. **Grazing periods** during the growing season must be kept as short as possible to avoid repeated defoliation of the most palatable plants. In practice, no more than 7 days and preferably 5 days or less, should be used, with periods over 14 days amounting to prolonged overgrazing.

2. **Rest periods** following use must be provided to allow palatable plants time to recover, replenish root reserves and achieve optimum growth. In practice, 6-8 weeks should follow between grazings, with 7 weeks proving about optimum for most veld types.

3. **High stocking density** is desirable to achieve evenness of grass utilization, chip up capped soil surfaces and trample unused plant material to the ground as a litter cover.

4. **Animal performance** is generally optimum when stock are permitted to freely select diets which provide both quantity and

quality of herbage at the right height and stage of maturity. Division of the herd into 2 or more groups and their movement on a follow-through pattern of grazing, can be used to achieve optimum animal performance.

5. **Minimized handling of animals** is essential in order to afford stock sufficient time period for maximum herbage intake. This is most suitably achieved by the provision of grazing cells, with centralized handling facilities in a cartwheel and hub paddocking arrangement.

Length of Rest Periods

Considering the needs of the grazing animal, emphasis should be placed on the length of the rest period in addition to the above criteria. Nutrient flow through standing live plants is a very dynamic process and long rest periods prevent animals from harvesting portions of these nutrients. Some plants are short lived and may only be available for periods of 6-8 weeks or less, and leaves of grasses only live for about 30 days. Therefore, during periods of rapid plant growth it is important to shorten the grazing cycle to 30-40 days to obtain maximum animal performance.

Since short (30-40 days) rest periods may limit the rate of plant succession on badly depleted ranges, two alternatives may be considered to benefit plant succession. Selected pastures may be skipped during the grazing cycle to allow reproduction or establishment of plants during critical periods. The second alternative is to lengthen the grazing cycle during critical periods. However, this may have undesirable effects. Pastures will be grazed longer during the critical period, and longer rest periods may reduce forage quality and thus animal production.

Need for Data on SDG

At the present time caution is advised with respect to application of SDG on rangelands. There is a noticeable lack of information on diet selection, nutrient intake, and animal production data from most grazing systems used on ranges and especially from SDG. Even more important is the lack of any quantitative, statistically sound data evaluating the animal production responses to SDG under a range of stocking rates. With the claims which have been made for two-fold and greater increases in stocking rates over very short periods following implementation of SDG, it is urgent that these relationships be evaluated properly in long-term studies. Since it is generally agreed that stocking rate is one of the most important factors affecting productivity and profitability of ranching operations, it is difficult to understand why more research has not been directed towards the response of grazing systems under different stocking rates.

HANDLING STRESS AND FACILITIES DESIGN AND LOCATION

Concentration of animals and frequency of movement are low under deferred rotation and rest rotation grazing systems, thus animal stress from handling is generally low. Conventional facilities and animal management practices are usually adequate for operation of these types of grazing systems.

Design of Facilities

HILF and SDG systems concentrate animals much more and require much more frequent handling of livestock than continuous grazing. The opportunity for animal stress resulting from increased handling and larger herd size is much greater than under continuous, deferred rotation or rest rotation grazing. In order to cope with animal handling problems under SDG and possibly under HILF grazing systems, it may be necessary to redesign pastures, pens, and water locations.

Savory (1979) developed a cell design for SDG with pens, water, and feed in a central location with pastures radiating from the center (Fig. 6-30). This design allows easy movement of livestock between any two pastures in the system. Since the central water facility serves all pastures, animals do not need to completely reorient themselves when they move to a different pasture. This is especially important during calving and lambing.

One of the concerns frequently expressed about the cell design is the use of wedge-shaped pastures with centrally located water and feed. Under conventional management systems this will probably result in large sacrifice areas around the center. However, with SDG the frequency and intensity of defoliation are controlled such that sacrifice areas are practically eliminated. Experiments have indicated that excessive utilization around the center is not a problem under SDG (Vaughn-Evans, 1978).

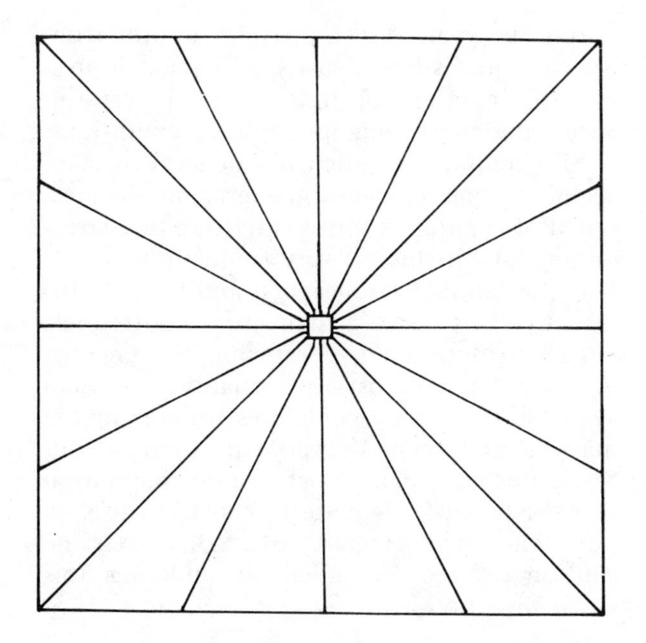

Figure 6-30. Hypothetical example of a 16 pasture grazing cell with water and working facilities centralized. This type of pasture design may be used with short duration grazing to reduce animal stress from handling and facilitate management.

The design of the cell center should allow for adequate holding and working pens, and other functions such as feed and equipment storage may be included (Savory, 1979). The center should be designed to discourage livestock from using it as a resting or bedding area. Animals should enter the center for water and/or feed and then move back into the pasture. If the center is too large, livestock will use it for resting and bedding. This will result in a build-up of dung and probably reduce the effectiveness of the system in promoting grazing distribution.

The physical location of the cell center is a very important decision since it is the site of intensive activity and investment. Factors such as soils, vegetation, drainage, topography, exposure and management considerations should be evaluated carefully prior to selection of the site for the cell center.

The size and kinds of facilities to be included in the center will vary with each cell. The kinds and classes of animals to be worked, the size and number of herds, and the kind of operations to be performed should be considered. Facilities should allow for flexibility of operation and for future changes in programs with minimal cost. One particular point is to allow for increases in herd size of up to 2-3X conventional stocking rates. Feed and pen space should be calculated and adequate drinking space and reserve storage of water provided.

Fences that allow calves and/or lambs to move between pastures have advantages in reducing the potential for separation of the young animal from its mother during moves. It is not necessary to build expensive, many-strand fences for the internal partitions because SDG is designed not to place nutritional stress on the animals. Individual animals that habitually cross fences should be culled quickly to avoid spoiling the rest of the herd (Savory, 1979).

Frequency of Animal Handling

The frequency of animal handling varies greatly among different types of grazing systems. Under continuous grazing, range livestock may be handled 4-5X a year for calving, working calves, weaning, and one or two additional times. Deferred rotation and rest rotation grazing require few additional moves and frequently these may be coordinated with regular livestock working schedules. However, HILF and SDG do require more handling of livestock. A HILF system with 8 pastures, one herd and a 125-day grazing cycle would result in animals being moved to different pastures 24X/year. A SDG system having 16 pastures, one herd and an average grazing cycle of 48 days would move livestock to different pastures 122X during a year. These systems contrast sharply to the Merrill deferred rotation grazing system [4-3; 12:4 mo.] in which a herd is moved to a new pasture once/year.

Stock handling techniques must be evaluated carefully, and care should be exercised much as would be used with a dairy herd. Practices used by many ranches such as chasing livestock with horses and dogs, using electric prods, and general rough handling of stock cannot be tolerated under HILF or SDG. While animals recover from these stresses under extensive grazing management, they will cause significant reductions in animal production under intensive grazing systems.

CONCLUSIONS

Many grazing systems that have been studied have relatively little benefit over continuous grazing with respect to animal or vegetation production and many reduce animal performance significantly. Failure of the systems to adequately control frequency and intensity of utilization on individual plants has been the single most important factor limiting plant and animal response to conventional grazing systems. Conventional grazing systems that use low stock densities result in a very low frequency of defoliation on a high proportion of plants. Thus, these plants mature and lose much of their nutritive value and develop large differences in palatability. The high average maturity of the vegetation is a major factor limiting animal production and carrying capacity on many grazing systems.

SDG gives the manager a high degree of control over both frequency and intensity of defoliation. Both can be adjusted at any time during the year. It also provides an opportunity to adjust stock density and grazing pressure to maintain optimum animal performance under changing forage conditions.

SDG requires a much higher level of management than conventional grazing systems, but it also requires more care to reduce stress on animals because of very frequent handling. Facilities should be designed and located differently for SDG than for continuous, deferred-rotation or rest-rotation grazing.

There are almost no quantitative data available to provide guidelines for making the many management decisions necessary under SDG. Because of the cost of implementation and the lack of adequate information to support the management of SDG, extreme caution is recommended in applying this grazing method.

Literature Cited, Part A

Allison, C.D. 1978. Ph.D. Thesis, Texas A&M Univ., College Station.

Allison, C.D. and M.M. Kothmann. 1979. Proc. West. Sect. Amer. Soc. of Animal Sci. 30:174.

Anderson, D.M. 1977. Ph.D. Thesis, Texas A&M Univ., College Station.

Arnold, G.W. 1960. Aust. J. Agr. Res. 11:1026.

Asplund, J.M. and W.H. Phander. 1972. J. Animal Sci. 35:1271.

Balch, C.C., D.A. Balch, V.W. Johnson and Jill Turner. 1953. Brit. J. Nutr. 7:212.

Brazle, K.F., L.H. Harbers and C.E. Owensby. 1979. J. Animal Sci. 48:1456.

Bryant, F.C. and M.M. Kothmann. 1979. J. Range Mgmt. 32:412.

Buentello, J.L. and W.C. Ellis. 1969. Proc. 24th Ann. Tex. Nutr. Conf., p. 167.

Chapman, H.D. et al. 1972. J. Animal Sci. 34:373.

Cook, C.W. 1956. 17th Ann. Faculty Res. Lect., Utah State Univ.

Cook, C.W. 1966. Utah Agr. Exp. Sta. Bul. 461.

Cook, C.W., R.D. Child and L.L. Larson. 1977. Range Sci. Dept. Series. No. 29, Colo. State Univ., Ft. Collins.

Cook, C.W. and L.E. Harris. 1968a. Utah Agr. Exp. Sta. Bul. 472.

Cook, C.W. and L.E. Harris. 1968b. Utah Agr. Exp. Sta. Bul. 475.

Cook, C.W., L.A. Stoddart and L.E. Harris. 1954. Utah Agr. Exp. Sta. Bul. 372.

Cunningham, M.D., F.A. Marty and C.D. Merilan. 1964. J. Dairy Sci. 47:382.

Davies, I. 1976. J. Agr. Sci. 87:25.

Deinum, B. 1966. 10th Int. Grassl. Congr., Helsinki. p. 415-518.

Dollahite, J.W. and J.M. Armstrong. 1974. Tex. Agr. Expt. Sta. L 1042, College Station, TX.

Duble, R.L., J.A. Lancaster and E.C. Holt. 1971. Agron. J. 63:795.

Ellis, W.C. 1978. J. Dairy Sci. 61:1828.

Ellis, W.C. and H. Lippke. 1976. Tex. Agr. Exp. Sta. Res. Monog. RM-6C, p. 27.

EPA. 1973. Proposed Criteria for Water Quality, Vol. I: Water for Livestock Enterprises. U.S. Environmental Protection Agency. Washington, D.C.

French, M.H. 1956. Empire J. Exp. A. R. 24:128.

Gorham, P.R. 1964. In: Algae and Man. Plenum Press, New York.

Harris, L.E. et al. 1963. J. Animal Sci. 22:51.

Hart, R.H. 1978. Proc. 1st Int. Rangel. Congr., p. 547.

Hart, R.H. et al. 1976. J. Range Mgmt. 29:372.

Hubbard, W.A. 1975. Canada Dept. Agr. Pub. 1390.

Jeffery, H. 1971. Aust. J. Exp. Agr. Animal Hus. 2:397.

Johnstone-Wallace, D.B. and K. Kennedy. 1944. J. Agr. Sci. 34:190.
Karn, J.F. and D.C. Clanton. 1977. J. Animal Sci. 45:1426.
Knight, R.W., W.H. Blackburn and L.B. Merrill. 1980. 33rd Annual Mtg., Soc. Range Manage. (abstr) p. 35.
Kothmann, M.M. and R.T. Hinnant. 1976. Proc. Ann. Mtg., Soc. Range Manage. (abstr.).
Kothmann, M.M. and Muhammad S. Kallah. 1978. The Southwestern Nat. 23:347.
Kothmann, M.M. and G.W. Mathis. 1971. 24th Ann. Mtg., Soc. Range Manag. (abstr.) p. 35.
Kothmann, M.M., G.W. Mathis and W.J. Waldrip. 1971. J. Range Mgmt. 24:100.
Kothmann, M.M., G.W. Mathis and W.J. Waldrip. 1978. Proc. 1st Int. Rangel. Congr., p. 606.
Kranz, J.J. and R.L. Linder. 1973. J. Range Mgmt. 26:263.
Laksesvela, B. and A.N. Said. 1978. Animal Prod. 14:49.
Larcher, W. 1975. Physiological Plant Ecology. Springer-Verlag, New York.
Marten, G.C. 1978. J. Animal Sci. 46:1470.
Marten, G.C. and R.N. Anderson. 1975. Crop Sci. 15:821.
McCartor, M.M. and F.M. Roquette, Jr. 1976. Tex. Agr. Expt. Sta. Res. Monog. RM-6C, p. 325.
Merrill, L.B. 1954. J. Range Mgmt. 7:152.
Moore, J.E. and G.O. Mott. 1973. In: Antiquality Components of Forages. CSSA Special Publ. 4, Madison, WI, p. 53.
NRC. 1969. United States-Canadian Tables of Feed Composition. Nat. Acad. Sci., Washington, D.C. Publ. 1684.
NRC. 1970. Nutrient Requirements of Beef Cattle. Nat. Acad. Sci., Washington, D.C.
NRC. 1974. Nutrients and Toxic Substances in Water for Livestock and Poultry. Nat. Acad. Sci., Washington, D.C.
Osbourn, D.F., R.A. Terry, G.E. Outer and S.B. Cammell. 1974. Proc. XII Int. Grassl. Congr. Moscow, Vol. III, p. 374.
Petersen, R.G., H.L. Lucas and G.O. Mott. 1965. Agron. J. 57:27.
Peterson, H.V. and V.T. Heath. 1963. J. Soil Water Cons. 18:103-108.
Radeliff, R.D. 1970. Veterinary Toxicology, 2nd ed. Lea and Febiger, Philadelphia.
Rainwater, W.A., W.C. Ellis, and W.M. Oliver. 1974. J. Animal Sci. 38:214 (abstr.).
Riewe, Marvin E. 1976. Tex. Agr. Expt. Sta. Res. Monog. RM-6C, p. 169.
Rittenhouse, L.R., C.L. Streeter and D.C. Clanton. 1971. J. Range Mgmt. 24:73.
Shirley, R.L. 1970. Proc. Nutr. Coun. Ann. Mtg. Amer. Feed Manuf. Assoc., May:23-25.
Shirley, R.L. et al. 1957. Q. J. Fla. Acad. Sci. 20:133.
Short, H.L., R.M. Blair and E.A. Epps, Jr. 1973. J. Animal Sci. 36:792.
Short, H.L., R.M. Blair and C.A. Segllquist. 1974. J. Wildl. Mgmt. 38:197.
Sinnott, E.W. 1960. Plant Morphogenesis. McGraw Hill Co., New York.
Stoddart, L.A. 1960. J. Range Mgmt. 13:251.
Taylor, C.A., Jr., M.M. Kothmann, L.B. Merrill and Doak Elledge. 1980. J. Range Mgmt. (in press)
Taylor, J.C. 1966. 10th Int. Grassl. Congr., Helsinki. p. 463.
Thorton, R.F. and D.J. Minson. 1973. Aust. J. Agr. Res. 24:889.
Thorton, R.F. and N.G. Yates. 1968. Aust. J. Agr. Res. 19:655.
Tilley, J.M.A. and R.A. Terry. 1963. J. Brit. Grassl. Soc. 18:104.
Tukey, H.B., Jr. 1970. Amer. Rev. of Plant Physiol. 21:305.
Underwood, E.J. 1971. Trace Element in Human and Animal Nutrition, 3rd ed. Academic Press, New York.
USDI. 1968. Water quality criteria. Sect. IV, Agricultural uses. Rep. Nat. Tech. Advis. Comm. Secr., U.S. Gov. Print. Office, Washington, D.C.
Vallentine, J.F. 1971. Range Developments and Improvements. Brigham Young Univ. Press, Provo, Utah.
Van Niekerk, D.G.H., D.W.W.Q. Smith and D. Oosthuysen. 1967. Proc. S. Afr. Soc. Animal Prod. 6:108.
Van Soest, P.J., R.H. Wine and L.A. Moore. 1966. 10th Int. Grassl. Congr., Helsinki. 10:438.
Webster, J.E., J.W. Hogan and W.E. Elder. 1965. Agron. J. 57:323.
Weeth, J.J. and A.L. Lesperance. 1965. J. Animal Sci. 24:441.
Willoughby, W.M. 1959. Aust. J. Agr. Res. 10:248.

Literature Cited, Part B

Acocks, J.P.H. 1966. Proc. Grassld. Soc. of South Africa. 1:33.
Allden, W.G. and I.A. McD. Whittaker. 1970. Aust. J. Agr. Res. 21:755.
Allison, C.D. 1978. Ph.D. Thesis, Texas A&M Univ., College Station.

Anderson, D.M. 1977. Ph.D. Thesis, Texas A&M Univ., College Station.
Arizona Interagency Range Committee. 1973. Grazing systems for Arizona ranges. S. Clark Martin and Charles R. Whitefield, Co-Chrm.
Bishop, A.H. and H.A. Burrell. 1975. Aust. J. Exp. Agr. Animal Hus. 15:173.
Corbett, Q. 1978. Rangeman's J. 5(6):201.
Gammon, D.M and B.R. Roberts. 1978. Rhod. J. Agr. Res. 16:117; 133; 147.
Goodloe, S. 1969. J. Range Mgmt. 22:369.
Hart, R.H. 1978. Proc. 1st Int. Rangeland Cong., Soc. for Range Mgmt., p. 547.
Hodgson, J. 1966. J. Brit. Grassld. Soc. 21:258.
Hodgson, J. and J.H. Ollerenshaw. 1969. J. Brit. Grassld. Soc. 24:226.
Hormay, A.L. and M.W. Talbot. 1961. USDA Prod. Res. Rep. 51.
Hormay, A.L. 1970. USDA, F. S. Training Text 4(2200).
Howell, L.N. 1978. J. Range Mgmt. 31:459.
Savory, A. 1979. In: Beef Cattle Sci. Handbook 16:375, 380. Agri. Services Found., Clovis, CA.
Savory, A. and S. Parsons. 1980. In: Beef Cattle Sci. Handbook 17:215. Agri. Services Found., Clovis, CA.
Tayler, J.C. and R.E. Deriaz. 1963. J. Brit. Grassld. Soc. 18:29.
Taylor, C.A., M.M. Kothmann, L.B. Merrill and D. Elledge. 1980. J. Range Mgmt. (in press).
Vaughan-Evans, R.H. 1978. Conex report, Rhodesia.
Wambolt, C.L. 1973. Montana State Univ. Ext. Bul. 340.

Chapter 7—Feeding and Nutrition of the Beef Breeding Herd

by Larry R. Corah

A practical year round beef cow nutrition program is generally built around some type of grazed forage in nearly all temperate regions where beef cattle are produced. In many regions native range provides virtually all of the required nutrients of the cow during the spring and summer months (Fig. 7-1). Native grass may even provide a high percentage of the needed nutrients during the fall and winter months, provided dry grass is available and weather conditions permit grazing. In other areas the cow herd nutrition program is built around seeded improved forages that may provide from a few months grazing to nearly a 12-month grazing program.

Figure 7-1. Cows and calves on native range.

Understanding the nutritional value of the grazed forage is the initial step in planning a cow herd nutrition program. Supplemental feeding is then used to compensate for the nutrient deficiencies of the forage at various times of the year. However, irregardless of the type of forage used, the nutrient requirements of the beef cow and the general guideline under which beef cows are fed remain fairly constant for most areas.

NUTRIENT REQUIREMENTS OF COWS

The general nutrient requirements for pregnant and lactating beef cows as recommended by the NRC (1976) are shown in Appendix Table 8. These guidelines serve as the basis of any cow herd nutrition program. However, any student of cow herd nutrition will realize that there are many factors that will alter the requirements of the cow. Thus, these values should only serve as a starting guide since they may need to be modified as indicated by new data or experience.

FACTORS INFLUENCING A BEEF COW'S NUTRIENT REQUIREMENTS

What are the factors that the beef cow operator must consider in determining the nutritional requirements of his cows? The following points are some of the most important things that should be considered in planning a feeding program for cows: stage of production–pregnant vs nonpregnant and lactating vs dry; age; physical condition and body weight; weather; breed; and specific area nutrient deficiencies.

Stage of Production

The stage of production that the beef cow is in has an important effect on nutrient requirements. There are various ways of looking at stage of production, but one of the best ways is to consider the 365-day beef cow year starting with calving and ending with the production of the next calf. If the beef cow is accomplishing what we want her to, she should produce a big, healthy calf every 365 days. The following table breaks the beef cow year up into four distinct nutritional periods.

Table 7-1. The 365-day beef cow year by periods.

Period 1	Period 2	Period 3	Period 4
80 days (postcalving)	125 days (pregnant and lactating)	110 days (mid gestation)	50 days (pre-calving)

Period 1. Period 1 is the 80-day period after calving when the cow is lactating at her highest level while trying to maintain a high

level of calf growth. In addition to this, the cow must undergo uterine involution, start recycling and rebreed during this period. Obviously, to the beef cow this period is her most important nutritional period. It is also during this period that many beef producers fail to feed their cows properly. In a spring calving situation quite often the cows are out on pasture in which the new grass is quite adequate in protein; however, it has a high moisture content and as such the beef cows cannot consume enough dry matter to meet their energy requirements.

Table 7-2 shows a 1100 lb cow's nutrient requirements during this period. It should be noted that the cow's energy requirements are fairly high through this period (13-15 lb of TDN) but also during this period protein is especially important to the cow because of her high level of lactation (Table 7-3). If the cow is poorly fed during this period of time, it will affect milk production, calf growth, conception rates, and percent of the cows cycling during the breeding season.

Period 2. During this period the cow should be in the early part of pregnancy while still lactating and maintaining a calf. It is also during this period that the cow should be gaining weight and laying on some energy reserve to prepare for the winter months, assuming a spring calving situation.

Table 7-2. NRC requirements for a 1100 lb beef cow during four stages of production.

Item	Periods 1	2	3	4
TDN, lb/d	13-15*	11-12	8-9	10-11
Protein, lb/d	2.0	1.6	.8	1.0
Dig. protein, lb/d	1.2	.9	.45	.55
Calcium, g/d	27	24	13	15
Phosphorus, g/d	27	24	13	15
Vit. A, IU/d	24,000	24,000	20,000	24,000

*Depends on milking ability, age and condition.

Most beef cows will be in a declining stage of lactation during this period and, as such, their nutritional needs are not as high as immediately after calving (Table 7-2). If the cow is poorly fed during this period of time, it will primarily affect her milk production and subsequently her suckling calf's growth rate (Table 7-4). Nutrition during this stage generally will not have any adverse effect on her early developing fetus.

Period 3. This is the period that follows the weaning of the calf and is referred to as mid-gestation. Basically, during this period of time the beef cow must primarily maintain her developing fetus. During this period the beef cow's nutrition needs are at the lowest level of any stage of the year (Table 7-2) and not much above maintenance.

When beef producers are looking for a period of time when they can save feed and "rough" the cows some, this is the period when this can be done with the least likelihood of undesirable results. It is during this period that some of the low quality forages such as straw or crop residue work very well (Billingsley et al, 1976). It is also at this time, assuming the cow was fairly fleshy entering this period, that she can afford to lose some weight up to possibly 10-15% of her body weight.

Period 4. This period is the second most important period during the beef cow year and again is a period when many of the producers fail to feed the cows as well as they should be fed. During this period 70 to 80% of the total fetal growth occurs and in addition the cow is preparing for lactation. If a beef cow operator has suffered through a

Table 7-3. Effect of protein levels pre- and postcalving on the milk production of first calf Angus heifers.[a]

Level of Protein	Weight of heifer precalving	Milk production, lb/day: 60 days postcalving	Milk production, lb/day: 150 days postcalving	Calf gains: 0-75 days	Calf gains: 75-150 days
High - 0.22 lb/100 lb BW	871	14.9	10.1	1.2	1.5
Med. - 0.16 lb/100 lb BW	836	13.2	10.0	1.3	1.4
Low - 0.07 lb/100 lb BW	710	11.0	5.7	.8	1.1

[a]From Bond and Wiltbank (1970)

Table 7-4. Effect of energy levels post calving on reproductive performance.[a]

Energy fed postcalving	Showing estrus by 90 days after calving, %	Not showing estrus during the breeding season, %	1st service conception, %	Overall conception, %
High - 16 lb TDN	95	0	67	95
Low - 8 lb TDN	86	14	42	77

[a]From Wiltbank et al (1962)

rough winter in which the cows have lost some weight, this last 50 days before calving is a period when adequate nutrition can do a lot to insure proper calf growth and maximum reproductive performance of the cow.

It should be noted that the cow should be fed well enough so that she is gaining weight during this last 50-day period. This is especially true in the cases where the cows are thin or have lost a considerable amount of weight during the winter. If the cow is poorly fed during this period, it will affect the birth weight of the calf by as much as 6-8 lb but even though the calf is lighter there will not be any appreciable change in the degree of calving difficulty (Bellows et al, 1972). Also, if the cow is on a poor plane of nutrition during this period, if the stress is severe enough there can be fetal loss through early abortion (Corah et al, 1975). Poor nutrition during this period will affect subsequent milk production and calf growth after calving (Bellows et al, 1972). In addition, there is evidence to show that it will affect calf survival at birth and calf health (Table 7-5). Nutrition in this period will have an important effect on cow reproduction by affecting the interval to first estrus which will cause a reduction in the percentage of cows cycling at the start of the breeding season (Wiltbank et al, 1962).

Table 7-5. Effect of energy levels precalving on cow and calf performance.[a]

Item	TDN fed/day	
	Low-low	Low-high
TDN fed from 100-30 d precalving	4.2	4.2
TDN fed from 30 d	10.6	4.2
Cow weight change, lb		
1st 70 d	-114	-119
Last 30 d	93	-23
Calf birth wt	66.8	58.7
Calving difficulty, %	9.5	9.5
Calves alive at weaning, %	100	71.4
Calves treated for scours, %	33	52
Cows' milk production, lb/d	12.1	9.0
Cows showing estrus by 40 d precalving, %	48	38

[a]From Corah et al (1975)

Age of the Beef Female

The age of the beef animal has a major influence on how the animal should be fed. Bred heifers about to produce their first calf should be handled separately and differently than the rest of the cow herd (Dunn et al, 1969). First, the beef producer should have some weight goals with the bred heifers. Assuming they are heifers of British breeding, he should strive to have them weighing 600-650 lb at breeding time and 850-900 lb at calving time. To achieve this, these heifers will need about 8.5 lb of TDN, especially during that period after breeding. It should also be remembered that these animals are in a growing stage and, as such, protein is very important with this type of animal. They should receive 1.6 lb of crude protein or 1 lb of digestible protein/day (NRC, 1976).

After calving these first-calf heifers also merit special consideration. First-calf heifers should be bred to calve 3-4 weeks prior to the rest of the cow herd. This will allow them more time to rebreed. Another management practice that should be followed, if feasible, would be to feed these first-calf heifers separately from the rest of the cow herd after calving. This gives these animals two advantages. First, they can be fed a higher plane of nutrition to be sure they rebreed and, second, it eliminates these younger animals being bossed around by the older cows in the herd. In some cases it may be beneficial to also maintain the 3-year-old cows as a separate group to prevent reproductive problems and continued physiological maturation of these beef females. Cows over 3 years of age and older can be maintained as one group.

Condition of the Beef Female

The most common method used by most cattlemen to determine whether extra roughage or supplement is needed with a cow herd is to evaluate the condition or fleshiness of the cows. Evaluating condition of cows at key times is an important part of a successful nutrition program.

The amount of energy that will be needed during the winter period is closely related to the condition of the cows as they enter the winter period. Cows that are in good condition after summer grazing can often lose 50 to 100 lb during the winter with no adverse effect on later cow productivity, provided the bulk of this weight is not lost the last 1-2 months precalving. Cows entering the winter period in a thin condition should gain weight during the winter months.

A key time to evaluate cow condition is 50-80 days precalving. Cows which are thin at calving will be slower to start cycling (Table 7-6), thus reducing conception rates, have weaker calves at birth and lighter calves at weaning (Corah et al, 1975). Cows that are thin at calving should be fed levels of energy and protein exceeding the NRC (1976) recommended levels to insure proper reproductive performance. Cows entering the breeding season in a thin condition generally respond more to flushing (feeding extra energy) than do fleshy cows.

Table 7-6. Effect of body condition at calving on reproductive performance.[a]

Body condition at calving	No. cows	Days postpartum, % cycling	
		60	90
Thin	272	46	66
Moderate	364	61	92
Good	50	91	100

[a]From Whitman (1975)

Cows maintained in an over-fat condition have a reduced percent calf crop, wean lighter calves and their longevity in the herd is usually shorter (Pinney et al, 1960). Excess condition at calving time is often associated with fall calving cows that are maintained on summer pasture (Fig. 7-2).

Figure 7-2. Upper; cows (at weaning) entering the winter period in excellent condition. Lower; an excessively fat cow that is likely to have poor reproductivity. Courtesy of D.C. Church.

Cow Size or Weight

Another factor that influences nutrient requirements is the size of the beef cow (NRC, 1976). As commercial cow herd operators strive to increase weaning weights, cow size has increased both due to genetic selection of calf growth and through the introduction of European breeds that have heavier mature weights.

The ideal cow size is not easily determined but will vary from one cow-calf operation to the next. However, as cow size increases, nutrient intake must increase, but the increase is not a one-to-one liner increase. Thus, a 1100 lb cow does not have a 10% greater nutrient need for maintenance than a 1000 lb cow. Rather, the extra nutrient needs will be less than 10% as the feed requirement for maintenance is proportional to metabolic weight ($BW^{0.75}$). The increase in energy requirements due to cow size is shown in Table 7-7.

Beef cows will expend considerable energy in grazing activities. Cow size or weight will

Table 7-7. Digestible energy requirement for mature beef cows for maintenance.[a]

Body weight, lb	Mcal for maintenance	Mcal for 1 mile travel
700	10.14	.74
800	11.21	.84
900	12.24	.95
1000	13.25	1.05
1100	14.23	1.16
1200	15.19	1.26
1300	16.13	1.37
1400	17.05	1.47
1500	17.96	1.58
1600	18.85	1.68

[a]From Brownson (1977)

affect energy requirements expended for activity with larger cows using more energy (Table 7-7).

Weather Conditions

Environmental conditions, particularly temperature, have a big influence on cow nutrient requirements. Heat and dust will alter cattle nutrient requirements but cold weather is a stress condition that has a great effect on much of the cow-calf regions of North America.

Cold is a stress factor that alters the energy requirements of the cow. Factors such as amount of hair, wind, humidity, age of animal, condition of animal, and breed all will influence how much the temperature affects the cow's energy requirement (see Ch. 16, Vol. 2). However, most feeding standards do not provide means of adjusting animal requirements as affected by these environ-

Table 7-8. Estimated critical temperature for beef cows.[a]

Hair coat description	Critical temperatures
Summer coat or wet	15°C (59°F)
Fall coat	7°C (45°F)
Winter coat	0°C (32°F)
Heavy winter coat	-7°C (18°F)

[a]From Ames (1978)

mental or genetic factors. The critical temperature at which cold weather affects beef cattle is shown in Table 7-8.

The amount of insulation created by hide, hair coat and air interface will influence how temperature affects increased energy needs. In general for each 1 degree (F) drop below the critical temperature, there is a 1% increase in the energy needs of the beef cow. It is important to note that if there is any change in protein requirements of cows during cold weather, it is a reduction in the amount of protein needed when expressed as a percent of the diet (Ames, 1976); this is a reflection of greater feed (energy) consumption, not of a lower protein requirement.

Breed

The requirements used for beef cows usually relate to the British breeds such as the Hereford, Angus, Shorthorn or reciprocal crosses of these breeds. As breeds from other countries have been introduced into the USA cattle industry, new nutrient requirement standards are needed.

Since many of these breeds produce more milk and are larger, the nutrient requirements for these types of cattle will be 10-25% higher

Table 7-9. Effect of breed on response to winter supplementation on native grass.

	Breed of cow				
	Hereford		Holstein		
Level of winter supplement	Moderate	High	Moderate	High	Very high
30% Protein supplement fed during winter, lb	279	570	312	570	869
Weight change for the year, %	3.3	2.3	-4.4	-1.3	.7
Milk produced/d, lb	13.6	13.0	25.1	27.5	26.8
Calf weaning weight, lb	585	580	715	704	695
Cows conceiving, %	92.2	96.2	50.0	72.2	87.0

than for conventional British breeds. If a producer is using dairy breeding in his beef operation, the nutrient needs of these cattle will be 20-30% higher to achieve the same level of reproductive performance as with British breeds (Wyatt et al, 1977).

Specific Area Nutrient Deficiencies

The major nutrients affecting cow herd productivity are energy, protein, Ca, P, and vitamin A. However, in certain regions of any country, due to variations in soil types, climate and type of grass, specific trace minerals such as I, Cu, Co and Se may be a problem (see Ch. 5, Vol. 2). In other regions, excesses of certain trace minerals will create toxicity problems. Once the specific problem for a particular region is determined, proper feeding of trace minerals may be a very inexpensive way of maintaining a high level of production.

CRITICAL NUTRIENTS IN COW HERD NUTRITION

Energy

In quantitative terms energy is the nutrient that is most important in a cow herd nutrition program and also comprises the major feed expense. The role of energy in a cow herd nutrition program has been well documented in many research trials over the years. It plays an extremely important role in reproductive performance, calf weaning weights, and in the overall productivity per cow. In most cattle operations the bulk of the energy consumed by cows comes from grazed forage or by feeding harvested roughage.

The energy requirement of the beef cow during gestation is actually fairly modest, especially during mid-gestation. During this time cows can often be maintained successfully on dry grass, corn or milo stalks or other low quality roughage.

During the last two months of gestation, 70-75% of the fetal energy requirement for growth occurs (Table 7-10). If the cow is in good condition at this time, some of the energy can be derived from conversion of body fat to utilizable energy. If cows are thin, extra supplemental energy must be fed or calf birth weight, calf vigor, and calf survival will be reduced (Corah et al, 1975). Energy deprivation will reduce birth weight by 4-8 lb, but no reduction in calving difficulty occurs (Bellows et al, 1972). Reduced energy precalving delays the onset of estrual activity postcalving (Wiltbank et al, 1962) and reduces the percentage of cows conceiving in the breeding season.

Table 7-10. Digestible energy required for fetal development.[a]

Month of pregnancy	Mcal required	
	Daily	Monthly
1	---	---
2	---	---
3	---	---
4	.10	3.05
5	.20	6.10
6	.40	12.20
7	.80	24.40
8	1.48	45.14
9	2.75	83.88
	Total	174.77

Following calving the energy requirements of the cow increase by 17-50% depending on the milk production of the cow (NRC, 1976). Inadequate energy during this period can lower weaning weights by 15-25 lb and reduce conception rates by 10-25% (Wiltbank et al, 1962). The amount of energy for the various activities of the postpartum cow are shown in Table 7-11. These data were derived from studies with beef cows in partial or complete confinement.

Protein

The second most expensive nutrient in a cow herd nutrition program is protein. Protein, particularly, plays a major role during lactation. Likewise, it plays a major role by affecting the appetite in that it alters the level of forage that animals will consumer and, as such, alters the level of energy they will take in. Minimal levels of protein recommended by NRC during mid gestation will depress roughage consumption. More protein is needed by the rumen microorganisms for a high rate of fermentation than is needed by the cow's tissues (see Ch. 4, Vol. 2).

Another important reason to consider protein is that it is often the nutrient that is most likely to be purchased in a typical cattle operation. Unfortunately, in the cattle industry in many cases, we put too much emphasis

Table 7-11. Digestible energy requirements for a 1000 lb (454 kg) cow and her 500 lb (227 kg) weaning calf.[a]

	Mcal for cows					Mcal for calf		Total Mcal	
Month[b]	Maintenance	Travel[c]	Reprod.	Milk prod.	Total	From milk	From pasture	Daily	Monthly
1	13.25	2.10	---	6.85	22.19	3.84	1.42	23.61	720
2	13.25	2.10	---	7.41	22.76	4.16	3.34	26.10	796
3	13.25	2.10	---	7.98	23.33	4.48	5.08	28.41	867
4	13.25	2.10	---	7.98	23.33	4.48	6.97	30.30	924
5	13.25	2.10	---	7.41	22.76	4.16	9.11	31.87	972
6	13.25	2.10	---	5.70	21.05	3.20	11.78	32.83	1001
7	13.25	2.10	.10	3.42	18.87	1.92	14.74	33.61	1025
8	13.25	2.10	.20	---	15.55	---	---	15.55	474
9	13.25	2.10	.40	---	15.75	---	---	15.75	480
10	13.25	2.10	.80	---	16.15	---	---	16.15	493
11	13.25	2.10	1.48	---	16.83	---	---	16.83	513
12	13.25	2.10	2.75	---	18.10	---	---	18.10	552
Annual totals	4850	768	175	1426	7218	800	1599	8817	8817

[a]From Marion and Riggs (1972)

[b]Calf born 1st mo.; cow bred 3rd mo.; calf weaned at the end of 7th mo.; and cow is dry next 5 mo.

[c]Average 2 miles daily.

on protein and not enough emphasis on level of energy fed to a group of cows. Some of the common mistakes in feeding protein to beef cows are discussed.

Over-Feeding During Mid-Gestation. A typical 1000 lb cow of average producing ability will need only 0.8 to 1 lb of crude protein during the middle part of gestation and yet in many cases producers will feed a roughage of fair quality during this period and yet feed a protein supplement when, in fact, it is not needed.

Under-Feeding Protein After Calving. When a cow calves, her requirements for protein double. For a cow producing 11 lb of milk, her protein requirements are 1.9 lb of crude protein after calving. But, when that cow produces 22 lb of milk, the level of protein needed is increased by 2.7 lb (Table 7-3).

Misuse NPN or Urea. Urea is a very cheap source of nitrogen and in many cases can be fed successfully to cattle, particularly feedlot steers. Yet in most cow herd nutrition programs when forage is often limited or forage is of low quality, urea is poorly utilized because of a lack of available energy (Table 7-12). When urea is utilized with a set of beef cows under these conditions, there is often a negative response to the high-urea protein supplements causing an increase in weight loss and subsequently a reduction in weaning weights and reproductive performance. In order for urea to be successfully utilized, it must be accompanied by adequate energy which is quite often not the manner in which beef cows are fed.

Table 7-12. Effect of urea on weight loss of cows wintered on dry native grass.[a]

	Supplement Fed	
Item	All natural 30% protein supplement	30% protein supplement with half of nitrogen from urea
Supplement/d, lb	2.5	2.5
Cow wt loss during the winter*	-284	-337
Estimated % the protein from the urea utilized by the cow		31%

[a]From Wright and Totusek (1975)

*Weight loss includes weight loss at parturition

Calcium

Because cows consume the major portion of their diet as roughage, a Ca deficiency in cow herd rations is not common. Even roughages such as dry grass, corn stalks and other crop residues, with the possible exception of cereal straws, usually provide adequate Ca for the beef cow. When cows calve, their Ca requirements nearly double and possibly at this time some response to supplemental Ca might be observed.

Phosphorus

The main supplemental mineral needed by beef cows is P. Cows are often maintained on forages such as mature grass or crop residues which contain low levels of P. A P deficiency is most likely to develop in cows maintained over long periods on mature dry grass.

Research (Short and Bellows, 1971; Ch. 4, Vol. 2) has indicated that P deficiencies will result in reduced appetite and impaired fertility which is often associated with reduced ovarian activity (Table 7-13). P-deficient cows have reduced appetites, may lose weight, milk less and may even show signs of depraved appetite such as chewing on bones, stones or dirt.

Table 7-13. Effect of phosphorus intake on first service conception rates of heifers and cows.[a]

	Level of phosphorus fed		
	Less than required	0-20 g above requirement	20-40 g above requirement
No. animals	97	233	198
1st service conception, %	50	58	70

[a]Hignett and Hignett (1951)

In certain areas of the USA, such as the region from Texas to California, more of a response to P is observed. In contrast, in other regions, little or no response to P has been observed in research trials (Call et al, 1978).

P would appear to have its greatest importance to the beef cow starting at least 60 to 100 days prior to calving and then 100 days post-calving on through the breeding season. P can be self-fed but more often it is fed when mixed with salt or in commercial protein supplements.

Salt

Salt should be supplied to cows at all times on a free choice basis. The requirements of mature cattle for salt is less than 28 g/head/d, but intake will vary greatly and is often influenced by the rest of the ration being consumed (Rich et al, 1976).

Vitamin A

Unlike other vitamins which are synthesized in the digestive tract of cattle, carotene or vitamin A must be supplied in the ration. Since vitamin A is accumulated during summer grazing and stored in the liver, a deficiency is not commonly observed in beef cows.

In some herds small calf crops, weak calves, retained placentas and other non-specific symptoms have been believed to be associated with vitamin A deficiencies (see Ch. 9, Vol. 2). However, these symptoms may often be more related to other nutrient deificiencies other than vitamin A. A 10-year Oklahoma study showed that cows could subsist up to 43 mo. on low-carotene rations before showing deficiency symptoms (Pope et al, 1961).

Since vitamin A is one of the most inexpensive nutrients, there is no justification for short changing the cow herd. Vitamin A can easily be supplied using mineral mixes or protein supplements as carriers, using injectable vitamin A, or feeding good quality legume hay.

NUTRITIONAL DEVELOPMENT OF REPLACEMENT HEIFERS

The replacement beef heifer is an extremely important part of the cow herd operation as she represents what hopefully will be a genetic improvement in the herd by replacing a lower producing or non-producing cow. The management of this heifer from birth to her first calf is an important period in order that she will develop into a productive cow that is successfully maintained in the herd over a period of years. In most areas of the USA replacement heifers are calved at two years of age. All consideration in this section will be based on developing replacement heifers to calve at that age. The development of this replacement heifer breaks down into three phases: pre-weaning; weaning to breeding; breeding to calving.

Table 7-14. Age and weight at first heat of heifers wintered at two rates of gain.[a]

Breed of heifer	Winter gain, 0.5 lb/day		Winter gain, 1.0 lb/day		Difference	
	Age at puberty, mo.	Age at puberty, lb	Age at puberty, mo.	Age at puberty, lb	Age, mo.	Wt, lb
Angus	13.1	518	11.2	572	1.9	54
Hereford	15.5	594	13.6	665	1.9	71
Shorthorn	13.7	500	10.9	544	2.8	44
Ang X Her	13.2	550	11.8	628	1.4	78

[a]Wiltbank et al (1966)

Pre-Weaning

It is important that a replacement heifer weight at least 400 lb by weaning, and it is also important that this growth is normal skeletal growth and not growth constituting a considerable amount of fat. Research has shown that creep feeding suckling heifer calves actually hinders their performance later as mothers because of deposition of fat in the developing udder.

Weaning to Breeding

In a well-managed cow operation heifers are bred 2-4 weeks prior to the cows to insure they have adequate time to rebreed. In order to have a high percentage of the heifers bred early, this means that 80 to 90% of heifers will need to have reached puberty by the time they are 12 to 13 mo. old. A sound nutrition program is the key to getting the job done.

Once a heifer is weaned she will need to grow at the rate of 1-1¼ lb/day until breeding at 13 to 15 mo. of age. This will mean that the heifer will need to gain approximately 250 lb during this period of time in order to insure that she will be weighing 600 to 650 lb, which is the necessary weight for heifers to have shown first heat (puberty). When heifers are not fed to gain a pound a day after weaning, it will take longer to reach puberty, causing a lower percentage of the heifers to be cycling at the start of the breeding season (Table 7-14).

In order to achieve this desired level of growth, a proper blend of protein and energy is extremely important. If heifers are light at weaning time, additional weight must be gained to insure they are cycling when breeding occurs as yearlings. This is illustrated in Table 7-15.

Table 7-15. Effect of weaning weight on heifer reproductive performance.[a]

Item	Heifers fed together: Light at weaning	Heifers fed together: Heavy at weaning
Weaning wt, lb	376	475
Age at weaning, d	187	204
Daily gain, weaning to breeding, lb	1.25	1.47
TDN/head/d, lb	9.7	9.7
Age at puberty, d	423	404
% Cycling at start of breeding season	60	90
% Pregnant	60	80

[a]From Varner et al (1977)

Breeding to Calving

The final phase of heifer development is making sure the heifer continues to grow from breeding to calving. Heifers that weigh 600 to 650 lb at breeding should gain another 200 to 250 lb or more before calving. This means that these heifers will have to gain ¾ to 1 lb/day from breeding to calving. Generally, with spring-calving cow herds, these bred replacement heifers will remain on summer grass from breeding until fall. If the quantity and quality of grass is adequate, they should easily achieve a gain of from 1 to 1½ lb/day. In the fall they often are fed harvested or stored feed until calving, possibly in some cases using the stored feed as a supplement to native grass. During this period of time it is important to remember what nutrients the bred heifer really needs. Protein is important as the heifer is continuing to gain weight;

however, protein is not nearly as important as energy. Heifers not properly developed will have more difficulty at calving and be slower to cycle and rebreed with their second calf.

FEEDING THE HERD BULLS

Proper nutritional development and maintenance of herd sires is an important segment of a well managed cow herd operation.

Young Bulls

Once bull calves are weaned it is important that proper growth and development be continued similar to the development of replacement heifers. This is especially important if the bulls will be used as herd sires as yearlings. An excellent management practice used by many purebred breeders is to evaluate the genetic potential of the bulls by feeding the bulls in a controlled test for 140 days. Rations used for bull evaluation often are high-energy rations consisting mainly of grain, however, high roughage rations can also be used.

When young bulls are purchased directly off a bull evaluation test or at a purebred sale, they often are carrying excess finish and present a distinct management problem. It is important that these highly conditioned bulls be "let down" from the time they are bought until they are turned out with the cow herd (Fox, 1972). During this let down period the bulls must be adapted gradually from the high-energy rations to a high-forage diet similar to what they will consume when out with the cow herd. During this period, use of bulky grains such as oats works well; proper exercise of the bull is also important.

Yearling bulls can be used successfully with the cow herd for breeding up to 20 or 30 cows with no adverse effect on later breeding as a mature bull (Corah et al, 1978). If yearlings are used, it is best if they only are used over a short breeding season (40-60 days) and then upon removal from the cow herd they should be separated from mature bulls and fed a ration fortified in protein, energy, minerals and vitamins to insure proper growth and development. Crushed oats, either alone or in combination with other grains, makes an excellent bull feed.

Mature Bulls

Once bulls have reached full maturity, their nutritive needs decrease and weight maintenance is all that is necessary. However, when bulls are removed from the breeding pasture in a thin condition, adequate energy should be fed to restore body condition. Bulls should always be conditioned 1-2 mo. prior to being turned out with the cow herd. Both overfat and over thin bulls may have reduced libido and fertility. Old bulls or thin bulls being used may benefit from supplementation during the breeding season. This can be accomplished by daily or every other day hand feeding of small amounts of grain to the bulls out on pasture.

Winter maintenance rations for mature bulls can be built around hay or silage with grain fed only to maintain proper condition. Bulls should always be allowed adequate bunk space and provided pen space to allow adequate exercise.

WINTER FEEDING SYSTEMS

Use of Hay

In many geographical rations, hay is the main source of feed used in wintering beef cows. Even in regions where cows are utilizing mature, dry winter grass, use of hay is common either as a source of supplemental energy and protein or as the main ration during periods of snow cover.

The quality of grass hay may be quite variable, depending on the species of grass, stage of growth at cutting, method of storage and degree of weathering during harvest. A good management practice is to analyze the hay and use the lowest quality hay during early winter and then save the higher quality hay for later winter when the cows' nutrient needs are higher. For the gestating cow, grass hays that contain over 6% protein can serve as the sole source of feed with only supplemental P, vitamins and salt needed. During mild winter weather, 16-20 lb of average quality grass hay will maintain mature cows. During extreme cold, 20-26 lb of hay may be required.

Legume hays, such as alfalfa and clover, serve as an excellent source of feed for beef cows; however, the cost of this type of hay often prevents use as the main roughage. Legume hays should be used after calving or as a protein source for cows on dry winter grass or for cows being fed low-quality roughage. When legume hay is being fed as a source of protein, it can be fed only once or twice per week with results equal to daily feeding. A common and successful practice is to feed low-quality roughage two or three days and alfalfa or clover hay the third or fourth day.

Use of Silage

Corn or sorghum silages are excellent roughage for gestating and lactating cows, but because they contain more energy per unit of dry matter than grass hay, they should not be fed free choice. Silage (65% moisture) at a rate of 30-40 lb will maintain a gestating cow while 50-60 lb of silage will provide adequate energy for a lactating cow. Because most corn or sorghum silage contains 7% to 8% crude protein (dry basis), extra supplemental protein is needed, especially when fed to lactating cows. Legume silages can be used similar to legume hay with beef cows.

Utilizing Crop Residue Material

One of the more effective ways of reducing wintering costs of the cow herd is to utilize crop residue material such as corn or sorghum stalks, cereal straw, beet tops, etc. Grazing of the crop aftermath is usually the cheapest way of using residue provided weather will allow grazing. With the exception of straw, most residue fields can be grazed for 30-60 days with no supplemental feeding needed except for minerals and vitamins. During this period the cows will often gain weight. When lactating cows are grazing residues, extra supplemental protein, energy, minerals and vitamins are definitely needed.

More crop aftermath material is now being harvested for winter feeding since new forage harvesting equipment is available. Because much of the crop aftermath contains over 30% moisture after grain harvest, care should be taken to dry the residue properly so mold and spoilage will not occur. Many of the residues such as corn stalks, sorghum stalks, and beet tops can be ensiled. Ensiled residue can serve as the only source of feed for gestating cows if it contains 5% protein (dry basis) and is self-fed (Kimple et al, 1977). Dry, harvested crop aftermath can be fed to cows grazing residue fields or grazing dry grass in a supplemental manner. Using residue in this manner will often maintain gestating cows, but supplemental protein is often needed from mid-winter on (Vetter, 1974).

Cereal or grass straws are satisfactory winter roughage and they provide more useful energy during cold weather than indicated by NE values. When cereal straws are self-fed, cows will eat 16-20 lb (Acock et al, 1979). Because most cereal straws contain only 3-5% crude protein (dry basis) and this protein is poorly digested, extra supplemental protein must be fed, but it does not have to be fed daily to the cows. When straw is fed as the main roughage, feeding alfalfa hay once weekly may be as successful as daily feeding (Acock et al, 1979).

Wintering on Dry Grass

A common program in much of the geographical region where beef cows are raised is to winter on dry native grass or grass-forb-browse ranges. The duration and intensity of winter grazing will be determined by the amount of available forage and weather conditions that permit grazing.

The nutritive value of the dry grass will be influenced by the native grass type, time of year and degree of weathering. Grass type will greatly influence the winter feed value of the grass as shown in Table 7-16.

Hand-clipped samples are often used to indicate feed value, however, this is misleading since cattle tend to feed selectively. When using esophageal fistulas, researchers have shown that cattle will selectively consume a

Table 7-16. Comparison of protein content in short grass range (blue grama) and bluestem range (clipped samples).

	Crude protein % DM basis	
Month	Blue grama[a]	Bluestem range[b]
January	3.7	2.8
February	4.2	2.6
March	5.0	2.5
April	5.0	2.4
May	9.0	15.0
June	10.2	8.6
July	7.9	5.7
August	7.5	5.9
September	5.0	4.1
October	5.0	3.7
November	4.5	3.5
December	4.0	3.1

[a]Data from Colorado Range Livestock Forage Analyses Handbook.

[b]From Smith et al (1959)

Figure 7-3. Cows on a snow covered winter range.

Figure 7-4. Cows being fed cubed supplement on winter pasture.

diet that is higher in crude protein and energy than is present in a random clipped sample. By feeding selectively cattle are capable of compensating somewhat for nutrient deficiencies. Through selective feeding cows are often able to maintain weight on native grass without supplemental feeding up until mid-winter, provided an adequate quantity of forage is available. In most cases protein is the first limiting nutrient. However, forage shortages also can create an energy deficiency.

One of the difficulties in building a supplement for cows wintering on native forage is determining the forage intake of the cows.

Table 7-17. Expected intake of range forage (if available) by a 1000 lb beef cow.

Month	% Body weight
January	1.5
February	1.5
March	1.5
April	1.6
May	2.6
June	2.7
July	2.2
August	2.2
September	1.9
October	1.8
November	1.7
December	1.6

Based on studies done at Oklahoma and Colorado, the preceding table of expected intake of native grass was developed. It should be realized that these are only estimated intakes and factors such as plant crude fiber, lignin and crude protein will influence intake as will the body condition of the cows. Fall calving cows wintered on dry grass will also have a higher forage intake than gestating cows.

Feeding Supplemental Feed

Protein. In deciding on how much protein to feed cows, three factors must be considered: the amount of protein that is needed, the percent protein in the supplement that is needed, and whether urea can be used. Factors influencing the amount of protein needed by the cow at various times have already been discussed. However, a common dilemma for the cow-calf manager is often whether to feed a 40% protein supplement versus a 20% supplement or some other level that can be purchased. When protein is the limiting nutrient, the greatest response in improved cow performance is achieved by feeding a protein supplement with a higher percentage of protein. In many cases this is also the cheapest way of supplying a unit of protein.

In many cases for cows on low quality roughage both protein and energy may be deficient. Under these conditions many cow-calf producers are successfully feeding protein supplements containing 14-20% crude protein but containing a TDN content of >70%.

These are fed at a higher level than for the more concentrated protein supplement. It is a common practice to start supplementing on native range using a supplement containing 14-18% protein during late fall-early winter and change to a 20-24% protein supplement as calving approaches. Supplemental proteins are usually pelleted or cubed unless it is feasible to feed in a bunk.

Although urea is a more inexpensive form of N, it must be fed in rations that contain adequate levels of energy in order to get good utilization of urea N by rumen microorganisms. Poor utilization is made of the urea by cattle on low quality roughage. Some research trials show only 25-50% of the N from the urea being utilized (Wright and Totusek, 1975). When cows are in moderately good condition and are either gaining or maintaining weight, protein supplements can contain some urea with performance sometimes being equal to feeding natural protein supplements. Use of liquid supplements, which usually contain a high percentage of protein from NPN sources, has become more popular because of the convenience they offer (Fig. 7-5). When roughage of adequate quality and quantity is fed, they have often worked fairly well. However, lack of control on intake and poor use of urea may make the cost of this supplement higher than desired or realized.

In some situations protein can be fed to cows on alternate days or even once weekly with as good a results as daily feeding (Acock et al, 1979). Use of winter cereal pastures or winter fescue or ryegrass make excellent sources of protein for brood cows.

Figure 7-5. One example of a lick wheel feeder for liquid supplements. Courtesy of D.C. Church.

Energy. In most cases energy is supplied to cows in some form of harvested or grazed forage. In some cases grain may actually serve as a cheaper source of energy for cows because of high hay prices or convenience created by feeding grain. Few problems are encountered when grain is used. If high levels are fed, it should be fed daily. When only energy is needed, feeding grain is nearly always cheaper than feeding a protein concentrate. Although excess protein is converted to usable energy, it is usually expensive to do so.

Minerals. A practical mineral for a cow-calf operation is one that contains at least 10-12% P, since P is usually the limiting mineral when forage is being fed. Many excellent commercial minerals mixes are available or a simple mix of 50% salt and 50% dicalcium phosphate or bonemeal is equally successful. To be effective some intake of the mineral is obviously necessary and certain P sources such as bonemeal are less palatable than other sources. Minerals and vitamins may also be fed successfully as part of the protein concentrate.

Methods of Feeding. As previously discussed every other day or even more infrequent feeding of protein supplement is successful unless the supplements contain high levels of urea. Self-feeding of supplements can be controlled by mixing loose salt with the feed or using commercial intake limiters. As a guide to determining how much salt to add to control intake, mature cattle will usually eat about a 0.1 lb salt/100 lb of body weight. This will vary with the palatability of the supplement and the rest of the ration being fed (Rich et al, 1976). Intake of salt-controlled supplements is variable from one animal to the next, a disadvantage for this method of feeding. Adequate water must always be available when salt is used to limit intake to prevent salt toxicity. Controlled experiments in several states have failed to show any harmful effects upon later productivity of the cows from proper use of salt-concentrate mixes.

Protein blocks may be a convenient way of feeding protein. They are popular in many of the cow-calf producing regions. Various methods of controlling intake are used including inclusion of salt or hardness of the block.

Separating Cattle by Age

Whenever possible, cattle should be segregated into groups by age and sex. This allows the nutrition program to be specific for that age or type of cattle and helps control some of the dominance effect during feeding. Replacement heifers should be maintained as a separate group and, if possible, first-calf heifers and even second-calf cows should be maintained as a group separate from the mature cows. If possible, young bulls should be separated from mature bulls.

Another management practice that helps in planning a nutrition program is to have the cows calving over a short 60-90 day calving season. Long calving seasons lead to many of the cows being poorly fed as well as resulting in an uneven calf crop with the late calves being especially light at weaning.

SPECIALIZED CATTLE PROGRAMS

Confinement or Semi-Confinement Cow Systems

Increases in prices of grazing land coupled with desires to intensify cow herd production has lead to interest in total or semi-confinement of the cow herd. This type of management system will require a greater investment in labor, facilities and equipment while saving on the investment in land.

Research trials conducted in many states have shown clearly that the productivity of cows in total confinement or semi-confinement was similar to cows maintained on grass. No apparent differences were observed on calf crop percentage, weaning weights, cow longevity and cattle health. Certain management practices such as early weaning of calves and creep feeding are much more feasible and economical with confined cows. Nutrient requirements of confined cattle are similar to grazing cattle except maintenance requirements are slightly lower because of reduced activity. Drylot cows will usually not show the weight fluctuations of grazing cows as the rations fed are more controlled.

In the future systems of semi-confinement may be more feasible than total confinement. Under this system cows would graze residue fields during the fall and winter and be fed roughage in confinement during the spring and summer months. Under an intensive production system such as this, a larger number of cows can be maintained without reducing acreage devoted to crop production.

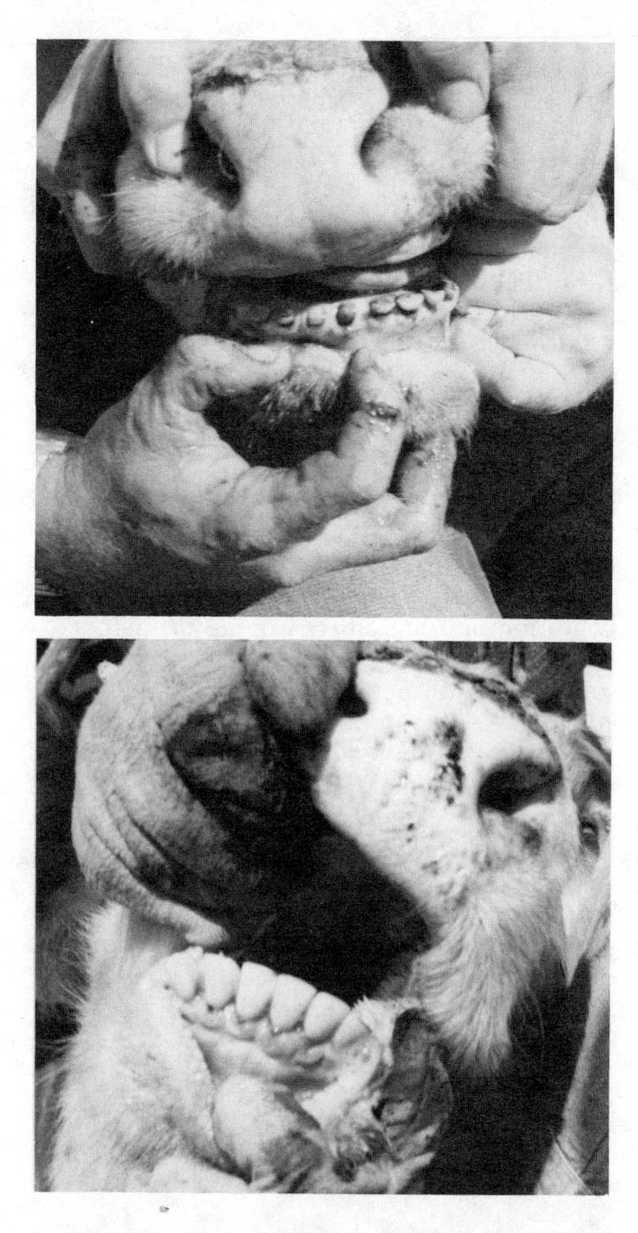

Figure 7-6. Upper: normal tooth wear in a 13-yr. old cow grazing on native range in Texas. Lower: note very little tooth wear on a cow of the same age. This cow was fed harvested feed in confinement. Courtesy of P.T. Marion.

Purebred Cattle

The nutrient requirements of purebred cattle are identical to those of cattle maintained on a commercial operation except that many purebred breeders may want to maintain slightly more condition on their cattle to improve the appearance at shows and sales. This can be accomplished by feeding more energy. However, care must be taken so the cattle do not carry excess condition which may reduce later productivity. Use of creep rations and higher levels of supplementation are common in purebred operations.

References Cited

Acock, C.W., J.K. Ward, T.J. Klopfenstein and I.G. Rush. 1979. Nebraska Beef Cattle Report No. EC 79-218.

Ames, D.R. 1976. In: First Int. Symposium, Feed Composition, Animal Nutr. Requirements and Computerization of Diet. Utah Agr. Exp. Sta., Logan.

Ames, D.R. 1978. Kansas Agr. Expt. Sta. Rpt. Progress 320:94.

Bellows, R.A., L.W. Varner, R.E. Short and O.F. Pahnish. 1972. J. Animal Sci. 35:185 (abstr).

Billingsley, R.O., D.D. Lee, Jr. and J.R. Males. 1976. J. Animal Sci. 42:1367 (abstr).

Bond, J. and J.H. Wiltbank. 1970. J. Animal Sci. 30:438.

Brownson, R. 1977. Great Plains Beef Cattle Handbook Factsheet No. 1951. Oklahoma State Univ., Stillwater.

Call, J.W. et al. 1978. J. Animal Sci. 47:216.

Corah, L.R., T.G. Dunn and C.C. Kaltenbach. 1975. J. Animal Sci. 48:819.

Corah, L.R. et al. 1978. Kansas Agr. Expt. State Rpt. Progress 320.

Dunn, T.G., J.E. Ingalls, D.R. Zimmerman and J.N. Wiltbank. 1969. J. Animal Sci. 29:719.

Fox, F.W. 1972. Commercial Beef Cattle Production. Lea & Febiger, Philadelphia.

Hignett, S.L. and P.G. Hignett. 1951. Vet. Res. 63:603.

Kimple, K., M. McKee and G. Fink. 1977. Kansas Agr. Expt. Sta. Rpt. Progress 291.

Marion, P.T. and J.K. Riggs. 1972. In: Digestive Physiology and Nutrition of Ruminants. Vol. 3. Practical Nutrition. 1st ed. O & B Books, Inc., Corvallis, Oregon.

NRC. 1976. Nutrient Requirements of Beef Cattle. Nat. Acad. Sci., Washington, D.C.

Pinney, D. et al. 1960. 34th Oklahoma An. Livestock Feeders Day Report.

Pope, L.S., F.H. Baker and R.W. MacVicar. 1961. Oklahoma Expt. Sta. Bul. B-578.

Rich, T.D., S. Armbruster and D.R. Gill. 1976. Great Plains Beef Cattle Handbook Factsheet No. 1950, Oklahoma State Univ., Stillwater.

Short, R.E. and R.A. Bellows. 1971. J. Animal Sci. 32:127.

Smith, E.F., V.A. Young, L.A. Holland and H.C. Fryer. 1959. J. Range Mgmt. 12:306.

Varner, L.W. et al. 1977. J. Animal Sci. 44:165.

Vetter, R.L. 1974. Iowa State University Beef Cow Research Report AS-395.

Whitman, R.W. 1975. Ph.D. Thesis, Colorado State University.

Wiltbank, J.N. et al. 1962. J. Animal Sci. 21:219.

Wiltbank, J.N. et al. 1966. J. Animal Sci. 25:744.

Wright, J. and R. Totusek. 1975. J. Animal Sci. 40:194.

Wyatt, R.D. et al. 1977. J. Animal Sci. 45:1120.

Chapter 8—Feedlot Cattle

by F.N. Owens and D.R. Gill

INTRODUCTION

Nutrition of feedlot cattle entails a broad variety of interlocking disciplines. Economics dictate what type of cattle and what type of ration is fed. Feedlot records are essential to project gains, costs and break-even prices from cattle weights and feed intake records. Environmental factors can alter performance drastically and are often overlooked by feedlot managers. Ration formulation for feetlot cattle on a least cost basis, described in Ch. 5, considers not only general requirements for energy, protein, vitamins and minerals, but also adjustment for factors influencing growth potential such as stress, cattle age, cattle background and frame size and specific growth stimulants and feed additives. Least cost gains, not just least cost rations, are of primary concern, so extra allowances are often provided as a safety factor. Proper mixing and delivery of the correct amount of this ration to the cattle is probably the greatest problem in cattle management in feedlots. Calling and delivering feed as well as trouble-shooting cattle problems often control success or failure of a feedlot. These general areas will be presented in detail in this chapter.

PRODUCTION COSTS

Cattle are fed for one reason–the cattle feeder believes he can make a profit by converting feed and labor into beef. Relative costs of cattle, feed and money determine the type of ration, the type of cattle and the length of feeding which is most profitable or least unprofitable. The following sections describe some of the factors involved in the complex gamble of cattle feeding.

Cost of gain can be subdivided into the costs of cattle, feed and feedlot charges. Consider cattle costs first. This can be subdivided into cost of transport, commission, interest, death loss, and medical treatment.

Figure 8-1. Cattle on feed in a large commercial feedlot in the southwestern USA. Photo courtesy of D.C. Church.

CATTLE COSTS

Transport, Commission and Shrinkage

Most cattlemen calculate cost of gain for cattle by comparing the "layed in" cost to the final sale price of the cattle. Layed in cost is defined as cattle cost plus commission and freight divided by the purchase weight. For example, 700 lb steers, bought through an order buyer at $50/100 lb live weight, transported 300 miles and shrinking 8% in transit will cost $55.87/100 lb (Table 8-1) when layed in. The cost of shrink increases with cattle price, while freight and commission costs increase with weight. Cost of shrink with different amounts of shrink are shown in Table 8-2. The greater the shrink, the greater the time for recovery of shrink and the greater the death loss.

Interest

In the past, interest cost seldom represented over a few ¢/lb of gain, but with high cattle prices and interest rates, this cost can grow considerably. Daily interest costs on different investments at various interest rates are presented in Table 8-3. With a cattle investment of $500 and a simple interest rate of 16%, interest amounts to 22¢/day. With a gain of 2.6 lb/day, interest cost would equal 8.5 ¢/lb of gain. This table demonstrates the

Table 8-1. Cattle cost: effects of freight, commission and shrink.

Purchase price, $/cwt	Cost after shrink adjustment,[a] $/cwt	Freight,[b] $/cwt	Commission,[c] $/cwt	Layed in cost, $/cwt	Layed in minus purchase price, $/cwt
130	141.30	1.02	.50	142.82	12.82
120	130.43	1.02	.50	131.95	11.95
110	119.57	1.02	.50	121.09	11.09
100	108.70	1.02	.50	110.22	10.22
90	97.83	1.02	.50	99.35	9.35
80	86.96	1.02	.50	88.48	8.48
70	76.09	1.02	.50	77.61	7.61
60	65.22	1.02	.50	66.74	6.74
50	54.35	1.02	.50	55.87	5.87
40	43.48	1.02	.50	45.00	5.00
30	32.61	1.02	.50	34.13	4.13

[a]Shrink of 8% used for calculation.
[b]Cattle shipped 300 miles, freight at $1.50 per mile, 44,000 lb load purchase weight.
[c]Typical commission charge.

Table 8-2. Cattle costs: effect of shrink.

Purchase price, $/cwt	Freight, $/head	Commission, $/head	Shrink, %	Cost after shrink adjustment, $/cwt	Layed in cost, $/cwt	Difference, %
80	.85	.50	2	81.63	83.01	3.01
80	.85	.50	4	83.33	84.74	4.74
80	.85	.50	6	85.11	85.11	5.11
80	.85	.50	8	86.96	88.42	8.42
80	.95	.50	10	88.89	90.39	10.39

Table 8-3. Interests costs: effect of principal and interest rate cost per day.

Principal, Amount, $	Interest rate, %			
	6	10	14	18
200	3.33	5.56	7.78	10.00
300	5.00	8.33	11.67	15.00
500	8.33	13.89	19.44	25.00
700	11.67	19.44	27.22	35.00
900	15.00	25.00	35.00	45.00

Table 8-4. Interest cost: effect of rate of gain.

Daily interest cost,[a] ¢	Rate of gain, lb/d	Interest cost/lb of gain, ¢/lb
.1389	0.5	27.78
.1389	1.5	13.89
.1389	1.5	9.26
.1389	2.0	6.94
.1389	2.5	5.56
.1389	3.0	4.63

[a]Based on a principal of $500 and an interest rate of 10%.

Table 8-5. Cattle cost adjusted for death loss.

Apparent cattle cost, $/cwt	Death loss, %						
	0.5	1	2	3	4	5	6
	. . . . True cost, $/cwt.						
20	20.10[a]	20.20	20.41	20.62	20.83	21.05	21.28
30	30.15	30.30	30.61	30.93	31.25	31.58	31.91
40	40.20	40.40	40.82	41.24	41.67	42.11	42.55
50	50.25	50.51	51.02	51.55	52.08	52.63	53.19
60	60.30	60.61	61.22	61.86	62.50	63.16	63.83
70	70.35	70.71	71.43	72.16	72.92	73.68	74.47
80	80.40	80.81	81.63	82.47	83.33	84.21	85.11
90	90.45	90.91	91.84	92.78	93.75	94.74	95.74

[a]Adjusted layed in cost.

importance of shopping for the lowest interest rate. Interest cost/lb of gain depends on rate of gain as shown in Table 8-4. At high interest rates, the economic benefit of rapid gain is greater. High interest costs prohibit long periods for growing feeder cattle prior to finishing.

Death Loss, Medical Costs, Taxes and Insurance

The cost of death loss increases as cattle prices increase or as cattle increase in weight. Table 8-5 shows the layed in price of cattle adjusted for anticipated death loss. Actual costs usually exceed the amounts indicated in the table because feed and medical costs are usually incurred before an animal dies. Medical costs vary widely but generally range from $2-5/head. Taxes and insurance costs also vary but normally fall in the range of $1-3/head.

Feed Costs

Feed costs include the cost of buying, transporting, storing, processign and delivering commodities to the feed bunk. Many cattle feeders buy grains at harvest in an attempt to buy at the lowest annual price. They often overlook the hidden costs of storage, shrinkage and interest. Table 8-6 shows the cost by month of a bushel of corn bought at $2.80 on November 1 and shrinking 3% during storage and handling. Note that after 6 mo., the interest, storage and handling charges increased the corn cost by 49¢/bu. Over time, interest and storage costs accumulate so that the cost of grain and thereby the cost of shrink increases considerably. These costs for storage must be balanced against the fluctuating cost and availability of grain purchased on a more continual basis.

Each 10 ¢/bu increase in grain price is considered by most feeders to equal about $1/cwt increase in feed cost of gain. The data in Table 8-6 illustrate that feed prices should be evaluated at least monthly, since hidden costs (interest, handling and shrinkage) make it possible for a feedlot to sell stored feed to cattle owners for less than cost.

Relationships Between Feed Efficiency and Total Cost of Gain

Light weight calves generally have feed efficiencies considerably superior to yearling cattle. Feed costs for feeding a steer from 500 to 900 lb are generally about 75% that of a steer fed from 700 to 1,100 lb. This greater feed efficiency and reduced feed cost may be offset by a higher non-feed cost (interest and death loss), as younger cattle generally have lower daily gains than yearling cattle.

Feed Costs of Gain

Estimates of cost of gain due to feed alone for cattle fed high energy rations can be estimated from Table 8-7. For these calculations it is assumed that the energy sources in the ration are priced similar to corn grain and the total ration is considered to be all corn since supplements usually cost more/lb than corn while roughages cost less. Feed conversions of 7.2 lb DM/lb gain were used to develop this

Table 8-6. Cost of corn purchased at $2.80/bu in November, and stored for various times.

Month	Charges					
	Cost, $/bu	Interest,[a] $/bu	Storage,[b] $/bu	Handling,[c] $/bu	Cost, $/bu	Cost + 3% shrink, $/bu
November	2.80	---	---	.15	2.95	3.04
December	2.80	.0364	.02	.15	3.01	3.10
January	2.80	.0437	.04	.15	3.03	3.12
February	2.80	.1094	.06	.15	3.12	3.22
March	2.80	.1435	.08	.15	3.17	3.27
April	2.80	.1818	.10	.15	3.23	3.33
May	2.80	.2193	.12	.15	3.29	3.39
June	2.80	.2585	.14	.15	3.35	3.45
July	2.80	.2970	.16	.15	3.41	3.51
August	2.80	.3372	.18	.15	3.47	3.57
September	2.80	.3780	.20	.15	3.53	3.69
October	2.80	.4180	.22	.15	3.59	3.70
November	2.80	.4598	.24	.15	3.65	3.76

[a] 15% simple interest
[b] Charged at 2¢/bu/month
[c] Charge of 15¢/bu

Table 8-7. Feed only portion of cost of gain.

Steers		Heifers	
Weight range	Corn factor	Weight range	Factor
800-1200	.1616	800-1500	.1569
700-1100	.1522[a]	700-1050	.1550
700-1200	.1616	700-1000	.1500
699-900	.1331	600-950	.1522
600-1000	.1369	600-900	.1484
600-1100	.1500	600-850	.1426
500-900	.1284	500-900	.1535
500-1000	.1331	500-850	.1484
500-1050	.1426	500-800	.1426

[a] Example: For a steer fed from 700 to 1100 lb, feed cost should be approximately 0.1522 times the cost of a bushel of corn as delivered to the bunk. If whole shelled corn at the bunk costs $2.45, the total feed cost/lb of gain will be $2.45 x .1522 = 37.3¢.

table. The values apply to initial and final weights and desirable environmental conditions.

These relationships were derived from feedlot data from Great Plains feedlot cattle which had received an estrogen implant or fed an estrus inhibitor and re-implanted when needed. No provision is made for death loss or shrink. If catter are fatter at the start of the feeding period, feed costs need upward adjustment. Feed processing, interest, and shrinkage must be included in the price of corn.

Margins

Complete updated feeding budgets need to be maintained for cattle of various types and weights to determine break-even cost and calculate possible profit/head/day for cattle feeders. Relationships change constantly as cattle, feed and non-feed costs shift. A negative margin exists when fed cattle sell for less per pound than feeder cattle and a positive margin occurs when fed cattle sell for more per pound than feeder cattle. Economically, costs of feeding, not merely initial and final cattle prices, need to be considered to determine the feeding margin. At times, cattle feeders can buy weight cheapter than they can put in on by feeding cattle. This situation encourages diligent cattlement to buy heavy cattle to feed for only the shortest period of time to finish cattle for market. In the reverse situation, where cattle are "cheapened back" while on feed or one has a "positive feeding margin", selection should be for feeder cattle which will gain a lot of weight but yield desirable carcasses with a good feed efficiency.

Feedlot Charges

In almost every situation cattle gains are most economical when rate of gain is maximum. Considering only interest and margin costs, it becomes apparent that low rates of gain are not feasible commercially. In addition, gross feedlot margins average 20-25¢/day in 1980 and probably will increase with energy cost. Although and farmer-feeder may feed only one group of cattle per year and does not directly pay a daily yardage charge, costs of interest and labor plus risk should encourage him to strive for maximum rates of gain by cattle. The challenge of the beef industry is to increase rate of gain at a pace fast enough to keep beef a competitive food.

FEED INTAKE AND FEEDLOT PERFORMANCE

Two factors are often confused when feed intake is considered. These are the short term feed intake or meal size (palatability) and appetite or feed intake over a sustained period. Adding a sweetening agent such as molasses to a low-protein diet may improve palatability, but such addition will have little long-term effect on feed consumption unless it alters animal metabolism. Palatability, as used here, refers to the preference for a feed or ration whereas appetite or feed intake refers to an improvement in intake over a longer feeding period.

A number of factors play roles in control of feed intake (see Ch. 11, Vol. 2). Depending upon ration composition, these factors may place a "ceiling" on feed consumption.

Rumen Fill

Bulk fill in the rumen limits feed intake with higher roughage rations. Intake is thereby related to the rate of passage from the rumen. More easily digested roughages ferment more rapidly and the residue passes out of the rumen faster while low-quality roughages or certain nutrient deficiencies reduce rate of fiber digestion and passage and thereby limit the amount of feed cattle can consume. When starting feedlot cattle on feed, bulky rations are commonly fed to prevent acidosis. Digestive stress and acidosis problems can be extreme when concentrate rations are consumed readily by new cattle. During the time period when concentrate levels are increased, growing calves must switch from bulk fill regulation of roughage rations and adjust to chemostatic intake regulation.

Chemostatic Control

The general relationship between energy density, rate of gain and feed conversion efficiency is presented in Figure 8-2. Rate of gain increases with energy concentration in the ration to a plateau. Below this plateau, bulk fill limits intake. On the plateau, chemostatic factors control intake. Daily gain may decline as the concentrate level approaches 100%. The roughage:concentrate ratio or energy density at which this plateau begins is influenced by a number of variables. In general, a higher level of roughage can be fed and still attain maximum daily gains with younger, lighter and/or thinner cattle than in older, heavier and/or fatter cattle. As cattle progress through the feeding period, the level of roughage must be decreased if maximum daily gain is to be obtained. Higher levels of roughage can be fed with higher quality roughages than poorly digested roughages, since rapid digestion reduces bulk fill. The most desirable rate of gain, the flat portion of the curve for finishing cattle, would be dictated largely, but not exclusively, by the cost and availability of roughages relative to concentrates. In general, feeders in the Corn Belt prefer to feed nearer the left hand side (point a) and High Plains feeders to the right hand side (point b) on the flat portion of this curve. For light weight growing cattle, a lower daily gain may produce a more desirable carcass grade. Whether such cattle should be backgrounded in a feedlot or on pasture depends on economics.

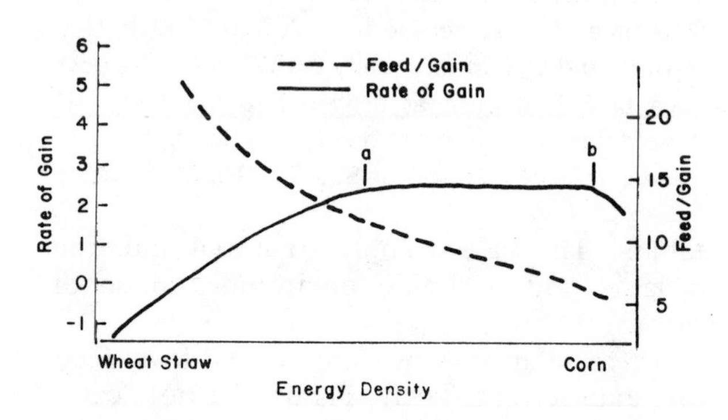

Figure 8-2. Effect of energy densities on rate and efficiency of gain.

Efficiency of Feed Use

The amount of feed required for maintenance of an animal increases with live weight. Only feed supplied above this maintenance requirement is available for weight gain. Although the maintenance requirement increases with weight, a 1,000 lb animal does not have twice the maintenance requirement of a 500 lb animal, since a 500 lb steer requires 4.55 Mcal of NEm per day and a 1,000 lb steer requires 7.65 Mcal of NEm. The requirement increases at the ¾ power of the body weight which reflects the relative protein content or lean mass of an animal. Since more feed is required for maintenance as feedlot cattle increase in weight during the feeding period, efficiency of feed used declines and cost of gain increases as feedlot cattle grow.

Efficiency and Increasing Finish

The amount of NE needed above maintenance (NEg) for a 400 lb steer to gain 1 lb/day is 1.25 Mcal. A 1,000 lb steer, however, requires almost twice as much NEg (2.49 Mcal) for a similar weight gain. Several factors are responsible. First, the 1,000 lb steer deposits considerably more fat and less water/lb of gain than the 400 lb steer. On a dry basis, fat contains almost twice as much energy as protein, although energy required to deposit fat appears lower than for protein. On a caloric basis, fat is more efficient to deposit than protein, but fat is deposited with only about 10% added water while protein is deposited with about 60% its weight of water. Due to the water retention, efficiency of weight gain is much greater with lean than fat deposition.

Efficiency of gain also changes with rate of weight gain. A 400 lb steer gaining 3 lb/day requires more than 3X as much NEg than one gaining only 1 lb/day (4.17 vs 1.25 Mcal). This is because the steer gaining 3 lb/day deposits more fat/lb of gain than one gaining only 1 lb/day. Since a higher proportion of the gain is fat, the energy requirement is greater.

Heifers require more feed/lb of gain than steers at the same weight. A 700 lb steer requires 1.91 Mcal of NEg to gain 1 lb daily, whereas a heifer requires 2.11 Mcal of NEg. Again, the heifer is depositing more fat/lb of gain than the steer and therefore requires more feed. Similar to the comparison with steers of different weight above, the heifer is nearer her mature weight than a steer at the same weight. Heifers of a large frame size and a high mature weight would have nutrient requirements most similar to steers of medium frame size and moderate mature weight. If heifers and steers are fed together, they might be pen fed in groups matched in this manner.

The chemical makeup of the diet, primarily the availability of energy, is of great importance in regulating feed intake with higher concentrate rations. Specific nutrients thought to be involved are discussed in detail in Vol. 2.

Other Factors

Excessive Concentrate. A small amount (5-15%) of roughage is included in the ration of feedlot cattle to aid in bunk management and avoid digestive disturbances. Added roughage decreases rate of feed consumption and increases salivary flow. The added saliva helps buffer rumen contents, aids ruminal mixing and enhances digestion within the rumen. Unfortunately, added roughage may reduce digestibility in the small intestine.

Physical Make-up of the Ration. Processing may influence total feed intake (see Ch. 3). Feeders generally prefer coarsely processed grains, even though undigested grain appears in the feces. The reason for feeding coarse grain is based on the premise that fine, floury feeds are less palatable. In Arkansas tests, where cattle were offered a choice of grain processed in different manners, they preferred the coarsely-processed material. Equally important may be a slower rate of passage for the coarse material from the rumen, with less rapid absorption of VFA and a greater appetite tolerance over a 24-h period.

Density or Bulk of the Diet. Sand or inert plastic materials have been added to steer rations to alter density of the ration. Intake of digestible DM with a high concentrate ration was unchanged by addition of sand or plastic although DM intakes differed. Odor, flavor and texture of the ration may alter palatability. In Texas tests, very high moisture levels (25-50%) reduced total DM intake. However, the dry, dusty nature of chopped and mixed rations is undesirable. Adding low levels of water, silage, fat, or sticky material of some type may slightly aid total feed intake of a dry ration. Overall, the chemical balance and nutrient content of the ration appear more important than taste or odor to the fattening steer.

A P deficiency can reduce feed intake and cause a depraved appetite. Depriving the ruminant of water also will reduce feed intake markedly. The practical importance of "gypsum" or unpalatable water on feed intake has received little attention, but a wise cattle feeder needs to monitor water as well as feed quality.

Time or sequence of feeding may also influence feed intake. Several studies have shown that feed intake increases with frequent feeding. Providing fresh, clean feed several times daily will sharpen appetites.

Figure 8-3. One example of a mixer-delivery truck normally used in large feed yards. Courtesy of Feedlot Management Magazine.

FEED INTAKE PREDICTION EQUATIONS

Rate of gain and feed efficiency for pens of cattle can be closely predicted from feed intake using the NE equations. Computer printouts in most large feedlots provide daily and cumulative estimates of gain and costs for each pen of cattle. Performance and efficiency can be predicted if feed intake of pens of cattle can be projected. The trick is to predict the feed intake. For cattle in the Great Plains, intake of 65 to 90% concentrate rations can be projected reasonably well based upon cattle feeder grade, initial skrunk weight, current weight, and ration DM. Formulas are given at the end of the chapter for predicting feed intake. Estimates of feed intake can easily be calculated on a daily or weekly interval with an iterative calculator or computer program. Together with NE values of the ration, these programs can predict weight gain, feed efficiency, total cost and break-even price at any period of feeding for the group of cattle being simulated.

When designing these equations, the amount of feed an animal will consume was found to be related to body type (BT). For calves, 1 is for the smallest frame type cattle and 5 is used for large frame Hereford cattle. A 2.5 was considered average. Yearling cattle, especially those from wheat pasture, usually eat more feed. For yearlings, body type values by weight groups would be: 600, 4.2; 650, 4.8; 700, 5.5; 750, 6.0.

Correct body type is essential for an accurate simulation of feed intake. The user should test each run by comparing average intake with past records of pens of similar type cattle and with daily feedlot management reports to ensure a reasonable simulation. Each 1 unit added to or subtracted from the body type score will change estimated feed intake for the entire feeding period by about 1 lb of DM daily.

This and other simulation models will prove helpful in projecting feedlot performance. Such models can remove some guesswork and assist in making economic decisions about cattle feeding profitability.

ENVIRONMENT AND STRESS

Climatic conditions that exist in an area have a major impact on the economics of beef production. A desirable climate gives individual feeders and beef producers an advantage in producing beef. Modifying the environment through structures or buildings may improve the competitive position of livestock feeders in certain regions of the USA. In recent years factors have been developed to estimate the impact of stress on animal performance and to evaluate the economic effects of environmental modification.

Mud and Wet Pens

Mud greatly reduces feedlot gains. Mud was produced by frequent sprinkling of pens at the University of California (Givens, 1969). Mud reduced rate of gain by 23% and increased feed to gain ratio by >20% in these studies. These trials had moderate temperature conditions and examined the impact of mud in the absence of high humidity and cool temperatures which often accompanies mud stress.

High Temperature

In the desert areas of the Southwest, summer temperature can reach 120°F. This temperature will greatly reduce gains of cattle of the English breeds. Modifications in feedlot construction can reduce the heat load and improve performance. Experiments at the Imperial Valley Field Station, near El Centro, California (Kelly et al, 1960; Johnson et al, 1974), have demonstrated that shades reduce the radiant heat load upon the cattle by as much as 50%. Shades do not lower air temperature and are not useful if the sky is heavily overcast, since the sky must be clear to serve as a heat sink for radiation from cattle under shades. Shades for cattle are not widely used except in Southern California and Arizona.

Cold Stress

Feedlot cattle in the warmer, dryer areas of the USA gain more rapidly and efficiently in the winter than do cattle in the northern areas. It is hard to separate the effects of cold, wind, moisture and mud which are usually concurrent problems. Adaptation of cattle to cold stress may be more important than cold itself. Southern cattle shipped into the high plains of Oklahoma and Kansas in October and November occasionally die of stress related to cold if severe conditions occur during the first month after arrival.

Feedlot cattle fed to gain rapidly with a thick, dry hair coat have a critical temperature well below -10°F in still air (see Ch. 16, Vol. 2). The combination of wind, cold and moisture will increase the critical temperature and increase maintenance requirements of cattle. Insulation loss due to rain can be readily regained after a shower by solar radiation and production of heat from the body. Insulation loss from lying in mud or water has a much longer duration due to reduced exposure for evaporation by radiation and convection. Maintenance NE requirements for feedlot cattle in north central Colorado plains by month are presented in Table 8-8.

Table 8-8. Estimated maintenance energy requirements of open-lot fed cattle in the north central Colorado plains.[a]

Month	NEm (Kcal/kg$^{0.75}$/da)
January	93
February	91
March	87
April	79
May	71
June	65
July	65
August	65
September	67
October	77
November	87
December	93
Mean	78

[a]Johnson and Crownover (1975)

Adjusting Rations for Heat or Cold Stress

Heat stress can be reduced by reducing the ratio of heat increment to net energy. Lofgreen (1973) improved winter feeding performance by formulating rations for a high heat increment, and improved warm weather performance by formulating for a low heat increment. The cost of feed selection on a heat increment basis may not always be justified by the improved performance, but some feeders may use some opportunities to increase the roughage level and reduce fat in rations fed during cold weather.

Breed of Cattle and Weather Stress

Breeds differ in response to hot and cold conditions. Cattle feeders frequently make the mistake of moving Brahma cross cattle from southern into northern feeding areas during the fall months. Lack of adaptation to cold and lack of protection from the cold frequently spells disaster. Similarly, British breeds of cattle do not perform well in the summer in the southwest desert. Seasonal cattle performance in feedlots should be improved by moving cattle the same direction as migrating birds.

Receiving Rations

Design of a receiving ration for a feedlot is complicated by two extreme types of newly arrived cattle. Calves on arrival have the problem of consuming enough energy to rebuild strength and make reasonable weight gains. In contrast, yearlings or cattle which have received less stress, will eat large amounts of feed rapidly and may subsequently suffer from acidosis and other digestive upsets. The best feedlot managers often fail to assess which eating pattern a set of cattle will follow on delivery. A 72% concentrate ration (Table 8-9) based on the work of Lofgreen (1977) produced economical weight gains with minimum health problems for stressed light weight calves. This ration probably would cause acidosis and death of non-stressed local yearlings eating large amounts of feed. The degree of stress must be considered in ration formulation.

Table 8-9. Composition of 72% concentrate receiving rations.

Ingredient	%
Alfalfa hay	23.0
Sudan hay	5.0
Rolled barley	49.5
Wheat millrun	3.0
Hominy feed	4.0
Linseed meal	4.5
Fat	3.0
Molasses	7.0
Limestone	0.5
TM Salt	0.5
Vitamin A	1,000 IU/lb of ration

A common practice in high plains feedlots is to offer newly arrived cattle a 60% concentrate ration together with some palatable grass hay for several days. Most feedlots will limit the daily amount of the concentrate ration to 12 lb of DM/head for the first few days and let the cattle consume as much hay as they please.

Management of the bunks the first few days is critical. Eating patterns established early will likely continue through the feeding period. It is desirable to have high and consistent (day-to-day) feed intake on all feedlot cattle. Offering too little or too much feed may start the cattle on a cyclic feed intake pattern which can only be broken by re-starting the pen. Cattle susceptible to respiratory viral infections can benefit from a longer time on lower energy rations, but overhead costs tend to force cattle feeders to move cattle rapidly to higher concentrate rations.

Nutrient Levels in Receiving Diets

Newly received cattle have special nutrient problems. Most are related to feed and nutrient intake. For example, a 400 lb stressed calf has a K requirement of at least 0.07 lb/day. If a calf ate 12 lb of DM, the K requirement would be met by a ration containing 0.6% K. However, if a calf only consumed 5 or 6 lb of feed, K supply is inadequate even though the percentage in the ration appears adequate. Cattlemen and nutritionists should consider nutrient allowances on an amount/head/day for stressed cattle. Particular attention should be given to protein and K, since these nutrients appear particularly important in recovery of normal feed intake.

Stresses in the Movement and Adaptation of New Feeder Cattle

About 2% of the cattle shipped in the USA die after arrival at feedlots or ranches. Two approaches have been proposed to solve this problem. The first involves broad-scale procedures prior to shipping and generally is termed preconditioning. The second involves intensive procedures after arrival. Since most shipped cattle change ownership a number of times prior to feedlot arrival and weight groups are assembled from a number of herds, identity of preconditioned cattle is usually lost. Since the degree of preconditioning (weaning, castration, dehorning, vaccination, parasite control) varies among producers, feedlot operators must intensively process all purchased cattle at arrival and cannot financially reward producers for preconditioning even if producers could be identified. If the producer retains ownership of cattle until slaughter, financial rewards should justify preconditioning.

Complex relationships exist between stress and viral and bacterial agents. Shipped cattle which die usually die as the result of a bactrial or viral infection to which the animal was predisposed by stress. Consequently, most feedlots have developed a systematic health-receiving program which is applied to new cattle.

PROTEIN

Protein requirements for maintenance and various rates of growth by feedlot cattle have been outlined in the NRC (1981) publication for feedlot cattle. Requirements should be visualized as daily *amounts* instead of diet *percentages,* since feed intake varies with ration composition and cattle type. In addition to the listed requirements which consider frame size, feed composition and feed intake, feedlot nutritionists should adjust for variability in protein content of feedstuffs, digestibility of fed protein, metabolic efficiency of utilization of various proteins and previous nutritional status of the cattle being fed. Some of these factors are difficult to assess, and many of the beneficial effects of added intact protein on animal health, bacterial efficiency in the rumen and starch digestibility are neither understood nor included in current estimates of requirements.

At normal prices, 1% protein above that required in the ration will increase cost of gain <0.5 ¢/lb. Consequently, a safety margin may be desirable despite the added cost since protein deficiency can markedly reduce digestibility and efficiency of feed use.

NPN Usefulness

Protein requirements for feedlot cattle can be subdivided into ammonia need for ruminal digestion and amino acid needs post-ruminally. Urea can be used to provide ammonia but only indirectly to supply amino acids. Several systems have evolved to estimate the amount of urea which can be added beneficially to a feedlot ration. Thumb rule maximums of 1% of the ration DM or 33% of the total protein are widely used. In feedlot rations, a lower maximum, 0.5%, is often used. The "Urea Fermentation Potential" (UFP; g of urea which can be usefully added to one kg of ration) can be estimated from TDN, protein content and ruminal protein degradability of a ration (see calculations at end of chapter). UFP values for common feedstuffs are presented in Table 8-10. One can calculate from ration composition whether in theory urea should be added (positive UFP) or not (negative UFP). Rations low in protein and high in energy content are best suited for urea use, but high moisture grains are less suited than dry grain rations for urea use.

Table 8-10. Urea fermentation potential of common feedstuffs.[a]

Feedstuff	% Crude protein	UFP
Alfalfa meal	16.3	-34
Barley grain	13.0	-1.6
Corn grain	10.0	11.8
Oats	13.2	-4.7
Milo grain	12.5	7.7
Soybean meal	51.5	-107.7
Urea (to UFP only)	280.0	1000.0

[a]From Burroughs et al (1975)

Management Factors for Urea Use

When only a small amount (<1 lb/head/day) of supplement is fed and the supplement is top-dressed onto the rest of the ration, some cattle may receive excesses and others will not receive enough of the supplement. For best results, all highly concentrated supplements, especially those containing urea, should be completely mixed with the total ration or at least with the grain portion of the ration. Mixing efficiency of equipment should be checked occasionally to be certain that the delivered ration does not have hot spots of concentrated ingredients. Rations which allow separation of fine from coarse material may also cause urea or unpalatable feed additives to become concentrated in the bottom of the feed bunk. This can lead to cyclic feed intake, erratic animal performance and waste of feed.

Cattle should be gradually accustomed to urea. Best results may be obtained with little or no urea in receiving rations with urea addition gradually over a period of 3 or 4 weeks. Since K, P and S are supplied by preformed protein sources, be sure that these are adequate when urea is fed as a substitute for preformed protein.

High urea supplements should not be fed to sick cattle. Ammonia toxicity is usually a result of uncontrolled supplement intake or feeding excessive quantities to sick or unadapted cattle.

Post-Ruminal Protein Needs

The post-ruminal protein supply consists of about half microbial protein synthesized in the rumen and half protein from the ration which bypasses destructive bacterial action with the rumen. Absolute protein requirements increase with rate of protein deposition in the carcass as well as with body size and maintenance functions. Rapid protein deposition corresponds with periods of fast rate of gain, especially for younger cattle, and with the use of growth stimulants such as estrogenic implants.

The percentage of protein required in the ration delcines as cattle mature. Ration protein can be reduced by changing to different rations through the feeding period when complete rations are fed. Protein reduction also can be achieved by feeding a constant amount of protein per animal daily in a supplement added to the grain. This system automatically reduces the percentage of protein as feed intakes increase. Such a system may prove particularly useful for newly received cattle since total feed intake at this time often is low and variable from animal to animal.

Protein Digestibility

Heat damage during storage of forages or processing of grains will reduce digestibility of protein by cattle. Such a depression can be monitored by pepsin digestibility or acid detergent fiber-N analyses. Bound protein is useless for the animal. The pepsin indigestible protein from feedstuffs which were used to establish the protein requirements of cattle probably ranged from 10 to 20% of the protein. Consequently, when indigestibility exceeds this value, extra protein must be added to compensate.

Tables estimating digestible protein requirements have been used in the past. These generally were calculated from crude protein requirements. Except when digestibility has been measured or when certain protein sources of questionable digestibility are fed (sorghum grain, animal protein sources, very low protein roughages), digestible protein has not proven superior to crude protein for formulation of rations. Some of the newer protein systems (metabolizable protein; net protein) may prove useful in the future as understanding improves of factors influencing ruminal protein bypass, microbial protein synthesis and protein needs for maintenance and gain.

Ruminal Bypass

Besides differences between protein sources, bypass of protein varies with bacterial factors in the rumen, which are influenced by fiber level in the ration and residence time in the rumen which is influenced by feed intake, roughage level, feed processing and feed additives. Proteins can be crudely divided into three bypass categories as follows: under 40%, casein, soybean meal, peanut meat; 40 to 60%, cottonseed meal, alfalfa meal, corn grain, brewers dried grains; 60 to 80%, meat meal, corn gluten meal, blood meal, feather meal, fish meal. Feeding of high bypass proteins will increase the use of and need for NPN by ruminal bacterial. Without such NPN addition, high bypass proteins may not increase the total supply of protein to the intestines. Performance benefits of increased protein bypass should be expected only when protein needs are high. Consequently, their benefit is greatest for young, rapidly growing, implanted calves and newly received feedlot cattle under stress.

Estrogenic growth stimulants increase protein deposition in the lean body mass by up to 37% (Byers, 1979). As about one-third of the protein used by growing cattle is used for protein deposition, this means that about 12% more protein would be required. But feed intake is increased about 6% by growth stimulants. This will increase the protein supply to the animal by 6%. This difference suggests that with implants, the protein requirements on a percentage of the ration may be about 6% greater for cattle with than those without growth stimulants. This would mean that if a 10% protein ration is suitable for non-implanted cattle, 10.6% would be recommended for implanted cattle.

Monensin feeding decreases feed intake an average of 10% without decreasing rate of protein deposition. This would suggest that an increased percentage of protein may be needed when monensin is fed. But monensin increases ruminal bypass of protein 15 to 30%. This increased bypass will reduce the requirement for protein and compensate for the decreased protein intake. Results from feeding studies to date indicate that the percentage of protein in the ration does not need to be increased when monensin is fed and indeed might be slightly reduces as monensin spares protein.

Feed Tag Interpretation

The term "protein equivalent" refers to the maximum amount of protein that can be synthesized in the rumen from the urea or NPN provided and is equal to NPN x 6.25. The amount that is actually synthesized depends upon a number of factors and will range from none with high-protein low-energy rations to near 100% with low-protein high-energy rations. Reduced intake is often observed the first month of urea feeding. This adaptation to urea may reduce performance in the short term. Higher urea levels may also depress gain when cattle are nearly finished. Since feed intakes are high at this time, an excessive amount of urea may be the problem.

VITAMIN SUPPLEMENTATION

Vitamin supplementation programs for feedlot cattle have centered on the fat-soluble vitamins since ruminal syntehsis of the B vitamins under normal conditions has been considered adequate to meet systemic needs.

Fat Soluble Vitamins

Vitamin injections have become part of the treatment scheme for newly arrived feeder cattle at most feedyards. Commonly, one million units of vitamin A are injected. Often, vitamins D and E are included in the injection mixture with no apparent nutritional reason. This amount of vitamin A, when stored for later release by the liver, is enough to meet total animal needs for a period of about 200 days. In addition, feedlot rations are typically supplemented with 1,500 IU of vitamin A/lb. This is about the requirement listed by the NRC (1976). Considering the vitamin A activity of corn grain and alfalfa products commonly fed, total intake exceeds the requirement considerably.

The other fat-soluble vitamins have received less attention. Vitamin D should be added to rations for cattle fed with no exposure to sunlight. The vitamin D requirement is 275 IU/kg of dry ration (NRC, 1976) and either D_2 or D_3 may be used. Sunlight and sun-cured roughages should meet the vitamin D requirement for other cattle. During fermentation and storage of high moisture grains, vitamin E activity is greatly reduced. However, experiments conducted to date have not demonstrated a performance benefit from vitamin E supplementation of feedlot rations with these or other normal rations. Vitamin K supplementation has proven beneficial only when moldy sweet clover was fed to cattle. In the past, internal hemorrhages of newly received cattle following castration or bruising was attributed to vitamin K deficiency and reduced prothrombin, but today the disorder appears associated with elevated thrombolysin.

Water-Soluble Vitamins

Four of the B vitamin have been tested recently in feedlot rations. In each case, it appears that the vitamin is altered in the rumen or the vitamin is altering ruminal or systematic metabolism instead of serving the classical vitamin role as a metabolic cofactor in tissues.

Polioencephalomalacia (PEM) of feedlot cattle has been attributed to a deficiency of tissue thiamin due to ruminal destruction of thiamin or synthesis of a thiamin antogonist in the rumen (see Ch. 9, Vol. 2). Elevated S intakes may accentuate the problem. Some of the symptoms—out-stretched neck, paralysis—are similar to one form of coccidiosis, but PEM cattle respond within minutes to an intravenous thiamin injection. Feed additives to prevent PEM have not been developed, although coating several B vitamins for ruminal bypass might prove useful.

Choline, unlike other B vitamins, does not appear to be synthesized within the rumen. Choline supplementation at 500 to 750 ppm increased gain and feed efficiency by 6% in two trials (Dyer et al, 1962; Rumsey and Oltjen, 1975), but several other feedlot trials, especially with higher roughage levels, have shown no benefit. Conditions under which choline might prove useful as a feed additive remain to be determined.

Niacin supplementation at higher levels (250 and 500 ppm) depresses feed intake and rate of gain of feedlot cattle. But lower levels (100 ppm) in 9 trials with feedlot cattle has increased gain and feed efficiency an average of 3.6%. Compared with the cost of supplementation (under 50 ¢/ton of feed), supplementation may be justified. Niacin might prove more useful with urea-supplemented rations or early in the feeding period during ration adjustment.

Injections of B_{12} have proven helpful for newly received cattle in several studies. Deficiencies of protein or certain minerals, ruminal dysfunction, or the stresses of ration adaptation or high concentrate may reduce ruminal B-vitamin synthesis. At such times, supplementation could prove beneficial. But other than injectables or special supplements for newly received cattle, benefit from B vitamin supplementation of feedlot rations has been small.

Table 8-11. Estimated macro-mineral requirements (g/day) for feedlot cattle gaining 2.2 lb daily.

Animal weight	Mg, g		K, g		Na, g		Cl, g	
	NRC	ARC	NRC	ARC	NRC	ARC	NRC	ARC
440	6.0	5.0	40	9-16	2.5	4.8	2.5	6.1
660	9.0	6.5	60	14-24	3.8	6.5	3.8	8.7
880	12.0	8.0	80	18-32	5.0	8.2	5.0	11.3

MINERAL NUTRITION OF FEEDLOT CATTLE

Specific functions of major and trace minerals in the rumen or tissues have been presented in detail in Ch. 4 and 5 of Vol. 2. Only special concerns about mineral nutrition of feedlot cattle will be reiterated here. Requirements for minerals as presented by the NRC (1976) or ARC (1965) are listed in Table 8-11. Values are cited (g/head/day) for cattle of different weights for K, Na and Cl and values for Ca and P and the trace minerals are given in Appendix Tables. When animals are consuming only small amounts of feed, the percentage of minerals in the ration needs to be increased. For many of these minerals, feedstuffs will meet much of the need. But for most commercial feedlot cattle, supplemental salt, Ca and P, K and a trace mineral package are included in the ration. Specific minerals are discussed below.

Salt (Sodium Chloride)

Extreme deficiencies or excesses of salt depress feed consumption. Since salt is deficient in most feedstuffs, it should be added to most feedlot rations. Salt with certain added minerals is commercially available in most regions of the USA. Need for added minerals will depend on local conditions (such as I in the goiter belt) and composition of the rest of the ration. For cattle, use of trace mineralized salt is cheap insurance. Concentrations of salt typically fed are 0.25 to 0.5% of the ration DM, although the amount can be decreased to 0.10% without reducing performance. Since salt is largely excreted in the urine, it accumulates in lagoons and will become concentrated as water evaporates. Excess salt also can increase soil salinity and reduce crop yields. Consequently, a lower salt concentration appears desirable in many feedlots.

Reduction of salt in a ration has two limitations. First, urine output is reduced when salt intake is decreased. More concentrated urine has been associated with an increased incidence of urinary calculi (see Ch. 13, Vol. 2). Secondly, when salt is used as the carrier of trace minerals, the amounts of trace minerals provided will decrease as salt concentration is reduced. Consequently, extra trace minerals may need to be added if the amount of trace mineralized salt in a ration is reduced.

When salt is added to the ration at concentrations above 0.10%, there is no nutritional need to provide free choice salt for feedlot cattle. However, free choice salt may speed restoration of intestinal motility following stress.

Calcium

Grains are low in Ca, so supplemental Ca is needed with high grain rations. Where alfalfa products are readily available and economical, this feedstuff adds Ca although its availability has been questioned. With growing cattle, a Ca deficiency causes rickets. But since most feeder cattle have a relatively mature skeleton, bone thinning and fragility rather than rickets would be more common symptoms of a chronic Ca deficiency. Ca deficiency also has been associated with reduced nutrient digestibility, especially for starch. Reduced ruminal digestion of starch, due to decreased buffering or retention time, may be responsible.

Although some tables list requirements for Ca as low as 0.38% of the ration, most feedlot operators use a minimum of 0.45% of the ration and concentrations up to 0.65% are not uncommon with high grain rations.

Supplemental Ca can be provided from limestone (38% Ca), together with Mg in dolomitic limestone, or with P from mono- or dicalcium phosphate, rock phosphate or bone meal (see Appendix Table 2).

A manufacturing by-product, cement kiln dust, being more readily soluble in acid than limestone, is potentially more available to a ruminant than other Ca sources listed. Contaminant metals (lead and cadmium) in kiln dust from some manufacturing plants limit its commercial acceptance. Limestone is typically the least costly source of Ca, but to maintain a desirable Ca to P ratio, use of the other products may be desirable.

Phosphorus

Cereal grains and most high protein feeds, except cottonseed meal and animal proteins, range from 0.15% to 0.65% P. The requirement of feedlot cattle is from 0.22 to 0.43%, so a small amount of supplemental P may be needed. In addition to the P sources which also provide Ca, mono- or di-basic sodium phosphate are often fed.

Calcium-Phosphorus Ratio

The ratio of Ca:P in bone is approximately 2:1. To meet requirements for Ca and P, a ratio of 1.3 to 1 is needed, but a wider ratio of up to 5:1 has proven neither detrimental nor harmful. Excessive Ca may decrease feed intake or add cost to the ration and has been implicated in phosphatidic urinary calculi problems (Ch. 13, Vol. 2).

Potassium

Feedlot rations containing intact protein usually contain >0.5% K. This compares with the requirement of 0.6% to 0.8% for feedlot cattle. For newly received cattle and possibly cold-stressed cattle, weight loss appears to be recovered more rapidly with 1% K in the ration. Although this weight may be largely gut fill or fluid retention, added fill may restore intestinal activity more rapidly and reduce morbidity and mortality of feeder calves. The chloride is the most commonly used supplemental source of K as the caronate is quite costly.

Sulfur

Normally present as a component of amino acids, S may need to be supplemented when NPN replaces supplemental protein. The cost and availability of supplemental S, listed from greatest to least, is amino acid sulfur, sulfite sulfur, sulfates (usually sodium) and flowers of sulfur. The latter two are most commonly used in supplements to maintain a ratio of N:S below 15:1 in cattle rations.

Magnesium

Although occasionally deficient for grazing cattle, Mg levels and availability appear adequate in most feedlot rations. Mg oxide at low levels depresses feed intake markedly. Excess Mg may be involved with an increased incidence of phosphatidic urinary calculi (see Ch. 13, Vol. 2).

Iodine

Deficiencies of I lower metabolic rate and can cause goiter. Requirements of 0.2-0.4 ppm are easily met if iodized or trace mineralized salt are fed as a source of salt.

Selenium

A Se deficiency causes white muscle disease of growing calves. Reduced pancreatic activity also has been reported. Soils and feedstuffs from soils in states bordering the Great Lakes are very low in Se whereas accumulator plants of the Great Plains concentrate Se from soil and can cause toxicity. Federal approval for supplementation of rations with Se to a level of 0.1 ppm has been granted. Supplementation to this level may improve performance of feedlot cattle in deficient regions.

The formula for one trace mineral mix in use in Great Plains feedlots is presented in Table 8-12. This mixture works satisfactorily with corn or milo grain rations.

Needs for Ration Analysis

Sometimes the cost of nutrient analyses far exceeds the cost of supplementing a ration with a nutrient. Cost for analyses of minerals

Table 8-12. Feedlot trace mineral mix.

Ingredient	Percentage[a]
$Fe\ SO_4\ 7H_2O$	30.25
$Ca\ CO_3$	27.85
$ZnSO_4\ H_2O$	26.75
$MnSO_4$	11.80
$Cu\ SO_4\ 5H_2O$	3.00
$Na_2Se\ O_3$	.87
$Co\ SO_4\ H_2O$	.25
$Ca\ (IO_3)_2\ 6H_2O$	.10

[a]When fed at 0.0025 % of the ration, the above mixture adds (in ppm) Zn 24; Fe 15; Mn 8.2; Cu 1.9; Co .2; Se .1; I .06.

Table 8-13. Analytical costs for minerals.

Mineral	Cost of supplement/ton of ration	Cost of analysis
Calcium	$0.15	$ 5.00
Phosphorus	0.20	5.00
Trace minerals	0.10	25.00+

can be compared with the cost of supplemental minerals in a typical feedlot ration to decide how frequently samples should be analyzed. Such a comparison is presented in Table 8-13. The low cost of the mineral supplements and frequent difficulty in interpreting analytical results with respect to availability of minerals present makes frequent analysis impractical. Only when specific problems in mineral nutrition are encountered does ration, blood and/or tissue analysis become practical.

IMPLANTS AND GROWTH STIMULANTS

Sex has a large influence on the feedlot performance and carcass composition of cattle. General differences in value of steers and heifers have long been observed by feeders and packers. Hormones inhibit or stimulate various physiological functions including growth rate and tendency to fatten.

Sex Influence

Bulls, steers and heifers differ in their rate, efficiency and composition of growth. At a comparable physiological age, these three rank in decreasing order for muscle and bone and increasing order for fat. These differences in composition account for some of the differences in rate and efficiency of growth between sexes and with age. Bulls generally gain faster and more efficiently than steers, particularly under full feed conditions, and produce carcasses with a higher percentage of lean meat. Bull carcasses generally have a lower carcass grade than steers due to less marbling.

Taste panel studies have varied in their evaluation of the flavor and tenderness of bull beef compared to steer beef. Beef from bulls slaughtered at less than 15 mo. of age has proven acceptable in flavor and tenderness. Bulls should be managed in such a manner that they reach market weight at as young an age as possible. Early castrated calves (1-3 mo.) have been compared with late castrated calves (7-8 mo.) in several trials. Generally, the late castrates show a slight advantage in preweaning gain but no advantage in postweaning feedlot gain. Performance of large groups of bulls assembled from various sources has been disappointing. This may be due to fighting and social problems in commercial feedlots. Although comparatively few bulls have been fed in the feedlot in the past, bull feeding may increase in the future due to more rapid growth rate and lean meat production. A portion of the difference between the performance of bulls, steers and heifers is attributed to hormonal differences.

Research workers have tested many naturally occurring hormones and other chemicals with hormone-like activity in an attempt to improve growth rate, feed efficiency and carcass traits of cattle. Various types of hormonal preparations used with feedlot cattle are discussed below by hormone class. Those approved for use are listed in Ch. 4 along with normal feedlot responses and withdrawal times.

Estrogens

Diethylstilbestrol (DES) had extremely wide use and consistently improved growth rate. Present FDA regulations forbid the use of DES in beef cattle in the USA. Estrogenic compounds can be fed or implanted at the base of the ear and have physiological effects at very low concentrations. Estrogens generally improve rate of gain by steers by 10-18% and feed efficiency by 6-9%. Compounds such as resorcinol lactone (Ralgro) and Synovex S are implantable substitutes for DES with similar effects on performance but at a higher drug cost. If fresh implants are not maintained, performance may regress to a point below that of animals never implanted.

Certain side reactions to estrogen treatment can occur. These include depressed loin, elevated tail head, enlarged mammary glands, riding and, in some cases, vaginal prolapse. Buller steers present management problems in feed yards. The severity of these side effects is related to rate of estrogen release and is minimal at the recommended levels. Damage or crushing of implant pellets is often believed to be involved with these reactions since broken pellets release the hormones more rapidly.

Androgens

Testosterone is presumably responsible for the rapid growth of bulls. This hormone has been tested as a growth stimulant. Results with testosterone have not been as favorable as estrogens for steers.

Hormone Combinations

A combination of progesterone and estradiol has been tested. Effects on growth rate and feed efficiency of steers compared favorably with DES. Estradiol and testosterone, combined in implants for heifers, have shown gain and efficiency increases equal to or exceeding that for DES.

Progestogens

Several synthetic progestogens suppress ovulation and estrus when fed to beef females. Melegestrol acetate (MGA) suppresses estrus and ovulation in feedlot heifers. A summary of trials in which MGA was fed to heifers during the feedlot period indicates an increase in gain of up to 11% and an improvement in feed conversion of about 5% over untreated heifers. Carcass grade is influenced very little by MGA treatment. Whether MGA exerts additional hormonal activity to increase growth rate of heifers is not known.

Effects of Implants on Feed Intake and Energetic Efficiency

Implanted animals show increases in rate and efficiency of gain which reflect an increase in feed intake of about 6%. A limited number of trials suggest that rate of gain response to implants is reduced when protein is deficient. This might be expected from the increased rate of protein deposition with estrogenic implants. Increased need for other nutrients due to estrogens has not been documented.

Feed Additives

A number of feed additives used in feedlot beef production have been discussed in Ch. 4. Costs and estimated incidence of feedlot use are shown in Table 8-14. Other additives with special uses include anthelmentics, coccidiostats, agents for control of foot rot, and specialty antibiotics, sulfas, electrolytes and buffers.

Use of most feed additives is closely controlled by the Food and Drug Administration. Uses and claims are under constant review. Thus, users need to keep up to date on current recommendations and limitations and follow label directions which specify level of use and withdrawal time intervals.

RATION HANDLING

Quality control (ration preparation, mixing, etc.) and feedbunk management (quantity control) are directly involved with obtaining economical performance from cattle. Feedbunk management represents the combination of complex factors involved in maximizing performance, minimizing digestive disorders, and keeping cattle on feed. Since both rate and efficiency of gain are directly related to nutrient intake, cattle feeders must strive for maximum feed intake of a consistently high quality ration.

Ration Management Factors

Fresh Feed. The cattle feeder should strive to have fresh and palatable feed in front of his

Table 8-14. Costs of using additives in feedlot rations.

Feed additive	Rate of gain, %	Efficiency of feed use, %	Approxi. cost/head day	1980 Estimated use, % of feedlot cattle in USA
Monensin (Rumensin)	0	4.5-11	1.5¢ at 30 g/ton	85-90
Antibiotics				
Tylan	3-5	3-5	0.85¢ for 70 mg	70-80
Oxytetracycline	3-5	3-5	0.25¢ for 75 mg	
Chlortetracycline	3-5	3-5	0.25¢ for 70 mg	
Bacitracin	2-4	2-4	0.25¢ for 50 mg	
Melengesterol acetate	10-12	5	1.2¢ for 0.4 mg	High percentage of heifers

cattle at all times. The feeder should avoid the use of substandard quality or spoiled feeds, as either will reduce feed intake. Mold growth, spoilage and mustiness can severely reduce feed intake. It is better to discard spoiled feed rather than risk reduced feed intake. As a rule, when poor quality feed reduces feed intake by 5%, gains will be reduced by 10%.

The length of time that feed may be left in the bunk before it becomes stale depends on moisture content, frequency of feeding, humidity and weather conditions and the specific combination of feeds in the ration. Eventually, all feed not consumed becomes stale and must be discarded. Feedbunks should be cleaned periodically, at least once each week.

It is not poor management to have the cattle clean up the bunks once a day, providing that cattle are not out of feed so long that they become restless or overeat when fed again. Clean water is equally essential as fresh feed. Many good cattle feeders clean waterers daily.

Control Variation in Nutrients and Quality. Moisture content of the feeds in a ration is the largest single source of variability when silage, haylage, green chop or high-moisture grains are fed. The only successful manner to formulate rations based on feeds which vary in moisture content is to formulate on a standard DM basis (such as 90% of 100%). When formulas are on a standard DM basis, switching from moist to dry feeds or vice versa is simple. Feed DM intake will remain constant and formula adjustments are easily made.

Figure 8-4. Two examples of feed delivery systems sometimes used by small feeders. Courtesy of Feedlot Magazine.

Select Only High Quality Silages and Roughages. Silage quality is variable, even when expressed on a DM basis. Factors which influence nutrient content include grain content, growing conditions, stage of maturity at harvest and variety. The grain content of corn silage is one of the most important factors affecting quality. Grain generally comprises 35-50% of corn silage DM, although the range is from zero to >50%. Varieties which produce high grain yields are usually preferable for silage. If silage is to be purchased, it is advisable to recommend specific varieties to potential suppliers in order to improve uniformity in the quality and grain content of silage. Another important factor is the stage of maturity at harvest. Corn harvested near the hard dent stage (about 65% moisture) produces the maximum digestible DM yield and such silage will have the minimum storage loss.

Though generally a reliable predictor of quality, grain content may be low for silage from drought-stricken corn. Silage from drought-stricken corn needs to be supplemented with more grain for cattle to reach equal energy intakes. Grain percentage can be estimated by separating and weighing grain

Figure 8-5. A facility designed for automated feeding of rations with high levels of silage and/or high-moisture grains. Courtesy of Feedlot Management Magazine.

from a dried sample. Starch or fiber content also are useful indicators of energy content of forage. When energy content is in doubt, it is desirable to conduct such analyses.

Hay crops harvested as hay silage also are variable in both protein and energy content and availability. The best indicator of hay digestibility is acid detergent fiber content. When hay is purchased, it is good to know when it was harvested and how much it was weather damaged. Protein and/or fiber analysis on forages aids in quality control and in determining supplement needs. The relative value of roughages must consider the reason for roughage addition to a ration. With some rations, a less digestible, more fibrous roughage is more valuable than one of high digestibility.

Questionable Feedstuffs as Only a Portion of the Ration. Damaged, aged or spoiled feeds can at times be economical ingredients with proper processing and management. Such feeds should be limited to only a portion of the ration in order to minimize the risk of reducing performance due to variability of such feeds. When fed as only a small portion of the ration, such feeds have less impact on ration variability and cattle performance. Care should be taken to supplement such feeds properly. Higher than normal levels of supplemental minerals, vitamins and protein will provide a wider margin of safety when such feedstuffs are fed.

Addition of inert or distasteful feeds to a ration at levels under 3% usually will not reduce cattle performance, but the true economic value of added feeds will be hidden by variability in animal performance. If such feeds are cheap, the total cost of the ration will be reduced. The digestibility and protein or energy value of such ingredients therefore must be evaluated carefully prior to use. Adding feeds or filler of questionable value merely to reduce ration cost is unethical. An ethical nutritionist can immediately justify use of each ingredient in a ration. Ration dilution with filler can be avoided if rations are formulated on a cost per unit of energy basis. Although some custom feeders calculate ration costs on a wet basis and dilute their ration with questionable feeds to compete for customers, the total cost of production, not simply cost per ton of feed, determines the profit or loss on a pen of cattle and long-term customer satisfaction.

Physical Condition of the Ration

Mixing. Several factors affect the physical condition, uniformity and palatability of the ration in the feedbunk. First, proper mixing will help improve palatability and avoid sorting. Mixing will often mask the flavor of less palatable ingredients such as high urea supplements, feeds containing weed seeds or mold and certain drugs. Care in formulating the ration and adjusting for moisture variation is useless if the ration is not uniformly mixed. Protein content, for example, can vary from 2-3% within a load of feed as a result of inadequate mixing. Adequate mixing is necessary so that every animal receives its intended daily level of nutrients and feed additives. The following guidelines are helpful to avoiding inadequage mixing:

1. Dilute additives and urea in a supplement to a minimum of 1 lb/head daily. If high protein supplements are used and/or only small amounts of supplemental protein are needed, they can be diluted and premixed with some ground grain before addition to the major portion of the ration. Proper mixing is more critical where high urea supplements, growth stimulants and other additives are being added with the protein supplement.

2. The physical form of the ration can be altered to reduce separation of ingredients. When using whole shelled corn, for example,

it is desirable to pellet the protein supplement. However, with finely ground grain, it may be desirable to have the supplement in a meal form, to add some molasses, or to use a liquid supplement to prevent separation. Large pellets in ground feeds may be sorted and rejected by cattle. In one case, switching from 3/8 inch to 1/4 inch supplement pellets prevented selective rejection of a monensin supplement by steers. In some rations the forage may need to be finely chopped to avoid separation. Mixing high moisture forage with dry grain or mixing a dry forage with high moisture grain helps prevent separation and improve uniformity of mixed rations. Where wind is a problem, high moisture feeds or small amounts of fat or molasses in the ration help reduce losses.

3. A higher roughage level compensates for inadequate mixing. When the ration is well mixed, 5-10% roughage (DM basis) may be enough to maximize NE intake and prevent digestive disorders. When the ration is not adequately mixed, it may be necessary to include 15-20% roughage to minimize founder and digestive upsets.

4. Obtain a chemical analysis on feed samples taken from the bunk periodically. These can be used to check the accuracy of formulation and mixing. Salt concentration is a simple check for mixing precision.

5. If problems with distribution of minor ingredients are encountered, try changing the order in which the ingredients are added to the feed wagon or truck. Distribution of the supplement is usually more uniform if the grain is added first and supplement second. These can be mixed for a short time before the roughage is added. In effect, this procedure is premixing grain and supplement before adding roughage.

Density. Cattle do not like a ration which cakes to the bottom of the bunk and they seem to prefer ingredients which have larger particle size. Rations of high moisture grain can form a hard, dense mass in the feedbunk after several hours or exposure to the weather and cattle saliva. A higher roughage level and high moisture grain or flaked grain helps avoid this problem. Likewise, certain types and forms of roughage are more suitable with finely ground grain. Feed in the bunk should have a loose texture. This can be checked by passing one's fingers through the feed in the bunk.

Moisture Content of the Ration. Each ration has an optimum moisture content. Cattle do not like dry, dusty rations. Likewise, when rations are too high in moisture, total DM intake may be reduced. Many studies have compared high moisture with dry corn in high concentrate rations. Intake is considerably reduced at corn grain moisture contents >30% compared with about a 5% reduction at moisture contents of 23-30%. This intake reduction probably is due to fermentation end-products since water addition at similar levels has usually not reduced feed intake. The optimum moisture level is affected by not only animal acceptability but also the ease of mixing and segregation of ingredients, wind, bunk life and the specific combination of feedstuffs fed.

Certain supplements for feedlot steers may be handled in a liquid form. A "Micromix" device suspends drugs or additives in water and sprays them onto mixing feed. Such a system aids in quality control and drug distribution and simplifies inventory control of micro-ingredients.

Complete protein-mineral supplements also can be provided in a liquid suspension to spray on feed or to feed in lick-tanks. Molasses or similar viscous liquids are used as suspending agents. Phosphoric acid often is included as a P source and an intake limiting compound. Instability of drugs, disuniformity of intake from day to day and animal to animal, settling of certain ingredients and high cost of transport of liquids are disadvantages. Low labor cost and simplicity of feeding are the major advantages.

Feed Combinations. Certain combinations of feedstuffs such as barley, wheat, oats and/or alfalfa hay contribute to digestive disorders. The following thumb rules can help avoid such troublesome combinations:

1. Include no more than 40-50% of the grain as wheat in a high concentrate ration. Limit rye to 20% of the ration.

2. Where barley and wheat are the major feeds available in addition to corn, proportions of 1/3 barley, 1/3 corn and 1/3 wheat work reasonably well.

3. If the ration is primarily barley or oats and alfalfa and problems with bloat are

encountered, replace part of the barley or oats with corn, replace part of the alfalfa with grass hay, and/or replace part of the alfalfa and barley or oats with corn silage.

4. If digestive disorders are encountered, make some changes in the combinations of feeds being used. For example, changing the proportion of grain and alfalfa may reduce the incidence of bloat.

Effect of Weather. Extreme, rapid changes in weather affect feed intake. During periods immediately preceding changes in weather, cattle abruptly increase their feed consumption and may develop digestive disorders. This will be followed by a period of reduced feed intake and possibly sickness. Feed intake also may be reduced during extended periods of bad weather. It seems advisable to add 10% extra roughage or limestone to finishing rations during periods of time when rapid and abrupt weather changes occur.

Effect of Lot Condition. Cattle should have a dry place to lie down close to feed and water. Cattle hesitate to wade through mud to the feedbunk. Studies from the University of California show that mud can reduce cattle gains over 20%. Cattle stand in a shed or behind the windbreak when winds are high rather than walk across the feedlot to the feedbunk. Earth mounds placed as a right angle to the bunk apron help cut crosswinds. Concrete aprons extending from the bunk some 5 feet into the pen also help eliminate mud between the resting area and the feed. When cattle are first placed on feed, space for all cattle to eat simultaneously should be provided to enhance observation of sick cattle and to allow easy access to feed for timid animals. Once cattle are on feed, bunk space per head can be reduced greatly. A minimum of 6 to 9 inches of bunk space for each animal is usually adequate.

Feeding Technique

Feeding System Needed. Fresh and platatable feed should be available for cattle at least 23 h of each day from the time cattle arrive in the feedlot until they are removed for slaughter. Each cattle feeder needs to develop a feeding pattern that will allow the bunks to be nearly cleaned up each day to avoid stale feed while at the same time preventing cattle from going without feed for more than 30-60 minutes. When refed after a longer period without feed, two things can happen. First, cattle may consume excess feed, go off feed and develop digestive disorders. Feeding cattle on a regular schedule helps the cattle feeder anticipate when to check bunks for higher than expected feed intake. A set schedule also allows the cattle to anticipate when fresh feed will be placed in the bunk and helps avoid restless behavior. Fresh feed appears to be a psychological factor which stimulates high feed intake.

Number of Times to Feed Each Day. Many cattle feeders believe that the more often fresh feed is fed, the more often cattle will come to the feedbunk to eat. Frequent feeding reduces the likelihood of both stale feed and extended periods without feed. But labor and equipment costs limit how many feedings can be justified. The major factors to consider are type of feed and weather. When high moisture feeds are used, cattle need to be fed more often than when dry feeds are used, particularly to reduce spoilage during hot weather. The feeding pattern may need to be changed with season of the year. In the winter, cattle eat more feed during the day than at night, so more feed should be placed in the bunk at the morning feeding. Conversely, in the summer, cattle eat more feed during the night or cooler hours of the day, so more feed should be placed in the bunk in the evening.

The most important man in the cattle feeding operation is the man who feeds the cattle. Clean bunks and lack of digestive problems are a sign of a careful feeder; spoiled feed caked over the bottom of the bunk indicates a careless feeder. Being able to anticipate the amount of feed needed each feeding is important in avoiding stale feed or having cattle out of feed for extended periods. Knowing how to get cattle on full feed and anticipate increases or decreases in intake to keep cattle on full feed requires skill and judgement; this is best developed by careful observation of cattle, close attention to bunk management and consistent quality control of the ration. Frequent ration adjustments to use least cost ingredients may prove counterproductive if such adjustments cause problems in bunk management.

Management of Self-Feeders. In many cases it is impossible to obtain equal performance from cattle fed with self-feeders to that attained from cattle fed in a bunk line by a skillful feeder. However, feeders who lack the skill or aptitude for bunk line feeding can do well using self-feeders. If the ration's physical form permits use in self feeders, reduced labor costs may make self-feeding practical.

Self-feeders must be inspected daily for bridging of feed as well as wasted and stale feed. Mixtures of urea, molasses, and dicalcium phosphates absorb water and at certain moistures and temperatures will bridge in self-feeders and may form blocks in mixers and storage tanks. Physical separation of the ration also limits usefulness of self-feeders. Accumulated fines in the bottom of the self-feeders will be rejected until cattle become quite hungry, often resulting in founder and bloat. Physical characteristics of the ration are more important when self-feeding than when bunk line feeding.

LEAST COST RATIONS

The following description applies to the output format generated by the Oklahoma State University Feedmix Program. This program or similar ones are available at most land grant universities. While the specific format and output of linear programs may differ in appearance, all show essentially the same information. The ration in the example below is for illustration purposes only.

The first page of output (not shown) gives the nutritional basis on which the formulation was computed. Any coefficient, price ($/100 lb) or restriction may be changed at the user's option. Fine tuning of ingredient composition based on local analysis of ingredients proves helpful.

The second page of output (Table 8-15) presents the least cost ration plus some additional information. Maximum and minimum constraints placed on ingredients are shown. When a maximum or minimum increases ration cost, the dollar cost per unit (1%) of the restriction is shown (restriction cost/unit). As shown in Table 8-15, 5% dehy alfalfa was forced into the solution. The minus sign before the .0051 signifies that decreasing this restriction would reduce cost. Each 1% dehy caused the price of this ration to increase by 0.0051/100 lb. In the case of barley, there is no minus sign on the restriction. This means that releasing the maximum limit on this ingredient would lower the cost of the ration.

The dollar cost of each restriction is provided so that a skilled nutritionist can re-evaluate his restrictions to determine if they are worth the cost they add to the ration. If the value of a feed decreases as it makes up a higher percentage of the ration due to palatability or handling and management problems which can be compensated by placing a higher price for the feedstuff, the same feed can be entered as a second feedstuff with the same (or adjusted) nutrient content, a higher maximum limit and a higher price.

Assigned cost/cwt is the cost of the commodity entered into the program. All prices and nutritional constraints must be on an equal DM basis. All data in these illustrations are on the basis of 90% DM. The cost of a commodity must include all costs for delivery of the feedstuff to the bunk. Special costs for delivery, storage, processing and mixing must be considered.

Incoming price is the price below which a commodity not included in the formula would enter the formula. For example, if corn at 90% DM were available at a price below $1.92/100 lb, it would displace some other feedstuff currently in this formula. This listing assigns values to feedstuffs that the feeder could pay relative to the feeds he is currently using. A check of the grain and feed markets can determine if a cattle feeder can buy a new feed for less than this incoming price, and judge whether and when a commodity has a place in his ration.

The lower prices listed for some included commodities are shown under the same column. If this commodity drops below the price specified, reformulation is useful since a new least cost formula will use a greater quantity of that item. Outgoing prices are costs above which an included commodity will be dropped from the ration. So long as

Table 8-15. Computer output.

Ingredient	Percent in ration	Max.	Min.	Restriction cost/unit	Assigned cost/cwt.	Incoming price	Outgoing price
Alf hay 24% fiber					100.00	1.75	
Alf hay 28% fiber					100.00	1.48	
Alf hay 34% fiber					100.00	1.13	
Alf dehy 17% pro.	5.00	10.00	5.00	-0.0051	2.30	1.79	***
Barley straw					100.00	0.59	
Cottonseed hulls	2.76				0.60	-53.72	0.86
Milo stover					100.00	0.73	
Oat hay					100.00	1.06	
Prairie hay					100.00	0.86	
Wheat straw		10.00			100.00	0.56	
Barley 46-48 lbs	40.00	40.00		0.0019	1.85	***	2.04
Barley light					100.00	1.93	
Corn dent No. 2					100.00	2.04	
Cotnsd meal O P	2.25				3.80	1.97	4.94
Fat-beef tallow		5.00			100.00	2.80	
Milo 9% Pro (C)					100.00	1.94	
Molasses, cane 90		10.00			100.00	1.33	
Oats	48.98		40.00		2.00	1.85	2.12
Soybean meal-sol					100.00	4.17	
Wheat-10% prot.		60.00			100.00	2.06	
Wheat mill run		20.00			100.00	2.21	
Dicalcium phosph.					4.50	0.66	
Ground limestone	0.51	1.00			0.75	0.43	6.17
Salt	0.50	0.50	0.50	-0.0057	1.00	***	***
Urea-45% N		0.60			100.00	13.41	

Total cost per cwt . . . 1.95

Note all weights and percentages are on an air dry basis.

Table 8-16. Nutritional specifications of this ration.

Ingredient	Amount in ration	Forced amount	Cost of one unit change	Valid range	
				Low	High
Est. net energy, megcal	65.00				
Energy, maint., megcal	77.03				
Energy, prod., megcal	51.00	51.00	0.0186	48.17	52.17
Crude protein, %	12.08				
Digestible protein, %	9.25	9.25	0.0659	8.23	11.81
Ether extract (fat), %	3.29				
Crude fiber, %	10.26				
Roughage, %	7.76				
Calcium, %	0.35	0.35	0.0084	0.16	0.53
Phosphorus, %	0.36	0.30	0.30		

commodity prices remain in the range between these incoming and outgoing limits, a rerun on the computer would give an identical least cost ration. Consequently, reformulation frequency should be dependent upon commodity price changes and incoming and outgoing limits, not calendar date.

Table 8-16 shows the nutritional specifications of the ration. Complete nutritional specifications of the mix are given in the column "amount in rations". Forced amounts are from the minimum or maximum levels constraints applied.

Cost per unit of change provides information on the cost to change nutritional constraints within the valid range. For example, the ration in Table 8-16 had a minimum digestible protein of 9.25%. If for some reason the nutritionist wanted to drop this level to 8.25%, the new ration would cost $0.0659 less/100 lb. Or, if he wanted to increase digestible protein from 9.25% to 10.25%, it would cost $0.0659 more. The range over which these statements are valid is also specified, in this case from 8.23 to 11.81% digestible protein. Using this information it is possible to re-evaluate the economic impact of a change in nutrient specification without rerunning the ration. Each nutrient restriction may be evaluated in terms of its dollar cost. For example, if the feeder felt that for performance, health or safety reasons, raising digestible protein 1% was worth 6.6 ¢/100 lb of feed, he can reformulate at a higher protein level.

Conventional linear programming is a very useful economic tool for the cattle feeder. Few feeders can affort to feed a ration which costs more than it must. Linear programming will pinpoint more economical feed ingredients and mixes. But least cost rations do not necessarily mean least cost gains for feedlot cattle since rate and efficiency of gain and non-feed costs are not all being considered together. Nevertheless, least cost formulation is a helpful tool to maximize profitability of cattle feeding programs.

FORMULAS FOR PREDICTING FEED INTAKE, PERFORMANCE AND PROTEIN REQUIREMENTS

All are set for weight and feed in kilograms

Energetic Relationships

A. Metabolizable Energy
   1. Digestible Energy (Mcal/kg) x 0.82 = Mcal/kg
   2. TDN (%) x 0.03616 = Mcal/kg

B. Net Energy
   1. Maintenance Requirement
      a. Calves $0.077 \times (W)^{.75} = $ Mcal/day
      b. Yearlings $0.0705 \times (W)^{.75} = $ Mcal/day

   2. Gain Requirement
      a. Steer Calves $(0.05272 \times \text{Gain} + 0.00684 \times \text{Gain} \times \text{Gain}) \times (W)^{.75} = $ Mcal/day
      b. Heifer Calves $(0.05603 \times \text{Gain} + 0.01265 \times \text{Gain} \times \text{Gain}) \times (W)^{.75} = $ Mcal/day
      c. NEg (Yearling Steer) $= (0.04871\,G + 0.00629\,G^2)\,W_{kg}^{0.75}$
      d. NEg (Yearling Heifer) $= (0.05161\,G + 0.01168\,G^2)\,W_{kg}^{0.75}$

   3. Gain Prediction (after subtraction of feed for maintenance)
      a. Steers $12.091 \times NEg/(W)^{.75} + .10157 - 3.854 = $ kg gain/day
      b. Heifers $8.891 \times NEg/(W)^{.75} + .06204 - 2.215 = $ kg gain/day

   4. NEm 77 / (antilog) (2.2577 – .2213 ME) = Mcal/kg
      NEg 2.54 – 0.0314 (antilog) (2.2588 – .2213 ME) = Mcal/kg

5. NE and ME Prediction From Gain

a. $NEm^1 = \frac{-(2.54\ NEm + 2.4178X) - (2.54\ NEm + 2.4178X)^2 + 4(NEg - 2.54\ X)(2.4178\ NEm)}{2\ (NEg - 2.54\ X)}$

b. $NEg^1 = 2.54 - \frac{2.4178}{NEm}$

c. $ME\ (Mcal/kg) = \frac{2.2577 - \log F}{0.2213}$

Feed Intake Prediction

$$K = .0954 \times \frac{NWT - 136}{4545} + \frac{FG}{220}$$

A. Daily Feed (kg) = $[[K \times W^{.75} - (\frac{W \times 2.2}{220}) - 500) \times .454)] \times .9] \times DM$

Feeder Grade

Calves	Small Frame	= 1
	Large Frame	= 5
Yearlings	280 kg.	= 4.2
	300 kg.	= 4.8
	320 kg.	= 5.5
	340 kg.	= 6.0

B. Daily Feed = KM $Wt^{.75}$ = kg dry matter/animal daily

K = 0.1 for cattle 360 kg
= 0.095 for cattle 360-480 kg
= 0.090 for cattle 480 kg

M = 1.0 for calves
= 1.1 for yearlings
= 1.17 for Holstein
= 1.00 for Helstein cross
= 1.00 for other breeds
= 0.91 for monensin
= 1.00 without monensin

Wt = Equivalent Weight

Nitrogen Relationships

1. Crude Protein (CP) Nitrogen (%) x 6.25 = % protein

2. Urea Fermentation Potential

a. .371 x TDN – .0357 x CP x DEG = g urea/kg feed

b. $31.64 - 3.558\ CP + \sqrt{237.8\ CP^2 - 237.8\ CP - .1778\ TDN^2 + 36.39\ TDN - 149.72}$

Where DEG = ruminal protein degradability as a decimal

3. Crude Protein Requirement

$F + U + S + G + M \; D \times BV$ = g protein/animal daily

$F = .03 \times$ dry matter intake
$U = 2.75W^{.5}$
$S = 0.2W$

G for cattle 250 kg = $192.3 \times ADG \times 0.8173$;
G for cattle = $710.8 - 103.8 \ln EBW + 39.06 \ln (EBG \times \text{mature } EBW)$

where EBW = .86 x live weight
EBG = .89 x daily live weight gain
mature EBW = .86 x mature live weight

or G for heifers = $ADG \times (242.2 - (1.322 + .2985\, ADG)W^{.75})$;
G for steers = $ADG \times (242.2 - (1.244 + 1.514\, ADG)W^{.75})$;

M = milk yield in kg x .0335
D = .66
BV = .73 for growing cattle
BV = .85 for mature cattle

Literature Cited

General Reference: Great Plains Fact Sheets, Cooperative Extension Service, Regional Extension Project GPE-7.

Special References:

ARC. 1965. The Nutrient Requirements of Farm Livestock. No. 2. Ruminants. Agricultural Research Council, Her Majesty's Stationery Office, London.
Byers, F.M. and R.E. Rompala. 1979. Ohio Beef Cattle Res. Prog. Rpt., p. 48.
Burroughs, W., D.K. Nelson and D.R. Mertens. 1975. J. Animal Sci. 41:933.
Dyer, I.A., R.R. Roa, J.D. Clark and J. Templeton. 1962. J. Animal Sci. 21:668.
Fox, D.G. and J.R. Black. 1977. A system for predicting performance of growing and finishing cattle. Mich. State University Res. Rpt. 328, p. 141.
Gill, D.R. 1979. Fine tuning management with computer-assisted decisions. Okla. Cattle Feeders' Seminar, p. C-1.
Givens, R.L. 1969. California Cattle Feeders Day, p. 58.
Johnson, D.E. and J.C. Crownover. 1975. Colorado State University Research Highlights.
Kelly, C.F., T.E. Bond, W.N. Garrett. 1960. California Agr. 14(9):11.
Lofgreen, G.L. 1973. California Feeders Day, p. 81.
Lofgreen, G.L. 1977. Arkansas Agr. Expt. Sta. Sp. Rpt. 50:15.
NRC. 1976. Nutrient requirements of beef cattle. Nat. Acad. Sci., Washington, D.C.
NRC. 1978. Nutrient requirements of dairy cattle. Nat. Acad. Sci., Washington, D.C.
Owens, F.N. and D.R. Gill. 1979. Amer. Soc. of Animal Sci. Abstr. (So. section), p. 16.
Reid, J.T. and J. Roff. 1971. J. Dairy Sci. 54:553.
Rumsey, T.S. and R.R. Oltjen. 1975. J. Animal Sci. 41:416 (abstr.).
Satter, L.D. and R.E. Roffler. 1975. J. Dairy Sci. 58:1219.

Chapter 9—Feeding Dairy Cows

by J.T. Huber

INTRODUCTION

As with many other phases of present technology, milk production potential of the modern dairy cow has surpassed the wildest dreams of cattlemen just a century ago. Table 9-1 illustrates the progress in average milk production in the USA from 1955 to 1978. Average milk yields have almost doubled from about 5,800 to 11,200 lb/cow during this period. More importantly, many well-managed dairy herds currently achieve annual yields of over 20,000 lb/cow, suggesting that some potential for increasing milk production still remains in most dairy herds. In the state of Michigan, in 1979, mean milk yields of cows on DHIA test were 15,379/cow and 30 herds achieved an average of >20,000 lb. Widespread genetic improvement of cattle made possible through artificial insemination is perhaps the most important factor responsible for the increased milk yields, but cows with more potential could not produce up to this level without improved feeding and management. Figure 9-1 shows a group of dairy cows from a high producing herd.

The dairy cow is one of the most efficient farm animals in converting feed energy and protein to human food. An average of 60-65% of her feed is derived from forages and by-product feeds which are poorly utilized by humans (see Ch. 1). In the case of ruminants it must be remembered that a relatively small percentage of the total feed they consume is edible by humans. In considering the proportion of the corn crop for which animals compete directly with humans, Bath (1977) stated that 50% of the dry matter (DM) is forage (leaves, stalks, cobs) utilizable by ruminants but not by humans. Of the grain, about 50% is inferior kernels and mill feed (not edible for humans), leaving only about 25% of the total energy in the crop suitable for human consumption. If this crop is fed to dairy cows, the result is a large positive balance in energy and protein edible to humans. This is particularly true if the N needs of the cow are partially met with nonprotein N (NPN).

Table 9-1. Milk production, production per cow and number of milk cows, 1955-78.[a]

Year	Milk production, 10^6	Milk per cow, lb	Number of milk cows, 10^3
1955	122,945	5,842	21,044
1960	123,109	7,029	17,515
1965	124,180	8,305	14,953
1970	117,007	9,751	12,000
1975	115,334	10,350	11,143
1976	120,269	10,879	11,055
1977	122,698	11,181	10,974
1978[a]	121,928	11,240	10,848

[a]Preliminary. From 1979 "Milk Facts," Milk Industry Foundation, Washington, D.C.

Figure 9-1. Group of high producing cows in a dairy herd in California fed a ration for high milk production.

Dairy cows in different countries vary greatly in the amount of milk they produce (Table 9-2). Many reasons exist for the variation in average yield/cow. Important factors are types and genetic potential of cattle, systems of feeding, care and management, health measures and disease prevention, milking systems, and climatic stresses.

Intensive dairying, as practiced in North America has resulted in the highest milk yields ever observed. Many positive factors are nearly optimized to achieve such high yields. Many other countries have great potential for increasing the productivity per

Table 9-2. World statistics of cattle milk production for 1976.[a]

	Milk animals, million	Milk production			Humans per milk cow
		metric ton, million	kg/animal	kg/capita	
World today	205	394	1922	97	20
Developed market economies	58	209	3603	274	13
North America	13	62	4769	261	18
West Europe	38	125	3289	342	10
Oceania	4	13	3250	774	4
Other[b]	3	9	3000	64	47
Developing market economics	83	54	650	27	24
Africa	16	5	313	15	21
Latin America	31	31	1000	93	11
Near East	11	7	636	35	18
Far East	25	11	456	10	45
Centrally planned economies	63	131	2079	102	20
Asia	7	4	571	4	132
East Europe and USSR	56	127	2268	347	7

[a]FAO Production Yearbook (1976)
[b]Israel, Japan and South Africa

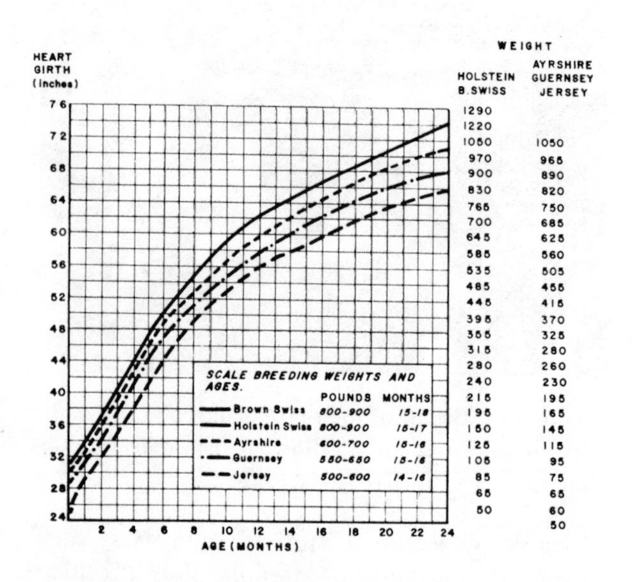

Figure 9-2. Recommended growth pattern of dairy heifers from birth to 24 months. From Hillman et al (1974).

cow without a substantial growth in cattle numbers.

PREPARATION OF HEIFERS FOR LACTATION

Minimizing calf mortality and raising strong, healthy heifers large enough to breed when they are 13-14 mo. of age are sound objectives for herd replacement programs. Raising of calves is discussed in Ch. 10, so this chapter will deal only with animals of 6 mo. of age and older. By 6 mo. the heifer is a functional ruminant so her nutritional needs can come mostly from forages. However, caution should be taken to insure that rations have adequate amounts of proteins, minerals and vitamins.

Feeding Levels

A desirable growth rate for heifers of large breeds from birth to 14 mo. is 750 g/day. At 14 mo. they should weigh 350 kg and be large enough to breed and calve without difficulties at termination of a normal gestation in which they would gain about 150 kg. Figure 9-2 illustrates recommended growth for dairy heifers from birth to 24 mo. of age. Because calves seldom average over 500 g/day gain to 4 mo. of age, a daily gain of 800 g/day is necessary from 4 to 14 mo. if the desired size is achieved by 14 mo.

Rations to support 800 g/day gain in dairy heifers need to be relatively high in energy. High quality forages (based primarily on grain silages with some hay) are satisfactory, but if energy value of forage is low, supplemental concentrate is recommended. Some liberties in feeding heifers can be exercised to use up low quality forages, but the practice should not be prolonged over a long period. Compensatory gain of heifers following periods of low energy nutrition tend to minimize the overall delay of growth if low energy rations must be used.

Retardation of growth under recommended levels (i.e. <0.7 kg/day) is unprofitable because it shortens the productive proportion of the heifer's life. On the other hand, acceleration of growth which fattens heifers (when growth rates exceed 1 kg/day) is undesirable because lifetime milk production and longevity decrease.

Cornell workers demonstrated that heifers raised from birth to first calving on 65 and 100% of recommended energy requirements produced about 8% more milk during 6 lactations than those fed 140% of normal (Reid et al, 1957). It was planned to breed all groups at 18 mo. of age, but many heifers in the low group had not started cycling. An important finding was that first estrus was determined more by the body size than the age of animal. Mean ages at first heat for the respective treatments were 20.5, 11.2 and 9.3 mo. Table 9-3 summarizes the performance of heifers on the 3 regimes. Note the large difference in body weights at first calving had essentially disappeared by second calving. Low groups consumed more feed during the first lactation to provide for greater weight gains after calving.

In an experiment with identical twins, Swanson (1960) compared rapidly grown heifers (which became fat) with those fed a normal energy level. At 2 years of age the fat twins were 32% heavier but produced only 85% as much milk as normals in first lactattion. Tissue slices of udders of fat twins showed less development of alveolar cells. The authors postulated that this was due to infiltration of fat into secretory tissue space. Fattened heifers have less udder development at first estrus and at breeding and take longer to conceive than those fed diets with normal energy (Pritchard et al, 1972; Larsen et al, 1974). Experiments in England (Little and Kay, 1979) demonstrated that accelerated growth and fattening during rearing rather than early breeding caused lower production. Fertility was not affected by treatment, but dystocia was higher in early-bred heifers, and assistance at first calving was necessary after breeding early.

Heifers bred to calve at 19 mo. exhibited a marked increase in calving difficulties and greater breeding problems (Wickersham and Schultz, 1963). First lactation milk yields of groups bred at 10, 14 and 18 mo. were 17% higher for the oldest than the two youngest groups of heifers, but differences were not significant. Data from the Wisconsin study are summarized in Table 9-4.

In conclusion, heifers should be fed to breed at 13-14 mo. so they will calve by 22-24 mo. For large breeds, gains from 3 to 14 mo. should be 800 g/day. Good quality forage balanced for needed minerals and vitamins is the desired feeding system, but grain should be supplemented if forage energy drops. Protein content of the total ration should be at least 12% (DM basis). Growing too fast and/or breeding too young results in less total milk produced, dystocia at calving, more breeding difficulties and decreased longevity.

Table 9-3. Influence of energy level from birth to calving on performance of dairy heifers.[a]

	Percent of standard		
	65 (low)	100 (medium)	140 (high)
Kg concentrate fed (to first calving)	294	412	1,091
Age at first heat, mo.	20.5	11.2	9.3
Body weights: (kg) first calving	384	482	548
second calving	661	584	630
third calving	620	626	672
Milk yields (6 lactations) (kg)	31,166	30,387	28,455

[a]From Reid et al (1957)

Table 9-4. Influence of breeding age on calving performance of heifers fed at normal energy levels.[a]

	Summary of performance data		
	Breeding age, mo.		
	10	14	18
Calving age, mo.	20	24	28
Services/conception	3.3	2.5	2.2
% conceived at first service	25	42	50
Calves died at calving	2	1	0
Living calves produced	6	10	9
% having calving difficulties	88	55	56
Birth wt of calves	86	84	83
Milk production, first lactation	4,491	4,603	5,356

[a]From Wickersham and Schultz (1963)

Table 9-5. Feed requirements for Holstein heifers on an alfalfa hay or corn silage system.[a]

Months of age	Milk[b]	Concentrate[c]	Hay[d]	Corn silage[e]
	....	.... kg	....	....
0 and 2	140	35	20	----
3 and 4	0	85	90	----
5 and 6	0	85	190	450
7 and 8		(60)	270	635
9 and 10		(60)	400	935
11 and 12	___	(60)	475	1170
Total to 1 year	140	205 (180)	1445	3190
13 and 14	0	(60)	535	1255
15 and 16		(60)	580	1360
17 and 18		(60)	635	1485
19 and 20		(60)	680	1590
21 and 22		(60)	720	1680
23 and 24		140	610	1435
Total 1 and 2 years		140	3760	8805
Total: Birth to 2 years	140	345 (480)	5205	11995

Total feed cost to 24 months.[f] Hay system = $462.75; corn silage system = $334.40

[a]Adapted from Hillman et al (1974). Clean, fresh water should be available at all times.

[b]Milk, colostrum or milk replacer (adjusted to 14% solids), or combination of these three.

[c]The 60 kg (in parenthesis) for every 2 mo. from 6 to 22 mo. represents 1 kg/cow/day needed only in hay rations to attain 800 g/day gain in heifers. Totals for year in parentheses are for extra corn given hay-fed heifers.

[d]Consumption of heifers on a hay system.

[e]Consumption of corn silage by heifers on corn silage system. Assume corn silage of 33% DM supplemented with NPN (urea or ammonia) to contain at least 13.5% crude protein with dicalcium phosphate and trace mineralized salt. Feed same as hay system up to 4 mo. Corn silage fed heifers do not need grain after 6 mo.

[f]Price per kg (¢): milk replacer, 38.6; concentrate, 14.3; corn, 11.0; hay, 6.6; corn silage, 2.2.

Feeding Systems for Raising Heifers

Protein Needs. The newborn calf grows best on a ration containing at least 20% protein, but by 3 mo. of age protein needs have decreased to about 14%, and 12% will give optimum growth after 6 mo. (Huber, 1973).

The calf can handle limited amounts of NPN by 6 weeks of age if eating over 750 g of starter daily and rumen function is established. The NPN can then be gradually increased to provide all of the supplemental protein in growing rations by the time the calf is 4 mo. of age. A maximum use of NPN is particularly profitable for corn silage rations because of their low protein content, but when heifers are fed mostly legume hay (15% CP±), supplemental N is not needed.

Hay vs Silage. Rations which provide nutrients for the desired growth rate are usually adequate in other respects. Hence, the decision on the type of ration to feed growing heifers comes down to economics. Total feed needed and costs for raising heifers from birth to 24 mo. of age were compared for an alfalfa hay or a corn silage system and are presented in Table 9-5. Currently, costs favor the corn silage ($344 vs $463) and support the trend in corn growing areas of feeding more corn silage to young dairy stock. Few dairymen feed heifers strictly on one or the other system, but will usually combine silage, hay and other feeds available on the farm. Heifers raised on hay and corn silage supplemented with needed protein and minerals are shown in Fig. 9-3.

Figure 9-3. Healthy heifers on a farm in Michigan.

ENERGY FOR LACTATION

Mismanagement of energy feeding in dairy rations is a costly error made by many dairymen. Undernutrition in early lactation, when propensity for milk production is greatest, and overfeeding during late lactation and dry periods are common occurrences on dairy farms.

Meeting Energy Demands in Early Lactation

The milk synthesis machinery of the mammary gland is at its peak for 6-8 weeks following parturition, but the cow's desire to consume nutrients to make milk is limited. Mean DM intakes for cows during the first few weeks of lactation seldom exceed 2.5% of body weight (BW) even though highly palatable rations are provided (Foldager and Huber, 1979). Postpartum intakes may climb to as high as 4% of BW between 6 and 16 weeks, but intake generally tapers off as milk yields decrease. Many cows attain peak milk production (40-55 kg/day) while intakes are still lagging behind; so to survive, they must mobilize body tissue. For cows yielding 35-55 kg daily, this creates an energy deficit which must be compensated for by mobilization of body fat (and some protein).

Fortunately, the cow in early lactation has a great capacity for conversion of body fat to energy for milk synthesis. Energy balance studies have shown that a deficit of as high as 20 Mcal/day (equal to 3 kg of body fat) for the first 3 weeks of lactation may be incurred by high-producing cows shortly after parturition (Flatt et al, 1965). The average loss of energy was 7 Mcal/day for the first 66 days of lactation for a group of cows consuming all they would eat of a ration containing 40% concentrate and 60% alfalfa hay. From 66 to 176 days, cows were essentially at energy equilibrium with a slight positive balance of 0.7 Mcal/day; this increased to 3.4 Mcal/day for days 176 to 292 (Flatt et al, 1969).

Considerations for maintaining high intakes of rations in early lactation cows are: (1) an adequate balance for necessary nutrients (especially protein, minerals and vitamins); (2) sufficient fiber to maintain optimal rumen function (15-17% ADF on a DM basis, depending on the type of forage fed). If corn silage is the main forage source (with limited hay), concentrate should not exceed 50% of the total DM, but up to 65% concentrate may be fed with mostly hay. (3) Texture and sweetness are important. Cows prefer a chewy texture (as with cracked and rolled grains or pellets) to one which is finely ground. They also respond to molasses, perhaps 4-6% in concentrate.

Restoration of Body Tissue

High producing cows will lose weight during early lactation when milk yields are greatest and appetite is lowest. Rapid weight losses cause metabolic disturbances which seriously impair productivity and profitability of cows for the entire lactation. Large cows can safely lose 100 kg of body fat during the first 70 days of lactation, amounting to about 15% of their post-calving weight. This weight reduction furnishes sufficient energy to produce 750 kg of milk. Smaller cows can lose a proportional percentage. Body weight losses, however, do not always reflect tissue losses in early lactation because of changes in gut fill and shifts in body composition (water filling adipose space, etc.).

Restoration of the body tissue converted to milk in early lactation should occur during mid and late lactation. If allowed ad libitum consumption, intakes are usually maximized by 50 days postpartum and most cows reach an energy equilibrium shortly thereafter. As milk production decreases, the intake of excess energy relative to needs allows for restoration of lost tissues.

Allotment of extra energy during the dry period is not generally recommended, because fat accumulated by non-lactating cows is less efficiently converted to milk (ca. 70% as efficient) than that gained during lactation (Moe et al, 1971). Moreover, there is greater danger of excessive fattening if dry cows are allowed to overconsume energy.

Source of Energy

Maximum consumption of energy is attained when 40-45% of the ration is made up of high quality forage and concentrate the remaining 55-60% (Spahr, 1977). As concentrate increases above 60% of the total ration, adverse effects such as off-feed, depressed butterfat and displaced abomasums increase. Physical state of the fiber is also important. Grinding or fine chopping of hay or silage to average particle sizes of <1 cm results in a feed which serves less effectively as a forage, and more of the above-mentioned problems might be expected.

Addition of fat to dairy rations is a possible method for increasing energy density without decreasing fiber to danger levels. Palmquist and Jenkins (1980) fed dairy concentrates containing 10% hydrolyzed fat without depressing milk production, milk fat or protein concentrations, feed intakes, or nutrient digestibilities. However, concentrate containing 10% tallow resulted in a significant reduction in milk yields and components. Added fat has not always improved milk yields and more research is needed before widespread use of the practice should be encouraged.

PROTEIN

Utilization by the Dairy Cow

The dairy cow is the most efficient ruminant in changing low quality protein from grains and forages into one of near optimal amino acid composition for humans such as in milk. This upgrading is largely accomplished through the bacteria and protozoa of the rumen (see Vol. 1). A large percentage of the natural protein which enters the rumen is deaminated (Fig. 9-4). The resultant ammonia is linked with specific amino acid precursors as altered by the microbes for synthesis of their structural protein. Thus, microbial protein provides the majority of the protein for absorption from the gut.

Casein, lactoglobulin and lactalbumen, which comprise 90% of the milk protein, are synthesized in the mammary gland from free blood amino acids. The remaining 10-15% comes from direct transfer of blood albumen into the mammary gland. A high producing dairy cow yielding 9,000 kg milk annually will synthesize about 280 kg milk protein. She will also produce about 20 kg of protein required by the fetal calf and associative tissues. Hence, she secretes as milk and calf protein a quantity equal to ca. 200% of the protein contained in her body. Thus, it is not surprising that the dairy cow's requirement for protein greatly exceeds that of non-lactating ruminants.

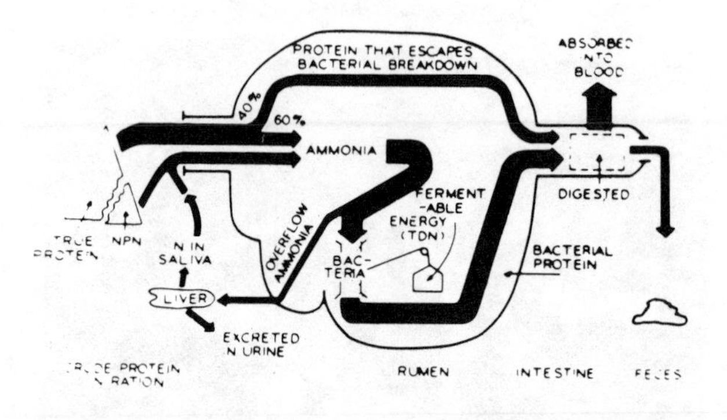

Figure 9-4. Scheme for protein and nonprotein nitrogen utilization by the ruminant (Satter and Roffler, 1975).

Methods of Protein Evaluation

In the past, protein requirements for dairy cows were often expressed as digestible crude protein (CP). However, digestible protein is highly correlated with total protein of the ration because the proportion as metabolic fecal N is fixed and makes up much of the total feed N (see Ch. 3, Vol. 2). Hence, digestible protein of a feed reflects poorly the absorbable amino acids available to the animal and has little functional significance for ruminants because rumen microbes reduce a high percentage of dietary protein to ammonia. Ammonia entering blood from the digestive tract has minimal value but would be considered totally digestible. Specific cases where digestibility reflects protein availability in ruminants are physical or chemical alterations such as overheating, fine grinding or excessive treatment with formaldehyde.

Systems which more directly estimate protein available for milk production by taking into account fluxes of the major N pools in the body are being developed. Such a system was proposed by Satter and Roffler (1975) who suggested a decreasing requirement for protein as lactation progresses. Their general recommendation is to feed 16% or more CP for the first 4 mo. of lactation and 12.5% thereafter. These workers suggest that protein needs are related to stage of lactation as well as milk flow. The rationale is that large amounts of body fat are mobilized by high producing cows early in lactation due to an energy deficit and that mobilized tissue is relatively lower in its protein contribution. However, data are scarce as to exactly how much body protein is available for milk synthesis if intakes do not supply sufficient energy and protein. New systems which base protein requirements for milk upon absorbed amino acids are still in the development stage, so recommended allowances in this chapter will be in terms of intakes of total CP.

Protein Needs for Lactation

Although it is agreed that protein demands decrease with decreasing milk flow, responses to high levels of protein in early lactation have been variable. Usually, where protein percentages >13% have increased milk yields in early lactation, energy intakes of cows were also stimulated (Sparrow et al, 1973; Grieve et al, 1974; Edwards and Bartley, 1979). In studies where energy intakes were as high for 13% as they were for 15-16% CP, milk yields were not stimulated by increased protein (Chandler et al, 1976; Foldager and Huber, 1979). Table 9-6 suggests modifications of the NRC (1978) total CP allowances for milk production based on the author's experience and opinions.

Digestibility of DM is affected by the protein content of rations. Huber and Thomas (1971) demonstrated that digestible DM increased from 56 to 69% by increasing dietary CP from 8.5 to 13.6% (of DM). About 38% of the decreased DM digestibility on the low protein ration was attributable to a decrease in the CP digestibility; the remainder was due to decreased absorption of other components of the ration DM. Tyrrell (1979) increased digestibility of corn silage rations by raising CP from 13 to 15%, but noted little further increase between 15 and 17%. Changes in intakes, milk yields and ration

Table 9-6. Suggested protein allowances for lactating dairy cows of varying milk yields (630 kg, mature cow).

Milk yield,	Feed intake		NRC allowance[a]		Author's allowance[b]	
kg/day	kg DM/day	% BW	g/day	% of ration DM	g/day	% of ration DM
15	16.3	2.5	1,745	10.7	1,956[c]	12.0[c]
25	19.5	3.0	2,565	13.2	2,486	12.7
35	20.8	3.2	3,385	16.3	3,218	15.5
40[d]	17.6	2.8	3,795	21.6	3,596	20.4
45	22.1	3.4	4,205	19.0	3,766	17.0

[a]Maintenance = 515 g/day, milk (3.5% fat) = 82 g/kg (based on NRC, 1978).
[b]Maintenance = 636 g/day, milk (3.5% fat) = 74 g/kg (based on author's experimental data over many years).
[c]Feed this group at least 12% crude protein to prevent depressed digestibility of dry matter and lowered intakes even though calculated needs are only 10.7%.
[d]This calculation is for a cow in very early lactation which peaks in milk yields before intake is maximized. She is obviously using considerable body fat, but the available protein reserve is not known.

Table 9-7. Response of lactating cows to increased protein supplementation in their ration (50% corn silage, 50% concentrate).[a]

	Crude protein in ration (%)			
	8.5	10.5	12.5	13.6
Source of supplement	none[b]	urea	urea & SBM	SBM
Milk yields, kg/day[c]	19.2	23.7	26.4	25.3
Dry matter intake, % BW[c]	2.46	2.65	3.05	2.81
Dry matter digestibility, %[c]	56.4	59.0	65.0	69.2

[a]Huber and Thomas (1971); (10 cows/treatment).
[b]Urea was added to corn silage at ensiling, soybean meal to concentrate at feeding.
[c]Linear increase with increased protein significant ($P < .01$).

Table 9-8. Marginal changes in milk, dry matter intake and estimated returns as a result of changing ration protein level with soybean meal (for cows in early lactation and capable of producing 7,000 kg or more).[a]

Change in ration protein level, DM basis, %	Daily increase in milk production, kg/cow/day	Daily increase in feed intake, kg/cow/day	Daily increase in Milk value[b]	Feed cost[c]	Feed cost[d]	Feed cost[e]
			$	$	$	$
10 to 11	1.91	1.36	.46	.22	.25	.30
11 to 12	1.50	.68	.36	.15	.18	.23
12 to 13	1.00	.41	.24	.12	.15	.20
13 to 14	.77	.27	.19	.10	.14	.18[f]
14 to 15	.59	.18	.14	.09	.13[f]	.17
15 to 16	.50	.13	.12	.09	.13	.17
16 to 17	.41	.13	.10	.09[f]	.13	.17
17 to 18	.27	.09	.07	.08	.12	.16

[a]Adapted from Satter et al (1979)
[b]Milk priced at $11/cwt.
[c]Cost of additional feed (mainly grain mix) is figured at $95/ton, and the cost of a 44% natural protein supplement, exclusive of added minerals and vitamins, is figured at $195/ton. It is assumed that replacement of 0.68 kg of the basal grain mix by 0.68 kg of 44% natural protein supplement will raise the total ration protein level by one percentage unit.
[d]The cost of additional feed is the same, except the 44% natural protein supplement is figured at $245/ton.
[e]The cost of additional feed is the same, except the 44% natural protein supplement is figured at $300/ton.
[f]This most profitable level of protein to feed. It would be advisable to feed at least this amount of protein, and perhaps one percentage unit of protein more. Some of the benefit from high protein feeding will accrue in later lactation, and will likely compensate for slight overfeeding of the maximum profit point in early lactation.

digestibility with increasing protein intake in lactating cows are given in Table 9-7. Minimum CP that a lactating cow should be fed so that energy utilization and DM consumption are not lowered appears to be 13-14%.

There is a diminishing stimulation of milk yields as CP is increased from 12 to 17%, and decreased milk production has been reported in studies where CP exceeded 19% (Satter et al, 1979). Because protein supply in dairy rations appears to follow the law of diminishing returns, it seems reasonable that protein be supplemented in accordance with income produced over the cost of the additional protein.

At high costs of protein supplement ($300/ton for soybean meal), it is doubtful that the maximum recommended feeding level for CP should exceed 15% (even in early lactation). Lower protein prices (soybean meal at $160/ton) may make a 17% ration profitable. Table

9-8 estimates changes in milk yield, DM intakes and returns from varying protein level of the ration.

NONPROTEIN NITROGEN (NPN) IN DAIRY CATTLE RATIONS

Economics of Feeding NPN

In the future world food demands may dictate that ruminants consume NPN to their maximum utilizable capacity, but the primary justification at present for feeding NPN to dairy cattle is economics. Generally, a reduced feed cost and increased profit result from inclusion of as much NPN as possible without depressing milk yields.

Urea (the principal form of NPN presently used) incorporation into ruminant rations has increased greatly in recent years. The primary reason for this increase is the upward spiral in the cost of natural protein supplements. In addition, research and extension programs showing nutritional benefit and increased profits from NPN use have stimulated use. To illustrate, in 1973 when soybean oil meal prices soared to over $300/ton and shortages were common, approximately one million tons of urea were incorporated into ruminant rations in the USA (Prebluda, 1974). The calculated savings were about $600 million in feed costs by substituting urea and corn for natural protein during that year.

Savings in Feed Costs for Dairymen

The profitability of substituting NPN and an energy source for natural protein supplements in dairy cattle rations depends on a number of factors. These are cost of the NPN source, relative price of the energy and the

Table 9-9. Economics of feeding NPN to dairy cows.

Price of shelled corn		Price of soybean meal, $/ton				
		160	200	240	280	320
$/bu	$/ton	¢/cow/day				
A. Savings in feed cost (¢/cow daily) from substituting 3.5 lb of shelled corn and 0.5 lb urea (12 ¢/lb) for 4 lb soybean meal[a]						
2.00	71	14	22	30	38	46
2.50	90	10	18	26	34	42
3.00	107	7	15	23	31	39
3.50	125	4	12	20	28	36
4.00	142	1	9	17	25	33
4.50	159	-2	6	14	22	30

Corn silage cost, $/ton[c]	Price of alfalfa hay, $/ton			
	40	60	80	100
	¢/cow/day			
B. Savings in feed cost (¢/cow daily) from substituting 35 lb corn silage and NPN for 14 lb of alfalfa hay[b]				
13	0	14	28	42
16	-5	9	23	37
19	-11	3	17	31
22	-16	-2	12	26
25	-21	-7	7	21

[a] To calculate savings per year for a herd, multiply the daily savings by the number of cows, then by 310 days. For example, a 23 ¢/cow daily savings ($3.00 corn and $240 soybean meal) would result in $14,460 annual savings for a 200-cow herd.

[b] 18.5% CP, early bloom.

[c] 30% DM corn silage, 47% as ears. The NPN would come from 0.55 lb urea or urea equivalent added to the total ration.

natural protein which is being replaced, and the animal response from substitution, usually reflected in changes in milk yields.

It is assumed that 8 lb of soybean meal are equal in net energy and CP to 7 lb of shelled corn and 1 lb of urea. Another basis of substitution is corn silage plus NPN (urea or ammonia) for alfalfa hay. The former comparison is valid for corn-growing areas and the latter for the western USA where alfalfa hay is the principal forage. It would take about 35 lb of corn silage (33% DM) and 0.55 lb of urea to replace the CP and energy in 14 lb of alfalfa hay (18.4% CP). Table 9-9 approximates savings in feed costs from substituting urea and corn for soybean meal or corn silage and an NPN source for alfalfa hay at varying ingredient prices.

As demonstrated, the potential profit from incorporation of NPN in dairy rations is great, but decreased milk yields from NPN diets could easily negate any benefit. At current feed and milk prices it would taken only 1-2 lb/day loss in milk to make NPN feeding unprofitable.

Factors Affecting NPN Utilization by Dairy Cattle

Energy. Precursors to amino acids are synthesized by rumen microbes from the fermentable carbohydrates. Cows fed rations low in energy and high in fiber are limited in the quantity of NPN they can utilize successfully. Animals often react negatively to high fiber rations supplemented with NPN.

Rations fed for high milk production generally contain sufficient available energy to support synthesis of microbial protein from NPN. For example, rations containing large amounts of corn silage are well suited for efficient NPN utilization. Reviews by Satter and Roffler (1975) and Burroughs et al (1975) suggest a strong dependence of NPN use on energy content of the diet and recommendations for NPN incorporation are based primarily on concentration of available energy in the ration.

Protein. Protein in dairy rations is often adequate without the added NPN, hence the NPN is of no benefit. The maximum dietary CP at which cows respond to NPN is still controversial, but it probably is not over 14-15% of the ration DM, even in high energy rations. Data of Jones et al (1975) and Kwan et al (1977) showed increased milk production when protein in complete rations was raised from about 14 to 16% with NPN. Supplementing rations of corn, soybean meal, corn silage and limited hay, which contain 11-12% natural protein, with NPN to increase CP to 14-15% is a proven practice in practical dairy operations (Ryder et al, 1972).

Quantity of NPN. In complete rations for lactating cows, urea (or urea equivalent as ammonia) should not exceed 1 to 1.2% of the DM. Total intake of urea (or urea equivalent) should be limited to 225 g/day for the average cow (weighing 650 kg). Addition of too much urea (1.7-3.0%) to complete rations of 11.5% CP decreased milk yields (VanHorn and Jacobson, 1971; VanHorn et al, 1975). These high urea intakes (300-500 g/day) were accompanied by depressed feed consumption. In contrast, Kwan et al (1977) raised CP of a complete ration from 11 to 14% by addition of 1% urea and increased milk yields 10% (from 31 to 34 kg/day).

Production Level. Satter and Roffler (1975) suggested that NPN is not indicated for cows early in lactation and that cows be supplemented with only natural protein for 14 weeks after calving. It is true that high yielders react more negatively to excessive NPN than low yielders (Dutrow et al, 1974), just as they are more severely affected by other dietary stresses. In contrast to the statement of Satter and Roffler (1975), a considerable amount of evidence shows that NPN feeding is compatible with high milk yields (Holter et al, 1968; Huber et al, 1975; Kwan et al, 1977; Clay et al, 1978; Murdock and Hodgson, 1978; Conrad and Hibbs, 1977; Foldager and Huber, 1979). Table 9-10

Table 9-10. Response of high producing cows in early lactation to NPN or natural protein.[a]

Item	Basal ration, 12-13% CP	Soybean ration, 15-17% CP	Urea ration, 15-17% CP
Milk yields, kg/day	30.64	32.67	32.34
DM intake, % BW	3.22	3.18	3.23

[a]Pooled from four recent studies comparing 54 cows per treatment starting 0-5 weeks postpartum and continuing from 9 to 20 weeks. Data from Kwan et al (1977); Murdock and Hodgson (1978); Clay et al (1978); Foldager and Huber (1979).

summarizes data from four recent studies showing a positive response to NPN in early-lactation cows receiving more than 12-13% CP.

Rumen Ammonia Nitrogen (RAN) Concentrations. Ammonia levels in the rumen will rise with increasing dietary protein, but they increase more rapidly on NPN diets than on natural protein. From in vitro studies, Satter and Slyter (1974) reported that microbial protein yield was maximized when RAN reached 2-5 mg/dl and Satter and Roffler (1975) proposed limits for NPN feeding based on these findings. Other data do not support such a low cut-off for protein synthesis in the rumen. With a similar in vitro system, Bull et al (1975) showed increased microbial protein until RAN was about 20 mg/dl. Barr (1970) demonstrated that microbial protein production was markedly increased by changing the substrate from grain plus urea to starea at levels of 100 mg/dl. In vivo studies suggest that production of microbial protein is not maximized until RAN reached 10-20 mg/dl (Hume et al, 1970; Miller, 1973). Mehrez and Ørskov (1976) reported recently that utilization of fermentable substrate by the rumen increased until RAN reached 23 mg/dl.

Protein deficient rations (10% CP) significantly reduced milk yields and resulted in RAN levels of 3-5 mg/dl, whereas, cows fed adequate protein were in excess of 8 mg/dl (Huber et al, 1976; Randel et al, 1975). It is the opinion of the author that many factors affect RAN levels in cows and they are often difficult to interpret. Using in vitro or in vivo RAN concentrations for predicting limits of NPN feeding is ill advised.

Adaptation and Deadaptation. High levels of NPN (>50% of the total dietary N) require up to 8 weeks for full adaptation, but gradually increasing by NPN-containing feed for 14 days is probably sufficient for adaptation of cows to quantities of NPN presently recommended. A simple and workable scheme is to feed 1/3 of the NPN feed during the first week, 2/3 for the second week, and a full-feed the third week. Feeding a high protein ration prior to feeding NPN facilitates adaptation, which apparently occurs through enhancement of urea synthesis by the liver due to ammonia absorption from the rumen. Adaptation minimizes dangers of ammonia toxicity and facilitates urea recycling into the rumen. Continued feeding of an NPN diet is necessary or deadaptation occurs in 3-5 days. Hence, additional adaptation will be required if NPN feeding is resumed.

Limits for Feeding NPN to Dairy Cattle. Data on the kinetics of urea utilization (Mugerwa and Conrad, 1971) suggest that the limit for urea addition was about 180 g/day for lactating Jerseys weighing 430 kg. Extrapolating this level to a 600 kg Holstein cow (metabolic weight basis) would give a suggested maximum intake of 245 g/day. The Ohio workers further proposed that limits for NPN intake should be expressed on weight/day rather than as a percentage of the ration.

Because most cows are fed in groups rather than individually, ration NPN levels should be gauged to those individual cows (which are the highest milk producers) consuming the greatest amount of feed. Even though average DM consumption for a total herd may be 3% of BW, studies suggest that the higher producers (6-16 weeks after calving) are consuming 3.5% of their BW. Hence, a 700-kg cow would receive 220-250 g urea if fed a complete ration containing 1-1.1% urea. When urea is blended with concentrate fed separately from the forages, the limit is 1.5-1.75% of the concentrate because of depressed intake when concentrates contain over 2% urea (VanHorn et al, 1967; Huber et al, 1967). For high moisture feeds or in warm, humid weather, a maximum of 1% urea in concentrates is indicated (Kertz and Everett, 1975).

Biuret takes about twice as long for adaptation as urea because the need for stimulation of biuretolytic activity by rumen microorganisms. However, deadaptation is just as rapid as with urea, occurring in only 3-4 days after withdrawal of biuret from the ration.

SUGGESTED PROBLEMS WITH FEEDING NPN

Toxicity

Consumption of high levels of urea (>45 g/100 kg BW) in a short period by unadapted cattle can be fatal, but adapted animals tolerate 2-3X that amount (see Ch. 17, Vol. 1). Mistakes, such as cows breaking into urea supplies, inadvertantly spilling urea on feeds, or miscalculating feeding levels often cause toxicity. Feeding NPN in complete feeds or mixing with grain silages minimize dangers even if errors are made. Modified forms of NPN such as Starea, dehy-100, urea-beet pulp or biuret, release ammonia at a slower rate and protect against toxicity.

Acute ammonia toxicity is treated by drenching cows with 10-12 ℓ of vinegar (5% acetic acid) as soon as symptoms are observed. Also, give half the original dose 2-3 h later when toxic signs reappear (Word et al, 1969). Acetic acid concentrations stronger than 10% are not advised because of injury to the esophagus.

Dairy cows which survive NH_3 toxicity do not suffer after-effects. Milk production, estrus cycling and other signs of well-being rapidly return to normal. The suggestion that cows will abort following ammonia toxicity has been discounted (Word et al, 1969).

Palatability

Concentrates containing over 2% urea result in depressed feed intakes by dairy cows even though animals might be physiologically adapted to tolerate higher amounts (VanHorn et al, 1967; Huber et al, 1968). Huber and Cook (1972) showed that intake depression of a urea-containing concentrate (cows consumed 320 g urea/day) was due to taste and not to ruminal or post-ruminal events. Wilson et al (1975) reported a marked depression in feed intake when the complete ration contained 2.3% urea (425-450 g/day). Urea also lowered DM consumption when administered through a rumen fistula, suggesting that large amounts of urea decreased intake by a mechanism other than taste. Chalupa et al (1979) demonstrated that decreased intake of urea-containing rations is a learned response.

Masking the bitter urea taste with molasses improved intake in studies by Huber and Cook (1972). Addition of urea to corn silage also reduces intake problems caused by urea, probably because about 50% of the N is consumed as ammonium salts and also the masking of the remaining urea by acids in silage (Huber et al, 1968).

Reproductive Performance

Some writers have suggested that feeding rations containing NPN lowers conception efficiency of cattle. To test this claim, 85,000 individual lactation records from Michigan DHIA herds were analyzed for calving intervals during the 5-year period of 1965 to 1969 (Ryder et al, 1972). Through careful interviews with herd owners and inspection of herd records, the amount of urea (or urea equivalent from ammonia) fed was ascertained. About 55% of the herds received NPN during at least 3 of the 5 years. Cows receiving no NPN had an average calving interval of 380 days, identical to that of cows fed NPN (Table 9-11). Little change occurred as NPN in the rations increased. Percent of cows culled because of sterility showed no relationship to the feeding of NPN. Erb et al (1976) reported increased abortions (14% of 37 calvings) in heifers raised on high urea diets. On the other hand, studies conducted by Clark et al (1970) with dairy heifers fed urea levels similar to those fed by Erb et al failed to show that the high NPN caused abortions or other reproductive problems.

Urea:Nitrate Relationship

Urea and nitrate have been fed alone and in combination. No synergism between the two toxicities was shown. Sebaugh et al (1970) reported milk yields of cows fed high nitrate

Table 9-11. Influence of urea intake on calving intervals, percent of cows sold as sterile and milk yields in Michigan DHIA herds 1965-69.[a]

Urea intake, g/day	None	1-370	1-60	61-120	121-180	180+
Average urea intake, g/day	0	79.8	36.1	90.5	146.7	219.5
Herd-year observations, no.	1,442	1,715	760	653	219	83
Calving interval, days[b]	380.4	379.9	379.4	380.7	379.8	377.8
Cows sold as sterile, %	2.15	2.4	2.4	2.4	2.6	1.71
Milk per cow, kg	5,937	5,867	5,676	5,893	5,982	5,977

[a] From Ryder et al (1972)

[b] Calculated by dividing the total number of individual cow observations for each herd by the total calving interval for each herd.

hay was greater for animals consuming supplemental urea than soybean meal, but initial toxicity symptoms appeared worse for the nitrate-urea combination. Virginia workers reported depressed feed consumption and lowered weight gains on a ration containing 1% KNO_3 and 1.7% urea compared to either fed singly. However, after intakes of feed were equalized, weight gains were not different (Lichtenwalner et al, 1973).

The possibility of nitrate enhancing a urea toxicity is meager. Rumen microorganisms do reduce nitrate to nitrite and the nitrite to ammonia, but the latter reaction is slow and very little of the original nitrate would be transformed to blood ammonia. Also, the yield of ammonia from nitrate is extremely low. Conversely, excess urea will not aggravate a nitrate toxicity, because the rumen is a reducing environment and oxidation of NH_3 to nitrite is impossible.

Method of Feeding NPN

NPN in Concentrate vs Complete Rations. Many dairy cows are still fed NPN through concentrates offered 1-2X daily. Holter et al (1968) showed no difference in milk production or breeding efficiency between urea or SBM groups fed concentrate twice daily. The CP of both rations was 14.6% (of DM), but removal of urea would have decreased ration protein fed the NPN group to 12.0%, a level too low for high producers.

In another study, Gardner et al (1975) started 132 Holstein cows at parturition on concentrates containing 1.5% urea, 14.8% SBM or 7.7% of a heat-processed corn-urea mixture (Golden-Pro). Daily milk yields for the first 18 weeks of lactation were highest for Golden-Pro (30.7 kg), next for urea (29.2 kg) and least for SBM (27.9 kg).

Complete rations allow for feeding a higher quantity of NPN than when added to concentrate fed at infrequent intervals. This is because distribution of NPN in complete feeds results in lower rumen ammonia and more efficient incorporation of the dietary N into microbial protein.

Modified Forms of Urea. In attempting to increase the level of NPN that dairy cows utilize efficiently, modified forms of urea have been developed. These are:

Starea. Kansas workers (Bartley and Deyoe, 1975) expanded grains with urea (Starea) and showed superior microbial protein synthesis, feed consumption and milk production compared to unprocessed grain plus urea. Slower rumen ammonia release and decreased toxixity were also observed.

Dehy-100. Conrad and Hibbs (1979) pelleted a mixture of 32% urea, alfalfa meal, dicalcium phosphate, sodium sulfate and sodium propionate which contained 100% CP. When fed as the only protein supplement to high-producing dairy cows for a full lactation, milk yields (8,000 kg), FCM, milk fat, and milk protein were similar to a ration containing soybean meal. Urea furnished about 40% of the total ration N in the Dehy ration (equal to 350 g urea/day).

Biuret. Biuret, a conjugation of two urea molecules, is tasteless, nontoxic and much slower than urea in ammonia release. This compound might appear to be the ideal NPN source for feeding ruminants. Superior growth compared to urea has been shown for high fiber but not production-type rations. Because of spillover in milk, biuret is not allowed for milk cows by the FDA.

Isobutylidene Diurea (IBDU). A slight advantage for IBDU compared to granular urea was shown, probably because of slower ammonia release in the rumen (Hemmingway et al, 1972), but more research is needed prior to widespread use. Like biuret, IBDU is not allowed for milk cows in the USA.

Ammonia and Urea in Silage. Silage is now widely used as a carrier for NPN in ruminant rations. Addition to silage distributes NPN intake over the entire day and, in the case of urea, masks the undesirable taste. Michigan studies conducted over a 7-year period resulted in slightly higher milk production (about 1 kg milk/day) for NPN-treated, compared to untreated silages when rations were of equal CP content. Addition of NPN to grain silages makes their protein content (DM basis) similar to that of the concentrate fed and allows greater flexibility in feeding cows differing amounts of concentrate for varying production, but still allows their protein needs to be met.

Table 9-12. Influence on performance of lactating cows of ammonia- vs urea-treated silage with or without urea in the concentrate.[a]

	Silage treatment[b]			
	Ammonia, %		Urea, %	
Concentrate urea, %	0	1.5	0	1.5
Dry matter intake, kg/day	19.11	17.80	17.98	17.32
Milk yields				
kg/day	20.9	20.7	20.8	20.0
Persistency[c]	86.3	85.3	84.7	78.2

[a]Huber et al (1980)
[b]16 cows per treatment
[c]Persistency = treatment/pre-treatment milk yields during the last 8 weeks of treatment

Figure 9-6. Cows consuming liquid protein supplement from a lick tank.

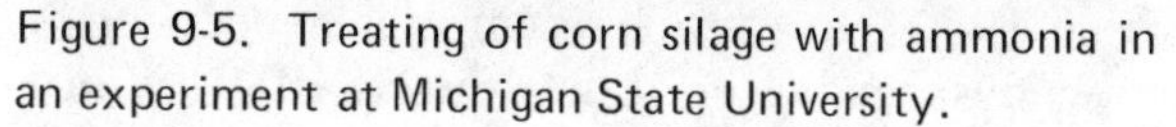
Figure 9-5. Treating of corn silage with ammonia in an experiment at Michigan State University.

Because ammonia N is the most economical NPN source and because the acids in corn silage render it an ideal substrate for ammonia, a program to evaluate ammonia additions to corn silage was initiated at Michigan State University in 1967 (Fig. 9-5). A summary of seven experiments involving 169 cows showed milk yields were 0.7 kg/day higher for groups fed ammonia than those fed control or urea-treated silages (Huber, 1973). The superiority of ammonia over urea-treated silage was supported by a study in which cows fed urea in concentrate and ammonia-treated silage produced more milk than those fed urea in concentrate and urea-treated silage (Table 9-12).

Compared to other silages, those treated with ammonia are higher in lactic acid and insoluble N and less apt to heat and spoil when exposed to air. The increased lactic acid is due to buffering by the NH_3. The increased insoluble N results partly from incorporation of ammonia into the insoluble fraction and partly to a decrease in proteolysis of the original plant protein (Huber et al, 1979). The increased stability is caused by an antifungal action of the ammonia or the ammonium salts (Britt and Huber, 1975).

NPN in Liquid Supplements. Liquid supplements have become an established method of furnishing NPN to dairy cattle. They conveniently supply cows with CP, vitamins and minerals and are usually palatable because of molasses. Most supplements contain 10-20% urea and some have up to 25% of their N as true protein derived from soubles from corn, fish or other sources.

Liquids are often mixed with forages or concentrates prior to feeding. They are also consumed ad libitum from lick-wheels (Fig. 9-6). With the lick-wheel, a key problem is an initial overconsumption, but after several days cattle regulate intake of liquid. The assumption that cows of different production levels, needing varying amounts of protein, voluntarily balance their rations with a free choice liquid is not well established, but the idea has some support (Braund, 1979). In the study summarized in Table 9-13, Braund and Steele (1971) demonstrated that cows fed 43% of their protein in the dry form ate more liquid supplement when allowed free choice

Table 9-13. Protein intakes of lactating cows offered varying amounts of dry protein and liquid protein supplement ad libitum.[a]

Item	Group		
	A	B	C
Protein in dry form			
Offered, % of requirement	50	100	150
Actual eaten, % of requirement	53	84	125
Liquid supplement consumed, kg/day	131	114	68
Total protein consumed, % of requirement	74	102	135

[a]From Braund and Steele (1971)

consumption from a lick wheel than cows eating 84 or 125% of their needs as dry protein. The intermediate cows (84% as dry) met their needs quite well with liquid, but liquid offered the high cows resulted in an even greater overconsumption of protein and obvious waste.

Feeding liquids to herd replacements and dry cows on low-protein feeds such as corn silage, corn stalks, or grass hay is often preferred to other methods of supplementation because of convenience. Mixing liquids with feeds of questionable palatability also increases DM consumption of such feeds.

EFFECT OF PROTEIN QUALITY ON MILK PRODUCTION

Availability of Amino Acids

There is some question as to the supply of essential amino acids for high producing animals such as dairy cows (see Ch. 3, Vol. 2). Attempts to gain information on this topic have involved infusions of amino acids or proteins into the small intestine of surgically prepared animals. Amino acids of most limited supply in blood relative to output in milk have been shown to be methionine, lysine, phenylalanine and threonine (Chandler and Polan, 1972; Vik-Mo et al, 1974; Clark et al, 1975). Milk protein production was stimulated more by methionine and lysine than other amino acids infused into the gut (Schwab et al, 1976). These were 70% as effective in increasing milk protein synthesis as casein.

Protection of Protein from Rumen Degradation

Approaches to minimize rumen degradation of protein have included physical alteration through heat and/or pressure; chemical treatment using formaldehyde, tannic acid, antimicrobial agents or other compounds which inhibit rumen proteolysis; or selection for specific protein supplements naturally resistant to rumen breakdown.

Chemical and physical treatments appear attractive, but they risk the danger of overprotection, making treated protein less digestible post-ruminally than untreated. Recent studies by Mielke and Schingoethe (1979) and Smith et al (1980) showed increased milk yields in cows fed extruded soybeans.

Selection of natural sources to reduce rumen protein degradation was effective in a Texas study (Majdoub et al, 1978). A patent granted to Braund et al (1978) showed that combining ingredients in dairy concentrate with reduced protein solubility resulted in increased milk production. Table 9-14 summarizes the data from one study. Basically, solubility was reduced by increasing brewers and distillers dried grains and decreasing corn gluten feed. Continued research on protein quality for ruminants should contribute to higher and more efficient milk yields, particularly in early lactation cows.

Table 9-14. Response of early lactation cows to rations of varying protein solubility.[a]

	Ration[b]			
	A	B	C	D
Crude protein, %	12.6	17.8	18.8	18.5
Natural protein, %	12.6	17.8	16.3[c]	18.5
Soluble protein, %	31.1	38.2	43.7	21.3
Milk yield, kg/day	26.6	29.1	28.5	31.8

[a]From Braund et al (1978)
[b]20 cows/treatment fed experimental rations from 5th to 15th week of lactation, all cows were fed similarly from 1 to 5 weeks.
[c]Contained 2.5% CP as urea

MINERALS

Feeding of minerals to dairy cattle is often handled poorly. Farmers are continually bombarded by mineral salesmen interested in placing their own product on the farm, regardless of ration needs. Consequently, the same minerals are added to the concentrate, to the forage, and allowed for free choice consumption in some herds. Excessive intakes of many minerals can be detrimental to productivity and longevity of cows.

MINERALS NEEDED FOR LACTATION

Daily requirements of minerals needed for a lactating dairy cow producing 30 kg milk daily are listed in Table 9-15. Recommended ration concentrations are also given. Because intake is directly related to milk yield, these ration percents generally apply to cows at all production levels.

A ration comprised mostly of corn grain and corn silage would need to be supplemented with most the minerals, but the alfalfa hay ration would need only a few. Specific schemes for supplementation should be developed, if analyses are available, for each individual ration depending on mineral content of feeds that are used. Otherwise, tabular values from official publications would be helpful.

Special Considerations for Mineral Feeding

Ca, P, Vitamin D and Milk Fever. Excessive intake and high Ca:P ratios, particularly in late lactation and during the dry period, have been associated with an increased incidence of milk fever (see Ch. 12, Vol. 2). The optimum Ca:P ratio for dairy cows varies with function (whether milking or dry) and type of feeds. From NRC (1978) standards, a ratio of 1.4:1 might be calculated. This ratio is probably satisfactory for dry cows, but it is too narrow for lactating cows which do best at ratios of 1.7 to 2.3:1 (Gardner and Park, 1973). Increased incidence of milk fever has been shown for ratios narrower than 1:1 and wider than 2.5:1.

Just as high Ca during late lactation and dry periods is detrimental to lactating cows, so also is excessive P in early lactation (Carstairs, 1978). Primparous Holstein heifers fed P at 138% of the NRC allowance for the first 12 weeks of lactation yielded an average of 2.7 kg less milk/day for the entire lactation than herdmates fed 98% of the suggested requirement. The reason for decreased milk

Table 9-15. Suggested mineral allowances and excesses relative to those allowances.

Mineral	Suggested allowance per day g or mg	Suggested allowance in ration[d] DM basis	Reported to cause problem[e] DM basis	Problem/allowance factor
Calcium (Ca)[a]	141 g	0.7%	1.4%	2.0
Phosphorus (P)[ab]	71 g	0.36%	0.6%	1.7
Potassium (K)[a]	150 g	0.75%	3.0%	4.0
Salt (NaCl)[ab]	60 g	0.3%	5.0%	16.0
Magnesium (Mg)[ab]	40 g	0.2%	2.0%	1.8
Sulfur (S)[a]	40 g	0.2%	0.35%	1.8
Iron (Fe)[ab]	1 g	50 mg/kg	1000 mg/kg	20.0
Zinc (Zn)[ab]	0.8 g	40 mg/kg	600 mg/kg	15.0
Manganese (Mn)[ab]	0.4 g	20 mg/kg	1000 mg/kg	50.0
Copper (Cu)[a]	0.16 g	8 mg/kg	100 mg/kg	12.0
Iodine (I)[a]	8 mg	0.4 mg/kg	10 mg/kg	25.0
Cobalt (Co)[a]	2 mg	0.1 mg/kg	10 mg/kg	100.0
Selenium (Se)[a]	2 mg	0.1 mg/kg	3 mg/kg	30.0
Fluorine (F)[c]	20 mg	1 mg/kg	20 mg/kg	20.0
Molybdenum (Mo)[c]	40 mg	2 mg/kg	20 mg/kg	10.0

[a]Generally need to supplement high corn silage rations.
[b]Generally need to supplement high alfalfa hay rations.
[c]Never supplement these to dairy rations.
[d]From NRC (1978) for a dairy cow producing 30 kg milk/day and weighing 630 kg.
[e]Mostly from NRC (1978). Some estimates from author's reading.

on high P was not clarified. Hence, indiscriminate and uncontrolled addition of several P sources to dairy rations should be discouraged.

Heavy supplementation of vitamin D on a routine basis is not desirable, but feeding massive doses (10 to 20 million IU/day) for 5 days before calving has reduced the incidence of milk fever, particularly in aged cows (Hibbs and Conrad, 1960). Caution was given to not extend the dosing period beyond 7 days to prevent problems of excessive intakes. The partially activated metabolite of vitamin D, 25-hydroxy-cholecaciferol (25 HCC), has effectively prevented milk fever in susceptible cows when it was administered at levels of 50,000-100,000 IU/day for 3-4 days prior to calving (Jorgensen, 1974). Recent reports suggest that 1,24-dihydroxycholecalciferol and 1-α-hydroxy-cholecaciferol may show better protection against milk fever than 25-HCC. As yet, none of the activated metabolites are approved for use by the FDA. Hopefully this approval is forthcoming.

In summary, the dairyman should feed P at 90-100% of the NRC standard (17 g for maintenance and 1.75 g/kg milk). Ca would then be calculated from the P intake so as to provide a Ca:P ratio of 1.7 to 2.3:1 for milking cows and about 1.5:1 for dry cows. When corn silage comprises the main forage, both Ca and P should be supplemented, but P without Ca would be indicated with rations high in legume forages.

Magnesium and Grass Tetany. In areas where dairy cows are grazed on lush, vegetative pastures in early spring, grass tetany and death have been observed (see Ch. 13, Vol. 2). The problem is usually aggravated by heavy N fertilization and high K levels in the grasses. Generally, Mg is less available in pastures than in grains or stored forages. Contrary to the other nutrients, Mg availability in forage tends to increase with maturity. Supplementation of cows grazing "tetanic" pastures with Mg (as the oxide or acetate) will usually maintain blood Mg at normal levels and prevent the danger of tetany.

Sulfur. With increased supplementation of NPN in ruminant rations, there has been a decrease in natural protein supplements which contain S. Also, the practice of heavier feeding of corn silage and less legume hay has decreased S intake by animals.

The S needs of cattle are often expressed in relation to the N content of the ration. For lactating dairy cows receiving 14-16% CP, 0.2 to 0.25% S is needed. This amount would result in the desired N:S ratio of 10 to 12:1. When corn grain and corn silage are the main ration components, S does not usually exceed 0.10%, so sufficient supplemental S is needed to essentially double original levels. Sulfate salts of K, Mg, NH_4, or Ca are satisfactory. Elemental S is less available than the salts (see Ch. 4, Vol. 2). Precautions should be taken to not exceed 0.35% S or a decreased ration intake might result (Bouchard and Conrad, 1973). When legume hay is fed in liberal amounts, supplemental S is generally not needed.

Trace Mineral Supplements

Some of the major feed crops (particularly corn and grasses) grown in certain regions of the USA contain an insufficiency of certain trace minerals to meet the requirements of animals fed primarily on these feeds. Recent approval by the FDA allows adding up to 0.1 ppm Se to dairy and beef rations. A problem related to low Se intakes on high corn rations is retained placentas in postparturient cows. Julien et al (1976) found that injection of 50 mg sodium selenite and 680 mg vitamin E approximately 14 days before calving effectively reduced the percent of placentas retained over 24 h. This practice has been effective in the field among herds receiving borderline levels of Se or vitamin E.

A practical way of avoiding mineral deficiencies in areas producing crops low in certain needed minerals (whether known or unknown) is to feed a trace-mineralized salt. The salt should be examined to ascertain that it contains sufficient amounts of the questionable minerals to cover any possible deficiency. For dairy cows the salt would be added at 1% of the concentrate, which generally makes up 40-50% of the total ration.

As a base, one might assume that dairy cows are consuming 0.5% salt. Hence, if it were desired to supplement the total ration with 0.1 ppm cobalt (the cow's requirement), the salt should contain 20 ppm or 0.002% Co. Similar calculations can be made for the other trace minerals contained in a salt.

Mineral Excesses

Adverse effects on milk production or health have been reported for macrominerals fed at levels of 1.5 (Ca and P), 2.0 (S), 5.0 (K and Mg) and 10.0 (salt) times the recommended allowance (Table 9-15). Margins of safety for microminerals appear wider than for some macrominerals (from 10X for Cu to 100X for Co), but these are highly variable. Moreover, the level at which many minerals are detrimental is not well-documented. Most studies have considered only weight changes, milk yields and overt health symptoms. Recent studies with iodine (Hillman and Curtis, 1980) have shown (at levels lower than those causing overt toxicity symptoms) shifts in white blood cell counts which might alter the animal's immunity and disease resistance mechanisms. The wide variation in concentrations at which the different minerals are detrimental to cattle, our lack of experimental data concerning these concentrations, and the subtle physiological changes that mineral excesses might stimulate, emphasize the need for more careful mineral distribution in dairy cattle rations than has generally been practiced.

Fluorine is a mineral which at low levels increases hardness of teeth enamel, but is of primary concern because of possible toxicity problems (see Ch. 14, Vol. 2). It was generally thought that dairy cattle could tolerate up to 30 ppm F without adverse effects (Shupe et al, 1972), but Hillman et al (1978) showed damaged teeth and bone lesions in cows fed rations of 15 to 20 ppm F. Cornell workers (Krook and Maylin, 1979) also reported F toxicity on Cornwall Island in cattle exposed to <40 ppm for prolonged periods (3 mo. of age to adult). Rations with <30 ppm (now required in Michigan) often require special processing of phosphate supplements. This results in more expensive mineral supplements.

Methods of Feeding Minerals

Free choice feeding of minerals has long been practiced by dairymen. Some think that if an array of minerals are offered, cows will consume according to needs. However, studies by Coppock et al (1972) and Pamp et al (1977) showed no relationship between the amount of minerals needed by cattle and free choice consumption.

Minerals should be force-fed by adding to the concentrate or forage or to a complete ration, which are consumed in relation to the animal's needs. Hence, the chance of mineral shortages or excesses for individual cows is minimized. Silages are ideal carriers of minerals needed in dairy rations and certain compounds (such as limestone) beneficially affect some silage fermentations. When cattle are grazing, there is usually no way to force-feed minerals, so free choice feeding is often the only alternative.

VITAMINS

Vitamin Needs for Lactation

Vitamins A, D and E must be furnished in the diet of ruminants. Dietary requirements for protein, energy and most minerals vary directly with milk production, but this is not entirely true for vitamins A, D and E (NRC, 1978). Even though these vitamins are secreted in milk (particularly A), there are no obligatory concentrations in milk and levels fluctuate in accordance with what the cow consumes. Hence, cows grazing early spring grass may produce milk that is 20X higher in vitamin A than later in the season.

It was generally throught that maintenance needs for these vitamins are sufficient for lactation and reproduction. However, the NRC recommended allowances for dairy cattle show slight increases during lactation and the last two mo. of gestation as compared to that suggested for growth.

A multitude of factors affect availability of vitamins A, D and E. Hartman et al (1976) injected vitamins A, D and E into 957 cows from 9 New York herds at drying-off and at freshening. The data showed no beneficial effect in 26 recorded measures of production, reproduction efficiency or herd health. Even though research has not clearly demonstrated a need for routine supplementation of these vitamins to dairy cattle rations, the addition of amounts equal to recommended daily allowances (7,000 IU A, 700 IU D, and 25 mg E/100 kg BW) is generally recommended. This practice provides cheap insurance against possible deficiencies.

Variations in Forages

Most of the commonly used forages are adequate in vitamins A, D and E if conditions are ideal, but many factors affect their availability from the forage. Some are: excessive heating during ensiling which destroys the biological value of A and E; storage of hay over winter, which reduces vitamin A to as little as 15% of original activity. Hay losses are accelerated if heating occurs due to baling too wet. Acidic conditions protect A and E and losses are minimized in well-managed silages. Sun-curing of hay increases vitamin D activity due to light irradiation.

Vitamin E Prevents Oxidized Flavors in Milk

Some regions of the USA have reported severe problems with oxidized flavors in milk. Maryland workers (King et al, 1967) showed that these flavors are prevented by feeding cows high levels of vitamin E (ca. 1 g/day). Even though the practice is expensive, it has been recommended for some milk markets during seasons when oxidized flavors are strong, and some major feed companies are now marketing dairy feeds with high enough concentrations of vitamin E to help control off-flavors in milk.

Niacin

Effects on Ketosis. Waterman et al (1972) reported an alleviation of symptoms of ketosis after oral dosing of cows with 160 g nicotinic acid in an 8-h period. Subsequently, a report from the same laboratory (Fronk and Schultz, 1979) showed that daily doses of 12 g niacin in early lactation cows until the milk acetone test was negative resulted in increased milk yields and blood glucose and decreased blood ketones.

Effects on Milk Yields. Michigan (Gubert and Huber, 1979) and Kansas (Bartley et al, 1979) workers showed that 6 g/day of niacin added to natural protein rations for milking cows resulted in about a 7% increase in milk yields; cows fed NPN did not respond to niacin. In the Kansas study, niacin stimulated intake of grain which explains in part the milk production response. In the Michigan study, intake tended to be higher for the niacin groups, but differences were not significant. An enhancement of rumen microbial protein synthesis was also attributed to the niacin additions (Bartley et al, 1979). Further research will be needed, but if a consistent benefit in milk yields is established, even though small, the low cost of added niacin would probably result in higher profits.

FEED ADDITIVES

Buffers

Rations high in concentrate and low in forage or of small particle size have been associated with metabolic disorders such as low milk fat, displaced abomasums, off-feed problems and, in extreme cases, lactic acidosis (see Ch. 17, Vol. 1). These disorders are usually accompanied by shifts in rumen fermentation to higher propionic and lower acetic acid with an accompanying drop in rumen pH from a normal of about 6.5 to the range of 5.0 to 6.0 (when milk fat % decreases) and on down to as low as 4.0 (in acidosis).

Addition of buffering agents such as Na or K bicarbonate, Mg oxide, dried whey or Na bentonite to high-grain, fat-depressing rations fed dairy cows have alleviated the butterfat depression in milk. The minerals (including those in whey) and bentonite act through restoring a more normal rumen fermentation. Also, Mg enhances mammary uptake of fatty acids from blood. The lactose from whey increases rumen butyrate and mammary uptake of β-OH butyrate.

Kentucky workers (Erdman et al, 1978) recently showed that 1.5% $NaHCO_3$ aided cows in adjusting to a 60% concentrate, 40% corn silage ration during the first 8 weeks of lactation. The buffer increased DM intakes and milk fat percentages. Adding MgO (0.8%) was not as effective as $NaHCO_3$; $NaHCO_3$ and MgO were no more effective than $NaHCO_3$ alone.

A recent report supported the usefulness of $NaHCO_3$ in adaptation of cows in early lactation to high energy rations (Kilmer et al, 1979). Cows receiving 0.7% of their total ration (60% corn silage, 40% concentrate) as $NaHCO_3$ consumed about 20% more DM during weeks 1 and 2 postpartum and produced significantly more milk (31.6 vs 28.7 kg/day) than controls.

Limestone and other Ca sources added to high corn silage-corn grain rations buffers the post-ruminal digestive tract (as indicated by higher fecal pH values), resulting in an improved starch digestibility. Practical benefits of increased milk production resulted from the limestone additions (Wheeler and Calvert, 1979). However, great variability in effectiveness of the different limestones has been shown. This appears to be related to particle size (Wheeler, 1979).

Methionine Hydroxy Analog (MHA)

This precursor to methionine was reported to increase milk fat percent and improve milk yields in high producing lactating cows. Others failed to show a response from methionine addition to lactating rations. Chandler et al (1976) reported studies from herds of 7 universities where cows were fed 12.5 and 15% protein with and without MHA and showed the only significant effect of MHA was to increase butterfat percent in cows on the low protein ration. More recent studies from several stations (unpublished data) showed that MHA consistently increased fat percent and FCM yields when added to high concentrate rations early in lactation.

Organic Iodine Compounds

Ethylene-diaminedihydroiodide (EDDI) is fed to dairy cows for prevention of footrot, improvement of breeding efficiency and other health-related reasons. More than 50 mg/day of EDDI results in an excess of I passing into milk, possibly causing jeopardy to humans drinking that milk (Hemken, 1979). Levels of as high as 300 mg I/quart of milk (ca. 20X the recommended maximum) have been reported from individual farms where excessive use of iodinated compounds occurred.

To avoid excessive I in milk, lactating cows should be treated with organic iodide only if absolutely necessary for controlling footrot. Treatment levels should not exceed 50 mg/day. Non-lactating animals can receive up to 400 mg/day if needed (Hemken, 1979).

Iodinated casein (thyroprotein) has been used at times to stimulate milk production (see Ch. 4). Full lactation data show that the additive does not improve milk yields over controls. Schmidt et al (1971) fed 10 g thyroprotein daily after 50 days postpartum and noted a 7% increase in milk for the first 2 weeks, but from 5-10 mo. of lactation, milk was lower than controls. Treated cows had lower total milk, body weights and general condition, but higher heart and respiration rates, services per conception and calving intervals. Because of its transient effects, thyroprotein is not recommended for lactating cows.

FEEDING PREGNANT DRY COWS

Dry Period Needs

The dry period serves for rest of cows and regeneration of mammary tissue. This period is necessary for a continuance of milk synthesis at optimum levels. The length of time a cow should remain dry has been questioned. Butcher (1980) reported that cows dry 20-30 days produced 421 kg less milk during lactations just prior to and after being dry than cows dry for 50-60 days. Cows dry 90 days or more were 184 kg lower than the 50-60 day group.

Nutrient needs for the dry cows are greatly reduced compared to those for even moderate milk yields. Assessment of body condition when turned dry in order to adjust energy intakes to achieve the desired fatness at parturition is suggested. A safe guide is to have cows in good flesh at calving, but not "patchy" fat. According to suggested allowances (NRC, 1978), energy needs for gestation during the last 2 mo. of pregnancy (while the cow is dry) are about 30% higher than maintenance; protein needs are increased 80%, Ca 60%, P 40% and vitamin A is not changed. If allowed to satisfy their appetites with high energy rations, most dry cows will grossly overconsume compared to their needs. Unfortunately, this is what happens on many dairy farms.

Problems with Overfattening

Extremely fat cows (Fig. 9-7) are very susceptible to ketosis, parturient paresis, mastitis, displaced abomasum, retained placentas and poor appetite (see Ch. 12, Vol. 2). The "fat cow syndrome", characterized by hyperlipemia and fatty livers, results in decreased resistance to pathogenic challenges because of an impaired synthesis of white blood cells (Morrow et al, 1979). Table 9-16 lists metabolic diseases encountered in cows calving on one dairy herd over an 8-mo. period. Approximately 1 year previous to the recorded dates, a reproductive problem occurred in this herd which increased the dry period of many cows. Cows which were dry for long periods had a tendency to become excessively fat. Also, energy furnished the herd prior to making ration adjustments was about 50% above that recommended by NRC. The mean white cell count for 8 cows with fat cow syndrome sampled just before ration changes were made

Table 9-16. Disease conditions and losses from the fat cow syndrome in a 600-cow Holstein-Friesian dairy herd during a 4-mo. period before and after treatment and preventive procedures were initiated.[ab]

		Disease condition				Losses	
Time	No.	Milk fever	Ketosis	Retained fetal membranes	Mastitis	Sold	Died
		%					
Before prevention (2/1-5/31)	120	5	38	62	6	3	25
After prevention	120	2	3	13	2	2	3

[a]From Morrow et al (1979)

[b]Prevention consisted of decreasing intakes of energy and other nutrients from 150% to 100% of the recommended allowance (NRC, 1978) during late lactation and while dry. The white blood cell count from 8 cows diagnosed to have fat cow syndrome averaged just 2,800 mm^3; whereas, normal white cell counts are 11,000-13,000/mm^3.

Figure 9-7. Excessively fat cow just prior to parturition on a dairy farm in Michigan.

was about 33% of normal (2,800 vs 8,100 cells/mm^3).

For large breeds, maintenance of the same body weight the cow possesses when turned dry, plus about 70 kg for the calf and placental growth, appears to be a desirable objective. Body condition of a cow is often difficult to determine. Some cows which do not appear excessively fat suffer from the fat cow syndrome. Body weight gains are not always satisfactory for monitoring fattening because of possible changes in body composition or gut fill. However, the experienced dairyman is often the best judge of when a cow becomes too fat. A high percentage of animals which are so obese that ribs are not visible and hipbones barely protrude above the plane of the back will suffer post-parturient problems. Cows which are dry for long periods (>60 days) because of miscalculated breeding dates or cessation of lactation, tend to become excessively fat. Separation of these cows from those in late lactation of normal dry period length is often necessary to avoid overconsumption of energy.

Preparing for Lactation

Before calving, dry cows and prepartum heifers should be fed a ration similar to what is fed the milking cows. Preparation for heifers should take place over 30-45 days. For cows, 14 days is probably sufficient. Concentrate intake should not exceed 1% of BW during this period. In one study ad libitum intake (10-15 kg/day) of concentrate for 21 days prior to calving was detrimental to mature cows and heifers, causing increased mastitis, metritis and less economical milk production compared to groups receiving only forage prepartum (Emergy et al, 1969).

On the day of calving, the cow has little desire to eat. Thereafter, feed should be increased slowly. Increasing concentrate at too rapid a rate will cause digestive upsets. The type of ration fed in early lactation should be similar to that received during the last 2-3 weeks of the dry period. One system suggested for building up energy intake of the fresh cow is to offer daily about 4 kg of concentrate just after calving and increase at 0.7-1 kg/day to a level commensurate with the cows' increasing needs. For most cows it will take 10-14 days to build up to the desired concentrate intake. Adjusting intakes of individual cows under conditions of group feeding is difficult, so close scrutiny of early lactation animals housed in groups is encouraged.

Additional measures are employed by some dairyment to prepare cows for calving. Injections of vitamins A, D and E are often given for protection against possible deficiencies of these vitamins at this critical time. The injection would certainly be cheap insurance if there is danger of the ration being marginal in these vitamins such as would occur when feeding heat-damaged forages. Vaccination of the dry cow to protect the newborn calf against viral diseases (such as the Reo and Corona types) has been quite effective in decreasing calf mortality in herds infected with these diseases (Hymas, 1979). The ideal time for vaccination is when the cow is turned dry. This allows for the build-up of immune titers in the mother's blood before calving. These antibiotics are incorporated into colostrum for protection of the calf.

Most feed companies market a dry and freshening ration, which is higher in fiber, vitamins and certain minerals than the regular herd ration. It is the author's opinion that a special ration is not usually needed and payment of a higher price than for the regular herd ration is not justified.

For prevention of calf death loss, it is important that cows be placed in an individual stall, away from the remainder of the herd just prior to calving (about the time the udder starts to enlarge). Studies by Ferris and Thomas (1976) showed much higher calf mortality (27 vs 8%) when cows were allowed to calve in group housing compared to individual pens. The higher death loss was associated with poorer sanitation and wetter bedding (67 vs 35% moisture) in the group pens.

INFLUENCE OF FEEDING ON MILK QUALITY

Fat

Feeding high-concentrate, restricted-roughage rations to lactating cows results in a marked depression in the fat content of the milk. This decrease has been associated with increases in rumen propionic acid and decreases in acetic acid. Such rations result in decreased mammary uptake of acetate and β-hydroxybutyrate (βHBA) from the blood (Huber et al, 1969). Theories proposed for the decreased fat are: diminished supply of acetate to the udder; less blood fat uptake because higher rumen propionate production increases blood glucose which stimulates fat movement to adipose tissue and diminishes the supply for milk synthesis; and depressed availability of βHBA (Emergy, 1980). The fat depression syndrome resulting from high-grain, low-forage rations probably has multiple causes rather than a single cause.

Feeding of whole cottonseed (2-3 kg/day) increases the fat content of milk and is common practice in western states. Even though some writers have reported a similar effect from feeding toasted soybeans, the results are less consistent than with cottonseed. Direct incorporation of saturated fat into lactating rations raised milk fat content, while unsaturated oils had a depressing effect (Davis and Brown, 1970). Rations very low in protein ($<$10% of the DM) depress milk fat (Huber and Thomas, 1972), but increasing protein above requirements does not raise fat above normal (Huber and Boman, 1966).

Protein

The high concentrate rations which depress milk fat also elevate milk protein but the gain in protein (about 10%) is not as great as the loss in fat (up to 40%). This divergence is even greater from an energetic standpoint. To explain the increases in milk protein on high-grain, low-roughage rations, English workers demonstrated that intraruminal infusion of propionic acid raised milk protein. They concluded that rations which increased rumen propionate elevate milk protein production (Rook and Line, 1961).

As mentioned, several buffering agents alleviate the decrease in milk fat on high concentrate rations. Studies by Yousef et al (1970) showed that the increased milk protein is still obtained even though fat depression is corrected and there is a shift toward lower rumen propionate. These data support those of Hotchkiss et al (1960) who reported an increase in milk protein due to higher energy level which was separate from the increased observed because of low fiber.

Furthermore, low energy intakes will decrease milk protein concentrations despite the grain to forage ratio.

Consumption of large quantities of protein will not raise milk protein content (Foldager and Huber, 1979). However, infusion of amino acids into blood (Yousef et al, 1970) or casein into the abomasum (Vik-Mo et al, 1974) increase milk protein. The mammary gland responds to higher concentrations of certain amino acids than normal rations supply because of rumen degradation of protein.

Lactose and Minerals

Concentrations of these components in milk are more stable than fat and protein and ar not greatly altered by dietary changes. However, rations very low in energy will cause milk lactose to drop and result in milk which measures below the legal standard in freezing point depression (Huber and Boman, 1966).

Glucose is the blood precursor for synthesis of milk lactose, but when glucose levels in blood varied between 45 and 75 mg/dl, milk lactose did not change (Storry and Rook, 1961). In contrast, Boman and Huber (1967) noted an elevation in the lactose-mineral fraction in milk in cows fed grain ad libitum and only 0.9 kg forage/day compared to those fed normal amounts of forage and concentrate. The increase in milk lactose was associated with a rise in blood glucose from 50 to 76 mg/dl.

Milk Flavors

Milk with no distinct natural flavors (a bland taste) is preferable for drinking. A "grassy flavor" will usually occur when cows commence grazing pasture in the spring of the year. This flavor can be practically eliminated if cows are removed from pasture and fed dry feed prior to milking. At start, cows would be turned to grass shortly after the AM milking and returned to dry feed after 30 min. A gradual increase in time of grazing is suggested and by 2 weeks cows can pasture up to 3 h prior to milking.

High Cu levels in milk catalyze milk fat oxidation, resulting in an oxidized flavor. Experiments to reduce susceptibility to oxidized flavors by attempts to bind Cu (Astrup, 1965), or of pteridylaldehyde to inhibit xanthine oxidase (Astrup, 1963) did not increase flavor stability. Oxidized flavors were also increased by feeding 400 mg thyroxine/day for 3 days (Astrup, 1964). Conversely, goitrogens reduced oxidized flavors.

Air contamination (caused by high risers in milk lines), exposure to sunlight or cows eating certain plants (such as onions) also cause off-flavor milk. As mentioned previously, a decrease in oxidative rancidity in milk has been achieved by feeding high levels of vitamin E.

FORAGE SYSTEMS

Quality of forage often makes the difference between profit and loss in a dairy feeding program, so it is imperative that every effort be exerted to maximize nutritional value and palatability in home-grown and purchased forages.

The Hay Crop

Alfalfa is the main hay crop fed to dairy cows in north America, although mixtures of legumes and grasses or just grasses may be predominant in some areas. Maturity at

Table 9-17. Estimated value of milk per acre from hay harvested at different stages of maturity and 2, 3 or 4 cutting system.[a]

	First cutting date		
	May 20 Vegetative	June 1 1/10 bloom	June 20 Full bloom
Number cuttings	4	3	2
Yield hay, T/yr	4.2	5.5	4.1
Milk/acre, kg	5,289	6,288	3,677
Milk value per acre, $ ($11/cwt)	1,282	1,524	891
Advantage over June 20, $	391	633	---

[a]Adapted from Bean et al (1968)

harvest has a profound influence on the quality of the hay crop. For highest milk production and profit to the dairyman, legumes should be harvested at about 10% bloom, perennial grasses (orchardgrass, timothy, ryegrass) just prior to budding and summer annuals (such as sudan and sudax) at 0.9 to 1.3 m in height. Table 9-17 shows a marked advantage in terms of profits and milk/acre of harvesting alfalfa for hay on June 1 (1/10 bloom), compared to earlier or later.

Cubing of alfalfa hay is commonly practiced in the western USA because of ease of handling. Many studies have shown favorable results with dairy cows fed cubes. Anderson et al (1975) reported higher intakes (12.8 vs 10.4 kg/day) and milk yields (23.8 vs 26.4 kg/day), but lower percent butterfat (3.10 vs 3.40%) with cows fed cubes compared to long hay.

To avoid weather damage many dairymen feed haylage as the primary forage. Direct cut haylage undergoes excessive fermentation due to high moisture and is less palatable to cows than the wilted. Wilting haylage to 30-45% DM is recommended. Wilting to above 50% DM results in poor packing and excessive heating. Protein and energy become less digestible and needed vitamins are destroyed. Heat-damaged haylages are readily detectable by increased acid detergent insoluble N (Yu and Thomas, 1976).

Figure 9-8. Harvesting and transport of hay cubes in central California.

Corn and Small Grain Silages

The optimum time to harvest corn for silage is when starch deposition in kernels is complete. This is characterized by the appearance of a small black layer where the kernel attaches to the cob. This stage (33-37% DM) is best for yield of nutrients, preservation in silos and intake by dairy cows (Table 9-18). As noted, milk yields of cows fed the hard dough silage (33% DM) was significantly higher than that of cows fed soft dough (25% DM).

Small grain silages (barley, wheat, rye, oats and mixtures) are lower in energy than corn silage (ca. 60% TDN), but have effectively maintained milk production in trials with lactating cows. Cutting small grains at the dough maturity yields maximum energy/acre, but double cropping often requires harvest in the boot stage for early removal of the crop. Even though field yields were lower at boot than dough stage, acceptability, digestibility and milk production by dairy cows was satisfactory (Polan et al, 1968). Poor results have been obtained from feeding small grain silages harvested in the milk stage of maturity (between boot and dough), but the reason for this is not clear (Huber, 1980).

Pasture and Green Chop

Dairy cows turned to lush, succulent pasture in early spring will generally show a surge in milk yields and a loss of body weight. As pasture quantity and quality diminish, milk production will also decrease, particularly if supplementary energy is not provided to cows. Additional concerns with pasture in intensive dairy systems are: damage to the

Table 9-18. Influence of maturity on the feeding value of corn silage for lactating dairy cows.[a]

Maturity	DM of silage, %	Total DM as ears, %	Milk yield, kg/day	Silage DM* intake, % BW	TDN in silage, % of DM
Soft dough	25	37	17.2	1.95	68.2
Medium dough	30	47	18.4	2.13	68.4
Hard dough	33	51	19.1	2.30	68.0

[a]Huber et al (1965)

*Cows were fed only corn silage (ad libitum) and soybean meal (1 kg/9.2 kg milk).

forage crop due to trampling by cattle; energy expenditure required by animals to harvest the pasture; and off-flavor milk produced by grazing cattle.

In many areas grazing is still the principal method of harvesting forages for dairy cattle. The shortage of fossil fuels will probably cause a return to more pasturing of dairy cows. Additional information on proper management of pastures to maximize nutrient supply and quality under varying conditions of management is needed.

Forage Particle Size

Particle size of forages exerts a profound influence on utilization of nutrients by cows. Forages cut too fine (of average particle size <1 cm), whether hay, haylage, corn silage or small grain silages, exit the rumen more rapidly than larger sizes, and cause a shift in rumen fermentation towards increased propionate and a decreased acetate (Miller et al, 1969). Depressed butterfat (Table 9-19) as well as an increased incidence of displaced abomasums (Coppock, 1974) have been associated with finely-cut forages. An average size of 7-10 mm for forage particles should avoid these problems (Odell et al, 1968), but minimums might be greater for denser feeds. Conversely, silage particles that are too large result in poor packing during fermentation and increase the chances of spoilage.

Table 9-19. Effect of particle size of corn silage on performance of lactating dairy cows.[a]

	Silage particle size	
	Normal, >10 mm	Recut, <3 mm
Milk yield, kg/day	27.2	27.0
Butterfat, %	3.5[b]	3.0[b]
Milk protein, %	3.4	3.4
DM digestibility, %	66.0	70.0
Rumen Ac:Pr, molar %	2.4	2.0

[a]Miller et al (1969)
[b]Normal > Recut ($P < .05$)

FEEDING CONCENTRATE TO THE DAIRY HERD

Grains

Corn is the grain fed in largest quantity in dairy rations in the USA, but there is widespread use of barley in the western states and sorghum grain is common in the south. Wheat does not usually go directly into dairy rations unless prices are low, but large amounts of wheat by-products from the flour milling industry are used. Most studies comparing response of cows to the different grains have found little difference between them when fed on an equal energy basis.

Figure 9-9. Harvest of high moisture corn in Michigan.

Milk yields and butterfat percent of cows fed high-moisture corn (whether ensiled or chemically treated) have been equal to dry shelled corn when fed on an equal DM basis (Clark, 1975). However, when high-moisture corn made up over 50% of the ration DM, some depression in butterfat has been reported. Moreover, totally fermented rations comprised of high-moisture corn and corn silage resulted in lower milk yields than when hay or dry corn were included as a major ingredient (Clark and Harshbarger, 1972). Advantage of high-moisture corn over dry corn are savings in drying costs and less field loss due to earlier harvest in the fall. Fig. 9-9 shows the harvest of high moisture corn on a Michigan farm.

Delivery of Concentrate

It had been presupposed by many dairymen that cows would not enter the milking parlor or milk out effectively if concentrate was not fed during the milking operation. This notion has been proven erroneous and a large proportion of the dairies do not furnish concentrate while cows are being milked. Studies comparing milk production and mastitis of cows fed concentrate in the milking parlor, as opposed to lot feeding before or after milking, have shown superior results when cows are not fed in the milking parlor, once they are adapted to the lot-feeding system (Bath, 1977). However, some dairymen are furnishing a token amount of concentrate in the parlor to improve cow movement during milking.

Processing of Concentrates

Coarse grinding, cracking or rolling are preferred to fine grinding of grains fed in meal form. Even though finely ground materials are more digestible, they cause dust and cake-up while being eaten, resulting in a decreased palatability to animals. Adding molasses or liquid supplements to finely ground grains will ameliorate this problem.

To prevent separation of ingredients having particles of varying sizes and weights, a large proportion of the commercial dairy concentrates are pelleted. Pelleting is particularly beneficial when mixed concentrates are transported in bulk form. Cows respond favorably to pelleted concentrates.

Steam flaking of grains increases feed utilization and gains in fattening cattle (see Ch. 4), but may depress butterfat if added at high levels to lactating rations, although there are some reports to the contrary. Cooking or other heat treatments of grains have also reduced milk fat. However, cooking of whole soybeans prior to inclusion in ruminant rations increases their palatability.

COMPLETE RATIONS

As dairy herds have increased in size, the feeding of complete rations to groups of cows has become a common practice. The main reason for this trend is to save labor, but some nutritionists feel a superior feeding program often results because each morsel of feed consumed by the cow is nutritionally balanced. Supporting the superior nutrition role was the experience of a large dairy of the author's acquaintance which recently adopted a complete ration program. There was an increase of about 1,000 kg/cow in milk yields during the first year after changeover to the complete ration.

Precautions in Preparing Complete Rations

The dairyman will often mix his home-grown silage (corn and/or hay crop) with purchased or home-grown concentrate in a mixer wagon or truck prior to feeding (Fig. 9-10). Hay is fed separately or chopped and incorporated into the mix. Precautions for complete rations are: 1) insure sufficient time for complete mixing, particularly the added minerals, vitamins and protein supplement. This is very important or cows will receive an unbalanced mix which could be detrimental to production or even health. 2) Assure that particle size of forages is not below recommended minimums (7-10 mm). 3) Maintain effective fiber high enough to avoid milk fat depression and associated disorders (i.e., 15-17%), depending on the source of fiber used. 4) Vary the energy concentration of the mix in accordance with milk production of the cows. The generally accepted maximum for percent concentrate mixed with hay crop forage is 60-65%. It would drop to about 50% if forage is predominantly corn silage (because of the grain in silage).

Figure 9-10. Preparation of a complete ration for feeding dairy cows in southern California.

Grouping of Cows Receiving Complete Rations

Grouping cows according to stage of lactation and milk production has worked successfully for many dairymen (Fig. 9-11). A practice often used is to place unbred animals in one group which include all cows up to 60 days after calving. Other cows are divided according to milk production. Dry cows and springing heifers should comprise an additional group. Where facilities permit (in larger herds), first- and second-calf heifers would be separated from mature cows to allow feeding for extra growth. The nutrient concentrations of complete rations fed the different

Figure 9-11. Group-fed cows in a large dairy in southern California. Courtesy of D.C. Church.

groups is according to group needs. Energy and protein are usually supplied at 10-15% in excess of the group's average requirement to accommodate the higher producers of the group. Complete rations fed to cow groups might vary from 0-65% concentrate, depending on the type of forage and needs of cows. Allowing sufficient bunk space and keeping feed before cows at all times often determines the success or failure of a group feeding operation. Feeding grain in the parlor to cows housed and fed in groups is no longer considered necessary, but cow movement may be improved if grain is available as a coaxing device.

Where facilities do not permit physical separation of cows, the use of magnet feeders, electronic gates or transponder feeding units allows consumption of extra concentrate by high producers (Fig. 9-12). Magnet feeders are usually the simplest and most economical of the three, and are widely used on dairy farms in the USA. Milk yields and profits can be boosted by these systems which provide extra feed to high producers. However, if they are not well-managed by controlling cows which have access to them, they might prove quite unprofitable.

NUTRITIONALLY RELATED PROBLEMS WITH DAIRY COWS

This section lists nutritionally related problems which have not been discussed in sufficient detail in other sections. They are only three of many which the dairyman is continually facing. What seems ironic is that many of these problems (such as ketosis and infertility) become more apparent in the best herds of highest milk production, perhaps because of greater stress on the cows. More

Figure 9-12. Operation of a magnet feeder on a dairy farm in Michigan.

detail on some of these problems is given in Ch. 18 or in Vol. 2 of this series.

Ketosis

Ketosis is a metabolic disease occurring more frequently in high than low producing dairy cows. It is due to the cow's failure to adjust to the heavy drain of nutrients placed upon her in periods of high milk yields, particularly early lactation. This is when energy output usually exceeds energy intake, regardless of the quality of the ration fed.

Ketosis is not usually fatal to cows, but causes a great loss in milk production and profits. Many of the cows in good herds will undergo a subclinical form of ketosis (with blood ketones up to 10 mg/dl and glucose down to 40 mg/dl), but most recover spontaneously without a severe loss of milk or appetite. During this period the fat % in milk is extremely high.

In cows suffering clinical ketosis, the onset of the disease is gradual as is its disappearance after treatment. If the disease is allowed to go unchecked, fatty livers may develop; this problem is often irreversible.

Normal recommendations for good feeding and management best protect cows against ketosis, but several precautions suggested by Schultz (1968) are: avoid excessive fatness at calving time; increase concentrates gradually after calving; cows should be fed some concentrate before calving; avoid abrupt ration changes up to 6 weeks after calving; provide sufficient fiber in ration to optimize energy intake (15-17%); use propylene glycol for prophylaxis and treatment of problem herds. Weekly milk ketone tests (from 1-6 weeks postcalving) will tell which cows are ketotic; maximize feed intake with a ration balanced for protein, vitamins and minerals in which palatability factors (texture, sweetness, etc.) are optimized.

Milk Fever

This metabolic disease occurs around parturition and is due to a failure in the Ca homeostatic mechanism, which is severely stressed by the onset of milk synthesis (see Ch. 12, Vol. 2). The disease is precipitated by a decrease in blood Ca from normal levels of 8.5-11.0 mg/dl down to <5 mg dl. When blood Ca drops to <5 mg/dl, it is usually accompanied by paresis which is fatal unless treated. Treatment consists of IV injection of Ca-Mg borogluconate. Aged cows and certain dairy breeds (Jerseys) are more susceptible to the disease than others. The relationship of Ca, P and vitamin D intakes to incidence of milk fever have been mentioned.

Jorgenson (1974) lists several measures tested which prevent the disease. These are: feeding prepartal diets low in Ca; adjustment of the dietary Ca:P ratio; feeding acidic diets, mineral acids or ammonium chloride prepartum; short-term administration of 90-100 g of calcium chloride daily; feeding massive doses of vitamin D prepartum; prepartum administration of metabolites of vitamin D_3 as mentioned in the mineral section.

Infertility

Rank deficiencies of certain nutrients such as vitamin A, Zn, I, and energy have caused impaired reproductive function, but the effects of excesses, marginal levels of nutrients, or nutritional imbalances are still obscure. It is doubtful (author's opinion) that most dairy rations are sufficiently imbalanced to cause infertility.

Feeding affects certain hormones (i.e. I feeding increases thyroxine), and many of these hormones interact with one another. Some of these interactions have been identified, although most are still to be discovered. Thus, it should not be surprising that many of the effects of nutrition on reproduction are still to be elucidated.

As mentioned previously, feeding NPN according to recommended practices does not decrease reproductive efficiency in dairy cows. However, energy deficiency in high producing cows at the time they should conceive is probably responsible for many reproductive failures. The literature is replete with proof to show that cattle should be in a positive energy balance for effective conception. Solution to the problem is not easy because peak milk production, lowered intakes and breeding often occur simultaneously in the same cow.

Feeding regimes such as inadequate forage (causing more displaced abomasums), high grain prepartum and fattening rations have been associated with increased metritis, which decreases conception rates. Overheating of silages destroys vitamins A and E and may cause infertility in some herds if rations are not corrected for these losses (Huber and Hillman, 1973).

FORMULATING DAIRY RATIONS

Forages are usually fed free choice to dairy cattle so concentrate or grain rations are formulated to provide the needed energy, protein, vitamins and minerals not furnished in the forage. A detailed discussion on ration formulation is given in Ch. 5, so this section will only present examples of formulated rations and important guidelines to follow for dairy animals.

Manual Calculations

A ration for a 650 kg cow producing 40 kg milk daily is illustrated in Table 9-20. The NE_l, protein, and P are in accordance with suggested allowances, but TDN is higher because of overevaluation of the productive energy in forages by the TDN compared to the NE_l system. Ca exceeds NRC (1978) levels in order to bring the Ca:P ratio to 2:1. Vitamin A is also in excess, mainly because of the high carotene content of the alfalfa.

Guidelines to be followed in planning lactation rations are given. 1) When balancing for groups, use the average milk production plus 3-7 kg/day to calculate requirements; 2) DM intakes vary with milk yields and size of cows, but range from 2-4% of BW. See NRC (1978) and Table 9-6 for estimates of maximum

Table 9-20. Formulation of a ration for a lactating dairy cow.[a]

	Nutrients					
	NE_1 Mcal	TDN kg	CP kg	Ca[b] g	P g	Vitamin A 1000 IU
Suggested allowance[b]						
Maintenance	10.30	4.53	.515	22	18	50
Milk	27.60	12.16	3.280	104	70	--
Total	37.90	16.69	3.795	126	88	50
Provided by forages (DM basis)						
Alfalfa hay (6 kg)	7.80	3.48	1.032	75	14	204
Corn silage (7 kg)	11.13	4.90	.560	19	14	126
Total from forages	18.93	8.98	1.592	94	28	330
Need from concentrate	18.97	7.81	2.203	32	60	--
Concentrate ingredients (DM basis)						
Ground shelled corn (6.41 kg)	13.01	5.14	.641	1	11	6
Soybean meal (3.25 kg)	6.04	2.63	1.592	12	24	0
Dical (0.133 kg)	---	---	---	37	25	0
Limestone (0.080 kg)	---	---	---	32	--	0
Total from concentrate	19.05	8.79	2.233	82	60	6
Total furnished	37.98	17.77	3.825	176	88	336

[a]650 kg BW, producing 40 kg of 3.5% milk. Total DM consumed = 22.79 kg or 3.51% of BW. Crude fiber = 16.9% of DM. Also add 0.7% trace mineralized salt to this ration.
[b]Ca increased over suggested allowance to give Ca:P ratio of 2.0.
[c]Suggested allowance and feed composition according to NRC (1978).

intakes; 3) DM intakes from roughages normally vary from 1.4 to 2.0% of BW, depending on quality. To estimate roughage intake for a specific ration, see NRC (1978; page 54). 4) For first and second lactations add 20 and 10%, respectively, to maintenance requirements to allow for growth. 5) Crude fiber content of the ration should be at least 15% to prevent problems of displaced abomasums and depressed butterfat. 6) Ca:P ratio should be between 1.8 and 2.3:1 for lactating cows and between 1.5 and 2.0:1 for dry cows. 7) The dry cow should receive principally forage. According to NRC (1978), energy (NE_1), protein, Ca and P needs are increased 30, 80, 40 and 40% above the maintenance allowance. However, these might be varied depending on the condition and history of the cow.

Formulation of concentrates will vary in accordance with the forages fed. Table 9-21 lists several rations containing varying amounts of alfalfa hay and corn silage (with or without NPN treatment). The rations are balanced for 27.2 kg of milk, which would be about 4.5 kg higher than the DHIA herd average.

Computer Balancing of Rations

At a nominal fee through their own State Cooperative Extension Service there are now available to most dairymen systems for computer formulation of dairy rations. Bath (1975) presents a good discussion on the use of computers in dairy cattle feeding.

These computer programs are designed to formulate rations of least cost. In some states a forage testing service is closely linked to the ration formulation program so as to enable dairymen to build a ration specific to the feeds on hand at a particular time. Generally, a contact by the farmer to his County Agricultural Agent is all that is necessary to start on the program. Similar services are also available to dairymen through certain feed companies and private consultants. In fact, almost all of the feed companies by economic necessity, employ a least-cost, computerized

Table 9-21. Feeding programs using varying amounts of excellent quality alfalfa (18.4% protein) and corn silage. Balanced for 27.2 kg milk/day (600 kg cows).[a]

Item					with 4.5 kg urea (2.5 kg NH_3) per ton corn silage		
Forage per cow per day							
Hay, kg	11.8	9.1	6.8	4.5	4.5	2.3	0
or							
Haylage, kg	21.0	16.0	12.0	8.0	8.0	4.0	0
Corn silage, kg	0	14.0	18.0	23.0	23.0	25.0	30.0
Concentrate mix (kg/1000 kg)							
Shelled corn	895	795	732	677	813	626	566
Soybean meal	89	185	248	300	161	200	254
Oats	--	--	--	--	--	150	150
TM salt	6	7	7	7	7	7	6
Dical phosphate	10	12	12	12	14	13	12
Limestone	--	--	--	5	5	5	10
Protein, %	12.2	15.7	17.4	20.0	14.8	16.5	19.0
Protein, % of total ration DM	14.3	14.3	14.7	15.2	15.3	15.7	16.0
Concentrate/cow/day for 27.2 milk, kg	10.4	8.2	8.2	8.2	8.2	8.6	9.0

[a]Adapted from Hillman et al (1976)

system for formulating commercial concentrate mixes.

Howard et al (1968) evaluated performance of cows receiving two least-cost rations (a least-cost constant ration based upon the previous year's prices and a least-cost variable ration which was reformulated biweekly). Both contained up to 1% urea and a conventionally formulated control ration with no urea. No significant differences in milk yields or other production parameters were noted between the rations, but there was a trend towards lower production on the least-cost variable ration, perhaps because of the abrupt changes made biweekly. Feed costs were reduced on both least-cost regimes. About 60% of the decrease in feed costs were attributed to the incorporation of urea.

Literature Cited

Anderson, M.J., G.E. Stoddard, C.H. Mickelsen, and R.C. Lamb. 1975. J. Dairy Sci. 58:72.

Astrup, H.W. 1963. J. Dairy Sci. 46:1425.

Astrup, H.W. 1964. Husdyrbruhsmøtet. Kontoret for Landbruksforskning, Oslo.

Astrup, H.W. 1965. Meldinger fra Norges Landbrukshøgskole 44:26:1.

Barr, G.W. 1970. Ph.D. thesis. Kansas State Univ., Manhattan.

Bath, D.L. 1975. J. Dairy Sci. 58:226.

Bath, D.L. 1977. Hoard's Dairyman, Nov. 25, 1977, p. 1349.

Bartley, E.E., J.S. Bartley, D. Riddell, and A.D. Dayton. 1979. Kansas State Univ., Dept. Anim. Sci. Ann. Rept. NC-115 (Exp. 2), Manhattan.

Bartley, E.E. and C.W. Deyoe. 1975. Feedstuffs 47(30):45.

Bean, B., J.W. Thomas, D. Hillman, and E.J. Benne. 1968. Michigan State Univ., Dept. Dairy Sci. Mimeo No. 169.

Boman, R.L. and J. T. Huber. 1967. J. Dairy Sci. 50:579.

Bouchard, R. and H.R. Conrad. 1973. J. Dairy Sci. 56:1276.

Braund, D.G. 1979. Large Dairy Herd Management. Univ. Presses of Florida, Gainesville.

Braund, D.G., K.L. Dolge, R.L. Goings, and R.L. Steele. 1978. Pat. No. 4, 188, 513. U.S. Patent Office, Washington, D.C.

Braund, D.G.and R.L. Steele. 1971. 2nd Progress Rept. Cooperative Res. Council, Ithaca, NY.
Bringe, A.N. and L.H. Schultaz. 1969. J. Dairy Sci. 52:465.
Britt, D.G. and J.T. Huber. 1975. J. Dairy Sci. 58:1666.
Bull, L.S., W.G. Helferich, T.S. Hollenshade and T.F. Sweeny. 1975. Proc. 13th Conf. on Rumen Function, Chicago, IL.
Burroughs, W.D., D.K. Nelson and D.R. Mertens. 1975. J. Animal Sci. 41:933.
Butcher, K.R. 1980. Dairy Processing Center, NC State Univ., Raleigh.
Carstairs, J.A. 1978. Ph.D. thesis. Michigan State Univ., East Lansing.
Chalupa, W., C.A. Baile, C.L. McLaughlin and J.G. Brand. 1979. J. Dairy Sci. 62:1278.
Chandler, P.T. et al. 1976. J. Dairy Sci. 59:1897.
Chandler, P.T. and C.E. Polan. 1972. J. Dairy Sci. 55:709 (abstr).
Clark, J.H. 1975. Proc. 2nd Internat. Silage Res. Conf. Nat'l. Silo Assn., Inc., Cedar Rapids, IA. p. 205.
Clark, J.H. 1978. Univ. of Ill. NPN Research Data (personal communications).
Clark, J.H. and K.E. Harshbarger. 1972. J. Dairy Sci. 55:1474.
Clark, J.L. et al. 1970. J. Animal Sci. 31:961.
Clay, A.B., B.A. Buckley, M. Hashbullah and L.D. Satter. 1978. J. Dairy Sci. 61 (Supp. 1):170 (abstr).
Conrad, H.R. and J.W. Hibbs. 1979. J. Dairy Sci. 62 (Supp. 1):79 (abstr).
Coppock, C.E. 1974. J. Dairy Sci. 57:926.
Coppock, C.E., R.W. Everett and W.G. Merrill. 1972. J. Dairy Sci. 55:245.
Davis, C.L. and R.E. Brown. 1970. In: Physiology of Digestion and Metabolism in the Ruminant. Oriel Press, Newcastle-Upon-Tyne, England.
Dutrow, N.A., J.T. Huber and H.E. Henderson. 1974. J. Animal Sci. 38:1304.
Edwards, J.S. and E.E. Bartley. 1979. J. Dairy Sci. 62:732.
Emery, R.S. 1980. Michigan State Univ. Dairy Sci. Dept. Mimeo.
Emery, R.S., H.D. Hafs, D. Armstrong and W.W. Snyder. 1969. J. Dairy Sci. 52:345.
Erb, R.E. et al. 1976. J. Dairy Sci. 59:656.
Erdman, R.A., R.L. Botts, R.W. Hemken and L.S. Bull. 1978. J. Dairy Sci. 61 (Supp. 1):172 (abstr).
FAO Production Yearbook. 1976. FAO, Rome, Italy.
Ferris, T.A. and J.W. Thomas. 1976. Michigan State Univ. Agr. Expt. Sta. Res. Dept. No. 271.
Flatt, W.P., C.E. Coppock and L.A. Moore. 1965. Eur. Assoc. Anim. Prod. Publ. 11:121.
Flatt, W.P., P.W. Moe and L.A. Moore. 1969. Eur. Assoc. Anim. Prod. Publ. 12:123.
Foldager, J. and J.T. Huber. 1979. J. Dairy Sci. 62:954.
Fronk, T.J. and L.H. Schultz. 1979. J. Dairy Sci. 62:1804.
Gardner, R.W. and R.L. Park. 1973. J. Dairy Sci. 56:385.
Gardner, R.W., R.A. Zinn and R.A. Gardner. 1975. J. Dairy Sci. 58:777 (abstr).
Grieve, D.G., G.K. Macleod and J.B. Stone. 1974. J. Dairy Sci. 57:633 (abstr).
Gubert, K. and J.T. Huber. 1979. J. Dairy Sci. 62:(Supp. 1):78.
Hartman, D.A., R.P. Natzke and R.W. Everett. 1976. J. Dairy Sci. 59:91.
Hemken, R.W. 1979. Hoard's Dairyman, Nov. 25, p. 1473.
Hemmingway, R.G., J.J. Perkins and N.S. Ritchie. 1972. Feed and Farm Supplies 69:4.
Hibbs, J.W. and H.R. Conrad. 1960. J. Dairy Sci. 43:1124.
Hillman, D., D. Bolenbaugh and E.M. Convey. 1978. Michigan State Univ. Agr. Exp. Sta. Res. Rept. No. 365.
Hillman, D. and A.R. Curtis. 1980. J. Dairy Sci. 63:55.
Hillman, D. et al. 1974. Michigan State Univ. Ext. Bul. 412, E. Lansing.
Hillman, D. et al. 1976. Basic Dairy Cattle Nutrition, 2nd ed., Michigan State Univ. Ext. Bul. E-702. Coop. Ext. Serv.
Holter, J.B., N.R. Colovos and W.E. Urban, Jr. 1968. J. Dairy Sci. 51:1403.
Hotchkiss, D.K., N.L. Jacobson and C.P. Cox. 1960. J. Dairy Sci. 43:872.
Howard, W.T. et al. 1968. J. Dairy Sci. 51:595.
Huber, J.T. 1973. Dairy Science Handbook, Vol. 6, Agriservices Foundation. Clovis, Calif.
Huber, J.T. 1980. (in press). Nutrition Handbook Series, Vol. III. CRC Press, W. Palm Beach, Fla.
Huber, J.T. and R.L. Boman. 1966. J. Dairy Sci. 49:395.
Huber, J.T., R.L. Boman and H.E. Henderson. 1976. J. Dairy Sci. 59:1936.
Huber, J.T., H.F. Bucholtz and R.L. Boman. 1980. J. Dairy Sci. 63:76.
Huber, J.T. and R.M. Cook. 1972. J. Dairy Sci. 55:1470.

Huber, J.T., R.S. Emery, J.W. Thomas and I.M. Yousef. 1969. J. Dairy Sci. 52:54.
Huber, J.T., J. Foldager and N.E. Smith. 1979. J. Animal Sci. 48:1509.
Huber, J.T. and D. Hillman. 1973. Dairy Science Handbook, Vol. 6, Agriservices Foundation, Clovis, Calif., p. 103.
Huber, J.T. and J.W. Thomas. 1971. J. Dairy Sci. 54:224.
Huber, J.T., J.W. Thomas and R.S. Emery. 1968. J. Dairy Sci. 51:1806.
Huber, J.T. et al. 1967. J. Dairy Sci. 50:687, 1241.
Hume, I.D., J. Moir and M. Sommers. 1970. Aust. J. Agr. Res. 21:283.
Hymas, Theo. 1979. Personal communications.
Jones, G.M., C. Stephens and B. Kensett. 1975. J. Dairy Sci. 58:689.
Jorgensen, N.A. 1974. J. Dairy Sci. 57:933.
Julien, W.E., H.R. Conrad, J.E. Jones and A.L. Moxon. 1976. J. Dairy Sci. 59:1954.
Kendall, K.A., K.E. Harshbarger, R.L. Hays and E.E. Ormiston. 1966. J. Dairy Sci. 49:720.
Kertz, A.F. and J.P. Everett, Jr. 1975. J. Animal Sci. 41:945.
Kilmer, L.H., L.D. Muller and P.J. Wangsness. 1979. J. Dairy Sci. 62 (Supp. 1):231 (abstr).
King, R.L., F.A. Burrows, R.W. Hemken and D.L. Bashore. 1967. J. Dairy Sci. 50:943.
Krook, L. and G.A. Maylin. 1979. Cornell Vet. 6:(Supp. 8):70.
Kwan, K. et al. 1977. J. Dairy Sci. 60:1706.
Larsen, J.B. et al. 1974. 25th Ann. Mtg. Eur. Assoc. Anim. Prod., 27 pp.
Lichtenwalner, R.E., J.P. Fontenot and R.E. Tucker. 1973. J. Animal Sci. 37:837.
Little, W. and R.M. Kay. 1979. Animal Prod. 29:131.
Majdoub, A., G.T. Lane and T.E. Aitchison. 1978. J. Dairy Sci. 61:69.
Mehrez, A.Z. and E.R. Ørskov. 1976. Proc. Nutr. Soc. 35:40 (abstr).
Mielke, C.D. and D.J. Schingoethe. 1979. J. Dairy Sci. 62 (Supp. 1):74 (abstr).
Milk Facts. 1979. Milk Industry Foundation, Washington, D.C.
Miller, C.N., C.E. Polan, R.A. Sandy and J.T. Huber. 1969. J. Dairy Sci. 52:1955.
Miller, E.L. 1973. Proc. Nutr. Soc. 32:79.
Moe, P.W., H.F. Tyrrell and W.P. Flatt. 1971. J. Dairy Sci. 54:548.
Morrow, D.A., D. Hillman, A.W. Dade and H. Kitchen. 1979. J. Amer. Vet. Med. Assoc. 174:161.
Mugerwa, J.S. and H.R. Conrad. 1971. J. Nutr. 101:1331.
Murdock, F.R. and A.S. Hodgson. 1978. J. Dairy Sci. 61 (Supp. 1):183 (abstr).
NRC. 1978. Nutrient Requirements for Dairy Cattle. Nat. Acad. Sci., Washington, D.C.
O'Dell, G.D., W.A. King and C.W. Cook. 1968. J. Dairy Sci. 51:50.
Palmquist, D.L. and T.C. Jenkins. 1980. J. Dairy Sci. 63:1.
Pamp, D.E., R.D. Goodrich and J.C. Meiske. 1977. J. Animal Sci. 45:1458.
Polan, C.E., T.M. Starling and J.T. Huber. 1968. J. Dairy Sci. 51:1801.
Prebluda, H.J. 1974. Professional Nutritionist, Fall, p. 6.
Pritchard, D.E. et al. 1972. J. Dairy Sci. 55:995.
Randel, P.F. et al. 1975. J. Dairy Sci. 58:1109.
Reid, J.T. et al. 1957. J. Dairy Sci. 40:610.
Rook, J.A.F. and C. Line. 1961. Brit. J. Nutr. 15:109.
Ryder, W.L., D. Hillman and J.T. Huber. 1972. J. Dairy Sci. 55:1290.
Satter, L.D. and R.E. Roffler. 1975. J. Dairy Sci. 58:1219.
Satter, L.D. and L.L. Slyter. 1974. Brit. J. Nutr. 32:199.
Satter, L.D., L.W. Whitlow and K.A. Santos. 1979. Proc. Distill. Feed Res. Council. 34:77.
Schmidt, G.H., R.G. Warner, H.F. Tyrrell and W. Hansel. 1971. J. Dairy Sci. 54:481.
Schultz, L.H. 1968. J. Dairy Sci. 51:1133.
Schwab, C.G., L.D. Satter and A.B. Clay. 1976. J. Dairy Sci. 59:1254.
Sebaugh, T.P., A.G. Lane and J.R. Campbell. 1970. J. Animal Sci. 31:142.
Shupe, J.L., A.E. Olson and R.P. Sharma. 1972. Clin. Tox. 5:195.
Smith, N.E. et al. 1980. J. Dairy Sci. 63 (Supp. 1):153 (abstr).
Spahr, S.L. 1977. J. Dairy Sci. 60:1337.
Sparrow, R.C. et al. 1973. J. Dairy Sci. 56:664 (abstr).
Storry, J.E. and J.A.F. Rook. 1961. Biochim. et. Biophys. Acta 48:610.
Swanson, E.W. 1960. J. Dairy Sci. 43:377.
Thomas, J.W. 1971. J. Dairy Sci. 54:1629.

Tyrrell, H.F. 1979. Ann. Rept. at NC-115 Conf. (Nutrition of the High Producing Cow), Manhattan, KN.
VanHorn, H.H., C.F. Foreman and J.E. Rodriguez. 1967. J. Dairy Sci. 50:709.
VanHorn, H.H. and D.R. Jacobson. 1971. J. Dairy Sci. 54:379.
VanHorn, H.H. et al. 1975. J. Dairy Sci. 58:1101.
Vik-Mo, L. et al. 1974. J. Dairy Sci. 57:869, 1024.
Waterman, R., J.W. Schwalm and L.H. Schultz. 1972. J. Dairy Sci. 55:1447.
Wheeler, W.E. 1979. Feedstuffs, October 1, 1979, p. 31.
Wheeler, W.E. and C.C. Calvert. 1979. Abstr. 71st Ann. Mtg., Amer. Soc. Animal Sci., p. 418.
Wickersham, E.W. and L.H. Schultz. 1963. J. Dairy Sci. 45:544.
Wilson, G., F.A. Martz, J.R. Campbell and B.A. Becker. 1975. J. Animal Sci. 41:1431.
Word, J.D. et al. 1969. J. Animal Sci. 29:786.
Yousef, I.M., J.T. Huber and R.S. Emery. 1970. J. Dairy Sci. 53:734.
Yu, Y. and J.W. Thomas. 1976. J. Animal Sci. 42:766.

Chapter 10—Feeding and Nutrition of Young Calves

by D.C. Church, A.D.L. Gorrill and R.G. Warner

INTRODUCTION

General management practices (housing, environmental control), health care and feeding of dairy calves are often neglected by dairy managers, particularly in dairies which do not raise their own replacement animals. Generally, more emphasis is placed on the milking herd, since inadequate management and nutrition shows up immediately in the quantity and quality of milk produced.

Baby calves have more exacting needs than older animals and they are more susceptible to pneumonia and other respiratory diseases and to a variety of microorganisms which may cause severe diarrhea, morbidity and death. Consequently, death losses may be quite high at times. One report indicated that average losses in New York ranged from 16 to 27%, losses being higher in lager herds (Hartman et al, 1974). Fortunately, information is at hand to allow adequate performance and limited mortality if the available information is put into practice.

Nearly all of the information that has been developed on calves has been obtained on young dairy calves. Only a few scattered reports have been published on beef calves or on buffalo calves (Asian sources). The reason, of course is that fluid milk, in most economies, brings a high enough price so that it is usually cheaper to feed some or all of the dairy calf's diet in the form of a milk substitute (milk replacer). Consequently, information has been developed to gain a better understanding of nutrient requirements and on utilization of feed ingredients fed in or along with milk replacers.

Because of the relatively high price which may be obtained for fluid milk, the trend in developed countries in recent years has been to restrict the feeding of milk and/or milk replacer and to wean at very young ages. Many calves have been weaned successfully at 3-4 weeks of age, provided management practices have been good.

Information on digestion and absorption of nutrients, nutrient requirements and nutrient deficiencies of preruminants (young ruminant animals that have not yet developed normal rumen function) has been covered in detail in Ch. 10 of Vol. 2 of this series. Additional information is available in reviews such as those by Appleman and Owen (1975) and Roy and Stobo (1975). Other publications providing nutritional and management information include those of Woelfel and Gibson (1978) and Hoard's Dairyman (1979).

NUTRIENT REQUIREMENTS

Recent reviews of nutrient requirements of calves include that in Ch. 10 of Vol. 2 and by the NRC (1978). The recommendations of the NRC are presented in Appendix Table 16.

The energy and protein requirements of young calves are dependent on the presence or absence of a functional rumen as shown in Table 10-1, since rumen function lowers efficiency of utilization of high quality dietary ingredients such as casein and lactose from milk or milk replacers. In addition, dietary requirements of protein will be affected by digestibility and quality and by content of NPN compounds. Rate of growth will, of

Table 10-1. Estimated energy and protein requirements of a 50 kg calf.

	Requirements		
		Maint. + daily gain	
Nutrient	Maintenance	500 g	1,000 g
Digestible energy, Mcal			
Preruminant			
Jacobson (1969)	2.33	3.99	5.65
NRC (1978)		4.01	
Ruminant			
Jacobson (minimum)	2.42	4.49	6.55
Roy (1964)	3.34	5.79	8.24
NRC		5.42	
Protein, g			
Preruminant			
DCP* (Jacobson)	31	113	195
Crude protein (NRC)		180	
Ruminant			
DCP (Jacobson)	50	135	220
DCP (Roy)		175	275
Crude protein (NRC)		198	

*DCP = digestible crude protein

course, have a marked effect on nutrient needs.

Energy requirements suggested from several sources are given in Table 10-1. The rate of gain will, of course, have a marked effect on requirements. The NRC publication, unfortunately, does not give tabular values for different rates of gain for calves of a given weight fed liquid diets.

The optimum protein:energy ratio is probably dependent on several factors such as rate of gain (and amount of fat deposition) and on age (see section on milk feeding). Jacobson (1969) calculated that calves should receive a digestible protein (g):digestible energy (Kcal) ratio of 1.35 and 1.29 for gains of 500 and 1,000 g/day, respectively. Donnelly and Hutton (1975, 1976) fed calves milk-based diets with 16-32% CP and sufficient energy (17.4 or 21.6 MJ) for gains of 610 or 830 g/day. They observed that increasing CP increased carcass water and protein and decreased fat content. The reverse effect was observed from feeding the higher energy level. On the lower energy level growth peaked at 22-25% CP and on 25-32% CP on the higher energy level. Thus, these data and others suggest for replacement animals growing at moderate rates that 20-22% CP should be sufficient.

Although it is known that preruminant calves require the usual essential amino acids (except arginine), very little information is available on calves. Likewise, the requirements for essential fatty acids are not known. The needs for the various minerals and vitamins are known to some degree for milk-fed calves (see Ch. 4, 9, 10 of Vol. 2), but with less certainty for those fed milk replacers. Recommendations for milk replacers are given in Appendix Table 16.

COLOSTRUM

Colostrum, the first postpartum mammary secretion, is considered essential for survival of the newborn calf under most circumstances. The composition of colostrum differs markedly from normal whole milk (Table 10-2). Colostrum contains a high concentration of proteins, particularly immunoglobulins. With each subsequent milking there is a rapid decline in total proteins, dropping from about 14 to nearly 4 g/100 g by the 4th milking.

Table 10-2. Comparative composition of colostrum (first 24 h after calving) and of Holstein milk.

Constituent	Colostrum	Milk
Fat, %	3.6	3.5
Non-fatty solids, %	18.5	8.6
Protein, %	14.3	3.25
Casein, %	5.2	2.6
Albumin, %	1.5	0.47
β - lactogolbulin, %	0.80	0.30
α - lactalbumin, %	0.27	0.13
Serum albumin, %	0.13	0.04
Immune globulin, %	5.5-6.8	0.09
Lactose, %	3.10	4.60
Ash, %	0.97	0.75
Ca, %	0.26	0.13
Mg, %	0.04	0.01
K, %	0.14	0.15
Na, %	0.07	0.04
P, %	0.24	0.11
Cl, %	0.12	0.07
Fe, mg/100 g	0.20	0.01-0.07
Cu, mg/100 g	0.06	0.01-0.03
Co, μg/100 g	0.5	0.05-0.06
Mn, mg/100 g	0.016	0.003
Carotenoids, μg/g fat	24-45	7
Vitamin A, μg/g fat	42-48	8
Vitamin D, IU/g fat	0.9-1.8	0.6
Vitamin E, μg/g fat	100-150	20
Thiamin, μg/100 g	60-100	40
Riboflavin, μg/100 g	450	150
Nicotinic acid, μg/100 g	80-100	80
Pantothenic acid, μg/100 g	200	350
Vitamin B_6, μg/100 g		35
Biotin, μg/100 g	2-8	2.0
Vitamin B_{12}, μg/100 g	1-5	0.5
Folic acid, μg/100 g	0.1-0.8	0.1
Ascorbic acid, mg/100 g	2.5	2.0
Choline, mg/100 g	37-69	13

Colostrum is an excellent source of vitamins, particularly A and E. The relatively high intake of vitamin A by colostrum-fed calves is generally considered to provide additional protection against an invasion of disease organisms. Colostrum is generally considered to have a laxative effect on calves, but Roy (1964) reported the opposite effect. The rate of passage of the meconium was less in calves fed colostrum than in those fed normal milk.

A number of recent publications have summarized the literature on colostrum immunity and its impact on the health of the neonatal calf (Metzsar, 1978; Bush and Staley, 1980; Roy, 1980; Stott, 1980). The following is a brief distillation from these papers.

The newborn calf derives little immunity from its dam and is born essentially devoid of protection from indigenous microbial challenges. This is in contrast to the human, who is born with passive immunity. Colostrum contains an ample supply of immune globulins (Ig), vis., IgM, IgG and IgA, which have been synthesized by the dam in response to environmental challenges. These intact proteins can be absorbed through the intestinal microvilli, transported via a tubule system to the basal cell and released into the lymphatic and, hence, the vascular system. A non-heat coagulable colostral whey protein appears to be necessary for maximum absorption. This absorption provides temporary immunity for the calf. Absorption essentially stops by 24 h and Ig is no longer released into the lymphatic system. This process is not abrupt, beginning shortly after birth but proceeding most rapidly from 12 to 24 h postpartum.

Colostrum deprived calves usually die, and it is clear that the better the level of immunity attained the better the chances for survival. Management is a factor. In herds with a low mortality rate, fewer calves with low Ig levels die than in herds with high mortality rates. The amount of Ig actually consumed during the first few hours of life is crucial. It is suggested that the calf should receive 300-400 g of Ig during the first 24 h for complete protection. This level usually can be provided by 7 kg of colostrum fed during the first 24-30 h of life in 4 equal feedings, with the first feeding given as soon as the calf can be coaxed to drink. Other recommendations suggest 2 feeds of first milk colostrum fed by the first 4-14 h of life such that the intake equals 10% of the calf's live weight (see Fig. 10-1).

It has been shown that bacteria become established within 5 to 8 h in the small intestine and, if present prior to colostrum ingestion, will impair Ig absorption. Good management practices other than timing and adequate intake have been effective in increasing blood Ig levels. Suckling and keeping the calf with the dam increases serum Ig (Stott, 1980). For calves that won't nurse, stomach tubing is an effective technique. Recently, the use of a lyophylized colostral immunoglobulin preparation has been shown to improve the health of calves when used prophylactically or therapeutically (Brownstein et al, 1979).

A recent study (Brignole and Stott, 1980) utilizing 983 calves, clearly shows the vagaries

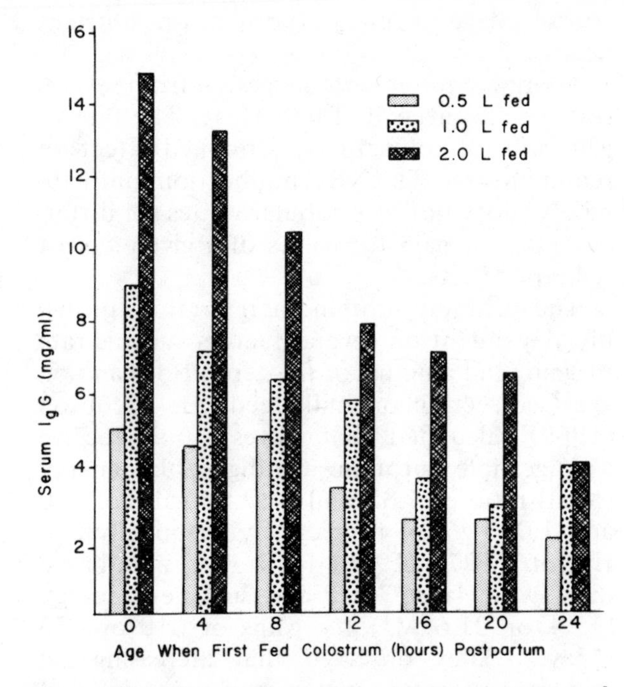

Figure 10-1. Effect of time after birth and amount of colostrum fed on serum Ig in young calves. From Stott et al (1979).

one sees when interpreting Ig absorption and response. About 42% of the calves failed to absorb appreciable Ig when left to nurse their dam for 24 h. After separation from their dam, they were fed 1 ℓ of pooled colostrum during the next 24 h. Most calves showed an increase in Ig following the second day's feeding, but about 12% did not. Of these, 13% died compared with 3.9% of all calves. Overall, the Ig levels of all the calves which died were two-thirds that of those which survived. This study clearly suggests (a) allowing calves to nurse their dam deprives 40% of the calves adequate immunity; thus hand feeding seems warranted; (b) feeding colostrum on day two helps improve blood levels in many calves and also fosters a respectable level of livability even among calves that fail to absorb much Ig.

Several simple field tests have been developed which are helpful in identifying those calves which failed to absorb adequate Ig. While these tests certainly are useful in identifying management problems, there is serious doubt that skillful calf feeders would be willing to pay a premium for identifiable colostrum-fed calves, as they seem capable of managing over the deficiency.

USE OF SURPLUS COLOSTRUM

The use of excess colostrum for calf feeding has long been advocated as a means of conserving salable milk. Modern cows produce from 40-50 kg (first 6 milkings), which is far more than their own calf can consume. Colostrum averages 14-18% solids and on a solids basis is equal to about 56 kg of whole milk, or enough to feed one calf for 4-5 weeks. Assuming the bulls are sold, most of the heifers could be raised on colostrum alone.

Foley and Otterby (1978) have summarized the extensive research on the characteristics, value and procedures for utilizing excess colostrum as fresh, frozen, refrigerated or fermented feed for calves. Freezing preserves it indefinitely, but requires special facilities and handling. Cold ambient temperatures are effective for a time, but since fermented colostrum is an excellent feed, some souring on warm days is in no way objectionable. The product can be fermented in plastic containers and is wholesome for a month or longer. In warm temperatures, a preservative will help prevent putrefaction, i.e., 0.7% acetic acid or 0.1% formaldehyde (all wt/wt).

If fed on an equal solids basis, colostrum is fully as effective as milk (Table 10-3), produces no more (or less) diarrhea and is usually accepted well. It is usually fed diluted from 1:1 to 1:3 with warm water to facilitate feeding and, perhaps, calf acceptance. It is a very effective means of utilizing an otherwise unsalable and highly nutritious product.

Table 10-3. Daily gain and starter consumption of calves fed colostrum or milk.[a]

Treatment	Daily gain, kg 0-30 days	Feed intake, kg		Water consumed, kg/day
		Liquid	Starter	
Fed 8% of BW				
TM-8[b]	0.13	3.23	0.15	
UM-8[b]	0.14	2.85	0.16	
C-8[c]	0.13	2.94	0.13	
N-8[d]	0.10	3.24	0.12	
Fed 10% of BW				
TM-10[b]	0.09	4.16	0.13	2.5
UM-10[b]	0.11	4.08	0.15	2.4
C-10[c]	0.18	4.04	0.17	2.7
N-10[d]	0.13	3.84	0.16	2.4

[a] From Keys et al (1980); calves were weaned at 28-30 days of age.
[b] TM, fermented mastitic milk from antibiotic-treated cows or UM from cows not treated with antibiotics.
[c] Normal colostrum diluted 1:1 with water.
[d] Fresh whole milk.

MILK FEEDING

Feeding whole milk is usually expensive when it is continued past the normal colostral period (first 6 milkings). Consequently, if whole milk is fed, it is usually limited in the amounts fed. In some areas it might be feasible to feed skim milk, but not so in most developed countries. In some undeveloped countries it is still the practice to allow calves to suckle for part of the day, either during the night or daytime, depending on when the owner prefers to milk. Information on feeding liberal amounts of whole milk and/or skim milk may be obtained from Roy (1970).

The milk requirements for calves to gain at different rates have been reviewed by Radostitts and Bell (1970) and Appleman and Owen (1975). When fed at a level of 8-10% of body weight/day, this amount will usually allow a daily gain of 0.3-0.4 kg/day to 3 weeks of age. A 45 kg calf requires about 4.8 ℓ of whole milk to gain about 0.5 kg. Larger amounts of milk may produce faster initial gains, but the response will often not be apparent by 12-16 weeks of age (Appleman and Owen, 1975).

Most studies indicate that satisfactory growth of calves can be achieved by feeding a total of 50 to 100 kg of whole milk if good quality, palatable starter is available from about 1 week of age. However, Holstein heifer calves have been fed as little as 30 kg of whole milk and weight gains to 15 weeks were 600 g/day (Gorrill, 1972). Calves fed lower levels of milk consumer more starter and, therefore, compensate for the lower nutrient supply from milk.

Feeding all calves of the same breed the same dialy amount of milk, regardless of size, will produce results equal to or better than feeding a certain % of body weight. The amount of milk fed can be increased at about 2 weeks of age and maintained constant thereafter (Table 10-4). Since small calves probably are physiologically and/or chronologically younger at birth than larger calves of the same breed, then he smaller calves should require as much or more total milk to supply their nutrient requirements before they could utilize solid feed to the same extent as heavier calves.

Table 10-4. Feeding schedule for Holstein calves fed milk or milk replacer once a day.[a]

Age, days	Whole milk, g	Milk replacer		
		Milk, g	Milk replacer, g	Water, g
2 - 4	1200 (twice)	1200 (twice)	--	--
5 - 7	2400	1200	200	400
8 - 14	2400	--	400	1400
15 - weaning	3500	--	600	1800

[a]From Gorrill (1972)

Whole milk is, of course, an excellent feed for young animals. During the first 2-3 weeks of life their protein requirement is high and milk provides an abundance of highly digestible protein of high quality. However, with increasing age, the N retention (of calves fed on whole milk) decreases. Roy (1964) showed that there was a reduction in N retention from 65 to 50% and a decrease in biological value from 78 to 63% between 4 and 10 weeks of age. These changes, in part, are a reflection of lower protein requirements/unit of energy supplied in the milk and changes occurring with age.

There is limited evidence that milk proteins can be improved by addition of methionine and, in some cases, by lysine (Robert, 1971; Butkevichene, 1973; Williams and Smith, 1975), although the practicality of this remains to be seen. It has also been demonstrated that weight gain could be increased and that milk protein was used more efficiently when glucose was added to whole milk milk. Increasing DE/g of digestible CP increased daily gains from about 825 to 1 kg/day, but higher levels of glucose or maltose caused severe diarrhea (Lister and Lodge, 1973; Lodge and Lister, 1973). Other work has shown that dilution of whole or skim milk with water tended to depress calf gains, although there was no effect on efficiency of energy utilization (Marshall and Smith, 1971).

Overall, milk supplies an adequate to excess amount of protein, lipids and of Ca and P. Milk is low in Mg (relative to Ca and P), and of the trace minerals, in Co, Cu, Fe and Mn. I and F may be low. Vitamins A and E are related to concentration in the dam's diet while vitamin D is proportional to the fat content of milk.

Normally, the young calf would receive enough of these limiting nutrients from colostrum to last until it starts to consume forage or other dry feed. Deficiencies are apt to occur only when the preruminant animal remains on a liquid diet for a longer time than nature intended.

Young calves will usually start to nibble on dry food by the time they are 7-10 days of age, especially if milk consumption is limited. Consumption is very low at first and is usually influenced by the amount of milk the calf receives and by the opportunity to eat dry feeds of a palatable nature.

MILK REPLACERS

A tremendous amount of research effort has been put into the development of milk replacers. They have been developed to the point that calf performance is essentially equal to that on whole milk and cost of production is considerably less. Consequently, milk replacers have been well accepted in most areasof North America and Western Europe. Reviews on this topic include those of Appleman and Owen (1975; general); Huber (1975), Ramsey and Willard (1975), Stobo and Roy (1978; proteins), and Raven (1970; fats). In addition, many individual research papers are available, of which only a few will be mentioned here.

One example of a milk replacer is shown in Table 10-5. This particular one contains only protein from milk. In one experiment with Holstein bull calves, the level of feeding was such that a 50 kg calf received 840 g air dry diet/day. Apparent N and gross energy digestibilities were 89 and 95%, respectively. Average weight gains over a 6-week period were 465 g/day.

In North America very few calf feeders prepare their own milk replacer since it requires careful handling of micro ingredients and because fats must be hemoginized well to produce an acceptable product. Nevertheless, some comments are in order on common ingredients used in current practice.

Proteins

Dried Skim Milk. Dried skim milk is the most widely used source of protein in milk replacers. Although good quality protein is derived from skim milk, the processing of milk to form the powder can have a marked effect on calf performance. For optimal calf performance the milk must be spray-dried,

Table 10-5. Composition of milk replacer for replacement calves.[a]

Ingredient	Amount
Skim milk, spray-dried	37.9 kg
Fat premix[b]	33.0 kg
Cerelose	25.3 kg
Mineral premix[c]	363 g
Vitamin premix[d]	697 g
Dicalcium phosphate	1.5 kg
Cobalt-iodized salt	1.0 kg
Aurofac-10	250 g
Dry matter	94.8%
Total fat, calculated	10 %
Crude protein, DM basis	20.9%
Gross energy, DM basis	4.58 Kcal/g

[a] Adapted from Gorrill and Nicholson (1969)

[b] Homogenized and spray-dried mixture of 30% stabilized lard (choice white grease) plus lecithin, and 70% skim milk powder.

[c] The premix contains: (in grams) MgO, 300; Na_2SeO_3, 0.055; $Al_2(SO_4)_3 \cdot 18 H_2O$, 26.4; H_3BO_3, 0.925; $Na_2MoO_4 \cdot 2H_2O$, 0.925; NaBr, 1.84; $CrCl_3 \cdot 6H_2O$, 1.0; $Fe_2SO_4)_3 \cdot nH_2O$, 22.0; ZnO, 2.5; $MnSO_4 \cdot H_2O$, 4.4; $CuSO_4$, 3.1.

[d] The premix contained: (in grams) vitamin A (325,000 IU/g), 3.1; vitamin D_3 (200,000 IU/g), 1.0; vitamin E (551 IU/g), 7.5; thiamin hydrochloride (USP), 1.0; riboflavin (53 mg/g), 18.8; niacin (feed grade), 2.0; pyridoxine hydrochloride (USP), 1.0; cobalamine (3,000 μg B_{12}/g), 1.33; pantothenic acid (66 mg/g), 38.0; choline chloride (50%), 600; menadione, 25.0; folic acid, 0.48.

using preheating temperatures of 77°C or less for a short period of time (15 sec). High, prolonged temperatures cause denaturation of the whey proteins, reduction of ionizable Ca, release of SH groups, poor clotting ability by rennet and reduced digestibility. The normal clotting of milk in the abomasum of the calf is altered, and greater quantitiesof undigested casein reach the upper small intestine. Heat damage of the milk can reduce calf weight gains by up to 30% during the first 3 weeks of life. There is a faster rate of build-up of infection in the calf barn, and a higher incidence of diarrhea and mortality (Roy, 1970; Emmons et al, 1976; Lister and Emmons, 1976).

Dried Whey. Whey is a relatively cheap source of proteins, minerals and vitamins. Although some reports suggest that >20% whey in milk replacers may cause diarrhea and poor performance, whey has been used as the only protein source in milk replacers (Toullec et al, 1969, 1971; Stewart et al, 1974; Volcani and Ben-Asher, 1974). Some information suggests that acid whey should be neutralized for optimal performance (Gorrill and Nicholson, 1972). Raven (1972) has shown that dried whey has a high digestibility and biological value for calves, although less than skim milk, when whey supplied half of the total protein.

Dried Buttermilk. Buttermilk may be used as part of or the entire protein source in liquid calf diets under certain conditions. Buttermilk is the only milk by-product of sufficiently high fat content (ca. 9%) to give nearly normal calf growth without the addition of extra fat. Davey (1962) reported that calves could be reared satisfactorily on buttermilk and concentrates or pasture, although incidence of diarrhea was higher than for calves fed whole milk. In new Zealand, veal calves fed dried buttermilk reconstituted to the same energy level as whole milk gained as well from 3 weeks to 91 kg as those fed whole milk (Fraser, 1961). High temperatures used to dry buttermilk may predispose calves to *E. coli* intestinal infection.

Non-milk Proteins. A wide variety of non-milk proteins have been tried in milk replacers. These include various soybean products, fish protein concentrate, fish meals of various types, liquified fish waste, single cell protein (bacteria and/or yeast) meat meal, blood meal and others. After reviewing literature on this topic, Stobo and Roy (1978) concluded that many of the non-milk protein sources, when used as partial or complete sources of N, may cause poor growth rates, digestive disturbances and mortality, particularly when fed to very young animals. Provided toxic factors are extracted or amino acid deficiencies rectified, they suggest that, by 3-4 weeks of age, between 20 and 40% of dietary protein can be used safely and that as much as 70% is feasible in some circumstances.

Fats

Dried whole milk from Holstein cows contains about 30% fat and 5.5 Kcal/g of gross energy. Therefore, milk replacers must contain appreciable amounts of fat to simulate milk. Fats which have been successfully used in milk replacers include tallow, lard, coconut oil, palm kernal oil, peanut oil, and hydrogenated marine fat. Calf performance has generally been better when lard rather than tallow was used. Tallow is less digestible and causes a higher incidence of diarrhea during the first 2 weeks. Mixtures of tallow (70%) and coconut oil (30%) result in better calf performance than only tallow (Raven, 1970).

Early trials using cottonseed, soybean or corn oil gave poor results. Hydrogenation of soybean oil improved calf performance, so many of the adverse effects of these oils have been attributed to unstable unsaturated fatty acids and can be prevented by the addition of antioxidants, including higher levels of vitamin E (Raven, 1970). More recently, it has been shown when calves were fed milk replacers with 25% of butter oil, lard or corn oil that fecal DM was lower in calves fed the corn oil (Gaudreau and Brisson, 1978). No doubt some of the early problems have been resolved by improvements in milk replacer formulas.

Although Veen (1974) has suggested that palm oil or coconut oil are preferable for young calves (<6-9 weeks old) as compared to animal fats, Bouchard (1977) found no difference in performance of calves fed tallow or tallow plus coconut oil. His data also showed that a commercial milk replacer with 21% fat was just as satisfactory (gain, feed efficiency) as milk replacers with 25 or 29% fat. A level of 35% fat resulted in lower performance.

Carbohydrates

The preparation of low-fat and relatively low-protein milk replacers dictates that large amounts of carbohydrates are incorporated as a source of energy and as a filler. Unlike the young of some species (i.e., baby pig), the baby calf can efficiently utilize only lactose and glucose. Starch is of little benefit to young calves and it is poorly digested by older calves. Sucrose is not utilized either because of a lack of enzymes from the gut (Burt and Irvine, 1970; Ch. 10, Vol. 2). When an ultrafiltrate of whey (to supply lactos) was used to replace whey in a milk replacer, Toullec et al (1977) observed that gain was reduced 16% and the incidence of diarrhea increased.

Huber et al (1968) concluded that the maximum level of starch should be 10% in milk replacers fed to calves under 3 weeks of age. Thereafter calf growth was similar on diets containing 9 and 27% starch. Normal growth of calves was obtained when 10% oat flour was incorporated in a milk replacer and fed from 3 days of age. However, the apparent digestibility of the energy in the oat flour was zero at 6 days of age and only 26% at 24 days of age (Radostits and Bell, 1970). Addition of cooked potato flour (7-21%) to milk replacers depressed growth rate of calves 11-25 days of age and performance at weaning (35 days) was similar. Thus, these and other data show that some starch can be utilized from milk replacers by calves about 3 weeks of age. This is an economical factor since starch usually costs less than glucose or lactose.

Other results indicate that a combination of processed starch and an amylolytic enzyme may be used in a milk replacer for very young calves (Morrill et al, 1970). Expanded sorghum grain plus a fungal amyloglucosidase gave a much greater blood glucose response than untreated or steamed grain sorghum plus enzyme (Fig. 10-3). Weight gains of calves fed milk replacers with the enzyme and either 11.8% expanded sorghum grain or 18% pregelatinized starch were similar to calves fed milk replacers containing glucose.

There is some question whether or not starch digestion by calves is due to enzymes produced by the calf or due to bacterial fermentation. Burt and Irvine (1970) postulated that precooked starch was more susceptible to bacterial attack, and the production of organic acids caused malabsorption and

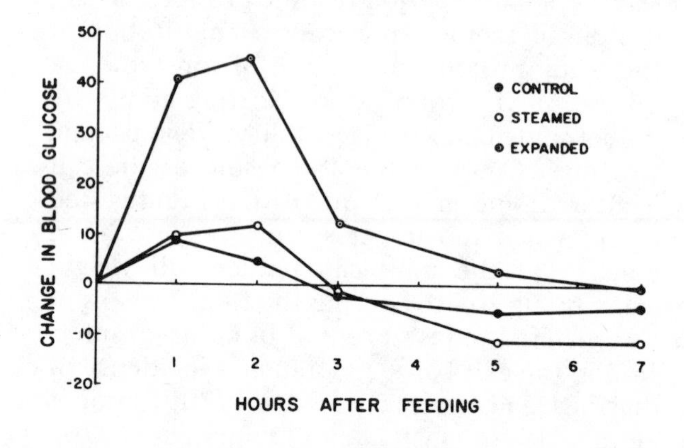

Figure 10-2. Effect of grain processing on availability of starch fed with an amylase to young calves as measured by changes in blood glucose (mg/100 ml) after feeding. From Morrill et al (1970).

diarrhea. Alcoholic fermentation of starch and glucose by yeast can also cause ethanol intoxication of calves (Abe et al, 1971).

Minerals and Vitamins

The need for supplemental minerals and vitamins in milk replacers depends, of course, on the basic ingredients. The content of these nutrients is not nearly so critical when feeding replacement animals as for veal calves since the former are either weaned at an earlier age or are fed solid feed along with the milk replacer, which provides an additional source of these nutrients. Essentially, no recent reports are available on these nutrients in milk replacers. Tabular data on recommended requirements are given in Appendix Table 16.

Antibiotics

Antibiotics are routinely added to milk replacers, although under ideal management and facilities they should not be necessary. Roy (1964) suggested that these conditions could be found only on farms where calves are home bred, colostrum intake is generous, the subsequent milk replacer is of good quality, the building is well ventilated, the number of calves in a unit is small, and the herdsman is experienced. When calves are brought in from unknown sources, antibiotics appear to be a necessity for survival (Lister and MacKay, 1970). The recommended treatment for these calves is 125 mg of tetracycline antibiotics/day for the first 3 weeks if colostrum was received after birth or 250 mg/day for the first 5 days followed by 125 mg/day for the next 16 days as a substitute for colostrum.

The tetracyclines are the only antibiotics which consistently stimulate growth of calves by 10-30%. The increased growth is generally associated with less diarrhea, particularly during the first few weeks of life. Aureomycin or bacitracin added at 23 mg/kg of milk replacer increased daily gains to 7 weeks of age by 34 and 42%, respectively (Lassiter et al, 1958). At 12 weeks of age, gains were increased by 20% by both antibiotics. Spiramycin fed at the rate of 11 mg/day to calves produced results equal to aureomycin fed at twice the level (Lassiter et al, 1963). Aureomycin has been used routinely at a level of 11 mg/kg of milk replacer in research trials with good results (Gorrill and Nicholson, 1969). In contrast with monogastric animals, penicillin seldom shows a growth response in calves.

The development of resistance to antibiotics by gastrointestinal microorganisms has been discussed by Roy (1964). He concluded that routine antibiotic feeding in conventional rearing of dairy herd replacements had not increased the incidence of diseases which were more difficult to control. However, under intensive systems of veal production, there was some evidence that routine antibiotic feeding may result in a decline in the effectiveness of the antibiotic. This may be offset in part by periodically changing the antibiotic fed.

FEEDING METHODS

Liquid vs Dry Feeding

Nutrients fed as a liquid are generally more efficiently utilized by the calf than the same nutrients fed dry. This difference can be attributed to less loss of nutrients by digestion in the gut as compared with rumen fermentation. Consumption of solid feed and rumen development have several advantages which offset the less efficient use of ingested nutrients. These include less risk of some bacterial infections and metabolic disorders, use of cheaper feed ingredients, and lower

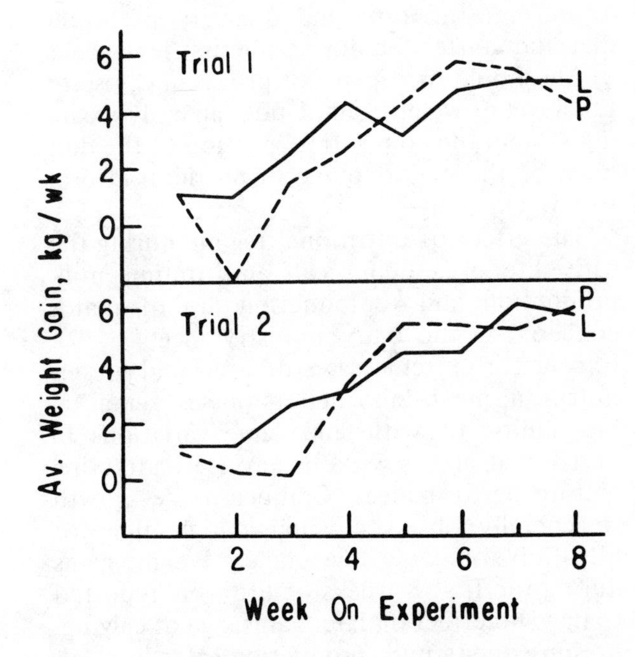

Figure 10-3. Cumulative weight gain of calves fed liquid (L) and pelleted (P) milk replacer. Trial 1 and 2 – milk replacer fed for 3 wk beginning at 1 and 2 wk of age, respectively. From Bush et al (1968).

labor requirements. About 1 kg weight gain/day can be obtained by either a liquid or dry feeding regime. Under these conditions the DM of liquid diets should not be more than about twice the cost of dry foods to be competitive.

Calves fed pelleted milk replacers (7.3 or 12% fat) at 1 or 2 weeks of age lost body weight during the first 1 or 2 weeks and then recovered rapidly (Fig. 10-3). However, liquid feeding produced greater total 8-week gains. When a high energy milk replacer (18% fat) was fed at 2 or 3 weeks of age, there was no difference between pellets and liquid in calf weight gains to 8 weeks of age. However, under an early weaning system, liquid feeding of milk replacer would be expected to give better calf performance with fewer management problems in feeding the replacer.

Greater overall efficiency of nutrient utilization at a reduced cost may be possible by feeding calves a combination of liquid and dry foods as is common in situations where calves are fed a liquid in combination with a dry starter concentrate.

Liquid Feeding Systems

Average daily gains of about 0.5 kg from birth to 8 weeksof age are generally considered satisfactory for dairy herd replacements. Therefore, the type and quantity of liquid diet and starter fed during the first few weeks of life should be capable of producing close to this level of weight gain. Under an early weaning system the concentrate portion of the diet must be the major source of nutrients after 3 to 5 weeks of age.

The effect of nutritional regime during the early life of a heifer calf and lifetime milk production and reproduction are of major concern to the dairy industry (see Ch. 9). Except for extreme cases of over- and undernutrition, most dairy heifers possess remarkable ability to withstand large variations in nutritional planes without adversely affecting mature performance. Compensatory growth is generally observed when a high plane of nutrition follows a low plane. Weight gains during the first 8 weeks of life were reported to have no effect on later gains, age of calving, or subsequent milk production (Martin et al, 1962). These workers suggested that calves should be kept healthy and thrifty during the early weeks of life without great regard for magnitude of gains.

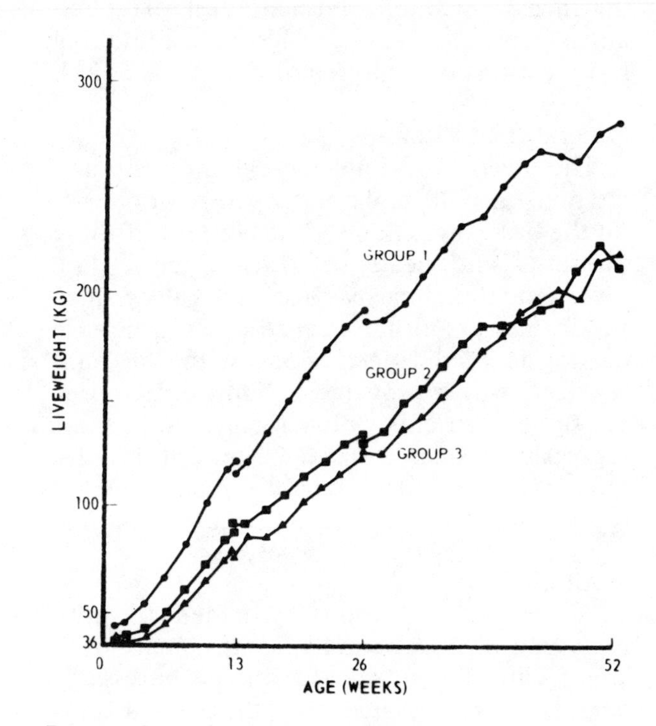

Figure 10-4. Cumulative weight gains of calves fed starter and a liquid diet either sucking the cow (Group 1), a commercial milk replacer (Group 2), or reconstituted skim milk powder (Group 3). All calves were weaned at 13 wk. From Wardrop (1966).

Some studies indicate that underfeeding of calves during the first few weeks of life has a permanent adverse effect on the animals and that compensatory growth does not occur. Results from one experiment are shown in Fig. 10-4. However, the relatively low nutritional plane after weaning (unsupplemented pasture) may not have allowed compensatory growth. Burt and Bell (1962) noted that calves which gained only 225 g/day during the first month of life showed no signs of recovery at 3 mo. of age. More recent work with identical twins suggests that underfeeding calves during the first 16 weeks of life will reduce mature body weights. Calves could tolerate shorter periods of undernutrition (4-8 weeks) with permanent effects (Everett and Jury, 1977).

Milk Replacers

Many research reports have shown that performance of calves was as good or nearly as good when fed a good commercial milk replacer compared with limited amounts of whole milk. Calves fed milk replacer generally appear normal, but may have rougher hair coats than calves fed whole milk during the first month of life.

There appears to be no advantage in feeding milk replacer ad libitum to calves during the first few weeks provided a palatable calf starter is offered free choice. Milligan and Grieve (1970) fed a low-energy milk replacer (6.5% fat) either ad lib during the first 3 weeks and restricted to 200 g/day during the 4th week, or restricted to 570 and 140 g/day during the first 3 and 4th weeks, respectively. Total intakes of milk replacer were 13.4 and 8.7 kg. Weight gains to weaning at 4 weeks were greater with the higher intake of milk replacer (0.32 vs 0.21 kg/day). However, both groups gained nearly 0.7 kg/day to 120 days of age. The calves on the restricted intake of milk replacer consumed more total starter by 4 weeks of age (6.5 vs 4.6 kg).

Once vs Twice Daily Feeding

Calves can be fed the liquid diet only once instead of 2X/day without adversely affecting overall performance (Mitchell and Broadbent, 1973; Galton and Brakel, 1976). The total daily allotment of liquid is given in one feeding, instead of the usual 2 equal feedings. For milk replacer, the amount of water for reconstitution is reduced. Consequently, a ready supply of fresh, clean water must be available to the calves at all times.

Calves fed a liquid diet 1X/day may be weaned at a slightly younger age (based on starter intakes) than calves fed twice (Table 10-6), but the incidence of diarrhea tends to be greater on 1X feeding. Daily starter intakes to weaning generally are greater, consequently feed efficiency tends to be lower (because of less efficient use of the starter). However, this reduced efficiency would be more than offset by the reduced labor requirement for 1X/day feeding. Furthermore, calves can be fed the liquid at any time during the day at the convenience of the feeder. However, calves should be fed at about the same time every day and should be checked at other intervals for any signs of ill health and treated accordingly. Under practical conditions, Williams (1968) suggests feeding liquid diets 2X daily for 2 to 4 weeks and then 1X daily until weaned. However, a more common farm practice is to feed 2X daily for 5 to 7 days and then go to 1X feeding.

Table 10-6. Performance of calves fed 1X or 2X daily.[a]

Item	1X daily	2X daily
No-calves	24	26
Birth weight, kg	40.2	34.3
Feed consumption, 39 days		
Milk replacer, kg	12.8	11.8
Starter, kg	4.9	5.4
Roughage pellets, kg	11.0	8.6
DM/kg gain, kg	2.27	2.04
Av. daily gain, 42 days, kg	0.29	0.28
$kg^{0.75}$	0.80	0.93

[a]From Galton and Brakel (1976)

Skipping Feedings

A further reduction in labor for feeding young calves may be possible by skipping one or more days of liquid diet/week. Omitting the Sunday feeding of liquid diet also facilitates weekend labor, particularly for large calf rearing units.

Results from some studies suggest that some feedings can be skipped for calves over 1 week (Wood et al, 1971) or 5 weeks of age (Boucque et al, 1971). Although this procedure can produce satisfactory results with less milk replacer and more starter consumption, not all calves will adapt to it and a higher level of management is required. Consequently, it is not a recommended practice.

Temperature of Liquid Diets

Attempts have been made to feed calves liquid diets that were lower tha normal body temperature. Reduced temperatures are particularly useful if large amounts of liquid are fed ad libitum, as in veal production. If some method of refrigerating the diet is used, then souring is not as great a problem compared with feeding warm liquids.

Calves are reluctant to drink cold milk from an open pail and some will starve rather than drink. This problem is minimal when the cold liquid is fed in nipple pails. Some studies have shown that voluntary intakes and weight gains were lower when calves were fed cold vs warm milk replacer (Taylor and Lonsdale, 1969). Flipot et al (1972) have shown that calves fed cold milk replacer (1.5°C) had lower gains and were less efficient than those fed replacer at room (18°) or warm temperatures (37°C). Mortality was also higher, particularly if the replacer was

Table 10-7. Effect of fluid intake and dry matter concentration of milk replacer on calves fed 1X daily.[a]

	Fluid intake, % BW/day			DM % in milk replacer		
Item	6	8	10	10	15	20
DM intake, kg/day	0.37	0.51	0.67	0.34	0.53	0.69
Water intake, kg/day[b]	2.33	2.10	2.75	1.57	2.15	3.40
Total water, kg/day	4.39	5.02	6.52	4.65	5.17	6.17
DM, % of water	9.3	10.7	10.7	7.8	10.7	11.9
Daily gain, kg	-0.01	0.13	0.32	-0.07	0.19	0.33

[a]From Jenny et al (1978)
[b]In milk replacer

restricted. Gray et al (1973) found that warm (35°C) allowed higher gains to 5 weeks although room temperature milk replacer (14-17°C) was just as satisfactory for older calves. Boeve and Weide (1972) obtained better performance in 1 of 2 trials with temperatures of 37 vs 18°C. Carcass quality was inferior in calves on 18° milk replacer. Granzer (1979) observed that abomasal pH decreased less after feeding cold milk replacer and attributed this to a reduced consumption.

Undocumented observations suggest that young preruminants do not consume cold milk replacer as rapidly as when it is warm. This should help to reduce nutritional diarrhea. Certainly, cold liquids will not spoil as rapidly and this fact cuts down on labor required for cleaning equipment. Most calf feeders now use milk replacer at room temperature after the first few days of feeding. At times the use of 0.1% formaldehyde has been recommended as this will also retard spoiling.

Solids Content

Normal cow's milk contains about 13% DM. The recommended dilution rate for many commercial milk replacers is about 11-12%. Some research studies suggest that higher performance may be obtained with about 15% DM (Pettyjohn et al, 1963; Gorrill, 1972). Jenney et al (1978) observed that both fluid and DM concentration influenced gain and consumption of DM and water in calves fed 1X daily (Table 10-7). DM concentration was more important in control of diarrhea than fluid intake.

In longer term studies with veal calves, Dammert et al (1972), Giessler et al (1973) and Stobo et al (1979) found that increasing DM from 12.5-14% (after 6-8 weeks) to about 20% resulted in increased DM consumption and daily gain, the values being 17 and 13%, respectively, in the work of Stobo et al. Consequently, the age of the calf and its intended disposition will affect the optimal amount of DM.

Liquid Feeding Devices

The use of open buckets vs nipple feeding has been thoroughly reviewed by Appleman and Owen (1975). Their conclusions are that calves learned to drink in about the same time (3 feedings) and that the major advantage of nipple feeding is a reduction in calves sucking on other calves. Nipple-fed calves consume feed at a slower rate, and nipple feeding is generally only recommended for veal calves which must consume large quantities of liquid diets for optimal performance.

EARLY WEANING OF REPLACEMENT CALVES

It is well accepted by nutritionists that healthy calves can be weaned successfully at 3-5 weeks of age. The advantages of early weaning include lower total feed costs, reduced labor and fewer problems with diarrhea and other digestive upsets. The success of early weaning is dependent on supplying a palatable, nutritionally adequate starter.

Abrupt vs Gradual Weaning

Abrupt weaning is the simplest procedure to follow. The full amount of milk or milk replacer is fed up to the day of weaning and then substituted entirely by water. Growth of calves has not been affected by abrupt weaning at 3 to 5 weeks of age, compared with the same total amount of liquid diet fed

Figure 10-5. Elevated metal calf stalls. Such stalls are favored by calf raisers because they facilitate individual attention and cleaning of the area, and because they help to reduce the spread of disease. Front view shows pails for feeding grain and water. Hind view shows open sides and metal floor slats. From Anonymous (1970).

and weaned gradually for 1 to 2 weeks. Good quality starter and water must be available prior to weaning if setbacks are to be avoided. Calves weaned abruptly will generally consume dry feed more readily after weaning. Some calves will not eat solid feed as long as a liquid diet is fed, and gradual weaning delays the time when the calf is forced to eat solid feed to survive. Some evidence indicates that calves suffer a greater setback when weaned abruptly at about 3 weeks of age (Bakker, 1968), but this has not always been the case.

Basis for Early Weaning

Calves have been weaned on the basis of age, body weight, weight gain, total intake of liquid diet and daily dry feed intake. In practice, a combination of two or more of these criteria may be involved. The simplest method is according to age, particularly where calves are fed in groups. Calves weaned at 21 and 42 days were essentially equal in weight gain at 42 and 84 days (Appleman and Owen, 1975). Weaning caused a sudden increase in dry food intake.

Weaning according to body weight has been recommended since this permits weaning at more nearly equal physiological ages (Gorrill, 1964). However, this requires that calves be weighed frequently or measured with a weigh tape. Weaning on the basis of starter intake has proven to be quite effective but, again, it is applicable only to individually fed calves (Fig. 10-5). It is commonly recommended that calves not be weaned until they are eating more than 1 lb (500 g) of starter/day. If weaning is delayed until consumption reaches 900 g/day, this will delay weaning for about 10 days for most calves with no apparent benefits compared to lower levels of dry feed

Table 10-8. Effect of weaning calves by starter intake and age on performance.[a]

Criteria	Age group at weaning, days		
	15-21	22-28	29-38
Average weaning age, days	18.9	24.4	34.1
Weight gain to weaning			
total, kg	5.6	7.2	11.8
g/day	386	349	378
180-day weight, kg	162	165	164
Feed intakes, kg			
liquid diet dry matter	5.7	8.9	14.3
total starter to weaning	3.8	5.0	6.9
Feed costs to 5 wk, $			
liquid diet, $0.44/kg DM	2.51	3.92	6.29
starter, $0.13/kg	4.20	3.30	1.52
total	6.71	7.22	7.81

[a]From Gorrill (1972)

intake (Leaver and Yarrow, 1972). A combination of starter intake and age has been used by Gorrill (see Table 10-8). This appears to work satisfactorily.

It should be pointed out that all calves don't respond the same and it is reasonable to expect that some should be fed liquids longer than others. Calves in individual pens generally learn to eat dry feed sooner from buckets than troughs, probably because they associate the former with milk. Once/day feeding usually stimulates starter intake. When calves are fed in groups, starter intake can be stimulated by feeding small groups of calves from a community trough, followed by placing starter in the trough as soon as the liquid is consumed. Providing water in a pail close to the starter also improved starter intake.

CALF STARTERS

Calf starters are feed mixtures made up of high quality concentrates which are intended to be eaten in dry form to supplement liquid feed. However, the term is also used to refer to concentrate feed given to early-weaned calves. A growing ration would probably be a more descriptive term for the latter.

The starter must be palatable to encourage consumption at an early age, and it must be a concentrated source of nutrients (low fiber) because of the small rumen capacity of young calves. Whereas nearly all of liquid diets will by-pass the rumen after swallowing, essentially all solid feed will go to the rumen and be subjected to the normal fermentative processes. The advantage in supply feed in dry form is that less expensive ingredients can be used and because labor costs are lower.

The nutrients required in starters will be affected by expected gain, by the amount and quality of ingredients in the liquid diet and nutrient consumption, if any, from various roughage sources. A brief coverage on this topic follows.

Protein Concentration

Data from a variety of trials on early-weaned calves have been reviewed by Kay (1969) and Morrill and Dayton (1978) have briefly reviewed data on younger calves. Analysis of these data and other research reports on this topic indicates that amount and duration of liquid feeding has an effect on response of calves to protein. The responses generally show little difference to (early) weaning when starters contained 12-18% CP, although Bartley (1973) observed that milk-fed calves ate and gained more when fed a pelleted starter with 22% CP vs one with 18% CP. NRC (1978) recommendations are for 16% CP. After weaning, results suggest that 15-16% CP (dry basis) is probably adequate (Leibholz and Kang, 1973; Morrill and Melton, 1973; Lee and McCoy, 1974; Henschel and Radloff, 1975; Winter, 1976; Wallenius and Murdock, 1977; Morrill and Dayton, 1978).

The response of early-weaned calves to protein levels is shown in Table 10-9. Note that feeding restricted amounts of starter increased hay consumption and gave improved performance on higher CP in the starter.

Energy Concentrations

As with protein, an optimum concentration of energy will depend on consumption of liquids. NRC (1978) recommendations for starter are 1.20 Mcal of DE/lb (1.01 Mcal ME or 80% TDN). For early-weaned calves, Kay et al (1970) fed diets with 2.4-3.1 Mcal ME/kg. DM intake and gain increased with each increase in energy.

Little emphasis has been placed on the protein:energy ratios in calf starters and the effect on peformance of early-weaned calves. Chandler et al (1968) varied energy levels in starters from 4.4 to 4.9 Kcal of GE/g (0-8% corn oil) and CP levels from 8 to 30% (dry basis) and fed these to calves from 8 to 18 weeks of age. Analysis of the data indicated that optimum calf response occurred with a starter containing 25% CP and 4.54 Kcal of GE/g. Feed intake decreased as the level of corn oil was increased, except for rations containing high protein levels (21 or 28% soy protein). With calves fed from 8-18 weeks, Traub and Kesler (1972) did not find any appreciable differences when diets were fed which contained protein:energy ratios (g digestible CP/Kcal of DE) ranging from 1:15 to 1:25. More recently, Daniels et al (1979) fed calves from 3-112 days (weaned after 56 days) and concluded that a ratio of CP(g):ME(Kcal) of 1:17 was optimal when compared to 1:14 and 1:23. Pre-weaning and post-weaning performances were the same on the 1:14 and 1:17 ratios.

Table 10-9. Effect of protein level in the starter and level of feeding on growth and feed intake by early weaned calves.[a]

	Ad libitum starter protein level,[b] %			Restricted starter protein level,[b] %		
	21	16	12	21	16	12
Period 5-8 weeks						
Weight gains, g/day[c]	790	790	700	750	790	730
Feed intake, kg						
Starter	38	36	39	33	35	35
Hay	2.5	2.9	3.6	3.4	3.7	3.6
Period 8-12 weeks						
Weight gains g/day[c]	840	850	790	840	810	750
Feed intake, kg						
Starter	79	76	77	56	55	56
Hay	7.7	8.2	7.4	13.1	11.7	10.8

[a]Adapted from Stobo et al (1967)
[b]Air-dry basis
[c]Adjusted for differences between treatment groups in mean intake of starter and body weight at the beginning of the period

Protein and Non-Protein-N Sources

Skim milk and other milk by-products were commonly used when starters were first being developed for calves. For the very young animal (up to 3 weeks), this is probably a good practice as these ingredients are highly digestible and will usually improve palatability in a starter (Bakker, 1968; Morrill and Dayton, 1974).

The current trend is to replace most if not all of the milk proteins with non-milk proteins. This is logical for weaned calves since the high quality of milk protein is negated to a considerable extent by rumen fermentation (see Ch. 13, Vol. 1). In the USA, soybean meal is the most common protein source. Others used widely include sunflower meal, cottonseed meal, peanut meal, fish meal (particularly outside North America), and meat and bone meal and single cell protein (bacteria and yeast). Bakker (1968) reported that calves preferred protein supplements in the following order: soybean meal, linseed meal, dried skim milk, fish meal, and meat and bone meal.

For ruminant calves there is no reason to believe that they cannot use any protein source that is satisfactory for older animals. There are a number of reports on starters including urea. Recent reports on this topic suggest that early-weaned calves can utilize urea, although growth rates tend to be lower as compared to soybean meal as a N source (Miron et al, 1968; Nelson, 1970; Winter, 1976; Morrill and Dayton, 1978). Addition of sulfur has been beneficial in some cases (Morrill and Dayton, 1978), and particularly when meat meal was a primary N source (Liebholz and Naylor, 1971; Liebholz and Kang, 1973).

Miscellaneous Feed Ingredients

In cafeteria-type selection studies, ground grains were preferred by calves in the following order: barley, wheat, rye, corn and oats. However, with whole grains, oats were preferred to barley (Bakker, 1968). When starters contained 75% sorghum grain, wheat or a mixture of the two, no difference was noted in acceptability by young calves but gains and efficiency to 90 days were markedly better for the starter containing wheat (Schuh et al, 1971). When barley was included in starters, steam-rolled barley resulted in improved performance over steam-cooked and flaked barley (Waldern and Fisher, 1978). Dried whey at a level of 10% has been shown to improve starter palatability (Morrill and Dayton, 1974), and sucrose and glucose have been ntoed to improve palatability and weight gains in young calves (Atai and Harshbarger, 1965), although Murphy et al (1970) observed no response to addition of brown sugar. Increased starter intakes have been noted with up to 10% molasses in the mixture, while higher levels may cause scouring (Bakker, 1968).

Added fats of various kinds in starter rations have not been shown to increase feed consumption, gain or efficiency (Miller, 1962; Gardner and Orme, 1971); however acidulated acids appear to give a good response (Waldern and Fisher, 1978).

Antibiotics

The inclusion of antibiotics in starters for early-weaned calves has generally increased feed intake, feed efficiency and growth (Preston, 1964; Roy, 1969). Levels of 20 to 40 mg of chlortetracycline, aureomycin or spiramycin/kg of starter have been used. Preston (1964) suggested that antibiotics in the starter had a protein-sparing effect, since a growth response was obtained only when protein intake was limiting. The greater calf growth response to adding antibiotic to the starter compared with adding it to the milk replacer for early-weaned calves was noted earlier.

Complex vs Simple Calf Starters

During the development stages of starters, a great many varied nutrient sources were included. Simple mixtures have since been formulated to reduce ingredient costs and enable farm mixing. Two examples of complex and simple formulas are shown in Table 10-10. There were no differencesin growth rates of calves fed the simple and complex starters. However, all formulas contained relatively high protein levels, especially those used by Milligan and Grieve (1970).

Texture of Calf Starters

Calves generally prefer starters containing coarse, flaky ingredients (Bakker, 1968; Kay, 1969). Coarsely ground or rolled grains stimulate greater intake at an earlier age than finely ground grain, even if the latter is pelleted. A higher incidence of digestive disorders also occurs with the finely ground ingredients.

Kay (1969) cited an experiment in which calves were fed 3 different starters: (a) coarse-mixed, containing flaked and crushed cereal grains and the protein-mineral-vitamin supplement as a meal; (b) same as (a) except the supplement was pelleted; and (c) a complete pelleted starter. Starter intakes between 55 and 95 kg live weight for calves given the 3 diets were 2.7, 2.4 and 2.1 kg/day, respectively. Growth rates were also greatest on starter (a). Pelleting starters has no beneficial effect on calves other than reduced feed wastage unless a high-roughage complete diet is fed (Gardner, 1967; Bakker, 1968), and it is more likely to result in parakeratosis in older calves, particularly if roughage is not fed (Tamate et al, 1978).

Table 10-10. Simple and complex calf starter formulas.

	Simple		Complex	
	% air dry basis			
	A	B	A	B
Ground corn meal		69.5		19.2
Cracked corn				13.7
Wheat	69.0		60.9	
Crimped oats				10.0
Oat groats			4.1	
Wheat bran				15.0
Alfalfa meal				5.0
Soybean meal	28.0	27.8		15.0
Linseed meal				7.5
Skim milk powder			10.0	
Whey powder				3.8
Fish meal			12.0	
Meat meal			1.2	
Stabilized animal fat			3.0	1.5
Sucrose			5.0	
Molasses			2.4	7.5
Ground limestone	0.5		0.2	0.6
Dicalcium phosphate	1.0	1.4		0.4
Trace mineral mix			0.15	0.05
Zinc sulfate			0.05	
Trace mineral salt		1.0		
Iodized salt	0.5		0.5	0.5
Brewers dried yeast	1.0			
Vitamin B complex mix			0.2	
Vitamin B_{12}, 20 mg/kg			0.1	
Vitamin A, IU/kg	1,760	6,600	4,400	6,800
Vitamin D, IU/kg	350	810	440	6,800
Aureomycin, mg/kg	33			
Terramycin, mg/kg		88	88	50
Crude protein (by analysis), %	23.4	21.2	23.5	19.3

[a]From Milligan and Grieve (1970)
[b]From Miller et al (1969)

ROUGHAGE FOR YOUNG CALVES

Although it is a common practice to provide hay to calves >2 weeks of age (other than veal), research results do not show much effect on performance, although Miller et al (1969) did note improved performance when

10% cottonseed hulls was added to a calf starter. Noller et al (1962) forced calves to consume hay by including it in a starter or giving it ad lib. Calves did as well when forced to eat hay as when given hay free choice. More recently, Murphy et al (1970) gave calves a choice of grass silage or hay and starter. Intakes of silage and hay DM were 49 and 50 lb, respectively, to 12 weeks. There were no differences in gain or DM consumption. Roy et al (1971) point out that veal calves given a good quality milk replacer ad libitum show very little desire to eat roughage an the carcasses from calves fed roughage tend to be inferior.

With early-weaned calves, silage has generally resulted in poorer performance of young calves than hay. Silage DM intakes are usually lower, presumably because of limited rumen capacity, although some of this difference might be due to lower quality of material in the silage. Brundage and Sweetman (1963) concluded that early-weaned calves should not be restricted to silage as the only roughage from 2 to 6 mo. of age. Thomas et al (1959) reported that wilted alfalfa silage plus supplement of hay (1% of body weight) or grain (1 kg/day) resulted in growth of heifers to 2 years of age nearly equal to those fed alfalfa hay. In other research with early-weaned calves, including either ground wheat straw (15%) or alfalfa hay (20%) in pelleted diets improved performance of the calves (Kang and Liebholz, 1973; Liebholz, 1975).

PASTURE FOR CALVES

Calves can be reared on pasture from 1-2 weeks of age (along with other feed) and, with few exceptions, performance has equalled or surpassed indoor rearing (Gorrill, 1964). The labor and inconvenience of hand feeding liquid diets to calves on pasture may be eliminated by feeding cold milk replacer ad libitum from nipples (Lonsdale and Taylor, 1969).

The quantity and quality of grass available is very important since calves are highly selective grazers and tend to starve themselves on unsuitable pasture. Digestibility of herbage DM by calves from 3 to 5 weeks of age has been reported to be about 75% and increased little with advancing age (Preston et al, 1957).

Rearing calves on pasture has generally resulted in reduced consumption and need for concentrates, compared with feeding hay indoors (Gorrill, 1964). If good quality pasture is available, no supplementary feeding is required after about 8 weeks of age, although feeding grain, such as oats, may increase weight gains during fall grazing (Gorrill, 1967).

The time of year when calves are put on pasture has a marked effect on performance. Roy (1969) noted that weight gains decreased as the time of year, when calves were born and put on pasture, advanced from spring to midsummer. Relatively poor growth of spring-born calves has been observed after putting them on pasture at weaning or at 15 weeks of age, compared with calves of the same age put on pasture at 1-2 weeks of age (Gorrill, 1964, 1967). The adverse effects have been attributed to a buildup of parasites on pasture, poorer quality pasture, and hot, humid conditions during midsummer.

VEAL PRODUCTION

There is a marked contrast in the rearing and management techniques required for replacement and veal calves. Replacement calves are fed a minimum amount of liquid diet plus solid feed and fed to attain moderate weight gains. Veal calves are generally fed only a liquid diet, and are fed to attain maximum gains to slaughter weight. Veal calves should gain 1 kg/day or more with a feed conversion of 1.4 to 1.5. Maximum gains of 1.4 kg/day or more are often obtained at the end of the feeding period. Limited amounts of diet should be fed during the first 2 or 3 weeks to reduce digestive upsets, but the calves should be on full feed by 4 weeks (Warner, 1970).

Commercial milk replacers for veal calves generally contain 15 to 25% homogenized fat and produce satisfactory calf growth and carcass quality. Formulas containing $<10\%$ fat, which are suitable for replacement calves, generally produce low weight gains and relatively low grade veal carcasses. Toullec and Mathieu (1969) indicated that fats rich in short-chain fatty acids produced greater growth early in life and fats with long-chain fatty acids at a later stage would enhance fattening of the veal calf. The fatty acid composition of carcass fat can also be altered by feeding liquid diets containing fats or oils with a particular fatty acid composition. Unlike the mature ruminant, the dietary fat by-passes the rumen and is not altered by rumen fermentation.

Milk replacer formulas require higher levels of protein than those recommended for replacement formulas. Weight gains and retention of Ca and N by veal calves were greater on a formula containing 26% protein rather than 19% (Roy et al, 1970). After 8 weeks of age, 20% CP was adequate in trials in Holland.

Although whole milk produces good veal calves, the protein:energy ratio may be too high for maximum calf performance. Lodge and Lister (1970) reported that weight gains of calves to 8 weeks were increased by 37% (877 vs 643 g/day) when 3% cream or an equivalent amount of energy from glucose was added to 3.5% milk. There was no obvious effect on carcass fat and lean. Roy et al (1970) observed that weight gains and N retention were not improved by increasing the fat content of milk replacer from 20 to 30%, but the fat content of the carcasses were higher on the higher fat level.

In many countries, particularly in Europe, "white veal" is considered a luxury meat and commands a high price. To produce white veal the calves must be moderately anemic, fast gainers and 12 to 14 weeks old. Consequently, mineral supplementation, particularly of Fe, of milk replacer formulas for veal calves is critical. Commercial Dutch formulas contain <3 ppm Fe, and the maximum Fe content of water used to feed veal calves should be 0.5 ppm to prevent red coloration of the meat.

The feed costs of veal calves raised on whole milk or high-fat milk replacers are relatively high compared with cereal-based concentrate diets. Gardner (1970) compared the performance and carcass characteristics of veal calves fed whole milk or fat supplemented grain rations (5% tallow) ad libitum. Weight gains to 108 kg live weight and feed efficiencies for calves fed grain or whole milk averaged 1.1 and 0.77 kg/day and 1.2 and 4.7, respectively. Dressing % of the respective carcasses were 55 and 61. The muscle tissue of the grain-fed calves was only slightly darker than for the milk-fed calves, and a taste panel did not detect any differences in veal from the two groups of calves.

MANAGEMENT OF THE YOUNG CALF

Roy has discussed management in detail in his books and various texts on dairy cattle provide considerable detail. Just to briefly summarize procedures, it might be useful to outline the practices used at a large dairy in Arizona, where calf death losses have usually been <2% (Fowler, 1976).

Calves are picked up in a modified golf cart soon after they are born. They are fed (in buckets) colostrum from several cows as soon as possible and to the extent of about 4 ℓ during the first day divided into 2 meals of 2 ℓ each. Iodine solutions are used to sterilize the navel to avoid infections, and the calves are given an oral dose of 10 ml of a commercial antibiotic and are vaccinated for *salmonella.* During the first 4 days of life the calf is fed on a night and morning feeding schedule; colostrum is fed for the first 2 days and whole milk for the 3rd and 4th days. By the 5th day they go on 1X/day feeding using 1 lb of milk replacer in 2.5 quarts of water (i.e., about 2.75 kg of a 15% DM solution). At the 5th day a concentrate mix is placed in front of the calf.

The calves start to eat solid food at an early age because they are hungry part of the time on the 1X daily feeding schedule. The calves are moved from individual stalls to corrals at 30 days of age where they continue to receive the milk replacer for another 10-15 days. Roughage (alfalfa silage or crushed alfalfa cubes) is mixed with the grain at this point. Using these procedures accompanied by special care in identifying and treating sick calves, this large operation (2,400 cows) usually has death losses <2%.

Literature Cited

Abe, R.K., J.L. Morrill, R. Bassette and F.W. Oehme. 1971. J. Dairy Sci. 54:252.

Ackerman, R.A., R.O. Thomas, W.V. Thayne and D.F. Butcher. 1969. J. Dairy Sci. 52:1869.

Appleman, R.D. and F.G. Owen. 1975. J. Dairy Sci. 58:447.

Atai, S.R. and K.E. Harshbarger. 1965. J. Dairy Sci. 48:391.

Bakker, K.J. 1968. World Rev. Animal Prod. 4:34.

Bartley, E.E. 1973. J. Dairy Sci. 56:817.
Betteridge, K., A.W.F. Davey and C.W. Holmes. 1976. N. Zeal. J. Agr. Res. 19:415.
Boeve, J. and H.J. Weide. 1972. Nutr. Abstr. Rev. 42:710.
Bouchard, R. 1977. Can. J. Animal Sci. 57:379.
Boucque, C.V., F.X. Buysse and B.G. Cottyn. 1971. Animal Prod. 13:613.
Brignole, T.J. and G.H. Stott. 1980. J. Dairy Sci. 63:451.
Brownstein, M.T., H.R. Conrad and K.L. Smith. 1979. J. Dairy Sci. 62 (Supp. 1):103.
Brundage, A.L. and W.J. Sweetman. 1963. J. Animal Sci. 22:429.
Burt, A.W.A. 1968. Animal Prod. 10:113.
Burt, A.W.A. and E.O. Bell. 1962. J. Agric. Sci. 58:131.
Bush, L.J., E. Coblentz, R.A. Rosser and J.D. Stout. 1968. J. Dairy Sci. 51:1264.
Bush, L.J. and T.E. Staley. 1980. J. Dairy Sci. 63:672.
Butkevichene, A.A. 1973. Nutr. Abstr. Rev. 43:516.
Chandler, P.T., E.M. Kesler, R.D. McCarthy and R.P. Johnston. 1968. J. Nutr. 95:452.
Dammert, S. et al. 1972. Zuchtungskunde 44:397.
Daniels, L.B. et al. 1979. Nutr. Rpts. Int. 20:107.
Davey, A.W.F. 1962. N. Zeal. J. Agr. Rex. 5:460.
Donnelly, P.E. and J.B. Hutton. 1975. N. Zeal. J. Agr. Res. 19:289.
Donnelly, P.E. and J.B. Hutton. 1976. N. Zeal. J. Agr. Rex. 19:409.
Emmons, D.B. et al. 1976. Can. J. Animal Sci. 56:317, 335, 339.
Everitt, G.C. and K.E. Jury. 1977. N. Zeal. J. Agr. Res. 20:129.
Flipot, P., G. LaLande and M.H. Fahmy. 1972. Can. J. Animal Sci. 52:659.
Foley, J.A. and D.E. Otterby. 1978. J. Dairy Sci. 61:1033.
Fowler, R.G. 1976. Hoard's Dairyman 121:1174.
Fraser, A.J.D. 1961. Animal Prod. 3:1.
Galton, D.M. and W.J. Brakel. 1976. J. Dairy Sci. 59:944.
Gardner, R.W. 1967. J. Dairy Sci. 50:729.
Gardner, R.W. and L.E. Orme. 1971. J. Dairy Sci. 54:802 (abstr).
Gaudreau, J.M. and G.J. Brisson. 1978. J. Dairy Sci. 61:1435.
Giessler, H. et al. 1973. Zuchtungskunde 45:45.
Gorrill, A.D.L. 1964. Can. J. Animal Sci. 44:235, 327.
Gorrill, A.D.L. 1967. Can. J. Animal Sci. 47:211.
Gorrill, A.D.L. 1972. In: Digestive Physiology and Nutrition of Ruminants. Vol. 3. 1st ed. O & B Books, Inc., Corvallis, OR.
Gorrill, A.D.L. and J.W.G. Nicholson. 1969. Can. J. Animal Sci. 49:305.
Gorrill, A.D.L. and J.W.G. Nicholson. 1972. Can. J. Animal Sci. 52:465.
Granzer, W. 1979. Tierernahrung und Futtermit. 41:197.
Gray, H.G., D.L. Pyton and R.S. Hinkson. 1973. J. Dairy Sci. 56:683 (abstr).
Hartman, D.A., R.W. Everett, S.T. Slack and R.G. Warner. 1974. J. Dairy Sci. 57:576.
Henschel, H.L. and H.D. Radloff. 1975. J. Dairy Sci. 58:741 (abstr).
Hinks, C.E., D.G. Peers and A.M. Armishaw. 1974. Animal Prod. 19:351.
Hoard's Dairyman. 1979. Calf Care and Raising Young Stock. Ft. Atkinson, WI.
Huber, J.T. 1975. J. Dairy Sci. 58:441.
Huber, J.T. 1969. J. Dairy Sci. 52:1303.
Huber, J.T.S. Natrajan and C.E. Polan. 1968. J. Dairy Sci. 51:1081.
Jacobson, N.L. 1969. J. Dairy Sci. 52:1316.
Jenny, B.F. et al. 1978. J. Dairy Sci. 61:765.
Kang, H.S. and J. Leibholz. 1973. Animal Prod. 16:195.
Kay, M. 1969. Rowett Res. Inst. Annual Rpt. 25:123.
Kay, M., N.A. MacLeod and M. McLaren. 1970. Animal Prod. 12:413.
Keys, J.E., R.E. Pearson and B.T. Weinland. 1980. J. Dairy Sci. 63:1123.
Lassiter, Ca. A., L.D. Brown and R.O. Thomas. 1963. Michigan Agr. Expt. Sta. Quart. Bul. 41:321.
Lassiter, Ca. A., L.D. Christie and C.W. Duncan. 1958. Michigan Agr. Expt. Sta. Quart. Bul. 45:631.
Leaver, J.D. and N.H. Yarrow. 1972. Animal Prod. 14:155, 161.
Lee, D.D., Jr. and G.C. McCoy. 1974. J. Dairy Sci. 57:651 (abstr).

Leibholz, J. 1975. Animal Prod. 20:93.
Liebholz, J. and H.S. Kang. 1973. Animal Prod. 17:257.
Liebholz, J. and R.W. Naylor. 1971. Aust. J. Agr. Res. 22:655.
Lister, E.E. and D.B. Emmons. 1976. Can. J. Animal Sci. 56:327.
Lister, E.E. and G.A. Lodge. 1973. Can. J. Animal Sci. 53:317.
Lister, E.E. and R.R. MacKay. 1970. Can. J. Animal Sci. 50:645.
Lodge, G.A. and E.E. Lister. 1973. Can. J. Animal Sci. 53:307.
Marshall, S.P. and K.L. Smith. 1971. J. Dairy Sci. 54:1064.
Martin, T.G., N.L. Jacobson, L.D. McGilliard and P.G. Homeyer. 1962. J. Dairy Sci. 45:886.
Metzgar, J.J. 1978. Ann. Rech. Vet. 9:177.
Miller, W.J. 1962. J. Dairy Sci. 45:759.
Miller, W.J., Y.G. Martin and P.R. Fowler. 1969. J. Dairy Sci. 52:672.
Milligan, J.D. and C.M. Grieve. 1970. Can. J. Animal Sci. 50:147.
Miron, A.E., D.E. Otterby and V.G. Pursel. 1968. J. Dairy Sci. 51:1392.
Mitchell, C.D. and A. Broadbent. 1973. Animal Prod. 17:245.
Morrill, J.L., R.K. Abe, A.D. Dayton and C.W. Deyoe. 1970. J. Dairy Sci. 53:566.
Morrill, J.L. and A.D. Dayton. 1974. J. Dairy Sci. 57:430.
Morrill, J.L. and A.D. Dayton. 1978. J. Dairy Sci. 61:940.
Morrill, J.L. and S.L. Melton. 1973. J. Dairy Sci. 56:927.
Murphy, M.J. et al. 1970. Irish J. Dept. Agr. Fish. 67:13.
Nelson, D.K. 1970. Feedstuffs 42(41):42.
Noller, C.H., I.A. Dickson and D.L. Hill. 1962. J. Dairy Sci. 45:197.
Preston, T.R. 1964. World Rev. Nutr. Diet. 4:121.
Preston, T.R., J.D.H. Archibald and W. Tinkler. 1957. J. Agr. Sci. 48:259.
Radostits, O.M. and J.M. Bell. 1970. Can. J. Animal Sci. 50:405.
Ramsey, H.A. and T.R. Willard. 1975. J. Dairy Sci. 58:436.
Raven, A.M. 1970. J. Sci. Food Agr. 21:352.
Raven, A.M. 1972. J. Sci. Food Agr. 23:517.
Robert, J.C. 1971. In: Proc. International Milk Repalcer Symposium. Nat. Renders Assoc. Des Plaines, IL.
Roy, J.H.B. 1964. Vet. Rec. 76:511.
Roy, J.H.B. 1970. J. Sci. Food Agr. 21:346.
Roy, J.H.B. 1970. The Calf. Management and Feeding. 3rd ed. The Pennsylvania State University Press, State College, PA.
Roy, J.H.B. 1980. J. Dairy Sci. 63:650.
Roy, J.H.B. and I.J.F. Stobo. 1975. In: Digestion and Metabolism in the Ruminant. Univ. New England, Armidale, NSW, Australia.
Roy, J.H.B. et al. 1971. Brit. J. Nutr. 26:353.
Schuh, J.D., W.H.Hale and C.B. Theurer. 1971. J. Dairy Sci. 54:801 (abstr).
Stewart, J.A., L.L. Muller and A.T. Griffin. 1974. Aust. J. Dairy Tech., June, p. 53.
Stobo, I.J.F. and J.H.B. Roy. 1978. World Animal Rev. #25, p. 18.
Stobo, I.J.F., J.H.B. Roy and P. Ganderton. 1979. J. Agr. Sci. 93:95.
Stott, G.H. 1980. J. Dairy Sci. 63:681.
Stott, G.H., D.B. Marx, B.E. Menefee and G.T. Nightengale. 1979. J. Dairy Sci. 62:1902.
Tamate, H. et al. 1978. Tohoku J. Agr. Res. 29:29.
Tayler, J.C. and C.R. Lonsdale. 1969. J. Agr. Sci. 73:279.
Thomas, J.W., J.F. Sykes and L.A. Moore. 1959. J. Dairy Sci. 42:651.
Toullec, R. and C.M. Mathieu. 1969. Ann. Biol. Anim. Bioch. Biophys. 9:661.
Toullec, R., C.M. Mathieu, L. Vassal and R. Pion. 1969. Ann. Biol. Anim. Bioch. Biophys. 9:661.
Toullec, R., J.L. Paruelle, J.Y Coroller and J.H. Le Treut. 1977. Ann. de Zootech. 26:29.
Toullec, R., P. Thivend and C.M. Mathieu. 1971. Ann. Biol. Anim. Bioch. Biophys. 11:435.
Traub, D.A. and E.M. Kesler. 1972. J. Dairy Sci. 55:348.
Veen, W.A.G. 1974. Nutr. Abstr. Rev. 44:208.
Volcani, R. and A. Ben-Asher. 1974. J. Dairy Sci. 57:567.
Waldren, D.E. and L.J. Fisher. 1978. J. Dairy Sci. 61:221.
Wallenius, R.W. and F.R. Murdock. 1977. J. Dairy Sci. 60:1422.

Wardrop, I.D. 1966. Aust. J. Agr. Res. 17:375.
Warner, R.G. 1970. Hoard's Dairyman 113:1123.
Williams, A.P. and R.H. Smith. 1975. Brit. J. Nutr. 33:149.
Winter, K.A. 1976. Can. J. Animal Sci. 56:567, 821.
Woelfel, C.G. and S. Gibson. 1978. Raising Dairy Replacements. Coop. Ext. Services, Vermont.
Wood, A.S., H.S. Bayley and G.K. McLeod. 1971. J. Dairy Sci. 54:405.

Chapter 11—Nutrition of Ewes and Rams

by R.M. Murray

INTRODUCTION

Sheep make a major contribution to world animal production with a present world sheep population of about 1000 million. The efficiency of sheep as converters of their feed to meat is much inferior to pigs and poultry (Table 11-1), largely because of the longer generation interval and small litter size. This results in a large overhead for maintenance of breeding stock that has to be carried by each slaughter animal. However, the non-ruminants are in direct competition with man for food, while sheep obtain almost their entire diet from grazing. It is the sheep's ability to live and produce on land unfavorable for other forms of agriculture that has made it so important in world agriculture.

Table 11-1. Efficiency of meat production.[a]

	Dry matter intake per	
Species	kg of carcass	kg of usable meat
Sheep	36	44
Cattle	30	35
Poultry	4.7	6.2
Swine	3.1	4.5

[a]From Owen (1976)

Sheep are responsible for three major products–meat, wool and milk. There are many breeds of sheep, and nearly as many forms of husbandry ranging from the transhumant and nomadic systems of southern Europe, North Africa and the Middle East through the extensive pastoral management of Australia, South Africa, South America and Asia to the intensive improved pastures of northern Europe, North America and New Zealand and the feedlot systems of the West and Midwest of USA. A comprehensive review of world systems of sheep production has been made by Owen (1976). The extensive husbandry of sheep in the pastoral areas of Australia has been described by Alexander and Williams (1973) while the intensive management of feedlot lambs in USA is described in some detail by Scott (1977). Throughout these management systems the central importance of pastures remains paramount. Therefore, nutritional management of sheep is one of adjusting peak animal demands to coincide with periods of greatest pasture production.

NUTRIENTS REQUIRED BY SHEEP

Sheep are primarily grazing animals which harvest their own food. This dependence on grazing while conferring economic benefits has the disadvantage that the animal must derive its nutrients from the diet accessible to it. This diet, being largely under the control of climate and other environmental variables, varies appreciably in both quantity and quality and can often be nutritionally inadequate.

CRITICAL NUTRIENTS

Wool and meat are mainly proteins. Meat may also contain appreciable amounts of fats and milk contains about equal amounts of protein, fat and sugar. Nutrition is concerned with supplying animal tissues with adequate substrates for the snythesis of their products. That is they require suitable sources of energy, proteins, minerals, vitamins and water.

Energy

A sufficient supply of energy to meet the maintenance and productivity needs of an animal is a basic requirement. Yet insufficient energy is probably the most common nutrient deficiency. Energy intake may be limited by either low forage availability or low forage quality. The significance of energy intake as a determinant of forage energy value is accounted for by the Nutrition Value Index (Crampton et al, 1960; see Ch. 5), which takes account of both ad libitum intake and an estimate of the available energy content of the feed.

Energy requirements of adult sheep depend on their level of production (see later section). At maintenance these energy requirements vary with the liveweight of the animal. The actual level of energy required to maintain live weight varies with the caloric concentration of the diet when ME is used as the base for computing requirements (Table 11-2).

Table 11-2. Metabolizable energy requirements (Mcal/day) for maintenance of adult sheep.[a]

Weight, kg	ME concentration, Mcal/kg DM			
	1.8	2.2	2.6	3.0
40	1.59	1.54	1.48	1.42
55	1.82	1.76	1.69	1.62
70	2.02	1.95	1.87	1.80

[a]From MAFF (1975)

Environmental factors have a major effect on feed intake. Although breeds differ in their ability to withstand environmental stress, food intake will be depressed by severe heat or cold stress. Moderate cold stress, however, enhances food intake, an effect commonly observed after shearing. The accessibility of feed and water obviously affects intake. Pasture density and pasture length will affect intake and also distance traveled and hence energy expended by the animal in harvesting its diet. Energy requirements of grazing sheep are 10 to 100% greater than stall-fed sheep.

Reproducing sheep require more energy than their non-producing counterparts. The requirement for energy during late pregnancy and lactation is 1.5X and 3X maintenance, respectively (MAFF, 1977). If it is considered that the energy requirement for grazing is 1.3X maintenance, then it is possible to calculate the average energy requirements for a reproducing ewe of any weight (Table 11-3). Insufficient energy intake by ewes will result in delayed maturity, reduced fertility, lowered milk production, poor wool production and an increased susceptibility to disease and parasitism.

Table 11-3. Metabolizable energy requirements for production by a 55 kg ewe.[a]

	Requirement, Mcal/day	
	Pen fed	Grazing
Maintenance	1.8	2.3
Pregnancy	2.7	3.5
Lactation	5.4	6.9

[a]From MAFF (1975)

Proteins

Proteins ingested by sheep are subjected to microbial attack in the rumen, which can result in considerable change in the composition of the amino acids absorbed from the small intestines. Therefore the quality of protein is not the critical factor in sheep nutrition, rather it is the quantity of protein in the ration.

The quantity of protein absorbed from the gut will increase both with protein content of the diet and the extent to which the forage is digested in the rumen. It can be seen, therefore, that protein availability per unit of forage consumed will decrease with advancing plant maturity, for not only does the protein content fall, but digestibility also falls. At maintenance sheep require dietary proteins to meet demands for wool production, for enzyme secretions within the gut, and to sustain the vital processes of the body. Protein required for endogenous N metabolism has been estimated to be about 2 g N/Mcal of maintenance energy.

The retention of N for growth has been estimated to be 2.5% of liveweight gain up to 40 kg liveweight and 2.4% from 40 kg to maturity (ARC, 1965). However, additional protein is required during pregnancy. The additional N retained in the growing conceptus has been estimated to be 0.18 to 0.24 g N/day in the 2nd month, 1.45 or 3.07 g N/day in the 4th month, and 4.96 or 7.40 g N/day in the 5th month of pregnancy, for ewes producing one 6 kg or two 5 kg twin lambs, respectively. The output of N in ewes' milk has been estimated at 9.7 g N/kg. These N retention figures represent the minimal requirements for protein, and do not take digestion losses into account. Recommended levels of protein (NRC, 1975) in rations for sheep are shown in Appendix Table 17, while Appendix Table 18 shows the daily protein requirements.

Minerals

Sheep require the same mineral elements as do other species. The requirements for essential mineral elements have been determined, and are shown in Appendix Table 19. Those apt to be critical include Ca, Mg, Na, P and S of the macrominerals and Co, Cu and Se of the trace minerals.

All surface and underground waters contain minerals in varying quantities, and these contribute significantly to the mineral nutrition of sheep (see Ch. 2, Vol. 2). However, in certain areas of the world the soils are deficient in specific minerals, and the pastures grown on these ranges are unable to meet the mineral requirements of sheep. Sheep producers are advised to determine the adequacy of their pastures before undertaking expensive and sometimes wasteful mineral supplementation programs.

Sodium and Chlorine. It is a common practice in many areas to offer sheep free access to trace mineral fortified salt licks. This undoubtedly is a cheap and efficient method of supplying minerals in areas where deficiencies of trace elements are known to occur. Sheep always relish salt and will take ad libitum about 10 g/head/day. However, there is little or no evidence to show that NaCl supplements are beneficial to grazing sheep; nevertheless, many farmers offer salt licks. Since these may serve to stimulate water intake and hence appetite, they can be useful in forage management.

Grains are low in Na and Cl, and when sheep are fed high grain diets, adding 1.0% salt to the ration is advisable. Salt is often used to regulate the intake of other mineral supplements such as N, P and S. Such mixtures can safely contain 10-50% salt as long as adequate water is available and iodized salt is not used.

Calcium and Phosphorus. Ca is seldom deficient in sheep diets except under conditions of prolonged high grain feeding. Pryor (1971) reports that wethers have been fed for 35 weeks on grain-only diets without depression in serum Ca levels. He suggests that Ca supplements may be useful in combating the acidosis which can accompany wheat feeding. When prolonged periods of dry lot feeding on high grain diets are contemplated, or when corn silage forms a major portion of the diet, addition of 1-1½% finely ground limestone in the ration is recommended. Generally, sheep seem to tolerate a high Ca:low P diet better than the reverse. The recommended Ca:P ratio is between 1:1 and 2:1.

P deficiency is widespread throughout the rangelands of the world. This is due primarily to the poor P status of many of the rangeland soils. However, the situation is further aggravated by the fact that most forages decline in P content with increasing maturity. Where P deficiencies have been shown to occur, supplements should be fed. Non-reproducing sheep require 2-3 g P/head/day, while pregnant and lactating sheep require up to 9 g P/head/day. Supplements should aim to provide at least half of the animal's daily requirements; that is at least 1-1½ g P/head/day.

The choice of supplement to provide P requires care, as recent evidence has shown some compounds of the mineral to be toxic (McMeniman, 1973). Detrimental performance has been recorded with phosphoric acid and sodium orthophosphate, while satisfactory results have been obtained using dicalcium phosphate, monoammonium phosphate and defluorinated superphosphate.

Magnesium. Grass tetany, a frequent problem in sheep grazing lush pastures, is a disease characterized by hypomagnesaemia. It is a serious problem on improved pastures in most temperate areas of the world. The disease is one of disturbed Mg metabolism rather than low pasture concentrations (see Ch. 13, Vol. 2). Nevertheless, it can be prevented by increased intakes of Mg and drenching with MgO to provide 7 g Mg/head/day is effective. Slow-release Mg bullets have been developed which, when given orally, lodge in the rumen, and supply Mg at sufficient levels to prevent grass tetany for 4-6 weeks.

Sulfur. Recent work has emphasized the importance of S in sheep nutrition. This element is necessary for the adequate utilization of protein and to sustain optimal fermentation within the rumen. An N:S ratio in the diet of 10:1 is recommended (NRC, 1975), and whenever N is fed to sheep to correct a protein deficiency, S should also be provided.

Trace Minerals. The metabolism of trace elements has been adequately reviewed in Vol. 2 and several other texts (Mills, 1970;

Underwood, 1977) and will not be discussed. Deficiencies in trace elements can be corrected readily in sheep by supplying mineralized salt licks, and only those elements of special importance will be considered.

Cu is one trace mineral of particular importance in many parts of the world. Not only can the element per se be deficient, but interactions with other minerals such as Mo and sulfate can precipitate Cu deficiency through interference with Cu metabolism. Further, excessive intakes of Cu can cause toxicity. Recommended dietary concentration of Cu is 5 ppm, however because of the danger of toxicity, supplementation recommendations must be cautious. While Mo excess can cause scouring symptoms which can be corrected by Cu supplements, deficiencies of Mo also occur and may result in Cu toxicity in forages with normal Cu levels (15-20 ppm).

Another trace element which may be deficient or occur at toxic levels is Se. There is a very narrow range between amounts required (0.1-2.0 ppm) and toxic levels (5+ ppm). Little can be done in situations of excess except to avoid or eradicate accumulator plants from the pastures. There is an interrelationship between Se and vitamin E and, when deficiency symptoms occur, it is recommended that both compounds be supplied to the sheep. Injectable Se solutions are available in most areas. If Se deficiency has been a problem, it is advisable to treat ewes 4-6 weeks before lambing and to treat newborn lambs. In some areas Se can now be added to salt. In the USA it is allowed at a level of 30 ppm. Premixes are also available for adding to concentrates.

There are many Co deficient areas throughout the world and more are being identified each year. The wasting disease caused by a lack of this element can be prevented inexpensively and efficiently for several years by oral administration of a slow-release Co bullet.

Vitamins

Vitamin deficiencies are virtually unknown in the normal feeding of mature sheep. Vitamin A or its precursor, carotene, is the only vitamin likely to be deficient, and then only in extreme cases of prolonged drought (1-2 years). Animals able to select green material from the pasture, or browse trees can maintain vitamin A reserves for long periods (Gartner and Anson, 1970). Rams, however, may be more susceptible to deficiencies than ewes, deficiencies may result in impaired fertility. Supplementation of breeding stock is often recommended, but this practice is probably unnecessary in most circumstances.

Water

An adequate supply of water is essential for the health and productivity of sheep. The provision of sufficient clean drinking water is a major consideration for the sheep producer. The requirements of sheep for water are modified by many factors including age, size, physiological status, environment, and diet.

There is a close relationship between water intake and food consumption. If water supplies are inadequate, food intake and productivity will fall. Non-reproducing sheep consume about twice the weight of water as of food dry matter (i.e. 2 ℓ/kg DM intake). This ratio is increased in pregnant ewes to as high as 7 ℓ/kg DM intake (Forbes, 1968). Lactating ewes require extra water for milk production and may consume in excess of 15 ℓ/day at peak lactation. Sheep will consume much more water in summer than in winter. The animals can derive a considerable proportion of their water needs from snow or dew, but consideration must be given to heating water supplies in snow-prone areas. Highly mineralized water is well tolerated by sheep, however intakes do increase with saline water. Productivity is not affected provided the salt does not exceed 1.2% (Wilson, 1975).

FACTORS AFFECTING THE NUTRIENT REQUIREMENTS OF THE EWE

With the majority of the world's sheep being kept at pasture, it is appropriate to confine our remarks to the grazing animal. The nutrients required by housed sheep are more easily defined and, while those of animals at range are similar, the demands of the latter are higher to allow for harvesting costs and overcoming environmental effects.

The nutrition of the lamb has been discussed in Ch. 12. This chapter considers the requirements of the ewe from 6 mo. of age, and deals first with the quantities sheep eat when they are in the different physiological states of maintnenace, growth, pregnancy, and lactation. Secondly it deals with the integration of demands of the breeding flock and supply of available pasture.

Maintenance

Adult sheep spend a high proportion of their life at maintenance. However, there are considerable fluctuations throughout a year in the live weight of producing ewes which are affected by pregnancy, lactation and availability of feed (Fig. 11-1). Ewes are generally offered pasture ad libitum and nutrient intake is determined by the quality of the forage. The maintenance requirements of ewes are shown in Appendix Table 18.

Growth

Feed requirements in the first year of life depend largely on the age of first mating. If the practice is to breed ewe lambs (7 mo. of age), then the diet must be designed to meet the requirements for growth as well as pregnancy and lactation. It is unlikely that pasture quality would be sufficient for successful mating of ewe lambs without some supplementary feeding. Under such management it is advantageous to breed the young ewes apart from the older flock to avoid unfair competition.

If ewe lambs are not mated, their requirements are much lower and good quality pasture is usually sufficient to maintain the desired growth rate. However, when pasture quality is low, growth of young ewes may be so low as to delay onset of puberty and ewes may not mate successfully until they are 2½ years old. Supplementation of ewes during their first 18 mo. may increase lifetime productivity significantly.

Flushing

Ewes on an increasing plane of nutrition at mating have higher ovulation rates than ewes on poor nutrition (Coop, 1966). This has led to the practice of increasing the energy supply to ewes before breeding (flushing). In most production systems weaning of the lambs is recommended about 4-8 weeks before mating. Cessation of lactation with weaning reduces the ewe's energy requirements, and even on

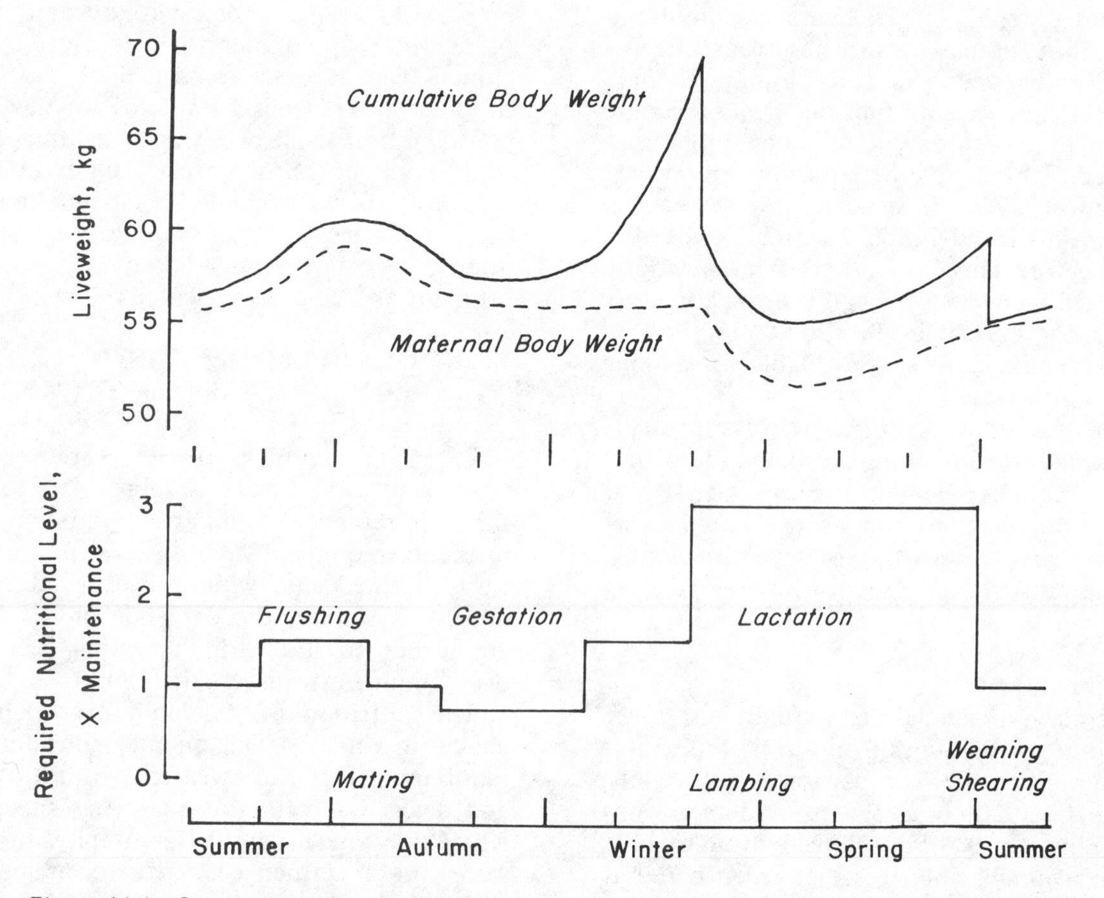

Figure 11-1. Seasonal variation in liveweight and nutritional requirements of a reproducing 55 kg ewe. After Jagusch and Coop (1971).

medium quality pasture, there will be considerable recovery of body condition. Flushing is usually accomplished by providing animals with fresh pasture, good quality hay or up to 0.25 kg grain/day. This special feeding usually begins 2-3 weeks before breeding and continues through the mating season.

Pregnancy

The nutrient demands of pregnancy (increase over maintenance) do not become significant until the 3rd mo. of gestation, when there is a marked increase in requirements. This increase is more pronounced with ewes carrying twin lambs. Appetite increases throughout pregnancy, but can decline sharply immediately prior to parturition. During this period the ewe may suffer a severe negative energy balance and pregnancy toxemia may result. Diets of high caloric density are required at this time if marked loss of maternal tissue is to be prevented. Owens (1976) suggests that it is impractical to try to meet the ewes' requirements in full at peak demand periods, and he suggests that ewes be allowed to make use of body reserves. Possible diet composition for pregnant ewes is shown in Table 11-4.

Protein requirements also increase during pregnancy from 50 g crude protein/Mcal ME during the early stages to 60 g/Mcal just before lambing. Recent research has suggested that during pregnancy amino acids may be required in excess of those which can be produced by microbial synthesis in the rumen, and that a supply of proteins which by-pass the rumen to be digested in the intestines may be necessary to achieve peak production (Ørskov, 1970). Leng (1976) has suggested that the real demand may be for glucose precursors, and that by-pass proteins and starch are used during these periods for gluconeogenesis.

Lactation

Voluntary food consumption by lactating ewes increases markedly to reach at least three times that of non-reproducing sheep. However, nutrient demand during early lactation is so great that it cannot be met from the diet, and loss of maternal tissue occurs. Ewes suckling twin lambs produce more milk than those suckling one lamb, and therefore have a greater lactation requirement. Nevertheless, it is generally impractical to segregate twin rearing ewes; rather it is better to prolong any period of supplementation.

EFFECT OF NUTRITION ON PRODUCTIVITY

Without doubt, of the factors influencing productivity, level of nutrition has the greatest effect. The current level of nutrition not only affects the current level of productivity, but through the ewe to the lamb can have far reaching effects on future productivity.

Reproduction

The fertility of the sheep flock is paramount. The greater the reproductive rate, the greater the number of animals available for

Table 11-4. Suggested composition of diets for 60 kg ewes in relation to requirements at various stages of production.[a]

Stage of production	Requirements to maintain weight: Mcal ME/day	Requirements to maintain weight: g CP/day	Voluntary intake, kg DM	Dietary composition: Energy, Mcal ME/kg DM	Dietary composition: Crude protein, % DM
Dry or early pregnant	2.0	100	1.5	1.8	9.0
Late pregnant	3.5	210	1.8	2.2	13.0
Lactating	5.7	340	2.5	2.2	13.0

[a]From Owens (1976)

sale and the greater the selection pressure which can be applied to the breeding flock to improve overall productivity of meat, milk, and wool.

The Ram. Young rams can enter the breeding flock as soon as they are big enough. Good nutrition is necessary to achieve early puberty and to maintain the mating ability of the ram. Level of nutrition not only affects libido but is a major factor in determining daily sperm production. While adequate protein is necessary in the ram's diet, it is energy intake which has the greatest effect on spermatogenesis. Supplementary feeding may be necessary to insure that rams are in good strong condition at mating. However, overfatness should be avoided in working rams as this can depress libido.

Rams have been shown to have a specific requirement for vitamin A for spermatogenesis. When green feed has been unavailable for several months, rams should be drenched with 1,000,000 IU vitamain A two months before and again just prior to the breeding season.

The Ewe. Fertility in the ewe is also determined by level of nutrition, and all aspects of the reproductive cycle are affected. Undernutrition is associated with delayed puberty, reduced conception rates, low twinning rates, and increased seasonal and lactational anestrus. Puberty in the ewe occurs while the animal is still growing; therefore nutrition has an enormous influence on age and liveweight at first estrus. Puberty has been recorded in Merinos as early as 4 mo. of age and at liveweights as low as 27 kg, but when nutrition is low and growth rates are slow, puberty may be delayed until 2 years of age. There is an interaction between the effects of age and weight on puberty. Data in Table 11-5 illustrate the variable reproductive performance of ewe weaners and the importance of early nutrition under grazing conditions. Sexual maturity can also be delayed, and there is a variable and often poor conception rate in young ewes although they have reached puberty. If ewe weaners are to be mated at 7 mo. of age, a high plane of nutrition is essential, and they should reach a target liveweight of 35 kg before mating.

Table 11-5. Age x liveweight interaction of the reproductive performance of young Border Leicester x Merino ewe weaners.[a]

Item	1968	1969	1970
No. of ewes bred	79	57	106
Age at breeding, months	7	7	9
Average liveweight, kg	35	44	34
% mating, of ewes bred	71	98	92
% pregnant, of ewes bred	13	16	63
% lambs born, of pregnant ewes	100	111	112

[a]From Tyrell et al (1974)

Poorly grown maiden ewes tend to have shorter estrous periods than adult ewes, less ewes mate and longer mating periods are required. The effect of nutrition on ovulation rates in maiden ewes (18 mo. of age) is shown in Table 11-6. An association between live weight and reproductive performance also exists for adult ewes. There appears to be a critical weight of 40-45 kg below which fertility is dependent on weight. That is, there is an increase in ewes showing estrus and lambing to first service as weight increases to the critical weight. Above this weight any increase in weight is associated with increases in ovulation rates. This is called the static nutritional effect. The proportion of multiple ovulations appears to be improved by a state of increasing body weight at mating. This is called the dynamic effect of nutrition. Although use is made of this phenomenon with the practice of flushing, results of research into the effects of body size, weight and condition on fertility are frequently confounded. Further, recent results implicate feed quality, although confounded in part by energy intake (Table 11-7). High protein supplements of lupin grain (35% crude protein) fed to Merino ewes

Table 11-6. Effect of liveweight on ovulation rate in a group of 18 month old maiden Merino ewes.[a]

Liveweight, kg	No. of ewes	Ovulations/ewe
25-29	3	1.00
30-34	24	1.00
35-39	52	1.04
40-44	16	1.13
45-49	13	1.38
50-54	3	1.67

[a]From Restall (1976)

for 5 weeks prior to breeding increased ovulation rates, pregnancy rates, and lambing performance. Nevertheless, Restall (1976) concludes that weight at breeding, irrespective of how it is achieved, is the critical factor. Any supplementation should aim first to get ewes above their critical weight (40-45 kg) and second to achieve the highest possible weight at breeding. A response in terms of extra lambs born/ewe lambing of 2½% has been found per kg increase in ewe weight over the range 40 to 60 kg (Coop, 1966).

Table 11-7. Effect of protein supplementation on ovulation rate of Merino ewes.[a]

Lupin grain/ ewe/day, g	Pasture source*	
	Sub. clover	Wheat stubble
	. . . ovulation rate . . .	
Nil	1.25	1.08
62.5	1.17	1.00
125	1.25	1.17
250	1.42	1.32
500	1.61	1.50

[a]From Lightfoot and Marshall (1976)
*50 ewes/treatment

The level of nutrition during pregnancy has a marked effect on the growth rate of the fetus and subsequent birth weight of the lamb. The extreme effects found in pen studies tend not to occur under grazing conditions. If ewes are losing body weight, then decreased lamb birth weights will result (Table 11-8).

Lambs with low birth weights have poor survival rates (Table 11-9), emphasizing the importance of adequate ewe nutrition during pregnancy. It can be seen in the table that there is an optimum lamb birth weight of 3-5 kg. Survival of heavy lambs can be jeopardized by problems at birth such as dystocia.

Poor nutrition during pregnancy will not only reduce wool production of the ewe, producing a tender fleece, but will also affect the lifetime wool production of her lamb (see section on wool). In the long run further loss of productivity from inadequate nutrition of the pregnant ewe can result from the subsequent reduction in milk production and decreased lamb growth rate. Onset of lactation may also be delayed in extreme cases, which may also jeopardize lamb survival. The level of milk production is also influenced to a large degree by the post-lambing plane of nutrition.

Table 11-8. Effects on lamb birth weights of level of nutrition during the last five weeks of pregnancy in grazing ewes.[a]

Level of nutrition	Live weight change of ewes, kg	Birth weights, kg	
		Singles	Twins
High	11.4	4.7	4.0
Medium	5.7	4.6	3.8
Low	0.0	4.4	3.3

[a]From Wallace (1960)

Table 11-9. Effect of birth weight on lamb survival.[a]

Birth weight, kg	% Survival	
	Singles	Twins
1.8	40	28
2.3	81	72
3.2	89	87
4.2	91	88
5.0	88	85
5.9	72	48
6.4	54	42

[a]From Holst and Killeen (1976)

Wool Production

A sheep's potential for wool production is determined early in its life. Before birth the initiation of the wool follicles is commenced, and the extent to which the lamb's genetic potential is achieved is determined by the level of nutrition of its dam during late pregnancy. Most follicles which will develop are initiated before birth. It is the level of nutrition which the lamb receives in its post-natal period which determines the number of the initiated follicles which will mature to become functioning, wool-producing follicles. This depends primarily on the milk yield of the ewe, which in turn depends on the ewe's plane of nutrition, both before and after parturition. The effects of pre- and post-natal

Table 11-10. Effect of pre- and post-natal nutrition on wool production at 2 years of age.[a]

Item	Treatment			
	High/high	High/low	Low/high	Low/low
Live weight at 2 yr, kg	53	48	49	44
No. of fibers, x 10^6	64	69	58	44
Fiber weight, μg	37	30	37	37
Wool growth, g/day	8.7	7.6	7.9	6.9

[a]From Schinckel and Short (1961); all lambs received the same nutritional treatment after 16 weeks of age.

nutrition on the productivity of the wool producing sheep are clearly demonstrated in Table 11-10. As can be seen from the table, nutrition in utero modifies the type of fleece produced. Although the lambs from ewes poorly fed to parturition, then well fed (low/high), produced as much wool as lambs from ewes fed well to parturition, then fed poorly (high/low), the fleeces of the former lambs were of poorer quality, containing fewer but coarser fibers.

Within the limits of a sheep's potential set by early nutrition, current level of nutrition, quantity and quality, has the greatest influence on wool production. When pasture growth is rapid, wool production is also rapid and in grazing sheep can reach levels as high as 3X that during the dry season. The limitation to wool production appears to be the sheep's voluntary feed intake, as experimental data indicate that production is linearly related to intake over the range of intakes investigated. However, the overall efficiency of wool production is low, with ca. 1 g clean wool being produced/100 g of food consumed.

The quality of the food consumed can also influence wool production, and it has been demonstrated that the amount of green pasture available has a major effect on wool production. Production per sheep increases as available green pasture increases, reaching a maximum at 300-450 kg available DM/sheep. Reducing the quantity of green material available per sheep by half (to 150-200 kg) only reduces production per head by 15-20%. However, the concurrent increase in stocking rate can result in considerable increase in wool production/ha.

Another nutritional factor affecting wool production is the availability to the animal of S-containing amino acids. Wool is a protein that contains about 15% S amino acids and which the grazing sheep must produce from a diet containing proteins of approximately 1% S amino acids. The degradation of dietary proteins within the rumen precludes the ration as a means of supplying sheep with high levels of S amino acids. However, quite large responses in wool growth can be achieved by experimentally supplying S amino acids directly into the intestines. This has led to a large research effort to develop methods of protecting dietary proteins high in S amino acids from degradation in the rumen, so that they reach the intestines unaltered. By-pass proteins are now available commercially, however the economics of their use requires careful consideration.

The nutritional stress of reproduction can have a major effect on the wool production of the breeding ewe. Breeding ewes will produce less wool than non-breeding ewes. The difference may be as small as 10% when nutrition is good, but when conditions are harsh, it can be as large as 25% or 35% for ewes rearing single or twin lambs, respectively.

Meat Production

Level of nutrition of the ewe through its effect on milk yield will affect the rate of growth of the lamb and hence subsequent meat yield. Nutrition of the lamb also affects carcass composition and carcass quality; these effects have been discussed in Ch. 13.

Milk Production

The ewe is an important dairy animal in many parts of the world. Milk production is important not only in these animals, but also in providing a foundation for the lifetime performance of all sheep. Lactation lasts 4-5 mo. in ewes. Milk yield is influenced by age, body weight, breed of ewe, stage of lactation, number of lambs suckled as well as the ewe's plane of nutrition. Peak yield, which normally is reached within 2-3 weeks post-partum, varies with breed, but is usually in the range 1-3 ℓ/day. The rate of decline which follows depends on breed, level of nutrition, etc., but generally production is still some 50% of peak 3 mo. after lambing.

Ewes' milk contains 1.1 Mcal GE/kg and is a concentrated source of energy, since fat may constitute up to 50% of the total solids. Protein is present in sufficient quantities to allow efficient use of this energy and to promote high lamb live weight gains (Walker and Norton, 1971). Ewes' milk is also adequate in vitamins and minerals.

The concentration of most components of milk quality, i.e. fat, protein and solids-not-fat, fall during the first 2-3 weeks of lactation, but then increase as lactation progresses (Peart, 1972). One notable exception is milk sugar, which increases in the early stages, reaching the highest concentration during peak yield, after which the lactose content gradually declines. The average composition for ewes' milk is 18.5% total solids which are composed of 38.4% fat, 30.7% protein, 26.0% lactose, and 4.8% ash.

Current level of nutrition is known to affect milk quality, and Ashton et al (1964) showed that ewes receiving supplementary concentrates in addition to grazing produce milk higher in solids-not-fat, protein and ash than ewes on a low plane of nutrition. Fat content is less affected.

Milk production is an energetically demanding process, and most ewes are unable to consume sufficient energy during lactation to prevent some loss in body reserves. Level of production varies not only with stage of lactation, but also is affected by the number of lambs suckled. Ewes nursing twin lambs produce 20-40% more milk than a ewe nursing one lamb. Table 11-11 shows the energy requirements of ewes suckling singles and twins in the first and second mo. of lactation. Gardiner et al (1964) have estimated that 65-83% of ME is converted to milk energy during the first 3 mo. of lactation. These workers estimate that NE for lactation was 1.3-1.4 x maintenance NE for ewes with single lambs and 1.7-1.9 x maintenance NE for ewes with twins.

While early work showed that severe undernutrition during pregnancy could lead to reduced milk yields by restricting mammary development, later field studies (Monteath, 1971) indicated that this was so only if the nutritional deficit was very large. However, milk production can be greatly influenced by plane of nutrition after lambing. An average ewe restricted to 2.4 Mcal ME/day will produce only about 1 kg milk/day while with ad libitum feeding of near 7.2 Mcal/day, up to 3 kg milk/day will be produced (Jagusch and Mitchell, 1971).

Table 11-11. ME requirements (Mcal/day)of ewes of differing live weights (kg) milking single or twin lambs, and milk yields (kg/day) during the 1st and 2nd month of lactation.[a]

		Month 1		Month 2	
No. of lambs	live weight	Milk yield	ME required	Milk yield	ME required
Singles	35	1.2	3.66	0.8	2.92
	55	1.4	4.26	1.0	3.42
	75	1.7	4.95	1.2	4.04
Twins	35	2.1	5.33	1.1	3.49
	55	2.3	5.98	1.3	4.04
	75	2.6	6.70	1.6	4.69

[a]After MAFF (1975)

FEEDSTUFFS APPROPRIATE TO SHEEP

Pasture has always been the traditional food of sheep and, of the world's domestic animals, the sheep is the most dependent on grazing for its food supply. However, feedlot sheep and housed flocks require complete diets formulated on conventional lines from available foodstuffs. The composition of many of the commonly fed feedstuffs is shown in Appendix Table 1.

GRAZING

Pasture

The importance of pasture to the world production from sheep cannot be overstressed. Pasture yield and nutritive value vary throughout the year depending on extent and pattern of annual rainfall, and from area to area depending on soil type, fertilizer applications and plant species present. The presence of pasture legumes in the sward increases the nutritive value of the pasture, and when pasture improvement programs are undertaken legumes should always be considered in the planting mixture.

The capacity of pasture to meet the requirements of the sheep flock depends on both yield and quality. Management must select mating times so that periods of high nutrient demand by the flock coincide with periods of peak forage supply. Young pasture growth is rich in protein (ca. 20%) and is highly digestible. At the end of the growing season plants mature and fiber content increases, while protein and mineral contents decline and nutritive value falls. In locations where rainfall is distinctly seasonal, pasture can be nutritionally inadequate for long periods. Frosts, dews and even light showers of rain cause extensive deterioration of dry season pastures through leaching of the soluble portions from the forage. Further deterioration can result from mold action.

The importance of forage fiber content has been discussed in detail by Weston and Hogan (1973). The capacity of a forage to provide energy depends largely on the extent to which it can be digested in the rumen, and this is strongly correlated with its fiber content. Further, nutritional deficiencies are often associated with high fiber, mature forage. Even when deficiencies in essential nutrients are rectified by supplementation, improvement in animal production may be slight, due to the inherently low fiber digestion.

Legumes occupy a place of particular importance in the feeding of livestock because of their high protein content. Alfalfa provides excellent grazing for sheep if careful management is used to prevent bloat. Rotational grazing systems have been developed for alfalfa dominant pastures, which have proved most successful. Clovers are other legumes of great value which are used extensively to enhance the nutritional quality of improved pastures.

Forage Crops

Crops may be planted specifically to provide temporary grazing when pasture yield or quality is low. Cereal crops such as oats, wheat and barley provide excellent grazing for sheep, whereas the taller plants such as grazing sorghums and corn are seldom used. Succulent crops such as rape, kale, turnips, etc., also find use in sheep nutrition, but care must be taken to avoid bloat.

Grain crop stubbles are of low nutritional value and should be used only by stock not being fed for high production. However, stubbles may provide roughage for dry ewes during the early stages of pregnancy.

Browse

Fodder trees and browse are generally adequate foods for sheep, but are for the large part out of reach of the animals. There are often considerable browse shrubs available on rangeland pastures, and these plants will form a variable proportion of the sheep's diet. Fodder trees seldom constitute the entire diet of sheep except where they are felled to provide feed during periods of drought (see later section).

FEEDSTUFFS FOR PREPARED RATIONS

Hay

The quality of hay depends on the stage of maturity at harvest, leafiness, color and lack of foreign material. A good hay should be low in fiber and high in crude protein, minerals and carotene. In this regard legume hays are superior to grass hays and preferable for feeding sheep. Hays containing more than 12% crude protein can be fed to sheep without the necessity for a protein supplement.

Silage

The nutritive value of silage depends on the nature of the crop ensiled, the chemical changes that occur within the ensiled mass and the losses of nutrients from the silo. Due to losses of soluble carbohydrates and true proteins, silage has less nutritive value than the original materials ensiled. Low-moisture legume-grass silage is an excellent roughage for dry sheep and can be fed on a self-feeding basis. Although adequate in protein, legume-grass silage should be supplemented with high energy feeds (grains) when fed to lactating ewes. Corn or sorghum silage, on the other hand, is usually low in N and requires supplementation with protein meals or legume hay. Silage is an excellent substitute for pasture, but when lactic acid concentrations are high, it is wise to restrict intake.

Cereal Grains

Most grains are satisfactory energy sources for ewes. When sheep are changed from pasture to high grain diets, the process should be gradual, taking several weeks in order to avoid acidosis and other digestive disorders. The proportion of grain in the final diet will depend on the energy demands of the ewes and quality of available roughage, but is usually about 50%. Grains can be fed either whole, cracked, rolled or flaked. However, no production advantage has been found from processing, except perhaps for corn.

Agricultural By-Products

By-products of the milling industry are useful stock feeds. Wheat bran and middlings are widely used in prepared foodstuffs, while rice bran and polishing are becoming available in greater quantities. Molasses is widely used in sheep nutrition, particularly in formula feeds and commercially prepared rations. Other by-products such as brewer's grains, cannery wastes, etc., provide good sources of nutrients for sheep, but their use usually depends on local availability. Cereal straws are seldom used for feeding sheep because of their poor nutritional quality. Straws may find use as a cheap source of roughage to prevent digestive disturbances when very high concentrate rations are being fed. The nutritional value of these by-products is given in Appendix Table 1.

Protein Sources

The oilseed meals—cottonseed, linseed, peanut and soybean—are all highly digestible and satisfactory protein concentrates for sheep. Animal protein meals such as blood meal, fish meal and meat meal are also suitable for inclusion in sheep rations but they are usually more expensive per unit of protein. The degree of heat treatment during manufacture of the various meals will affect the solubility of the proteins and hence their digestion in the rumen. These by-pass proteins may have an added effect through stimulation of appetite. Non-protein N sources such as urea can also be fed to sheep, but should be restricted to one-third of the total N in the diet.

FEEDING PRACTICES

Under extensive range management, sheep husbandry is concerned primarily with assuring adequate water supplies, treatment for internal and external parasites and protection from predators. Under these conditions additional food is offered only to the animals in times of drought to prevent loss of life. This is called feeding for survival.

Intensive management practices demand a level of productivity from the sheep which is beyond the potential of the pasture yield, and supplementary feeding is necessary to make up the deficit. Such supplementation is called feeding for production.

PASTURE MANAGEMENT

Pasture management involves the control of animal numbers and movements in order to maximize the productivity and profitability from animals, while maintaining the longevity and productivity of the pastures and the fertility and stability of the soils. Sheep production from rangeland depends first on the dry matter yield of the pasture, second on the quality of diet which sheep can select, and third on the energy which must be expended in harvesting this diet. Management can influence productivity through manipulation of stocking rates (see Ch. 6).

Factors Affecting the Productivity of Grazing Sheep

Pasture Utilization. Dry matter yield varies from extremely low levels in poor rainfall areas of the world to greater than 12 T/ha/year in favorable temperate environments. However, due to trampling, soiling or oversupply, only a proportion of this yield is harvested by the grazing sheep. Level of utilization can be as high as 90% for intensively stocked high producing pastures, but will be <30% on poor producing pastures where stocking rates are kept low to assure herbage is available during peak demand periods (Eadie, 1968). The more closely herbage availability follows nutrient demand by the flock (Fig. 11-1), the greater will be the degree of pasture utilization.

Selective Grazing. Sheep show a high degree of selectivity in their grazing habits and the degree of selectivity increases as stocking pressure decreases. The selective ability of sheep, therefore, has more significance in areas of sparse pasture production. While selection allows the animal the advantage of being able to obtain a diet of greater nutritive value than the average of herbage available, it can have a detrimental effect on the long term productivity of the pasture.

Overgrazing of range land in semiarid areas may not only change the botanical composition of the pastures but may also completely destroy the vegetation, thus producing desert areas.

Pasture Yield. The availability of pasture has a marked effect on the nutrient intake and energy expenditure of grazing sheep. As pasture availability declines, a sheep must graze longer and walk further to obtain sufficient energy to meet production demands. Sheep will increase grazing time from 7 h/day when pasture is abundant to a maximum of 10-11 h/day for dry sheep or 11-12 h/day for lactating ewes when herbage becomes extremely scarce. Lactating ewes will also eat some 32% faster than dry sheep in order to satisfy nutrient requirements (Arnold and Dudzinshi, 1967). In areas where pasture yields are generally low, the availability of drinking water becomes an important factor in determining the distance sheep must walk and, hence, productivity of the sheep.

Stocking Rate. The most important factor determining animal production from pasture is the stocking rate (sheep/unit of land). Stocking rate not only affects production/animal, but also production/ha. As grazing pressure increases, performance per individual declines but production/unit area increases to a maximum beyond which it declines due to deterioration of pasture yield and quality, and the effect on individual performance. As stocking rates approach those for maximum productivity, the costs of production for each added animal unit rise steeply until the cost of carrying the extra sheep exceeds the financial return from the extra product. Hence, the economic optimum stocking rate will be below that at which maximum production is achieved. The costs of carrying extra sheep are generally associated with increased fodder conservation and supplementary feeding. The stocking rate selected as the economic optimum by an individual producer will, therefore, depend largely on his idea of drought risk.

Pasture Management Systems for Sheep

The many management systems in use throughout the world have been described by Owen (1976). Further information is presented in Ch. 6.

Continuous Grazing. Continuous grazing is generally regarded as the simplest form of management. The sheep are allocated to an area of pasture and allowed to graze continuously. A set stocking rate may be used throughout the year, or the numbers of animals varied at times to modify the grazing pressure. However, the sheep remain together on a given pasture area for a prolonged period of time, with little interference in their normal behavior.

The stocking rate used will depend on the type of enterprise undertaken (wool production, lamb production), and the breed of sheep used. While local practice will indicate the stocking rate which might be used, the farmer should assess the productivity of his own enterprise to determine the optimum stocking rate for his particular farm.

Deferred Grazing. Deferred grazing may be practiced to allow the pasture to recover from a period of severe grazing pressure, to produce an accumulation of forage for grazing at a time when nutrient demands are high, or in preparation for fodder conservation. There is little conclusive evidence on the benefits of deferred grazing despite the many experiments which have been carried out (Bishop and Birrell, 1973). At low stocking rates conservation has no effect on animal production, while at a medium level of stocking, conservation can markedly increase production; at high stocking rates deferred grazing and conservation can decrease production considerably (Hutchinson, 1971).

Rotational Grazing. There are several forms of rotational grazing. All involve the division of the grazing area and the movement of the sheep from section to section at intervals. Rotational grazing systems should not be used with stock which must maintain a high level of feed intake such as pregnant ewes and fast growing lambs. Although rotational grazing is essential for the efficient utilization of dry land alfalfa dominant pastures, it has seldom been effective for other pastures (Robards, 1976). Nevertheless, rotational grazing is widely practiced in many countries, especially where pasture production is high.

Sheep are generally allowed 6-8 days in each field.

Strip Grazing. Basically a rotational system, strip grazing usually involves the use of portable electric fencing to confine sheep to an area of grazing; the sheep being moved to fresh pasture each day. The system aims to increase efficiency of pasture utilization by decreasing soilage and trampling. This system is seldom used today because of the high labor requirement.

Creep Grazing. Priority grazing may be given to lambs through a creep grazing system. The lambs are given access through gaps in a fence to pasture which is not available to their dams. This allows lambs to commence grazing and have access to the most nutritious pasture without competition from the ewes. The system can be used with a set stocked situation or incorporated into a rotational system. Creep grazing often requires considerable capital and labor input and has few if any advantages (Jordan and Marten, 1970).

FEEDING FOR PRODUCTION

Herbage varies in quantity and quality throughout the normal seasons from a highly nutritious diet to dead or dormant material of little value. Supplementary feeding is required to maintain production when pasture quality is poor. Fortunately, sheep can usually obtain a significant amount of their requirements from pasture, range or crop residues without the necessity of feeding complete diets in the drylot.

Supplementation for Production

Pasture production during the growing season can usually supply an adequate diet for the ewe during pregnancy and lactation but there is at least half the year in most environments when pasture growth is low or ceases. The forage which remains after the growing period is inadequate to support maximum production. The major concern at this time is increasing the nutritive status of the ewe to reach target live weights at mating, to insure high ovulation and conception rates. Flushing is most economically achieved with deferred grazing, although harvested hays, silages or prepared rations can be fed. Examples of prepared rations are given in Table 11-12. Supplements can be fed through labor saving self-feeders (Fig. 11-2) if the rations are designed to control energy intake.

Table 11-12. Rations of differing ME content, suitable as self-fed supplements for ewes and rams.

	Ration #		
Ingredients	1	2	3
Roughages	80	70	60
Grains	10	20	30
Protein meals	5	5	5
Urea	5	5	5
Minerals	1½	1½	1½
Energy content, Mcal ME/kg	1.7	1.8	2.0
Crude protein, % of DM	21	22	23

Figure 11-2. Farm-constructed self-feeders for supplements can save labor. Courtesy of D.P.I.

Self-Feeders. There are various types of self-feeders on the market, but suitable feeders can be constructed easily and cheaply on the farm. Self-feeders should be weather proof; have an internal construction which will allow the grain and roughage to flow freely; and should be of sufficient capacity to keep refilling to a minimum, yet small enough to be conveniently moved from site to site.

Drylot Feeding

It is the practice in many snow-prone countries to house sheep during the winter months. Housed sheep require a complete ration, as they are entirely dependent on the ration fed. These rations can be fed through self-feeders or handled mechanically. Self-feeding has the advantage of a lower initial capital cost, greater labor economy, and a reduced feeder space requirement. Most rations are suitable for self-feeding provided the components can be mixed uniformly to prevent selection. This requires that the roughage be coarsely ground which also eliminates most of the waste that occurs with long roughages.

Figure 11-3. Sufficient trough space must be provided to insure adequate feeding of all sheep when they are hand-fed. Photo courtesy of D.P.I.

Care must be taken when changing sheep from pasture to high grain diets. The process must be gradual to avoid digestive disturbances. This can be achieved within 2 weeks if dietary changes are made every second day (Table 11-13). When the roughage is coarsely ground and mixed with the ration, sheep can be put immediately onto diets containing some grain. Many operators prefer to start sheep on a diet of 70% roughage, changing the diet weekly and reducing the roughage content by 10% each time until a ration of the desired caloric density is being fed. Examples of rations of differing caloric content suitable for housed ewes are shown in Table 11-14. Adequate fresh water should be available at all times.

Figure 11-4. Hay may be provided through easily constructed wire racks. Photo courtesy of D.P.I.

Table 11-13. Feeding schedule for introducing sheep to high grain rations.[a]

Day	Grain-protein mix, %	Hay, %
1-2	0	100
3-4	20	80
5-6	40	60
7-8	50	50
9-10	60	40
11-12	70	30

[a]From Davis (1976)

Table 11-14. Suitable rations for self-feeding housed ewes.

Ingredients	Ration 1	2	3	4
Roughage	70	60	50	40
Grain	20	30	30	40
Protein meals	10	10	20	20
Minerals	1	1	1	1
Energy content, Mcal ME/day	2.0	2.2	2.3	2.5
Crude protein, % of DM	9.0	9.5	12.5	13.0

Pen Lambing. Pen lambing is a special situation in which production feeding of housed sheep is required. Ewes are housed for a few days before lambing, allowed to lamb in individual pens, and returned with their lamb to pasture a few days later. Additional capital expenditure and good husbandry skills are required. Results are generally successful, but careful nutrition is essential. As ewes are housed for such a short period, preconditioning to the production ration is required so that nutritional setbacks at this crucial period are avoided.

Accelerated Lambing

In the normal breeding calendar ewes are bred once per year. This allows the ewes adequate time to recover from pregnancy (5 mo.)

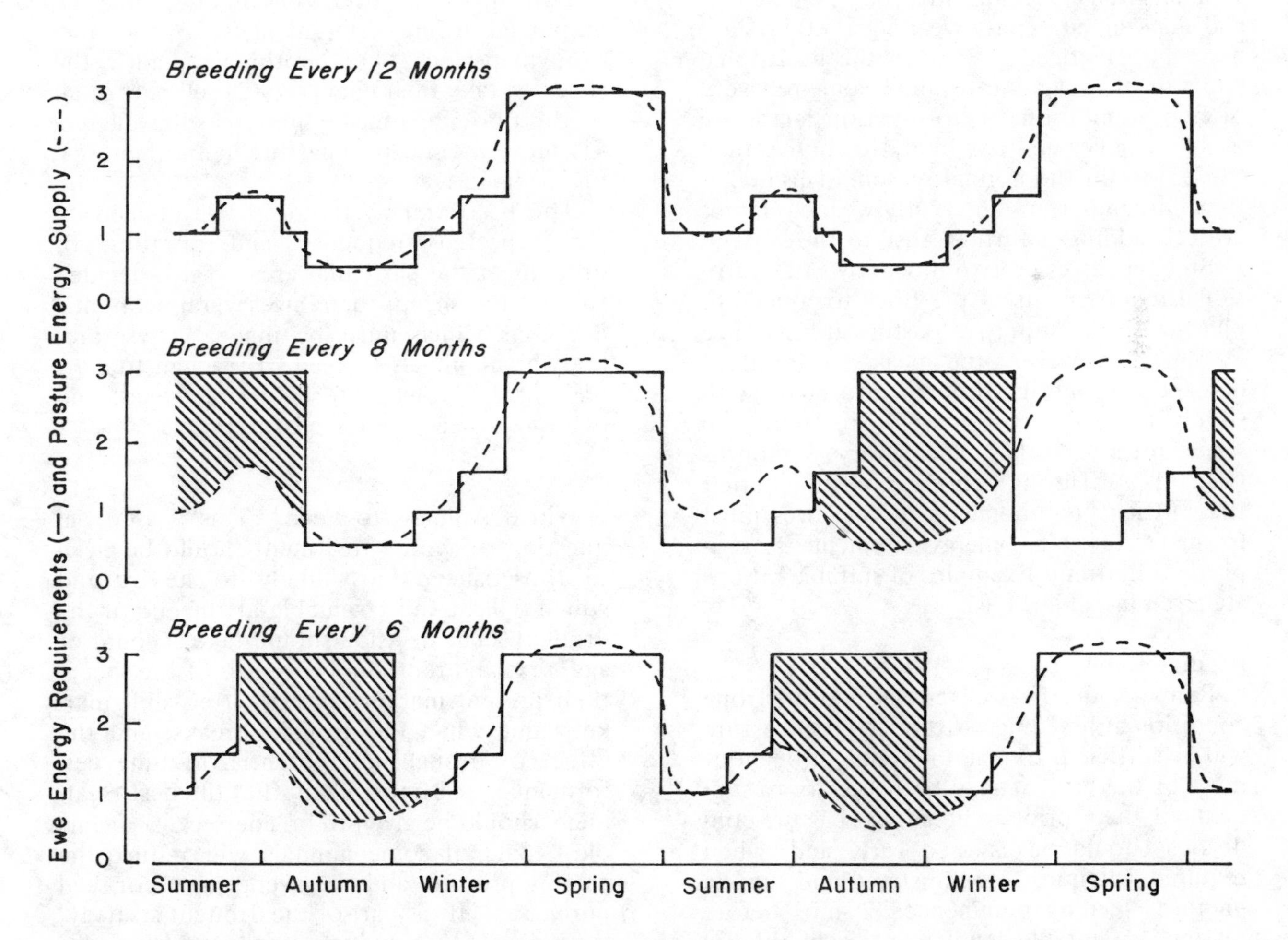

Figure 11-5. Energy requirement of ewes under various accelerated breeding regimes (—), and possible energy supply by pasture (- - -). Shaded areas show periods of energy deficit when supplementary feeding will be required.

and lactation (3-4 mo.). Considerable interest has developed in recent years in obtaining a lamb crop every 6-8 mo. Such acclerated programs do not allow the ewe a recovery period and place a large nutritional strain on the animal.

Accelerated lambing may be considered in order to take advantage of market fluctuations through a continuity of slaughter lambs or cheap seasonal labor supply, etc. Nevertheless, there will be increased costs and these must be taken into account in the overall economic assessment. Furthermore, outside the normal breeding season, the fertility of the ewes will be reduced. Hormone therapy may be required. Some breeds have better reproduction efficiency than others. Thus, consideration should be given to using these breeds when accelerated lambing is contemplated.

The nutritional implications of lambing every 6-8 mo. are enormous. High levels of supplementary feeding must be implemented and maintained. Early weaning at 80-90 days must be practiced to reduce the nutritional stress on the ewe and to allow some recovery of body condition. Out-of-season lambs are usually smaller and less likely to survive than those born in the normal season. This makes the husbandry of the early-weaned lambs critical, adding a further cost to the system.

Breeding ewes every 6 mo. allows the nutritional requirements of the flock to come into phase with the supply of pasture at least once a year. However, the peaks of nutrient demand can only be brought into phase with supply every 2 years or every 3rd lamb crop when breeding takes place every 8 months (Fig. 11-5). During periods of pasture deficit the nutrient requirements of the ewes must be met from high energy supplements supplied ad libitum. Examples of suitable rations are given in Table 11-12.

Feeding Rams

Rams should be well grown and in strong condition at breeding. Although good pasture will be sufficient to maintain rams throughout most of the year, it may be necessary to supplement them prior to breeding. Pasture conditions should be assessed early and, when required, vitamin A administered and supplementary feeding commenced 2 mo. before mating. A suitable supplement (Table 11-12) could be self-fed at 1-2 kg/head/day.

FEEDING FOR SURVIVAL

When rains fail, the survival of pasture fed sheep becomes the important consideration and it may be necessary to supply extra forage. Thus, various procedures have been developed to aid in survival.

Feeding Strategies

The onset of drought can be relatively sudden, as occurs when there is a complete absence of rain during the wet season, or it may be gradual, as when some rain falls but not in sufficient amounts to provide the quantity of pasture necessary to maintain the stock throughout the year. Rainfall patterns are generally well known in drought-prone areas, and whenever drought periods appear imminent a decision must be taken on the strategy to be adopted. It is necessary to have a plan. There are many drought strategies and one or all of the following practices may be employed during different phases of the same drought period: (1) do nothing and allow the sheep to take their changes; (2) sell part or all of the flock; (3) feed edible trees if available; (4) move the sheep to pasture in another area; or (5) hand feed.

The best strategy to adopt will depend on the expected frequency and duration of drought in the particular area. If it is decided to feed the sheep, there are several economic decisions which must be made. These are: (1) which animals to feed; (2) when to start feeding; (3) what feedstuffs to feed; and (4) how to feed these feedstuffs.

Which Animals to Feed. This is a crucial question. Favored treatment should be given to those sheep most likely to survive and which will be most valuable at the end of the drought. The decision should take account of age, sex, and reproductive status of the sheep, their present market value, their possible market value when the drought breaks, and the effect of drought on potential lifetime performance. It is most likely that the best treatment should be given to breeders 1½-5½ years old. Sale of some animals will reduce the grazing pressure and the overall need for feed purchases. If, as part of the drought strategy, it is decided to reduce stock numbers, this action should be taken early.

When to Feed. The feeding practice employed will depend upon the quantity and quality of herbage available. The producer must assess the contribution of the pasture and act accordingly. If there has been any rain, and if this has produced a green response in the pasture, sheep will selectively graze the new growth. There will be no need to provide supplementary feed until the new growth has been consumed and only the mature roughage from the previous season remains. This dry, standing roughage is low in protein, high in fiber and supplies only marginal quantities of energy. N-rich supplements should be fed to balance these inadequacies (see subsequent section on protein supplementation). When the available roughage is completely removed, consideration should be given to drylot feeding with high energy rations (see section on drought feeding).

What to Feed. Survival feeding is an expensive undertaking. Its financial success depends on careful evaluation of costs and returns of alternative programs. Quantities of feed required should be calculated for the number, type, live weight and physiological status of the sheep to be fed, the expected duration of the drought, and the costs of feeds available. The cheapest source of feed can be selected by calculating the cost per unit of ME at the feeding site.

Protein Supplementation

Vegetable or animal proteins are suitable supplements for sheep (see Appendix Table 1). Protein supplements should provide 20-30 g crude protein/head/day depending on the live weight of the sheep. Protein meals are often fed ad libitum mixed with salt to control intake to desired levels. Some difficulty may be experienced in starting sheep on protein meals of animal origin. This can be overcome by the use of molasses or corn and incorporating the protein meal slowly.

Non-protein nitrogen sources (urea, biuret, ammonium phosphate) have also been used as a source of dietary N. These supplements may be used to correct any protein deficiency and, in addition, aid the digestion of fibrous roughages which can be used as a source of energy. Urea is generally fed in combination with molasses in a ratio of 1:5 to 1:10. Some system is required to regulate intake of urea, which should aim to supply 10 g urea/head/day. This can be accomplished to some degree with lick wheel self feeders. If molasses is not fed, some form of sulfur should be given with urea to provide an N:S ratio of 10:1 in the supplement.

Drought Feeding

Drought always produces a financial loss. Survival feeding must be very carefully managed to minimize this loss. The energy requirements of the animals will be reduced if live weights are decreased to the lowest level compatible with survival. This critical live weight depends on the normal mature weight of the breed, but most sheep will survive a gradual loss of 25% of body weight. Furthermore, during the time of decreasing body weight, the tissue loss is metabolized to produce energy and this temporarily reduces dietary requirements. The minimum weight which adult sheep should be allowed to reach during drought feeding is 30 kg for fine wool Merinos, 40 kg for strong wool Merinos and 45 kg for Dorset Horn and Border Leicesters. Heavy mortalities may occur if sheep lose weight too quickly, therefore it is essential that drought feeding be started early. Energy expenditure should be minimized and this is best achieved by confining animals to yards. It is necessary to have some estimate of body

Figure 11-6. Dry lick supplements may be offered in cheap log troughs. Use has been made here of old railway ties. Photo courtesy of D.P.I.

weight. The energy requirements of sheep of differing weights can be obtained from the various published feeding standards (NRC, 1975; MAFF, 1975). Examples of the amounts of feed required for survival feeding of various classes of stock are shown in Table 11-15. From Fig. 11-3 it is possible to estimate the quantities required for sheep of differing weights and, hence, select the most economical feeds. Cereal grains are usually the cheapest energy source, but in many areas it may be less expensive to buy roughage than to buy and feed cereal grains.

Figure 11-7. Grains and roughages may be fed to sheep without using feeders, provided the ground surface is hard. Photo courtesy of D.P.I.

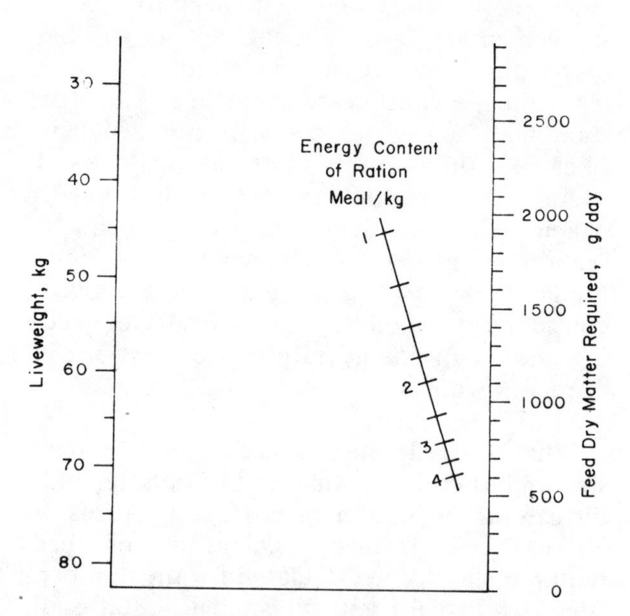

Figure 11-8. Dry matter quantities of rations of different ME content required to maintain ewes of variour liveweights. Adapted from NRC (1975).

Table 11-15. Quantities of energy required for survival by different classes of ewes, and quantities of feed which would provide this energy.[a]

Item	Energy requirement, Mcal ME/day	Factor of maintenance	Feed required, g/head/day, as fed					
			Corn	Wheat and sorghum	Oats	Alfalfa hay	Pasture hay	Corn silage
			Corn					
Dry ewes								
40 kg LW	1.2	1.0	420	440	470	760	820	3200
Pregnant ewes,								
last month	1.8	1.5	630	660	705	1140	1230	4800
Lactating ewes,								
first month	3.6	3.0	1260	1320	1410	2280	2460	9600
2-3 months	2.4	2.0	840	880	940	1520	1640	6400
Weaners,								
20 kg LW	1.0	0.8	340	350	380	610	660	2560

[a]From MAFF (1975)

Figure 11-9. Energy conservation during drought can be achieved by feeding ewes in yards. Photo courtesy of D.P.I.

Table 11-16. Feeding program for introducing sheep weighing 40 kg to a full grain ration.[a]

Day	Grain, g/head	Feeding frequency
1 and 2	50	daily
3 and 4	100	daily
5 and 6	200	daily
7 and 8	300	daily
9 to 11	400	daily
12 to 14	500	daily
15 to 21	1000	every 2nd day
after 21		every 3rd day

[a]After Davis (1976)

Whole grain diets can be fed successfully for long periods. However, not all sheep become accustomed to all grain diets. Poor feeders should be identified early and transferred to a "hospital yard" and given special attention. Some roughage is required by this group and the ration should contain 10-20% good quality hay. A suitable introductory program should take 3 weeks (Table 11-16). Feeding 2-3X/week gives better survival results than daily feeding once sheep are trained grain eaters (Davies, 1976). Sheep can consume several times their maintenance requirements and, when fed daily, competition from aggressive feeders will prevent shy feeders from obtaining their maintenance needs.

Pregnant ewes must receive their full energy requirements and must be fed daily if survival of the lamb is not to be jeopardized and pregnancy toxemia is to be prevented (Table 11-15). Ewes with lambs at foot also require extra food. Energy requirements increase in cold weather; this is particularly so for newly shorn sheep. Roughage provides a high heat increment during digestion and, therefore, has relatively more value during cold stress than indicated by its ME or NE value.

No benefit has been shown by grinding either grain or hay. Hay is best fed loose in hay feeders. However, good results have also been obtained by feeding hay loose on hard ground. Grain also can be fed by trailing onto hard ground allowing 100 m for every 1000 sheep fed. Where trough feeding is practiced, it is essential that adequate trough space be provided to allow all sheep to eat at once. If sheep have access to both sides of the trough, 5 sheep can be fed from 1 m of trough. The feeding procedure can be made quick, efficient and labor saving by mechanization.

Once the drought breaks it is important to keep feeding for a week or two until sufficient grass response has occurred. It is necessary to keep a store of feed on hand near the sheep for easy feeding during wet, boggy weather.

Feeding Edible Trees

Any non-poisonous trees or browse which sheep will eat can be used as drought fodder. The basic principle for successful use of woody plants is the same as all other drought feeding, that is to start before the sheep lose too much condition and to feed an adequate quantity. It is important that enough material is cut to provide a sufficient ration for the sheep. This can be easily estimated, as there should be some leaf left on the fallen trees when the next cutting is started. If this is not done, the sheep will eat woody twigs and in time develop fiber impaction of the rumen.

Most trees will provide a diet sufficient in protein, but digestibility of energy is low, and intakes must be maintained if adequate intakes of ME are to be achieved. Palatability of this woody forage can become seriously affected after 6 to 9 mo. of feeding. It may be necessary, as appetite falls, to supplement with some grain to maintain energy intakes. Mineral mixtures containing 50% salt provided ad libitum are also recommended to prevent mineral deficiencies and to stimulate water consumption and dry matter intake.

Figure 11-10. Fodder trees are readily eaten by sheep and can provide an economical drought reserve. Photo courtesy of D.P.I.

NUTRITIONALLY RELATED DISEASES

Nutrient imbalances can precipitate many diseases of sheep. The more important will be highlighted here.

Pregnancy Toxemia

Pregnancy toxemia, also referred to as ketosis and twin lamb disease, is associated with insufficient energy intake. There is a disturbance of carbohydrate metabolism resulting in reduced blood glucose and elevated blood ketones. It occurs in ewes in late pregnancy and is more common in ewes carrying twins, although it can occur in ewes carrying single lambs. It usually occurs in undernourished ewes, but can be precipitated in well-fed ewes by incorrect management.

The high-producing ewe is in critical nutritional balance, and management during pregnancy is most important. The ewes must be in good body condition at mating to ensure high conception rates, but nutrition should be reduced during the first 3 mo. of gestation. The plane of nutrition should again be increased during the last 2 mo. of pregnancy. Steps also must be taken during this period to minimize stress conditions such as transportation or changes in feeding that might reduce the plane of nutrition.

The disease is characterized by sluggishness, anorexia, staggering gait and nervousness followed by recumbancy and death. The purpose of treatment of clinical cases is to elevate blood sugar. This can be achieved by intravenous administration of glucose or oral dosing with propylene glycol or sodium propionate. The nutrition of all the animals in the flock should be improved by offering a high energy supplement.

Milk Fever

Milk fever characterized by hypocalcemia is another disease known as lambing sickness. The disease usually occurs in ewes in late pregnancy or early lactation, which are subjected to starvation or transport stress. Pregnant ewes fed high levels of cereal grains may also show signs of the disease, as may those grazing lush pasture or green crops. Sheep allowed to graze plants high in oxalate may also exhibit signs of hypocalcemia.

Affected animals become nervous, then weak and recumbant, and death follows. Treatment is by injections of calcium borogluconate, either intravenously or subcutaneously. Care should be taken at all times to avoid any stress to the pregnant ewe flock.

Grass Tetany

Grass tetany is another metabolic disease of ewes in early lactation. There is a sudden fall in blood Mg levels. Animals usually show violent nervous signs of galloping and convulsions. If untreated the disease is usually fatal. It is precipitated by a sudden change in environment or grazing lush pastures or green crops. The disease is easily confused with hypocalcemia, which may also be a concurrent disease. Treatment is seldom effective unless the disease is diagnosed early. Intravenous injections of Mg, either alone or in combination with Ca, have been successful. Careful management of the lactating ewe flock is essential for prevention of the disease.

Grain Poisoning

Whenever feeding with high grain rations is contemplated, extreme care must be taken to prevent excessive intake. Sheep, especially those which are hungry, must be graduallly conditioned to any change of diet. The disease usually occurs when sheep are first introduced to the drylot or to drought feeding. When excessive quantities of grain are eaten, fermentation is rapid and large amounts of lactic acid are formed within the rumen causing acidosis. Quantities of histamine are also produced and absorbed. There is ruminal stasis, impaction and dehydration. Treatment with oral $NaHCO_3$ solutions and perenteral antihistamines is usually successful. Fluid therapy may be necessary in extreme cases.

Fiber Impaction

Sheep forced to eat large quantities of browse, or which are fed fodder trees during drought may become impacted in the rumen with quantities of indigestible woody fiber. The animals show signs of acute ruminal dysfunction and lose weight rapidly. Treatment consists of placing the animals on high quality, laxative diets. This may not be successful in extreme cases. Feeding mineral licks and adequate water may be helpful. The condition can be prevented by assuring that sufficient forage or supplement is available so that the animals are not forced to eat the woody parts of the browse plants.

Urea Toxicity

Urea is a cheap and useful source of N for sheep. However, poisoning can occur when ruminants are fed large amounts of urea when they are not accustomed to it, or if they accidentally gain access to large quantities of urea, as when rations are poorly mixed. The toxic effects are due to the sudden production in the rumen of high levels of ammonia which is absorbed at a rate which exceeds the capacity of the liver to detoxify the ammonia back into urea. The excess ammonia then enters the peripheral circulation and toxicity occurs.

Unaccustomed animals may show clinical signs if they ingest as little as 15 g at once. However, once a tolerance has been developed, sheep can safely consume in excess of 60 g/day without apparent ill-effect, provided the urea is thoroughly mixed with the ration and the ration is consumed over a period of time.

Internal Parasites

Economic demands requiring increased productivity have led to increased grazing pressure on pastures. Associated with these higher stocking rates is the increased threat from internal parasites. Heavy infestations with these parasites can cause dramatic losses unless control steps are taken. The rationale of all worm control should be to break the cycle in two main places; inside the animal by anthelmintic treatment to prevent pasture contamination, and on the pasture by range management either through spelling pastures for long periods, or by grazing cattle alternatively with the sheep. When internal parasites are found to be a problem, advice should be sought on the optimum pasture management for the particular locality.

References Cited

Alexander, G. and O.B. Williams (eds.). 1973. The Pastoral Industries of Australia. Sydney University Press, Sydney.

ARC. 1965. Nutritional Requirements of Farm Livestock. No. 2 Ruminants. Agr. Res. Council, London.

Arnold, G.W. and M.L. Dudzinski. 1967. Aust. J. Agr. Res. 18:349.

Ashton, W.M., J.B. Owen and J.W. Ingleton. 1964. J. Agr. Sci. 63:85.

Bishop, A.H. and H.A. Birrell. 1973. Efficiency of grazing/fodder conservation systems. Proc. III World Conf. Animal Prod., Melbourne.

Coop, I.E. 1966. J. Agr. Sci. 67:305.

Crampton, E.W., E. Donifer and L.E. Lloyd. 1960. J. Animal Sci. 19:538.

Davis, C.H. 1976. In: Sheep Production Guide. Graziers Assoc. of N.S.W. MacArthur Press, Paramatta.

Eadie, J. 1968. 4th Rpt., Hill Farming Res. Organization.

Forbes, J.M. 1968. Brit. J. Nutr. 20:33.

Gardiner, R.W. and D.E. Hogue. 1964. J. Animal Sci. 23:935.

Gartner, R.J.W. and R.J. Anson. 1966. Aust. J. Exp. Agr. Hus. 6:321.

Holst, P.J. and I.E. Killeen. 1976. In: Sheep Production Guide. Graziers Assoc. of N.S.W. MacArthur Press, Paramatta.

Hutchinson, K.J. 1971. Herbage Abstr. 41:1.

Jagusch, K.T. and I.E. Coop. 1971. Proc. N. Zeal. Soc. Animal Prod. 31:224.

Leng, R.A. 1976. In: Reviews in Rural Science. II. From Plant to Animal Protein. N. England University Press, Armidale.

Lightfoot, F.H. and T. Marshall. 1974. J. Agr. W. Aust. 15:29.
Mills, C.F. (ed.). 1970. Trace Element Metabolism in Animals. E. & S. Livingstone, London.
McMeniman, N.P. 1973. Aust. Vet. J. 49:150.
MAFF. 1975. Tech. Bul. No. 33. H.M. Stationery Office, London.
Monteath, M.A. 1971. Proc. N. Zeal. Soc. Animal Prod. 31:105.
NRC. 1975. Nutrient Requirements of Sheep. Nat. Acad. Sci., Washington, D.C.
Ørskov, E.R. 1970. In: Proc. IV Nutr. Conf. Feed Manufacturers. J. & A. Churchill, London.
Owen, J.B. 1976. Sheep Production. Bailliere Tindall, London.
Peart, J.N. 1972. J. Agr. Sci. 79:303.
Pryor, W.J. 1971. Sheep Nutrition. 3rd ed. Queensland University Bookshop, Brisbane.
Restall, B.H. 1976. In: Sheep Production Guide. Graziers Assoc. N.S.W. MacArthur Press, Parmatta.
Robards, G.E. 1976. In: Sheep Production Guide. Graziers Assoc. N.S.W. MacArthur Press, Paramatta.
Schinckel, P.G. and B.G. Short. 1961. Aust. J. Agr. Res. 12:176.
Scott, G.E. 1977. The Sheepman's Production Handbook. 2nd ed. Sheep Industry Development Program, Denver, Colorado.
Tyrrell, R.N., N.M. Fogarty, R.D. Kearins and B.J. McGuirk. 1974. Proc. Aust. Soc. Animal Prod. 10:270.
Underwood, E.J. 1977. Trace Elements in Human and Animal Nutrition. 4th ed. Academic Press, London.
Walker, D.M. and B.W. Norton. 1971. Brit. J. Nutr. 26:15.
Wallace, L.R. 1960. Proc. Ruakura Farmers Conf. 13.
Watson, R.H., R. Rixxoli, A. Blackburn and D. Robinson. 1967. J. Agr. Vic. 65:420.
Weston, R.H. and J.P. Hogan. 1973. In: The Pastoral Industries of Australia. Sydney University Press, Sydney.
Wilson, A.D. 1975. Aust. J. Exp. Agr. Animal Hus. 15:760.

Chapter 12—Feeding and Nutrition of Young Lambs

by J.W.G. Nicholson

INTRODUCTION

There has been increased interest in the feeding and nutrition of young lambs in recent years. This has come at a time, at least in the USA, when sheep production is coming under increasing pressure because of various marketing problems and a gradually decreasing sheep population. Part of this interest is due to the fact that early weaning allows the ewe to be bred more than once per year, making it feasible to produce at least 3 lamb crops in 2 years. There is also greater interest in breeds of sheep such as the Finish Landrace, a breed that frequently produces litters of lambs, 4 or 5 being relatively common. Obviously, very few ewes could be expected to nourish successfully this many lambs so it becomes necessary to feed them in some other manner. Furthermore, the long successful use of milk replacers in feeding dairy calves has encouraged the use of similar products for lambs. This is true particularly when the ewe has twins or triplets or when a ewe is sick or dies after giving birth to a live lamb.

As a result of these various factors, a number of research papers have come out in recent years providing useful information on the practical nutrition and feeding of young lambs. However, this information is insignificant compared to that available on calves. It is likely, though, that much of the information on calves may be applicable to young lambs. In this respect, the reader is referred to Ch. 10 for additional information. A useful bulletin which gives a great deal of background information on the artificial rearing of lambs is that by Gorrill et al (1978).

PRENATAL NUTRITION

The influence of maternal nutrition on fetal lamb growth has been reviewed by Robinson (1977). It is clear that undernutrition during pregnancy reduces the weight of the fetus but there is no convincing evidence that it differentially affects the development of specific organs. During development of the fetus there is a linear relationship between the cube root of organ weight and days after conception (Richardson and Herbert, 1978). The only exceptions noted were in the development of the brain, cerebellum and spinal cord where weight gain tends to follow a sigmoid curve.

Although absolute growth rates are small in early gestation, it must be remembered that relative weight increases are exceedingly high. Robinson has pointed out that by 90 days of gestation, fetal weight is only about 15% of that reached at parturition only 8 weeks later. Weight increases at 90 days of gestation are approximately 6%/day while rates of about 20% are found earlier in gestation.

Wallace (1948) observed that undernutrition in pregnancy had a greater effect on weight of the placenta than on the fetus. It may be that the reduced fetal size due to undernutrition in early pregnancy is a consequence of the effect of undernutrition on placental function rather than from a shortage of nutrients per se. It is common practice to supplement the diet of ewes in late gestation. There is evidence (Bennett et al, 1964), at least for mature ewes, that adequate nutrition in late gestation will compensate for undernutrition in early gestation in terms of lamb birth weight.

Undernutrition in late gestation is likely to have more serious consequences than undernutrition in early gestation. This is because the lambs from ewes undernourished late in the gestational period are not as completely developed before being born. Not only are they born at an earlier stage of the differential growth and development process (Everitt, 1968), they are also born lighter in weight. Many studies have shown an inverse relationship between birth weight and lamb mortality. For example, Saunders (1977) found that 16.1% of lambs weighing under 3.6 kg were born dead or died within the first 7 days, compared with a loss of 4.7% of lambs weighing 3.6-5.4 kg and 3.5% for lambs 5.5 kg or more at birth.

From a survey of the literature, Gunn (1977) concluded that there are more persistent effects on adult body size and age at maturity when nutritive treatments are imposed prenatally and/or during the first few months of life, rather than when they are imposed at later stages of development. This was confirmed in his experiment when ewes were raised on a low or a high level of nutrition starting during the last 6 weeks of pregnancy and continuing until the ewes reached 12 mo. of age. When the ewes were 12 mo. of age, half the ewes raised on either level were switched to the other level of nutrition and maintained on it for 5 production years. Liveweight differences and lifetime lamb production were significantly affected by the rearing method. Gunn concluded that while reproductive potential is related to body size, it may also be influenced by the pattern of early growth during unidentified developmental stages which can be as early as 6 weeks before birth.

Cartwright and Thwaites (1976) have shown that the detrimental effects of undernutrition in late gestation on fetus development are accentuated by high ambient temperatures (42°C day and 32°C night). At these temperatures the ewe's energy requirements are increased while appetite is suppressed, resulting in decreased energy available to the fetus. This observation could have important implications for attempts at out-of-season breeding in hot climates.

Undernutrition during pregnancy can result in reduced size and maturity of lambs at birth with a consequent lower survival rate and, at least when the restriction is continued into the rearing period, reduced adult size and reproductive performance of ewes. Adequate nutrition in late gestation can largely compensate for undernutrition in early gestation. Robinson (1977) cited unpublished results which showed that gross overnutrition in early and mid gestation also can cause reductions of up to 40% in lamb birth weight.

The practical implications are that it is possible, perhaps preferable if ewes are fat, to restrict nutrient intake in early or mid gestation, but to do so in late gestation is detrimental.

THE NEWBORN LAMB

When the birth process has proceeded without undue delay and both ewe and lamb(s) are healthy, there is little need for intervention by the shepherd. However, when problems at parturition result in an exhausted ewe, or when a lamb is born weak and lethargic, it is often necessary for the shepherd to intervene to insure an early bonding of the newborn lamb with its mother. Bonding is facilitated by grooming of the neonate by the dam while it is still recumbent following its expulsion at birth. The struggles of the neonate to stand and its first unsteady steps stimulate the maternal instincts of normal ewes. A lamb which is too weak to stand may be ignored by the ewe, especially if there is a lively twin to claim her attention. An early and strong bond between mother and offspring has obvious advantages for survival. The first bonded of twin lambs tends to be nursed more successfully than the second (Fraser, 1976).

Lambs are born without blood serum antibodies (γ-globulins) against common infectious diseases (see Ch. 9, Vol. 1). Antibodies are acquired by the ingestion of colostrum soon after birth while the lamb is still capable of absorbing these proteins intact from the intestine. The lamb retains the ability to absorb antibodies for 24-48 h after birth, but the absorption mechanism begins to "shut down" soon after the lamb's first feeding (Doxey, 1977). It is important, therefore, that the first food the lamb receives is colostrum. Most shepherds try to keep a supply of frozen colostrum on hand to feed lambs from ewes which cannot supply their own. Cow's colostrum can be used as a substitute when ewes' colostrum is not available (Logan et al, 1978). These authors calculated that as little as 50 ml of colostrum/kg of lamb's weight would provide effective levels of serum antibodies. The colostrum used in establishing this level of feeding was from the first milking. It is known that the level of antibodies in colostrum decreases with subsequent milkings and more colostrum must be fed to achieve the same level of serum antibodies. The antibody titer of colostrum varies widely from ewe to ewe (Harker, 1977), but this is probably not too important provided the ewe has plenty of colostrum for her lamb(s).

NUTRIENT REQUIREMENTS OF YOUNG LAMBS

Information on the precise nutritional needs of young lambs is slowly accumulating. Much of this information is based on research which provides data on milk production and its composition by ewes nursing lambs or when young lambs have been fed known amounts of milk replacers of various types. This topic is discussed in detail in Ch. 10, Vol. 2.

The level of production and composition of milk from ewes is discussed in Ch. 11. In summary it can be noted that peak production of 2 to 2.5 kg/day occurs by week three post partum. Daily production declines progressively to less than half this level by week 10. Ewes nursing twins will generally produce about 20% more milk than those nursing single lambs, especially early in lactaion (Doney et al, 1979). While there is some variation between breeds in milk composition and lactational pattern, this is not of great significance for the breeds common in North America. Breeds selected for milk production, such as the East Friesland, have a more presistent lactation and do not show the increased fat content typical of common breeds in late lactation. The fat content of ewe's milk (ca. 8%) is higher than the content of protein (5.5%) or lactose (5%) and represents the main difference between ewe's milk and cow's milk.

The general concencus is that ewe's milk provides adequate nutrition for the lamb for the first 2-4 weeks at which time it begins to eat significant amounts of solid feed. Daily dry matter consumption ranges between 4 and 6% of the lamb's weight depending upon the milk production of the ewe and the number of lambs being nursed. Because of its high fat content and high digestibility, ewe's milk is a high-energy feed. Some trace minerals may not be supplied in adequate amounts by the milk. Trace mineral deficiencies are likely to be a regional problem and, where they are known to occur, it is necessary to supplement the diet of the lamb or the ewe. Deficiency areas have been identified in North America for Cu, Fe, I and Se.

The latest NRC (1975) requirements tables on sheep suggest some nutrient needs for early-weaned lambs weighing 10-30 kg. These are presented in Appendix Table 17. The authors of the NRC (1975) booklet point out that to achieve satisfactory results palatable high-energy, adequate protein diets are necessary. These younger lambs will not have the rumen development and the capacity to utilize less concentrated feeds considered normal for finishing lambs.

Milk Replacers

Milk substitutes have been used by many generations of shepherds to raise orphan lambs. Only in recent years has there been interest in rearing lambs on milk replacers as a planned procedure instead of an emergency procedure. Increasing interest in breeds such as the Finnsheep and Romanov which give birth to litters larger than the ewe can be expected to feed adequately and increasing interest in early weaning to facilitate early rebreeding have both led to research on milk replacers for lambs.

Much more research and experience have been accumulated with milk replacers for calves than for lambs. While the general principles may be similar for the two species, there are basic differences in the composition of ewe's milk and cow's milk. The major difference is a higher content of fat (65% of the energy in ewe's milk vs 50% in cow's milk) and a lower carbohydrate (lactose) content.

The first milk replacers formulated for lambs were patterned after ewe's milk and contained 30-35% fat on a dry matter basis. More recently there is a trend to using a lower fat content. For example, 24% fat in the dry milk replacer is now used in the standard milk replacer at the Animal Research Institute, Agriculture Canada, Ottawa, where several thousand lambs are reared artifically each year.

While butter oil is probably the best source of fat, many experiments have shown that lambs can utilize fat from many other animal and vegetable sources. Gibney and Walker (1977) showed that there is an interaction between the quantity and quality of protein and the source of fat in the efficiency of digestion of the fat. Provided the milk replacer contains adequate protein of good quality, such as that from spray dried skim milk, a wide range of fats is tolerated. However, if there is a reduction in the quality or quantity of protein, there is a reduction in the digestibility of the dietary fat.

Digestibility of the saturated fatty acids usually decreases with increasing chain length. The digestibilities of unsaturated fatty acids are usually higher than for saturated fatty acids of similar length. Thus Gorrill and Walker (1974) found that rapeseed oil containing 16 to 40% erucic acid (C 22:1) was less than 80% digestibile by young lambs, but other rapeseed oil which contained oleic (C 18:1) acid in place of the erucic acid was 94% digestible.

High-quality milk protein that has not been heat-damaged during drying is considered an ideal protein source for milk replacers. However, many experiments (see, for example, Gorrill et al, 1976) have shown that plant protein can replace part or all of the milk protein. Generally, animal performance will not be as high on replacers containing protein from sources other than milk. Nevertheless, they may be more economical, especially in cases where it is not necessary to achieve the maximum rate of growth.

Dove (1978) has shown that the proportion of essential to non-essential amino acids in the milk replacer is important as well as the absolute amount of essential amino acids. When the proportion of essential to non-essential amino acids varied very far from 1:1, there was a reduction in rate of growth and efficiency of utilization of both N and energy.

One reason that the growth of lambs may be lower on diets containing plant proteins in place of milk proteins may be the acid-base balance of the minerals. Walker et al (1978) point out that the mineral composition of ewe's milk is acid forming while that of cow's milk is base forming. When they added $NaHCO_3$ to replacers containing isolated soybean protein, the voluntary intake of the lambs increased.

Black et al (1972) found that for maximum N retention by 8 kg milk-fed lambs the protein should supply 25% of the GE and this value decreased to 12% for lambs weighing 30 kg. Penning et al (1978) showed that lambs fed a milk replacer containing only 30% of the energy as protein grew at a rate 42 g faster/day than those fed a milk replacer containing only 20% of the energy as protein. These authors also showed that the quality of the milk protein was important. Replacing heat-damaged milk proteins with undamaged protein resulted in 33 g more gain/day by the lambs. This response was obtained when the protein supplied 30% of the energy as well as when it supplied only 20% of the energy. Increasing the protein content or quality improved the efficiency of feed conversion to live weight gain and resulted in more protein and less fat deposition. Black and Griffiths (1975) found the N requirement/unit of energy intake was constant at 0.9 g N/MJ of ME for all milk-fed lambs irrespective of live weight when ME intake was 0.23 MJ/$kg^{0.75}$ per day. However, as ME intake/unit metabolic weight ($W^{0.75}$) was raised above this level, N requirement/unit ME intake increased for lambs weighing less than about 23 kg but decreased for heavier lambs.

The carbohydrate content of ewe's milk is low, and the newborn lamb can utilize only simple hexoses such as lactose and glucose. Provided the milk replacer is high in protein, the lamb does not need a source of carbohydrate to supply its blood glucose needs. If the level of protein is restricted, the inclusion of carbohydrate will improve N balance and liveweight gain. Gibney and Walker (1978) recommend the inclusion of the maximum level of glucose compatible with normal feces. It is well known that too much carbohydrate will cause diarrhea in lambs. These authors suggest the level of glucose should be <13.9 g/day/kg of liveweight. Others would set the limit at about 10 g.

MANAGEMENT OF YOUNG LAMBS

The bulletin by Gorrill et al (1978) outlines methods of feeding lambs on milk replacers and gives many useful management tips. Because it is essential that lambs receive colostrum, it is usually recommended that the lambs be left with their mothers for about 1 day. If they are left with their mothers for over 2 days, they are more difficult to train to an artificial teat and it may be necessary to hand milk the ewes to prevent udder damage.

When removed from their mothers, the lambs should be placed in a warm, dry pen, free of drafts and far enough away so that they cannot hear the ewes. If they are left for several hours without feed, they will usually take to the artificial teat more readily. Some writers have recommended up to 12 h fasting before introducing the milk replacer, but this would seem to be too long without feed for small, weak lambs. Milk replacer is usually fed at body temperature (39°C) for the first week, as young lambs do not take to cold fluids as readily as to warm. Once the lambs are accustomed to the artificial teat they can be changed to cold milk replacer which does not sour as quickly as does warm replacer.

Cold vs Warm Feeding

Other advantages noted for using cold milk replacers after the first week are more frequent nursing, with a smaller intake at each meal. This results in fewer digestive upsets and fewer problems with scours (Frederickson et al, 1971). The ingredients in the milk replacer are less apt to settle out of cold solutions. However, the big advantage in using cold replacers is the reduction in bacterial growth, consequently the need for less frequent cleaning of the feeding system and replacement of the milk supply. The addition of a low level of formaldehyde, as discussed in the section on problems encountered with milk replacers, will further reduce souring and the frequency of cleaning and sanitizing the feeding equipment.

If only a few lambs are raised on milk replacers, they can be fed by hand using any convenient bottle and nipple. Mechanical feeders are necessary to reduce the labor requirements when large numbers of lambs must be fed. These can range from simple heated reservoirs such as "The Lambsaver" (Fig. 12-1), homemade-gravity fed teat bars (Fig. 12-2), or elaborate systems in which refrigerated milk replacer is circulated to several pens (Fig. 12-3).

Teaching Lambs to Use Nipples

As noted earlier, it is easier to teach a hungry lamb to use an aritificial nipple and, also, it is easier to start with warm than with cold milk. To teach the lamb to drink, the nipple is palced in its mouth. If it does not start sucking immediately, its jaws can be moved by hand to stimulate sucking. To assist newborn lambs get their first meal, many ewes will gently nuzzle them on the rump. Some shepherds believe that placing a hand on the lamb's rump and pushing gently to imitate the action of the ewe, will stimulate sucking. With most lambs the first taste of warm milk from the artificial teat is all that is needed to start vigorous sucking. Only two or three training sessions are necessary with such lambs. The presence of one or two small lambs already trained to the artificial teat will help train new lambs. However, caution must be used in selecting the trained lambs. If they are much larger than the lambs being introduced, they may be adopted as substitute

Figure 12-1. The Lambsaver, a feeder used to dispense milk or milk replacer for young lambs. Courtesy of Sheep Breeder and Sheepman Magazine.

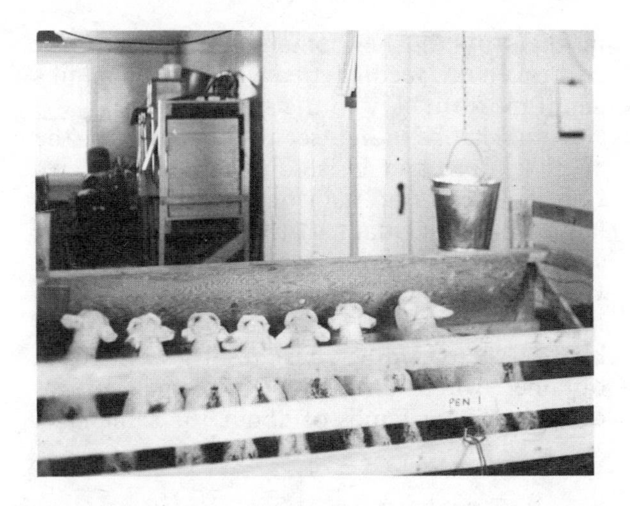

Figure 12-2. Simple, low cost, gravity-fed teat bar for feeding small groups of lambs. The bucket is attached by tubing to the teat bar.

Figure 12-3. Lambs weaned at one day of age and reared on cold milk replacer continuously to teat bar at the far end of the pen. Courtesy of the Director, Animal Research Institute, Ottawa.

mothers by the newcomers. A heat lamp or light over the feeding area will help the lambs adapt more quickly to the system.

Some lambs never seem to adapt to the artificial teat and insist on sucking the scrotum, navel or wool tags of bigger lambs. If a lamb does not adapt to the artificial feeding system in two days, it probably never will. Unless these lambs are separated into the individual pens and hand fed they will probably starve to death. Best results seem to be obtained when lambs are fed in small groups of 12 or fewer lambs of about the same size.

Amount of Milk Replacer Needed

Milk replacers are expensive, and so it is deisrable to limit the intake to the least amount compatible with acceptable levels of production. If the lambs are being reared for feedlot finishing or as breeding flock replacements, it is not necessary to push for high rates of gain at this early stage of development. However, if they are being raised for the light weight Easter lamb market, it may be desirable to feed higher levels of expensive milk replacer.

It is generally recommended that lambs be fed milk replacer ad libitum for the first 10-14 days. At this age they will not consume appreciable amounts of solid feed, and the intake of milk replacer is relatively low. Creep feed and hay should be available after the first day or two of age. If the milk replacer allowance is restricted after 10-14 days of age, the intake of dry feed is encouraged (Table 12-1; Owen and Davies, 1970). Lambs on the restricted intake of milk replacer consumed appreciably larger amounts of pelleted concentrate and were largely able to compensate for the reduced intake of milk replacer.

Duration of Feeding Milk Replacer

Lambs are being successfully weaned at an increasingly early age to reduce the high cost of milk replacers and the exacting labor requirements for feeding the liquid diets. There is no such thing as a right age to wean lambs. This depends upon the size of the lamb, the amount of dry feed being consumed per day and the type of dry feed that is offered. The digestive system of most lambs is not sufficiently well developed to wean them onto dry feed before 3 weeks of age. Likewise, most lambs under 11 kg in weight will have difficulty existing on dry feed alone. Therefore, it is probably best to base the decision of when to wean on several criteria. The lamb should be at least 3 weeks of age, weigh at least 11 kg and be known to consume dry feed.

A palatable, highly nutritious creep feed must be offered free choice both before and after weaning. As noted previously, restricting the intake of milk replacer after 10-14 days of age will encourage lambs to eat more creep feed and facilitate weaning. If possible, the whole group of lambs should be weaned at once. It is then possible to leave them in their familiar pen where they know the location of the creep feed. Fresh water should be available to the lambs from an early age.

Milk Replacer Formulas

High quality lamb milk replacers are commercially available from several companies. These are based on the use of relatively large amounts of dried whole or skim cow's milk. Dairy by-products, such as dried whey and

Table 12-1. Effect of initial live weight and milk allowance on lamb performance.[a]

	Milk replacer allowance			
	Large lambs		Small lambs	
Milk replacer fed	9 kg[b]	5 kg[b]	9 kg	5 kg
Initial live wt, kg	4.70	4.43	2.51	2.60
Daily gain to 15 kg, g/day	252	208	224	219
Total concentrate consumption to 15 kg, kg	7.28	12.52	8.67	12.77

[a]From Owen and Davies (1970)

[b]Values represent the milk replacer dry matter fed.

buttermilk, are popular ingredients. Many experiments have shown that lambs can utilize plant protein in place of part or all of the milk protein. The ad libitum intake and rate of gain is often lower on milk replacers containing plant proteins, and they are not widely used commercially. Many plant protein ingredients are difficult to get into solution and tend to settle out in mechanical fedding systems.

The deisrable proportions of protein, fat, and carbohydrates used in milk replacers were discussed in the section on nutrient needs. While it is possible for sheep raisers to make up their own milk replacer based on the information available, it is not advisable because of the availability of good commercial replacers. The necessity of homogenizing the fat (Gorrill and Nicholson, 1972) and of adding small, but essential, amounts of vitamins and minerals requires sophisticated equipment to get good results. Those who wish to formulate their own milk replacer are referred to the bulletin by Gorrill et al (1978).

The extent of dilution with water for feeding can range widely provided the lambs are fed ad libitum or at frequent intervals. Large (1965) found little difference in rate of gain of lambs fed 4X/day on solutions with 10, 15, 20 or 25% dry matter. If the milk replacers with <20% dry matter are fed less frequently, it is likely that dry matter consumption per day will be restricted. Most commercial firms recommend feeding their milk replacers at 15 to 25% dry matter. If the dry matter content of the milk replacer is over 15%, it is necessary to provide a supply of fresh water.

Problems Encountered with Milk Replacers

Of the problems that have been mentioned by various writers, diarrhea and abomasal bloat appear to be the most common ones that can be attributed to the diet. The use of too much carbohydrate or of an inappropriate carbohydrate is the usual dietary cause of diarrhea. The digestive system of young lambs is only able to use lactose, glucose, and galactose in any quantity (Gibney and Walker, 1978). Other carbohydrates such as starch and sucrose pass along the digestive tract unchanged and provide a substrate for the proliferation of undersirable bacteria. Another factor known to be very important with calves is the quality of the milk proteins used in milk replacers. Overheating during the drying process will result in denaturization of the proteins so that they do not produce a normal clot in the abomasum (see Ch. 10, Vol. 2). The end result of either feeding excess carbohydrate or poor quality protein seems to be that this allows bacteria such as *E. coli* to invade the upper intestine and multiply rapidly; the toxins produced apparently cause the diarrhea. The adverse effects of either of these two problems are exaggerated by feeding large amounts of the milk replacer at infrequent intervals.

Abomasal bloat may be related to this problem. In experimental work, it seems to be more common when formulas containing plant protein were fed twice a day but it also occurred on diets containing only milk protein. McQueen (unpublished) has isolated a microorganism, *Lactobacillus fermentum,* from bloating lambs and showed that it produces large amounts of gas when incubated with milk replacers. It was not controlled by low levels of the antibiotics commonly used in milk replacers, but it was greatly inhibited by low levels of formaldehyde. This observation had led to the development of a practical control procedure (Gorrill et al, 1978) involving the addition of 0.05% or 0.10% formalin in the liquid milk replacer. The 0.05% level is recommended for lambs fed cold milk replacer ad libitum and the 0.1% level for lambs fed milk replacer 2X/day. Levels of formalin higher than 0.1% in the milk replacer are unpalatable to some lambs and result in poor performance.

Formalin is a 37% by weight solution of formaldehyde. To obtain a 0.05% level it is necessary to add only 0.5 ml of formalin to 1 kg of liquid milk replacer. Twice this amount provides 0.1%. The formalin should be mixed in the liquid milk replacer some time before feeding, preferably overnight, to let the excess formaldehyde odor dissipate.

The addition of formalin at the levels suggested greatly retards souring of milk replacers fed ad libitum at room temperature. When used in teat bar feeding systems (Fig. 12-2), it is not necessary to dismantle the system for cleaning and sanitizing more frequently than once a week.

EARLY WEANING OF LAMBS

Early weaning is a term that has been used to define weaning of lambs at ages as young as

2 weeks or as old as 12 weeks of age. In the farm flocks of 50-100 years ago, the practice was to leave the lambs with the ewes on pasture throughout the summer. Lambs in range flocks are still commonly pastured with the ewes throughout the summer. Compared with this practice, 12-week weaning can be considered early weaning; however, the common practice on farm flocks now is to wean at 8-12 weeks of age. It was pointed out in the section on milk replacers that lambs will not consume and utilize significant amounts of solid feed before they are 2 weeks of age so this sets a minimum age for early weaning. For most efficient use of food by ewes and lambs, Ørskov et al (1973) suggest weaning at 33 days, but many factors can modify this recommendation.

Early weaning is used to advantage in a number of situations: where it is desirable to rebreed the ewe at an earlier time than usual; where there is some alternate use for the milk, such as in the cheese industry of Europe or the Near East; where the ewe has twins or larger numbers of lambs that she cannot adequately support; where the availability of high quality forage is limited and competition between ewe and lamb can be reduced by moving ewes to other pasture areas; or where stomach worm infection can be reduced by running ewes and lambs in different pastures (Brown, 1964).

Unless one of the above conditions exists there is little reason to wean lambs before 8-12 weeks of age. In the section on nutrient requirements, it was pointed out that the milk production of most ewes declined to less than half the peak production by 10-12 weeks after lambing. The ewe can obtain most of the nutrients needed for this level of production from good quality forages or pasture, feeds which are usually more economical than the high quality concentrates needed to feed the early-weaned lambs.

Results of the experiments reviewed by Brown (1964) generally indicate that lambs can be weaned with little growth setback by the time they are 8-10 weeks of age or that one of a set of twins can be weaned at 3-4 weeks of age and the remaining twin may be expected to grow at the same rate as a single lamb. It was suggested by some earlier writers that weaning of one of a set of twins might not have the desired results as each member of the set of twins tends to suck from only one teat. Recent observations (Hess et al, 1974) indicate that this is not a problem and the remaining lamb quickly learns to take advantage of the decreased competition for the available milk supply.

Age alone may not be a satisfactory criterion for determining when to wean lambs. Physiological age would probably be more satisfactory but is difficult to assess accurately. It is related to chronological age, weight and dietary treatments. Lambs that do not have access to solid feed before weaning will not develop a normal rumen and can be expected to suffer a growth check if suddenly weaned regardless of their age or size. On the other hand, Brown (1964) successfully weaned lambs at 2 weeks of age that had been on pasture with their mothers since birth and were fed a high quality concentrate and hay after weaning.

Protein Requirements

Recent research with ruminating animals has shown the importance of describing protein in terms of the amount of metabolizable amino acids available to the animal. Proteins that are readily soluble in rumen fluid are quickly deaminated by rumen microorganisms and furnish less metabolizable amino acids than others that are degraded to a lesser extent. The amount of metabolizable amino acids absorbed from the digestive tract is a function of the amount of dietary protein escaping degradation in the rumen, the amount of microbial protein synthesized in the rumen, and the digestibility of these proteins in the lower digestive tract.

Many of the experiments reported in the literature comparing different sources of protein for early weaned lambs have failed to consider the amount of dietary protein bypassing rumen degradation. Andrews and Ørskov (1970) have further criticized many earlier experiments for failing to control feed intake and for starting with animals that were already well grown (over 30 kg). In spite of these criticisms it is possible to draw general conclusions on the protein requirements of early weaned lambs.

It is obvious that the protein requirement (as a percentage of the dietary dry matter) is higher for young lambs than for older ones. Many research reports suggest that maximum rates of gain and feed efficiency will be attained at protein levels approximately 2 percentage units above those recommended by NRC. For lambs of about 10 kg liveweight, a level of 18-20% crude protein (CP) is needed, declining to 11-12% at 30 kg liveweight (Miller, 1968; Andrews and Ørskov, 1970; Gorrill et al, 1975; Cazes and Quackebeke, 1977). If one level of protein must be chosen for the entire growing period to 30 kg, a level of 16-17% of the dry matter is a commonly recommended compromise for lambs fed ad libitum (Andrews and Ørskov, 1970; Jordan and Hanke, 1970; Craxes and Quackebeke 1977).

Urea or other sources of non-protein N are not likely to be of much value in diets containing 16% or more CP. Below this level the value of urea will depend upon the amount of dietary protein degraded in the rumen and the amount of energy available to rumen microorganisms.

Table 12-2. Suggested protein levels allowing maximum gains for male and female lambs fed low, medium and high levels of cereal-based rations.[a,b]

Live wt of lamb		Feeding level[c]		
kg	lb	High	Medium	Low
		Indicated crude protein concentration, %		
20	44	17.5	15.0	12.5
25	55	15.0	12.5	12.5
30	66	12.5	10.0	10.0
35	77	12.5	10.0	10.0

[a]After Andrews and Ørskov (1970).

[b]Protein levels ranged from ca. 10 to 20%, dry basis.

[c]The high level was determined by Y = 4.24 - 0.03W, where Y = dry matter intake as a % of live wt (kg/day) and W = live wt (kg). The medium and low levels were 85 and 70% of the high level, respectively.

Protein-Energy Studies

Andrews and Ørskov (1970) have studied the gains of lambs from 16-40 kg when fed pelleted, cereal-based diets containing a range of protein from ca. 10 to 20% (dry basis) supplied from soybean meal. These diets were also fed (2X daily) at 3 different levels. Their suggested protein levels for maximum gains are shown in Table 12-2. Growth rate increased with higher protein concentrations as feeding level increased, giving a significant interaction. This confirms the generally accepted idea that protein requirements are influenced by feeding level. Andrews and Ørskov concluded that the overall optimum protein contents of diets for lambs between 16 and 40 kg were about 17.0, 15.0 and 11.0% for mean DE intakes of 3.0, 2.6 and 2.1 Mcal/day, respectively. Not only did the higher levels of protein increase rate of gain, but also increased the rate of protein deposition and decreased fat production.

Table 12-3. Effect of crude protein and energy level on performance of young lambs fed from 11 to 27 kg in body wt.[a]

	Daily gain and feed conversion								
	Period 1[b]			Period 2[b]			Overall		
	Gain,			Gain,			Gain,		
Item	kg	(lb)	FC[c]	kg	(lb)	FC	kg	(lb)	FC
Energy level, Mcal/kg[d]									
2.64	.23	(.51)	2.93	.29	(.64)	3.71	.26	(.57)	3.34
3.08	.24	(.53)	2.89	.29	(.64)	3.45	.26	(.57)	3.25
3.52	.21	(.46)	3.34	.24	(.53)	3.96	.22	(.49)	3.53
Crude protein, %									
14	.20	(.44)	3.02	.28	(.62)	3.72	.24	(.53)	3.50
17	.23	(.51)	3.24	.26	(.57)	3.74	.25	(.55)	3.35
20	.24	(.53)	2.91	.27	(.60)	3.66	.26	(.57)	3.26

[a]After Chipman et al (1971). [b]Period 1 was from the start through the 5th wk; period 2 from the 6th wk to the termination at 27 kg body weight. [c]FC = feed conversion = feed/gain. [d]Values equivalent to 60, 70 and 80% TDN, respectively.

The protein level of the diet should increase with the available energy level. Chipman et al (1971) self fed lambs (11 to 27 kg in weight) pelleted diets with ca. 14, 17, and 20% CP and with energy levels of 2.64, 3.08 and 3.52 Mcal/kg (ca. 60, 70 and 80% TDN). Some of their data are shown in Table 12-3. Gains during period 1 (first 6 weeks) were statistically better on the 17% CP diet than on the 14% diet, but there were no differences in the second period. With respect to energy levels, the high level resulted in less gain during both periods, indicating that about 70% TDN (or 3.08 Mcal/kg) was an adequate level for lambs of this age when fed pelleted rations.

By way of summing up the information on protein and energy, it seems quite certain that the very young lamb (3-6 weeks of age) needs more protein than older lambs when expressed as percent of the diet. A lamb on milk is used to getting a diet with about 28% of very digestible protein in the dry matter of milk. As consumption of solid feed increases, the protein needs can apparently be met with reasonably good rations containing ca. 18% CP. This amount can gradually be adjusted downward, the extent depending upon the ration energy concentration and its total intake. A very high energy concentration (ca. 80% TDN) will tend to depress intake, requiring a greater percentage of protein. Likewise, a very bulky diet will physically limit intake to the point that energy may be more limiting than protein. This applies particularly to rations in meal form. On the basis of information presented on the protein needs of older lambs (Ch. 13), it seems unlikely that the CP levels should be decreased below 16% (as fed basis) for lambs weighing 30 kg or less, depending upon what disposition is to be made of the lamb. If the lamb is being grown out for a replacement animal, then lower levels of protein may be acceptable for economic reasons.

With respect to energy, information at hand indicates that energy concentrations of 60-70% TDN are apparently adequate for lambs of this age, providing the mixture is a palatable one. Meal rations will need to have higher energy levels than pelleted rations (at least for those containing appreciable amounts of roughage) to obtain the same level of intake.

Amount of Roughage in the Diet

The form in which the roughage is supplied is at least as important as the total amount provided. The diet must contain enough rough material to prevent rumen parakeratosis —a build up of tissue on the rumen papillae which interferes with absorption. In this regard, chopped or long forages are much more effective than are ground forages and whole or rolled grains are better than finely ground grains. Fraser and Ørskov (1974) showed that there was no difference in daily weight gains of young lambs fed barley-based rations when the barley was either pelleted rolled barley, pelleted whole barley or unprocessed whole barley. Feed efficiency, organic matter digestibility and rumen pH were all somewhat higher for lambs fed the unprocessed whole barley. There would not seem to be any reason to go to the expense of processing grain for lambs. In the research work cited, the protein-mineral-vitamin supplement was pelleted and mixed with the whole grain for feeding.

Young lambs fattened intensively on high-concentrate diets may develop a soft subcutaneous fat that is unacceptable to the meat trade (Fraser and Ørskov, 1974; Ørskov et al, 1975; L'Estrange and Mulvihill, 1975). Ørskov et al (1975) showed that it was a greater problem with intact male lambs than with female lambs of similar age and weight fed the same diet. They related the incidence of soft fat to the molar proportion of propionic acid produced in the rumen. When the proportion of propionic acid in the rumen of male lambs was below 25%, all carcasses were acceptable; when it ranged from 25-30%, one out of 4 carcasses were unacceptable and over 30%, 3 out of 5 carcasses were unacceptable to the trade. The soft fat is related to a reduced level of stearic acid content and a higher proportion of odd-numbered straight- and branched-chain fatty acids, which are derived from propionic acid.

High concentrate diets favor the production of propionic acid in the rumen, while high roughage diets favor the production of acetic and butyric acids (see Ch. 3). If intact male lambs are being fed, it may be advisable to limit the amount of concentrate so that the molar proportion of propionic acid does not go above 25%. The ratio of concentrates to roughage required to produce this level of propionic acid will depend upon the quality of the forage—more digestible forages give higher proportions of propionic acid—and on method of processing—ground forages give higher ratios of propionic acid than do long forages.

Forages are usually less expensive sources of nutrients than grains so that it is desirable, economically, to feed the highest level of forage compatible with acceptable rates of gain. It might be expected that lambs could gradually consume relatively greater amounts of roughage as they gain in size and maturity. Even quite small lambs are able to increase their intake of diets diluted with forages so as to maintain energy intakes. For example, Owen et al (1967) found that lambs through the weight range of 11 to 34 kg could consume pelleted rations containing as much as 40% oat hulls without depressing daily gain. When similar rations were fed in the meal form, the addition of 20% oat hulls resulted in somewhat lower gains. Because of the interactions between quality of forages, form of processing and other diet ingredients, it is not possible to give precise amounts of forage except for specific situations.

Creep Feeding of Unweaned Lambs

Most sheep producers creep feed young lambs, particularly during the time that they are confined to dry lots. The benefit that may be obtained from creep feeding lambs in any given situation is impossible to state in general terms. The value of creep feeding will be dependent upon the relative adequacy of the milk supply for one thing and so will be of more value for twins than singles.

The acceptance of the creep feed is also a critical factor. Feeders must be designed and located so that lambs have ready access to it. One example of a portable creep feeder that can be used on pasture as well as in the dry lot is shown in Fig. 12-4. Another factor is the amount and quality of feed the lamb may obtain when the ewes are fed, and a 4th factor is the adequacy of the ewe's diet, particularly with respect to mineral supply.

Figure 12-4. An example of a simple type of portable creep feeder that can be used with lambs on pasture or in the feedlot. Courtesy of S. Dakota State University.

The value of feeding complex creep rations will depend upon the nutrient content of the cereals, protein supplement or other concentrates in a given formula. If the ewes are receiving adequate mineral nutrition, then a mineral source would be of less value in creep rations. Milk is an excellent food, but it is deficient, at best, in the trace minerals. However, if the milk supply is adequate, then the source and amount of protein supplement in creep rations will be less important.

Palatability of creep rations is important in order to obtain a high intake. Some feedstuffs that are particularly palatable to the young lamb include corn, barley, soybean meal, and wheat. Pellets, particularly small ones such as 3/16 in. (4.7 mm) are an appropriate size for young lambs. If the physical nature of the formula is not particularly appealing to a lamb, it is likely that pelleting will be of more value than when the formula is highly acceptable. Lambs will usually take readily to whole, cracked or rolled grains, but dusty and finely ground mixtures should be avoided as consumption of such mixtures will nearly always be low.

References Cited

Andrews, R.P. and E.R. Ørskov. 1970. J. Agr. Sci. 75:11.

Bennett, D., A. Axelsen and H.W. Chapman. 1964. Proc. Aust. Soc. Animal Prod. 5:70.

Black, J.L., G.R. Pearce and D.E. Tribe. 1972. Brit. J. Nutr. 30:45.
Black, J.L. and D.A. Griffiths. 1975. Brit. J. Nutr. 33:399.
Brown, T.H. 1964. J. Agr. Sci. 63:191.
Cartwright, G.A. and C.J. Thwaites. 1976. J. Agr. Sci. 86:573.
Cazes, J.P. and E. van Quackebeke. 1977. Ann. Zootechnie 26:286.
Chipman, G.H., K.R. Fredericksen, R.C. Bull and D.A. Price. 1971. Proc. West. Sec. Amer. Soc. Animal Sci. 22:215.
Doney, J.M., J.N. Peart, W.F. Smith and F. Louda. 1979. J. Agr. Sci. 92:123.
Dove, H. 1978. Aust. J. Agr. Res. 29:145.
Doxey, D.L. 1977. In: Perinatal Losses in Lambs, a symposium at Stirling University, February, 1975.
Everitt, G.C. 1968. In: Growth and Development of Mammals. Butterworths, London.
Fraser, A.F. 1976. Appl. Animal Ethology 2:193.
Fraser, C. and E.R. Ørskov. 1974. Animal Prod. 18:75.
Frederickson, K.R., D.A. Price and T.D. Bell. 1971. Proc. West Sec. Amer. Soc. Animal Sci. 22:199.
Gibney, M.J. and D.M. Walker. 1977. Aust. J. Agr. Res. 28:703.
Gibney, M.J. and D.M. Walker. 1978. Aust. J. Agr. Res. 29:133.
Gorrill, A.D.L., G.J. Brisson, D.B. Emmons and G.J. St-Laurent. 1978. Agriculture Canada Pub. 1507, Ottawa.
Gorrill, A.D.L., T.M. MacIntyre and J.W.G. Nicholson. 1975. Can. J. Animal Sci. 55:377.
Gorrill, A.D.L. and J.W.G. Nicholson. 1972. Can. J. Animal Sci. 52:477.
Gorrill, A.D.L., J.R. Seoane, J.D. Jones and J.W.G. Nicholson. 1976. Can. J. Animal Sci. 56:401.
Gorrill, A.D.L. and D.M. Walker. 1974. Can. J. Animal Sci. 54:411.
Gunn, R.G. 1977. Animal Prod. 25:155.
Harker, D.B. 1977. In: Perinatal Losses in Lambs, a symposium at Stirling University, February 1975.
Hess, C.E., H.B. Graves and L.L. Wilson. 1974. J. Animal Sci. 38:1313.
Jordan, R.M. and H.E. Hanke. 1970. J. Animal Sci. 31:593.
Large, R.V. 1965. Animal Prod. 7:325.
L'Estrange, J.L. and T.A. Mulvihill. 1975. J. Agr. Sci. 84:281.
Logan, E.F., W.H. Foster and D. Irwin. 1978. Animal Prod. 26:93.
Miller, E.L. 1968. Animal Prod. 10:243 (abstr).
NRC. 1975. Nutrient Requirements of Sheep. Nat. Acad. Sci., Washington, D.C.
Ørskov, E.R., W.R.H. Duncan and C.A. Carnie. 1975. Animal Prod. 21:51.
Ørskov, E.R., C. Fraser and J.C. Gill. 1973. Animal Prod. 16:311.
Owen, J.B. and D.A.R. Davies. 1970. J. Sci. Food Agr. 21:340.
Owen, J.B., D.A.R. Davies, E.L. Miller and W.R. Ridgman. 1967. Animal Prod. 9:509.
Penning, P.D., I.M. Penning and T.T. Treacher. 1978. J. Agr. Sci. 90:221.
Richardson, C. and C.N. Herbert. 1978. Brit. Vet. J. 134:181.
Robinson, J.J. 1977. Proc. Nutr. Soc. 36:9.
Saunders, R.W. 1977. In: Perinatal Losses in Lambs, a symposium at Stirling University, February 1975.
Walker, D.M., M.J. Gibney and R.D. Kirk. 1978. Aust. J. Agr. Res. 29:123.

Chapter 13—Finishing Lambs in the Feedlot

by D.C. Church

INTRODUCTION

Many lamb finishing operations in the USA are relatively high volume, low profit businesses which have feedlot capacities for thousands of lambs. A high percentage of lambs that do not go directly into the meat trade off the ewe come from these large feedlots. There are exceptions to this generalization, of course, where limited numbers of lambs are finished on many different farms or ranches. In the midwestern USA many lambs are finished, perhaps with some grain, by pasturing on winter wheat fields, while in some of the western states, particularly California, large numbers of lambs are finished on irrigated legume pastures.

Generally, the lambs that go to feedlots are those which come from range bands where grazing is relatively sparse or they are the tail-enders from farm flocks. Such lambs may have been born late and, thus, be undersized and insufficiently finished to go to the block meat trade. These lambs may also be poor doers for a variety of reasons. They may have been seriously sick at one time or another, and in some areas many are apt to be heavily parasitized with stomach or lung worms and liver flukes. Poor productivity by the ewe may account for lack of growth. Another reason could be that twins or triplets did not get enough milk. In addition, some may be deficient in one or more nutrients. Obviously, such lambs are not to be expected to perform as well as young, thrifty lambs that will reach market size and condition while suckling their dams. By the same token, it is to be expected that a relatively higher percentage of these lambs will gain very little in the feedlot and death losses may be high.

As suggested previously, the lamb feedlot business is a relatively low profit situation. The high death losses that may occur and the large numbers of poor doers partially account for this. In turn, these problems occur partly because sheep do not generally respond as well to medication and treatment as do cattle.

Although lambs will gain at about the same rate/unit of metabolic size ($W^{0.75}$) as steers, they are, generally, no more efficient in terms of feed conversion. Slaughtering costs/unit of edible meat are higher for sheep and the carcass yield is lower/unit of live weight than steers or hogs. There is more meat on the carcass that is difficult to merchandise, so lamb is nearly always at a competitive disadvantage with beef or other meats even if the carcass sells at a similar price. In addition, packers or meat buyers may readily discount carcasses they think are too large, and those that do not show a break joint (and are then classed as yearlings) are usually discounted very severely. Thus, lamb feeders, faced with many of these problems, must be rather efficient to survive and make a profit. Consequently, adequate nutrition, careful buying of feed, and proper management are, perhaps, more vital than for many other feeding situations.

For the reader who is interested in information from other sources, it might be noted that a variety of texts are available on sheep production. The Sheep Industry Development Program has prepared a loose-leaf booklet edited by Scott (1976) which may be a useful source and Jensen (1974) has published a book on sheep diseases which would be recommended.

NUTRIENT REQUIREMENTS

Protein

Of the major nutrients which are of concern, protein is probably of primary importance in most practical situations since it is an expensive ingredient to add. However, there is not unanimous agreement on how much is required for finishing lambs. The most recent NRC (1975) publication recommends 11% crude protein (dry basis) across weights of 30 to 55 kg for finishing lambs and 16% for early-weaned lambs of 10-20 kg and 14% for those weighing 30 kg (see Appendix Table 17, 18). For finishing lambs the expected gain is 200-250 g/day and 250-300 g/day for early-weaned lambs.

The author would prefer to go along with recommendations suggested by Preston (1966)

Table 13-1. Digestible protein requirements (in grams) as suggested by Preston.[a]

Body weight		Daily gains in g or lb							
		grams: 150	200	250	300	350	400	450	500
kg	lb	lb: 0.33	0.44	0.55	0.66	0.77	0.88	0.99	1.10
30	66	68	79	90	100	111	122	133	143
40	88	84	98	111	125	138	151	165	178
50	110	100	116	131	147	163	179	195	210
60	132	114	133	151	169	187	205	223	241

[a]Derived from Preston's (1966) formula, DP = 2.79 wt $kg^{0.75}$ (1 + 6.02G), where DP is the amount of digestible protein in g, wt $kg^{0.75}$ is the metabolic size, and G is the daily gain in kg.

which have been calculated and are presented in Table 13-1. Note in this table the amount of protein required for lambs of a given weight depends on the body weight as well as the expected gain. Preston's formula gives values that are somewhat similar to those suggested by the ARC (1965). Although the latter values are expressed as available protein and were derived from a formula which takes into account the amount of N excreted and the biological value of the proteins which are fed.

There is a fair amount of relatively recent evidence indicating that early-weaned lambs respond well when given moderate to high levels of protein in their diets (Andrews and Ørskov, 1970; Robinson and Forbes, 1970; Ørskov et al, 1971, 1972, 1976; Ely et al, 1979). Generally, early-weaned lambs have responded to diets with as much as 15-17% crude protein while higher levels may or may not give any additional response. Feed consumption may be expected to increase with protein level, at least on a high barley diet. For example, Ørskov et al (1971) observed that lambs ate more and were more efficient as protein increased (Table 13-2). Other researchers have reported similar findings. In subsequent research, Ørskov et al (1976) fed early-weaned lambs barley-fish meal diets with 12 or 20% crude protein. Lambs were slaughtered at weights ranging from 20 to 75 kg. Rate of gain was substantially higher with the 20% diets or when shifted from the 12 to 20% diets. Feed conversion was also improved on the 20% diet. Weston (1971) concluded that a level of about 18 g of digestible crude protein/100 g digestible organic matter was adequate for maximal rumen volatile fatty acid production of early-weaned lambs fed pelleted diets ranging from 11.7 to 19.1% crude protein. Comparable findings with respect to protein needs of early-weaned lambs have been reported by Ranhotra and Jordan (1966) and Jordan and Hanke (1970).

Table 13-2. Response of early-weaned lambs to increasing levels of protein.[a]

Protein in diet, %	Feed consumed, g[b]	Feed conversion	Gain, g/d
11.0	874	4.85	184
15.7	1012	3.88	248
19.4	1015	3.33	316

[a]From Ørskov et al (1971)
[b]Feed consumption averaged at weights of 20, 25, 30 and 35 kg.

With respect to data on older feedlot lambs, there are very few recent reports. A number of older papers show a favorable response by lambs when fed more protein than NRC recommends, although other work indicates that NRC levels may be appropriate at times. For example, Preston and Burroughs (1958), when comparing CP levels of 9, 13 or 17%, found maximum gains at the medium level with low-energy rations. When the energy level was increased, maximum gains were obtained at the 17% CP level and feed conversion was considerably improved. Later work by Preston et al (1965) and Preston (1967) indicated that rations with 11.7-13.5% DP (dry basis) gave similar performance in lambs fed mixed rations. Jones and Hogue (1969) found that 11.2% CP gave a better response than 8.4% CP. Other reports from various experiment stations (Rea et al, 1963; Church et al, 1966) indicate a favorable response by finishing lambs when CP levels of 14-16% are fed. More recently Craddock et al (1973) fed pelleted alfalfa-barley-corn diets with 10.5 or 13.5% CP and with 50 or 80% alfalfa. Gain (not significant) and feed conversion were improved about 8-10% by the higher protein level.

The author's experience with linear programmed pelleted self-fed rations fed to medium- and heavy-weight feeders is that gains and feed conversion are at maximal levels when rations contain about 14-16% CP (dry-weight basis). These studies were carried out with a variety of different feedstuffs and with rations which were formulated to have similar fiber levels and either isocaloric or different specified energy concentrations (Church et al, 1966, 1967).

It may well be that age and condition of the lamb, the type of diet, and time on feed may affect the response. Old crop lambs that have, essentially, made all of their growth would be expected to require less protein than young lambs. Malnourished or heavily parasitized lambs would, most likely, benefit by higher protein levels than unaffected animals. Increasing the amount of poorly digested fiber will increase N excretion and thus tend to increase N requirements. Pelleted feeds may be expected to pass through the GI tract at a more rapid rate than non-pelleted diets. If the digestibility is reduced as a result, this would tend to increase the need for protein. Another factor that will modify the need for protein, as expressed in percentage, would be the energy concentration of the diet. When the energy concentration is sufficient so that physical capacity of the GI tract is not limiting, then animal intake of dry matter will gradually be reduced if energy concentration is increased further. This results in the need for a greater concentration of protein, provided digestibility of protein does not increase as it frequently may with higher quality feedstuffs and reduced intake.

On the basis of these different considerations, it can easily be deduced that the exact needs of protein are difficult to determine for a given situation. Individual animal variation adds to the problem. Even though higher levels of protein may increase gain, from a practical point of view it might not always be feasible to try to obtain maximum gain if the cost of supplements is relatively high.

However, another important factor that should be considered is the effect of the finishing diet on the lamb carcass. Data published by Andrews and Ørskov (1970), Craddock et al (1973) and Ørskov et al (1976) show clearly that a relative protein deficiency results in greater fat deposition by the body tissues. Since many of our present day market lambs are probably over finished, particularly if self-fed to rather heavy weights, this fact alone might justify the inclusion of more protein in the diet than might be recommended on the basis of live weight gain or feed conversion. Therefore, the author will stick to his opinion that rapid-gaining lambs should receive 14-16% total protein in their diets. The nature of the N source, its solubility, and the amount degraded in the rumen may alter needs in a specific situation (see Ch. 3, Vol. 2).

Energy

The most recent NRC (1975) recommendations for energy concentrations for finishing lambs from 30-55 kg are for TDN levels of 64 and 67% for 30 and 35 kg lambs, respectively, and 70% TDN for heavier lambs. If expressed as Mcal of DE, the corresponding values are from 2.8-3.1 Mcal/kg of feed. For early-weaned lambs 73% TDN and 3.2 Mcal/kg of DE are recommended.

The energy concentration needed for maximum gains must be related to the type of feed preparation involved in the diet. There is ample evidence that finishing lambs do very well on high-roughage pelleted feeds (see Ch. 4, Vol. 3; Neale, 1958; Hartman et al, 1959; Perry et al, 1959; Weir et al, 1959; Menzies et al, 1960; Church et al, 1961; Woods and Rhodes, 1962). The evidence further shows that greatly increasing the energy content will not result in increased intake of appreciable amounts (see Ch. 11, Vol. 2). As an example of some data on pelleted rations, see Table 13-3. These and other data generally agree that maximum gains can be achieved with rations containing 75-85% alfalfa hay in pelleted rations, but satisfactory carcasses (US standards) can be produced on diets of 100% pelleted alfalfa. Feed conversion, of course, will usually be improved by increasing the amount of readily digestible materials such as the cereal grains. The cost of feeding pelleted rations may or may not be attractive.

Table 13-3. Effect of amount of alfalfa in pelleted rations on lamb performance.[a]

Item	Percent alfalfa hay in pellet[b]					
	100	90	85	80	75	65
Experiment 1						
Daily gain, g*	204	231		222	245	
Feed conversion	9.0	8.4		8.4	7.4	
Experiment 2						
Daily gain, g*	263		304		281	308
Feed conversion	9.0		8.9		10.0	10.0

[a]From Church et al (1961)

[b]Remainder of pellet made up of 5% molasses and barley.

*Data corrected for differences in GI tract fill.

It might be noted that lambs can be finished to satisfactory USA market conditions on energy levels that are lower than are usually considered satisfactory for steers. Pelleted 100% alfalfa rations, which will be on the order of 52-55% TDN, would take a long time to get a steer to satisfactory USA market conditon.

With respect to the amoung of energy needed in lamb finishing rations, Menzies (1968) has reviewed some of the experimental data. He points out that to get sufficient intake of ground-mixed rations into lambs for satisfactory performance, the ration must contain a higher concentrate level than is required in pelleted rations. This is, primarily, a matter of factors which affect passage of ingesta through the GI tract (see Ch. 7, Vol. 1). Particle size and density are two of the primary factors involved.

Menzies (1968) also pointed out that most of the data on lamb feedlot performance has been expressed in terms of gain, carcass yield, and feed conversion. Information from many different experiments indicates that an increase in the amount of ration concetrates will usually increase the carcass yield. For example, Preston and Burroughs (1958) found that increasing the energy level of lamb diets resulted in substantially increased yields and weight of kidney fat. Menzies (1968) presented data showing that high-energy diets resulted in much more back fat, and Andrews and Ørskov (1970) have shown that increases in the level of feeding gave a linear increase in carcass fat. Thus, it would seem that the commercial lamb feeder should be quite concerned not to feed high energy rations ad libitum in order not to get overfinished carcasses.

In the author's opinion, the NRC recommendations on energy are too high and would probably result in many over-finished lambs if intake was not restricted. The author would suggest TDN values of 55-60% for pelleted self-fed rations and, for mixed unpelleted rations self-fed, something more on the order of 63-65%. It is recognized that more rapid gain and especially improved feed conversion may be obtained with higher energy concentrations, but the overall effect of the carcass must be considered if sheep producers in the USA are even to hold, let alone expand, their current market position. When lambs are self-fed, higher energy levels would only be recommended for heavy, short-fed lambs (35-50 days) which might profitably utilize them or where feed intake is restricted for lambs fed for longer periods of time (60-80 days).

Menzies (1968) has listed some sources of information put out by various experiment stations (field day reports, etc.) relating to feeding lambs low-fiber (high-concentrate) rations. Other reports that have appeared in journals include those of Meyer and Nelson (1961), Andrews et al (1969), Vetter and Ternus (1971) and some of the British work with early-weaned lambs fed barley-fish meal diets. Examples of some rations that have been used are shown in Table 13-10. Information from these various sources indicates that gain of lambs may be as high as with more conventional rations and feed conversions, especially, may be considerably better. Lambs may be more subject to problems such as acute indigestion and enterotoxemia and more care will be required in formulation of diets with respect to minerals such as K and S as well as some of the trace minerals.

Mineral Requirements

NRC (1975) recommendations for Ca vary from 0.37% for light lambs to 0.26% for heavy lambs, and from 0.23% to 0.16% P. The author is not aware of any relative recent studies on the Ca and P needs of feedlot lambs other than a paper by Ullrey et al (1968) in which it was demonstrated that lambs 10-12 weeks of age initially required 0.26-0.29% Ca when they were fed pelleted corn-based diets. There are a number of papers concerned with Ca:P ratios in connection with studies on urinary calculi (see Ch. 13, Vol. 2 for details). It seems unlikely that a Ca deficiency *per se* would be a problem for feedlot lambs due to the short time span they are in the feedlot. With respect to P, a severe deficiency also seems unlikely, although one might develop in animals which were deficient before coming into the feedlot. The author's recommendations for Ca and P would be for about 0.35-0.40% Ca and 0.25% P, keeping in mind that that concentrations should be greater as the energy concentration increases and that too much P may cause more problems than too little. An excess of Ca is unlikely and, if it occurred, would not likely cause any problems other than to increase Zn requirements.

With respect to the other macro minerals, quantitative data of any recent nature on Mg are lacking. With regard to NaCl, Hagsten et al (1975) fed lambs 0, 0.25 or 0.5% added salt. Daily gain was not affected, but feed efficiency was lower by 3 and 26% for the 0 and 0.5% levels. They calculated that 0.2% added salt would be adequate for their ration which had the equivalent of 0.19% NaCl based on Na analyses. Studies with sheep on purified diets indicate a need of about 0.6% K, but finishing rations are unlikely to decline to this level, except with low-roughage diets which are not commonly fed to lambs in most feedlots. In the case of S, data (see Ch. 4, Vol. 2) indicate that a ration level of ca. 0.1 to 0.14% is adequate or that the N:S ratio should be about 10:1 (Moir et al, 1968). Analyses of most feedstuffs in the USA indicate that S content is apt to be above this level. Loggins et al (1971) demonstrated that 0.06% S added to a urea, corn, cob and cottonseed hull ratios resulted in a slight improvement in gain and feed conversion. No data were given on the S content of the basal diet. When feeding a barley, urea, mineral, and vitamin diet to early-weaned lambs, Ørskov and Grubb (1977) observed that addition of 0.4% sodium sulfate improved gain and feed efficiency. Adding 0.12% methionine hydroxy analog in addition had no effect.

With respect to the trace minerals, the reader is referred to Ch. 5 of Vol. 2 for details. Requirements indicated to be adequate for some of the trace minerals are (in ppm): Co, 0.1; Cu, 10; Fe, 40; I, 0.1; Mn, 20+; Se, 0.1-0.3; and Zn, 30-40.

From a practical point of view, the feeder needs to be mainly concerned with NaCl and with the Ca:P ratios. The NaCl need can easily be taken care of by self-feeding salt, and the Ca:P ration can be adjusted when formulating the rations or supplements to be fed. In the case of the trace minerals, Co, Cu, Fe, I, Mn, Se and Zn can be and usually are supplied in trace-mineralized salt. There may be areas where use of commercial mixes of trace-mineralized salt may not be recommended, especially if Cu toxicity is likely. However, in general, this is a sound practice to follow. In areas where Se is not an approved feed ingredient it can be provided, if needed, by intramuscular injections or oral dosing. However, lambs of the age commonly found in feedlots probably have rather low requirements, especially if an adequate intake of vitamin E is maintained.

In areas where Zn is apt to be deficient, it should be kept in mind that Zn requirements are increased as the Ca content of the diet increases. Cu requirements will be influenced by the consumption of Mo and sulfate, so that the actual requirement may be more or less than 10 ppm. A high level of either Mo or sulfate will increase the need for Cu. Conversely, if Mo and sulfate are very low, normal levels of Cu (10-15 ppm) may be toxic to lambs. Likewise, the needs for I may be increased by feeding material that has an appreciable amount of antithyroid compounds of one type or another, particularly plants in the cabbage family such as kale, rape, and cabbage.

Vitamins

Sheep are rather resistant to vitamin A deficiencies (see Ch. 9, Vol. 2), so it is unlikely, in the author's opinion, that deficiencies are apt to occur during relatively short stays in the feedlot. It is, of course, entirely possible that deficient lambs might come into the lot after a summer and fall on poor quality forage. Considering the low cost of providing vitamin A, little harm can be done by including it in the ration, particularly if the diet does not contain a good quality green roughage such as alfalfa hay. In addition, if lambs are fed on high-silage rations, the probability of a deficiency increases (Martin et al, 1968). NRC (1975) recommendations for carotene range from 0.8 to 1.1 mg/kg and, for vitamin A from 588 to 738 IU/kg of diet for finishing lambs. For early-weaned lambs the corresponding recommendations are 2-2.7 mg/kg of carotene and 1,417 to 1,821 IU/kg for vitamin A. Thus, something on the order of 2 mil IU/ton (2.2 mil· IU/metric ton) would meet these recommendations and should be more than ample in nearly every situation.

With regard to other vitamins, there is normally no reason to believe that vitamin E would be deficient in common feedstuffs used in lamb diets. The same applies to vitamin D if any green roughage is used. With the B-complex vitamins, the only problems indicated in the literature are with thiamin. Lambs sometimes develop polioencephalomalacia which responds to thiamin injections (see Ch. 9, Vol. 2). This condition is believed to be due to the development of thiaminase enzymes which degrade the thiamin or to the production of analogs of the thiamin molecule which block normal utilization of thiamin. Unless the condition occurs, there is no need to supplement thiamin routinely to feedlot lambs.

Fats

Ruminants, like other classes of animals, require minimal amounts of unsaturated fatty acids in their diet. However, there is no reason to believe that natural diets are not entirely adequate for lambs.

Water

There is little quantitative data on water needs of feedlot lambs. Data from a variety of sources (see Ch. 2, Vol. 2) indicate a need ranging from 1.4 to 5 kg of water/kg of dry matter consumed. This will be influenced by the fleece length, environmental temperature, mineral content of the feed and water, and consumption of protein, among other things. An average value might be 2-3 kg of water/kg of dry matter consumed. Thus, a lamb weighing 45 kg and consuming 1.5-2 kg of feed might be expected to drink 3-6 kg of water.

FEEDSTUFFS FOR LAMBS

Roughages

As in most other situations for ruminant animals, roughage should be considered first in evaluating which feedstuffs are appropriate in a given situation. This statement generally is more applicable for sheep than for cattle since growing and finishing lambs are capable of utilizing more roughage in their diets while still performing at maximal rates than is true for fattening cattle.

Alfalfa hay is highly preferred roughage for finishing lambs, although they will readily eat other forages such as well-cured clover or grass hays of various kinds. In Europe, particularly in the northern areas, dried grass is often used with good success. In North America, far more alfalfa is utilized than any other hay for feedlot lambs, partly because of its availability. When used in pelleted feeds, as much as 75-85% of the pellet can be alfalfa hay without reducing gain, although feed conversion will be reduced as the amount of the hay reaches very high percentages. Pelleting of grass or alfalfa hay results in appreciable increases in consumption and live weight gain as opposed to feeding long, chopped, or ground hay in a high-roughage diet (Wilkins, 1970; Wainman et al, 1972; Thomson and Cammell, 1979).

Offering long alfalfa free-choice with a high concentrate ground ration may not decrease lamb performance and, in some cases, may decrease costs (Chase et al, 1970). Some of the author's unpublished data on this topic are shown in Table 13-4. In this case the basal ration contained ca. 40% roughage and it was fed in pelleted form to heavy weight feeders. These data indicate clearly that lambs will do very well, at least in some situations, when consuming relatively large amounts of roughage in addition to a basal ration which produces a very satisfactory gain and feed conversion.

With respect to silage, data indicate that sheep utilize silage from mature corn plants to a greater degree than cattle (Colovos et al, 1970). Lambs show an obvious relish for high

Table 13-4. Effect of offering supplemental alfalfa hay and pellets to finishing lambs.

Item	Basal	+ad lib long hay	+0.23 kg pellet/day	+ad lib pellets
		Rations		
Daily gain, g	267	325	273	327
Daily feed intake, kg	2.00	2.26	2.22	2.37
Feed conversion	7.52	6.97	8.15	7.26
Relative cost of gain	100	89	107	94

quality silage and will consume 1.8-2.3 kg (4-5 lb) of silage/day along with some hay and supplemental concentrates. Menzies (1968) has reviewed older research that has been reported on lambs. He points out that lambs consuming 1.8-2+ kg of corn or sorghum silage, 0.45-0.7 kg (1-1.5 lb) of sorghum grain, 0.34 kg (0.75 lb) of alfalfa hay or pellets, and 0.04 kg (0.1 lb) of protein supplement will make gains on the order of 180 g/day (0.4 lb) at very reasonable costs. Silages fed with added concentrates or protein supplements have been used successfully by Illinois workers (Karr et al, 1965). In studies with silages made from wheat or barley plants at various stages of ripeness, Bolsen et al (1976) observed in one experiment that lambs gained more on sorghum silage than on any of 7 wheat silages. In a second trial differences were noted between barley or different varieties of wheat silage.

It might be pointed out that silage feeding may be a convenient method of limiting feed consumption without physically restricting feed intake. Provided the silage and concentrate can be well mixed, this is one means of gradually increasing grain consumption during the adaptation period.

There are many other roughage sources that are quite acceptable for inclusion in lamb fattening diets. Although comparative data on many of these have not been obtained under feedlot conditions, many, many digestion and metabolism trials have been carried out and voluntary feed intake has been reported in a number of instances. For example, Dinius et al (1970) found that material such as various wood products, verxites, clays and sugar cane bagasse could satisfactorily replace corn cobs (about 10% of ration) as roughage substitutes for lambs. Brugman et al (1969) have indicated similarly that sawdust and barley could be substituted for chicken litter and barley, and shredded paper has also been indicated to be useful feedstuff. Pelleted milo stover may be used to produce moderate rates of gain (Bolsen et al, 1975).

Straws have been used in some instances in finishing rations. In a pelleted ration 60% ryegrass straw produced acceptable gains in one instance (Church and Kennick, 1977). When lambs were fed a diet based on whole wheat and protein supplement with wheat straw added at 2, 7 or 13%, Weston (1974) concluded that 2% straw was satisfactory. Digestible organic matter consumption was not affected by the amount of straw given although dry feed consumption increased with each increase in straw in the diet. If the wheat was ground or pelleted, data suggested that more straw was beneficial.

Hydroxide- or ammonia-treated straws have also been used with success. In one instance lambs were fed diets with about 80% hydroxide-treated oats straw. Daily gain and feed conversions were 0.14 kg/day and 12.0 vs 0.18 and 9.5 for diets with similar amounts of dehydrated alfalfa meal (Javed and Donefer, 1970). In another instance it was shown that lambs performed satisfactorily (180 g gain/day; feed conversion of 9.7) when fed diets with 50% hydroxide-treated ryegrass straw (Church and Kennick, 1977). When lambs were fed hydroxide-treated rice straw along with alfalfa and barley, the treated straws produced an acceptable rate of gain and feed conversion when 36% straw was included, but including 72% straw reduced performance (Garrett et al, 1979). Treating corn cobs with hydroxide and then ensiling has been shown to result in increased gain of lambs vs hydroxide treatment only (145 vs 118 g/d). In this case protein supplements were fed in addition to the cobs (Rounds et al, 1976).

Cereal Grains

There is very little recent information in the literature in which the various cereal grains have been compared for finishing lambs, although there is an appreciable amount of older literature. Some of the papers that might be mentioned include a comparison of low-moisture corn vs ensiled high-moisture corn (Benjamin and Jordan, 1960), high-amylose corn (Preston et al, 1964), regular, waxy, and white sorghum grain (Nishimuta et al, 1969), waxy vs regular corn (Braman et al, 1973), and rolled barley vs whole and ground oats (Andrews and Ørskov, 1970). Data are also available where lambs have been finished on whole wheat and straw (Weston, 1974) or whole wheat-mineral combinations (McManus et al, 1972) under Australian conditions. In addition to these, there are a number of papers dealing with digestibilities of cereal-based diets; for example that of Sherrod and Albin (1973) in which information was published on waxy, floury, intermediate, regular, and bird-resistant sorghum grains, or that of Prigge et al (1976) which showed that lambs utilized N from high-moisture corn more efficiently than from dry corn.

Generally, it would seem that the tabulated values for DE, ME or TDN are appropriate measures for selecting the cereal grains for lambs. There may be differences in palatability and feed consumption due to the physical form or when high-concentrate diets are fed. The data of Andrews and Ørskov (1970) illustrate differences that may be observed. In their work they found that consumption of an oat diet was greater than with rolled barley, but digestible dry matter intakes were not different.

Some comment should be made about wheat, a feed that requires a longer adaptation period than most other cereal grains. Lambs are particularly susceptible to acute indigestion from over consumption of wheat, particularly newly harvested grain. If fed in high percentages in any type of diet, but especially in a high grain diet, a high rate of morbidity and death is likely to occur unless the adaptation period is relatively long (4 weeks+) or other measures are taken to prevent acidosis and rumenitis (McManus et al, 1972, 1973).

When lambs are given a choice of feedstuffs, the relative palatability may differ considerably. Data on this topic are shown in Table 13-5. Summarization of these data indicate that order of acceptance of these different feedstuffs to be: alfalfa > soybean meal = beet pulp > corn peas = barley = wheat mill run. In a conventional ration—where grains make up less than half of the ration—the relative palatability is probably of much less importance than in rations in which little roughage is fed.

Table 13-5. Relative intake (%) of different pelleted feedstuffs given in a cafeteria trial.[a]

Feedstuffs	Period number 1	2	3
Alfalfa hay	46.9	38.8	27.9
Barley	9.8	8.2	
Corn		18.7	
Wheat	7.4		
Beet pulp			24.5
Wheat mill run			7.8
Soybean meal	21.4	24.9	27.1
Peas	14.4	9.4	12.3

[a]From Goatcher and Church (1969)

The effect of feed processing and preparation methods has been covered in Ch. 3, so there is no need to duplicate that information here. Suffice it to say that lambs generally consume and digest whole grains to a greater extent than do cattle. This may be because they probably chew them more thoroughly. Barley is generally preferred to a greater degree when rolled, and utilization of small hard seeds as milo will usually be increased if the seed is cracked or rolled but other grains are not usually improved for finishing lambs by elaborate processing.

Molasses and Other Liquids

Molasses is commonly used in pelleted lamb finishing rations, partly as a binder. However, there is very little research on which to base any recommendations for use. Jordan and Hanke (1958) found that lambs consumed ca. 0.5 kg/d of molasses when fed free choice with shelled corn and bromegrass hay; performance was adequate. Merion et al (1965) found that lambs did well with 10 or 20% of cane molasses in their rations, but performance decreased when the rations contained 30 or 40%. More recently Zorrilla and

Merino (1972) fed 20% molasses and achieved a gain in lambs of 181 g/d; with 40% molasses gain decreased to 100 g/day. Hartnell and Satter (1978) fed weanling lambs a ration containing soybean meal which had been extruded with either 10% hemicellulose extract (Masonex) or 10% cane molasses and noted no differences in performance of the lambs. Likewise, Perry et al (1976) have determined that condensed soybean solubles (8% CP) could replace up to 10% of corn in pelleted ration without any appreciable effect on gain or feed conversion. Liquid whey (16% CP dry basis) appears to be a satisfactory feedstuff for sheep (Anderson, 1975), but no reports are available for feedlot lambs.

It might be noted that sheep do not have a particularly strong preference for sweets, thus molasses probably does not have the influence on palatability that it may have for other ruminant species. At any rate, data indicate that relatively large amounts of molasses can be used, if desired.

Fats

Animal fats (tallow, etc.) probably have a palatability value which might merit being included if rations are otherwise unpalatable and, if the price is in reasonable range. However, since lambs do not generally require high-energy diets for satisfactory performance, the general practice is to feed very little fat in commercial lots. Most of the experimental results have not been too encouraging.

Clapperton (1969) has reported that sheep were eager to consume linseed oil fatty acids. When fed at levels of 2%, he observed a substantial increase in consumption of rations made up of 50% grass pellets and 50% lamb creep pellets.

With respect to research on fattening lambs, Jordan et al (1958) noted that addition of stabilized tallow to soybean meal, which was included in the ration, resulted in increased gains in 3 of 4 experiments. When using pelleted rations, Perry et al (1959) observed that inclusion of 7.5% yellow grease reduced feed consumption and daily gain. The addition of large amounts of corn oil (15%) has also been shown to reduce gains, particularly on low-fiber rations. However, smaller amounts appear to be utilized more efficiently. Some of the author's unpublished data comparing tallow and corn oil are shown in Table 13-6. In another case lambs were fed high-roughage (93-99%) or high-concentrate (80%) rations with or without 6 or 8% hydrolyzed animal-vegetable fat (Johnson and McClure, 1972). Neither 6 or 8% of this product improved gains or feed conversion on either type of ration. One pen fed on high-concentrate with 6% corn oil was more efficient.

Table 13-6. Effect of added corn tallow on lamb performance.

	Fat added to ration, %			
Item	1	2	3.5	5
Tallow				
daily gain, g	223	197	197	195
feed intake, g/day	1819	1674	1592	1601
feed conversion	818	846	806	823
relative cost of gain	100	108	98	111
Corn oil				
daily gain, g	179	175	196	186
feed intake, g/day	1692	1637	1642	1479
feed conversion	936	934	839	796
relative cost of gain	100	94	88	82

If the responses in these experiments are typical of what might be expected with other rations and other situations, then it would appear that lambs may utilize corn oil more efficiently than tallow or hydrolyzed fat. The data in Table 13-6 show that feed conversion gradually improved as the corn oil was increased, as did the relative cost of gain; whereas with the tallow, feed conversion was not changed appreciably and intake was reduced somewhat more by tallow than by corn oil.

At least one feedlot study of a very limited nature has been done with protected fat (protected against rumen action by treating with formaldehyde). In this instance over a short-term study and with very limited numbers of lambs (3-4/treatment), feeding protected cottonseed oil (8%) appeared to increase daily gain and improve feed conversion (Shell et al, 1978).

Other Energy Sources

Numerous other by-product or waste materials have been fed to sheep or there have been reports on digestibility; however, very little information is available with regard to performance of lambs under feedlot

Table 13-7. Performance (3 trials) of lambs fed different protein sources.[a]

Item	Protein concentrate					
	Guar meal	Cottonseed meal	Soybean meal	Blood meal	Feather meal	Urea
Daily gain, g	200	280	290	250	250	260
Feed intake, kg/day	1.22	1.48	1.38	1.37	1.39	1.41
Feed conversion	6.10	5.29	4.76	5.48	5.56	5.42
Dry matter digestion, %	83.1	79.6	81.6	82.1	82.3	84.3

[a]From Huston and Shelton (1971)

conditions. Such materials would include citrus pulp, beet pulp, raw, cooked, dried and ensiled potatoes, spent coffee grounds, root crops, waste from food processing plants, and others. Other than for beet pulp, very little recent information is to be found on sheep. With beet pulp, it has been demonstrated that lambs do very well when fed only dried beet pulp combined with 1% urea and pelleted (Bhattacharya et al, 1975).

Protein Supplements

A wide variety of protein supplements has been used in practice for fattening lambs, generally with relatively little difference in the response. Some relatively recent data of Huston and Shelton (1971) are shown in Table 13-7 in which various protein sources were compared. In these experiments the rations were formulated so that the crude protein was the same (12.6%) in each ration. The authors noted that the lambs appeared to require some time for adjustment to the guar meal. The data also indicate that soybean meal resulted in the greatest and most efficient gains. Kromann et al (1966) have reported that dehydrated alfalfa produced the same response as cottonseed meal and Theurer et al (1968) found that soybean meal and corn gluten meal produced similar responses with feedlot lambs.

Other relatively inexpensive sources are frequently available to lamb feeders. Work with alfalfa and clover seed screenings and cull peas indicates that these products can be well utilized in lamb rations although pesticide residues may be a problem at times (Church et al, 1967). More recent data with cull beans indicate that as much as 12% of the diet could be fed to fattening lambs without any effect on performance; above 12% gain and efficiency were depressed (Doyle and Hulet, 1978). Treating dehydrated alfalfa meal with formaldehyde to protect it from rumen degradation did not result in any improvement over untreated meal, presumably because the net effect of treating with formaldehyde was similar to the heating during dehydration (Reynolds et al, 1978).

In the tropics cassava is a plant which has generated considerable interest for production for animal feed because of its high yield. In one instance when uncooked or cooked cassava-urea were fed with low quality hay, the results were better than feeding no protein supplement but less than if sesame-cassava meal was fed (Schultz et al, 1972).

Various liquid ingredients may be used commercially. Of these, ammonium lignin sulfonate has been used extensively for cattle. In one growth trial with sheep they were fed 2, 4 or 8% and compared with similar rations containing isonitrogenous amounts of urea. Weight gains were not different but there was a trend for reduced gain on the higher levels of the lignin sulfonate (Croyle et al, 1975). In a growth trial in which liquid whey was included in a liquid protein supplement, data suggested similar performance to a control ration containing soybean meal (Steeds et al, 1979). Older data where liquid protein supplements have been fed to fattening lambs suggest that performance was not as good as with dry supplements.

Non-Protein Nitrogen Supplements

There have been many, many metabolism and digestion experiments with sheep in which urea and other NPN sources have been studied. Although a wide variety of NPN compounds can be utilized safely, many of these are not economically feasible. Generally, data from a variety of sources (Hoar et al, 1968; Theurer et al, 1968; Amos et al,

1970; Huston and Shelton, 1971; Ørskov et al, 1971, 1974; Bhattacharya and Pervex, 1973; Shiehzadeh and Harbers, 1974; Church and Kennick, 1977; Britton et al, 1978) indicate that urea as the only supplemental source of N will not result in as rapid gains as a high quality protein such as soybean meal. Partial replacement will, however, often result in gain almost if not quite as high and, in some instances, more efficient feed conversion.

Other NPN sources that have some commercial application at times include biuret and ammonium phosphates. Although biuret is less toxic than urea, evidence indicates that an appreciable time may be required for sheep to adapt to it (Schaadt et al, 1966; Farlin et al, 1968). This fact plus its rather high cost do not provide any valid reason to recommend the use of biuret for finishing rations for lambs. In the case of diammonium phosphate, it is relatively unpalatable to sheep (Schaadt et al, 1966), so it would be best used at very low levels, if at all. With regard to other NPN sources, Ørskov and Grubb (1977) found that addition of ammonium sulfate to a barley-urea-mineral diet resulted in increased gains and improved efficiency of early-weaned lambs. Bolsen et al (1973) observed that ammonium sulfate plus urea was not as good as urea alone but Eng and McNeil (1975) stated that ammonium sulfate in combination with urea or cottonseed meal was quite satisfactory for lambs on high-energy finishing rations.

Non-Nutritive Additives

This subject has been covered in some detail in Ch. 4 of this volume, thus there is no need for much repetition here. Suffice it to say, with respect to the antibiotics, that compounds such as the tetracyclines have usually shown a small but relatively consistent response in terms of daily gain, fewer digestive disturbances, and less enterotoxemia. Appropriate levels in complete rations are on the order of 10-25 mg/kg (5-10 mg/lb). There is some relatively recent evidence that sulfamethazine (55 mg/kg or 25 mg/lb of ration), when given in combination with chloretracycline, may improve both daily gain and feed conversion as well as offering protection against enterotoxemia. Although monensin has been shown to improve feed efficiency of fattening lambs (Horton, 1979; Horton and Stockdale, 1979; Joyner et al, 1979), it is not an approved feed additive for lambs in the USA. Horton and Stockdale (1979) have observed that amprolium fed with monensin resulted in improved gain and efficiency.

With respect to buffers, some data indicate that Na bicarbonate may improve gain when included in rations at 2% (Woolfitt et al, 1964) and other information indicates that a mixture of Mg hydroxide and K bicarbonate prevented lambs from going seriously off feed when they were suddenly changed from a 60% roughage ration to a high-concentrate ration (Calhoun and Shelton, 1969). In a recent example, when lambs were abruptly changed from a ration with all hay to 92% concentrate, inclusion of 2% Na bicarbonate reduced lamb death loss from 12/pen to 2/pen (Dunn et al, 1979). Inclusion of a variety of bicarbonates or hydroxides for the total feeding period does not appear to be advantageous (Loggins et al, 1968; Shelton et al, 1969; Shelton and Calhoun, 1970; Dunn et al, 1979) and may result in a greater incidence of urinary calculi (Crookshank, 1966; Hoar et al, 1969; Dunn et al, 1979).

Sodium bentonite has also been investigated as a potentially useful feed additive. Huntington et al (1977) and Dunn et al (1979) observed an improved performance of lambs on high-concentrate rations during the first 2-3 weeks, but not much response later in the period except when fed with bicarbonate. Lambs tended to eat more consistently and has less diarrhea. Britton et al (1978) noted that bentonite added to diets with soybean meal-urea improved performance over similar diets without the bentonite.

HANDLING NEW ARRIVALS IN THE FEEDLOT

The average lamb which comes to the feedlot has been stressed in a variety of ways that are detrimental to its health and well-being. They have been rounded up, sorted, sometimes weaned, and transported over appreciable distances, often without feed or water. In addition to being exposed to new surroundings and strange feed and water, they have probably been exposed to diseases and parasites to which they may have no immunity, so it is no wonder that death losses may be relatively high.

It is beneficial to provide new arrivals with rest and a medium quality roughage and, perhaps something such as a limited amount of alfalfa pellets for a few days while they recover from these stresses. It may be difficult to get range lambs to drink, so water fountains that have water dripping or spraying to attract the lambs are often helpful. After a period of rest, most lamb feeders find it advisable to drench for internal parasites, depending upon the origin of the lambs. In some countries approved medicines are available for treating for liver flukes which may be a severe problem for lambs. Treatment for external parasites may also be advisable. Vaccination for enterotoxemia is often done, depending upon the type of ration to be fed, and in the USA, vaccination for soremouth may be advised.

Depending upon the time of year, most feeders will shear lambs. The lambs tend to gain more weight; the amount of fly strike is reduced during the fly season, and it is usually profitable to do so in terms of the wool obtained. The average lamb will have sufficient regrowth of wool to produce a no. 1 pelt by the time it is slaughtered. Shearing during wet, cold weather would not be advised, however, even if adequate shelter is available. In work in New Zealand, Elvidge and Coop (1974) determined that shorn lambs with no shelter and at a temperature of 16-17°C had an increased feed requirement during the next month amounting to 18%; the increase was 24% if not under shelter. If the temperature was at 7-10°C, the feed requirement of housed sheep increased by 46% and by 76-78% if no shelter was available.

STARTING LAMBS ON FEED

When starting lambs on finishing rations, most feeders want to get lambs to a full feed of concentrate as rapidly as possible. It must be remembered, however, that some time is required for adaptation, particularly for animals that have come to the feedlot from the range or other places where they have had no previous or recent exposure to feedstuffs containing large amounts of readily available carbohydrates. Following a change from grass to a ration with a high level of starch, one might expect that as much as 4-6 weeks may be required for reasonably complete adaptation of the rumen microorganisms. This particularly applies to rations containing wheat (see Ch. 17, Vol. 1) and probably to other concentrates as well. A sudden change to a high cereal diet may, and often does, result in acute acid indigestion as a result of lack of adaptation of the rumen microflora. Enterotoxemia may also be related to incomplete rumen digestion. In addition to adaptation of rumen microorganisms, some adaptation is undoubtedly required in the production of intestinal and liver enzymes. Unfortunately, little information is available on this subject. All of this is not to say that lambs may require 6 weeks to get up to full feed, but it is meant to point out that an abrupt change may present problems to the animal that take some time to overcome. Providing time and size of the animals is not critical, then a gradual increase in concentrate feeding is recommended.

Self-Feeding Pelleted Rations

With respect to getting lambs on feed, there are numerous possibilities. Where heavyweight feeders are involved, it is often advisable to get them on feed rapidly so that the carcass is not too large for the market. In starting lambs on feed, pelleted feeds are particularly useful. Lambs that have never seen any kind of feed other than pasture or range forage will readily take to pelleted rations with a minimal amount of problems such as acute indigestion and enterotoxemia. Lambs can be started on self-fed pelleted rations that have sufficient energy to provide an adequate finish, thus no further changes need be made in the ration during the finishing period. As pointed out in another section, pelleted rations require less concentrate than mixed rations to provide an adequate finish. Although the cost may not always be favorable because of processing expenses, the feeding of pelleted feeds is a highly preferred management practice.

Restricted Feeding of Concentrates

If pelleted feeds are not to be used or if only to get lambs started on feed, then it is advisable to start lambs on concentrates gradually. Concentrate intake may be limited by mixing with chopped hay, silage, cottonseed hulls, etc., with gradually increasing amounts of concentrates over a period of 2-3 weeks. Another method is to hand feed the concentrates, but this is hardly convenient in a large feedlot although it may be practical under farm conditions. With either of these methods, some lambs may still get sufficient concentrate to cause them trouble. Another possibility would be to include an unpalatable ingredient such as diammonium phosphate, sodium bicarbonate (see section on non-nutritive additives) or a high level of salt in the ration so that consumption is held down. The controlling agent can then be gradually reduced over a period of time to allow a gradual increase in consumption.

Self-Feeding High-Concentrate Rations

In the event that it is deemed desirable to increase the feeding of concentrates at a rapid rate or to self feed them, then it is advisable to allow for some means of reducing the incidence of acute indigestion or enterotoxemia that might otherwise occur. There are several possibilities here. One means would be to provide the concentrate in a physical form that is not exceptionally palatable during this initial period. For example, use of finely ground feeds as compared to rolled or cracked feeds that lambs might prefer. Other and probably more effective means include the addition of a relatively high level of antibiotics such as the tetracyclines in the ration, perhaps along with sulfamethazine, at levels of about 25 mg/lb (55 mg/kg of feed) for the first 2 weeks, after which time the antibiotic level would normally be reduced and the sulfamethazine would probably best be removed from the ration. A third possibility is to include 2-5% of Na bicarbonate or other buffers in the concentrate mix for the first 2 weeks of feeding. Buffers help to prevent the development of acute indigestion, although they may not greatly improve the rate of gain (see section on non-nutritive additives). All of these methods have worked well at times in the hands of experienced feeders.

FEEDING FACILITIES FOR LAMBS

Menzies (1968) has discussed some of the published information on facilities for lambs. Some recommendations indicate that lambs need on the order of 16-25 ft^2, each. However, this would depend on the particular facilities, the type of bedding used, if any, the use of mounded lots, etc. Kansas' recommendations are that not more than 500 lambs should be fed/lot. Evidence indicates that the density can be increased to about 4 ft^2/lamb where lambs are fed over slotted floors, but there is little information on this subject with large numbers of lambs. The high cost of developing confinement feeding facilities does not seem to be justified at the present time, except for small farm feeding operations where it may be feasible to construct the required facilities. Feeding over slotted floors has the advantage that re-infestation with internal parasites is less of a problem than with conventional feeding facilities.

Lamb feeders use a wide variety of feeding facilities. Some that are in common use include: fence-line bunks supplied by a feed wagon, various auger systems that take feed from a central supply to individual lots, feed bunks down the middle of a lot, or self-feeders that are particularly adapted to the use of pelleted feeds. Some of these are illustrated in Fig. 13-1 through 13-3. All have been used successfully in different situations, although the very large lots tend to use self-feeders or fence-line bunks more than the other methods.

The amount of feeder space required/lamb will depend upon the type of ration used, how frequently the lambs are fed, and the type of feeder. With self-feeders, normally not all of the lambs will want to eat at any given time, thus the space required may be as little as 10 lambs/foot of feeder space or less. When a system and ration are used that stimulates all lambs to eat at once, then about 1 ft of space/lamb may be required.

Salt-mineral feeders should be provided at convenient locations and should be sheltered from rain and snow if loose mixes are used. The number required/pen will depend on the number of lambs, of course, and on the design

Figure 13-1. An example of self feeder for pelleted diets. This type of feeder is often used in large commercial lots.

Figure 13-2. A simple feeder which can be used for both concentrates and roughages. Courtesy of C.S. Menzies.

Figure 13-3. A feeder similar to that shown in 13-2, except that feed can be supplied from an alley. Courtesy of C.S. Menzies.

of the feeder. If only 2-3 lambs can have access at once, then probably there should be one feeder/150-200 lambs, as a rule of thumb. Blocks are readily utilized, but loose mixes may be required in order to supply some mineral needs.

Shaded shelters should be provided for lambs subjected to hot summer temperatures and some type of a windbreak is highly desirable for winter feeding conditions. The amount of winter shelter needed will depend somewhat on the amount and nature of precipitation and on the shearing practices utilized.

Finishing on Pasture

In many areas it is a common practice to fatten weaned lambs on pastures of one type

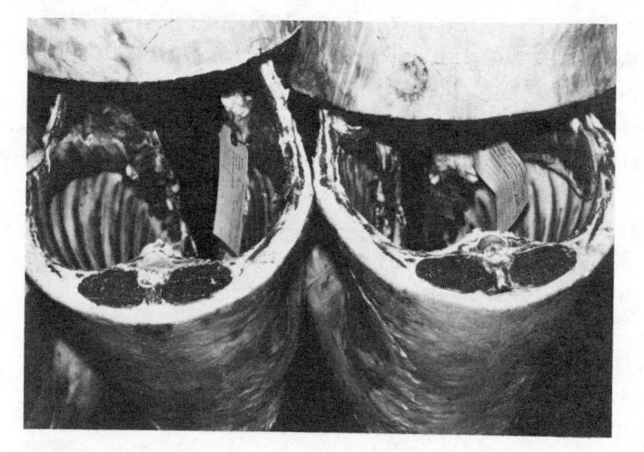

Figure 13-4. Effect of finishing on pasture. Left. Carcass of lamb fattened in dry lot. Right. Carcass of lamb finished to about same backfat thickness; this carcass is larger since the lamb did not fatten quite as rapidly. Courtesy of D.L. Thomas.

or another. Native or improved pasture species may be used. Reid et al (1978) demonstrated that early-weaned lambs grew more rapidly when pastured on bromegrass and orchardgrass than on tall fescue or perennial ryegrass (Figure 13-4), however no carcass data were presented. Other, older work on this topic has been reviewed by these authors. In addition to pasture species, winter wheat has often been used for fattening lambs with very good results (Noble et al, 1961). More recently, Ely et al (1979) have compared the response of early-weaned lambs fattened on pasture with and without supplemental feed to comparable lambs finished in a dry lot. The rations fed were corn, cottonseed hulls, soybean meal, and molasses base with either 13 or 16% crude protein. The unsupplemented pastured lambs gained significantly less than the other three groups (Table 13-8). Those fed concentrate on

Table 13-8. Performance of pasture-fed and dry lot-fed lambs.[a]

Treatment	Daily gain,	Feed consumed, g	Feed efficiency kg	Carcass fat, %
Pasture	156	---	---	23.9
Pasture + suppl.[b]	263	1.00	3.89	27.3
Dry lot, 13% CP	267	2.07	7.98	33.2
Dry lot, 16% CP	246	1.74	7.23	31.6

[a]From Ely et al (1979); data on lambs carried from an initial weight of 31.8 kg to a slaughter weight of 49.9 kg.
[b]Supplement same as feed given dry lot 13% CP.

Table 13-9. Examples of moderately high concentrate rations used for lambs.

Item	Texas rations[a]			Oregon rations[b]		
	1	2	3	4	5	6
	%					
Roughages						
Alfalfa hay		10	10	68.9	36.3	31.4
Cottonseed hulls	25					
Barley straw						5.2
Wheat straw					8.1	
Carbohydrate feeds						
Barley				20.0		
Beet pulp				10.0		16.1
Molasses	5				5.0	5.0
Oats		8.6	10			
Sorghum grain	50	60	74.2			
Wheat flour screenings					20.0	
Protein sources						
Alfalfa seed screenings					20.0	10.1
Cottonseed meal	19					
Cull peas					8.2	31.2
Soybean meal		17.9				
Urea			2.3		1.2	
Miscellaneous						
Antibiotic premix				0.1		
Limestone	0.5	1.75	1.75			
Salt	0.5	1.75	1.75	1.0	1.0	1.0
Dicalcium phosphate					0.15	
Approximate crude protein, %	12.7	15.9	15.1	12.4	17.0	16.7

[a]Examples from Huston and Shelton (1971)

[b]Pelleted linear programmed examples from Church et al (1967)

pasture required less than half of the dry feed/unit of gain. Carcasses of grass-fed lambs contained more protein and less fat and had lower quality grades but higher yield grades. The lambs fed in dry lot tended to be too fat at this weight, while those fed concentrate on pasture were more desirable overall. Thus, this procedure of feeding concentrate on pasture has considerable merit where it can be practiced.

EXAMPLES OF LAMB FINISHING RATIONS

Due to the fermentative and synthetic activity that takes place in the rumen, the ruminant animal is capable of adapting to a tremendous variety of feedstuffs and ration components. As a result, there is no one ration that is likely to be superior to many others. Consequently, it may be a futile exercise to show examples of rations that will result in adequate performance under practical conditions. Nevertheless, many readers, especially students, like to see examples, therefore this is the reason for including them in this chapter.

The take home lesson on this topic for the feedlot manager is to remain flexible since particle size, palatability and availability of a given feedstuff are extremely important factors in determining whether it is an acceptable ingredient. Almost any ingredient acceptable to livestock can be used in some manner and in some amount if it is desirable to do so. Low palatability can often be masked by other ingredients and, where selection is not allowed, palatability is of less importance as a regulator of voluntary feed consumption. Particle size may be effectively altered by

Table 13-10. Examples of high-concentrate rations used in experimental work.

	Iowa rations[a]		Illinois rations[b]			
Item	1	2	3	4	5[c]	6[d]
Feed ingredients						
Barley, whole						84.5
Corn, gr.					58.8	
Corn, whole shelled	71.35	94.05				
Corn, high moisture			96.0	98.0		
Corn cobs	10.0					
Alfalfa hay	10.0					
Cottonseed hulls					10.0	
Molasses	5.0	2.0			5.0	
Fish meal						14.0
Soybean meal			3.2		23.0	
Urea	1.0	1.0		0.6		
Limestone	0.5	1.3	0.4	0.4	0.7	1.5
Dicalcium phosphate	0.5					
Calcium sulfate			0.05-0.11	0.05-0.11		
Trace-mineralized salt	0.6	0.6	0.6	0.6	1.5	+
Vitamin A premix	1.0	1.0				+
DES premix	0.05	0.05				
Antibiotic premix			+	+	1.0	
Animal performance						
Daily gain						
g/day	227	222	186	209	267	
Feed conversion	6.32	5.49	5.89	5.49	7.23	

[a]Data from Vetter and Ternus (1971)
[b]From 1967 Illinois Sheep Day Report AS-646f
[c]Ely et al (1979)
[d]Ørskov et al (1976)

grinding, pelleting or mixing with liquids. Thus, availability, physical compatability with other ration ingredients, and cost are the critical factors to consider.

Some examples of high-roughage pelleted rations have been shown in Table 13-3 and one other (#4) in Table 13-9. When pelleted, these rations will result in very adequate performance, although the feed conversion will be in favor of those containing less hay.

Examples of sorghum-based rations (#1-3) are shown in Table 13-9. In these examples the amount of roughage varies from 10-25%. Also shown in Table 13-9 are some rations (4-6) based on the use of by-product feedstuffs such as alfalfa seed screenings and flour screenings, beet pulp, or barley. These 6 different rations will give the reader an idea of the variety of rations that can be used successfully to finish lambs.

As a further example some high concentrate rations are shown in Table 13-10. These range from 80 to 100% concentrate. Although performance has sometimes been quite good, particularly for feed efficiency, usually such high-energy rations have a number of undesirable things associated with them (discussed previously) and the author would not recommend them for these reasons.

EFFECT OF DIET ON LAMB MEAT FLAVOR

Although the rumen results in modifications of many chemical components which pass through it resulting in relatively uniform fatty acid composition in the tissues, among other things, some diets have been noted to have an influence on meat flavor or other quality characteristics. For example, with lambs fattening on pasture, it has been observed that rape forage may result in an unattractive foreign flavor and aroma (Wheeler et al, 1974). In another case meat from wethers grazed on pure stands of rape, vetch or oats was evaluated. Grazing on rape often produced a meat with a nauseating aroma and flavor, vetch gave an intense meaty flavor and oats resulted in meat with a pungent odor and flavor but less acceptable than from more typical pastures (Park et al, 1972). Formaldehyde-protected sunflower seed has been noted to produce a bland pork-like sweet flavor (Jagusch, 1975; Park et al, 1978) and dry diets containing 70% oats and 27% barley were considered to produce a less desirable flavor than diets with 19% barley, 8% oats and 70% hay (Vesely and Hironaka, 1975).

Yellow fat is objectionable in the meat trade. In one instance it was determined that this is a genetically-linked characteristic (Kirton et al, 1975). The sweet, pork-like flavor from feeding protected sunflower meal has been related to elevated levels of linoleic acid (Jagusch, 1975). Feeding high-energy diets results in less carcass protein and fatter carcasses (mentioned previously). In one instance carcasses of lambs receiving high-energy diets were judged to be physiologically older than those from lambs on lower energy diets. However, the meat from the low-energy diets was judged to be inferior in tenderness (Crouse et al, 1978).

In the meat trade, soft, oily carcasses are discriminated against and they have a less desirable flavor than those with firm fat (Campion et al, 1976). High concentrate feeding has been related to soft carcasses (Hartman et al, 1959; L'Estrange and Mulvihill, 1975), although whole oats have been shown to be better than whole barley (Ørskov et al, 1975). Field et al (1978) observed that high-corn diets (73-79%) resulted in more unsaturated fatty acids in subcutaneous fat than low-corn (36-38%), although they did not detect any differences in flavor or aroma due to nutritional treatments. They did observe that total polyunsaturated fatty acids decreased as weight increased and that panel scores for flavor were less desirable for rib roasts from ram lambs weighing 68 kg as compared to younger rams weighing 41 kg.

These data on relationship of diet to preferred flavor and aroma are rather incomplete, but indicate that diet can, in some situations, affect meat flavor. It is to be hoped that information on this topic will be expanded in the future.

MISCELLANEOUS MANAGEMENT PROBLEMS

Death losses in the feedlot may vary tremendously, depending upon the quality of lambs, but also on the severity of stresses imposed and on the level of applied management and nutrition. Pierson (1967, 1970) reported that death losses in Northern Colorado feedlots averaged about 2.5%, with enterotoxemia being the major cause. He has suggested that lambs should be vaccinated with *Clostridium perfringens* type D toxoid before arrival at the feedlot, where feasible. Some evidence indicates that a second vaccination will give additional protection. As was noted in the section on non-nutritive additives, the use of antibiotics and sulfas may help to reduce this problem as will the use of high-roughage pelleted feeds.

Urinary calculi may also be a very severe problem at times (Ch. 13, Vol. 2 for details). This can largely be prevented by proper formulation of the ration so that an excess of P is not present. Base-forming mineral supplements such as disodium phosphates, Na tripolyphosphate or dipotassium phosphate may result in an increased incidence of calculi, also. Thus, by controlling these factors, the problem can usually be prevented. In the event that calculi do develop, then the problem may be partially controlled by the use of acid-forming supplements such as ammonium chloride and phosphoric acid or additional Ca in the form of $CaCl_2$. When ammonium chloride is used, levels of 0.5 to 1% of the ration have been used with reasonable success.

References Cited

Amos, H.E., D.G. Ely, C.O. Little and G.E. Mitchell. 1970. J. Animal Sci. 31:767.

Anderson, M.J. 1975. J. Dairy Sci. 58:1856.

Andrews, R.P., M. Kay and E.R. Ørskov. 1969. Animal Prod. 11:173.
Andrews, R.P. and E.R. Ørskov. 1970. Animal Prod. 12:335; J. Agr. Sci. 75:11,19.
ARC. 1965. The Nutrient Requirements of Farm Livestock. No. 2 Ruminants. Agricultural Research Council. Her Majesty' Stationery Office, London.
Benjamin, W.E. and R.M. Jordan. 1960. J. Animal Sci. 19:515.
Bhattacharya, A.N., T.M. Khan and M. Uwayjan. 1975. J. Animal Sci. 41:616.
Bhattacharya, A.N. and E. Pervez. 1973. J. Animal Sci. 36:976.
Bolsen, K.K., L.L. Berger, K.L. Conway and J.G. Riley. 1976. J. Animal Sci. 42:185
Bolsen, K.K., G.Q. Boyett and J.G. Riley. 1975. J. Animal Sci. 40:306.
Bolsen, K.K., W. Woods and T. Klopfenstein. 1973. J. Animal Sci. 36:1186.
Braman, W.L., E.E. Hatfield, F.N. Owens and J.D. Rincker. 1973. J. Animal Sci. 37:1010.
Britton, R.A., D.P. Colling and T.J. Klopfenstein. 1978. J. Animal Sci. 46:1738.
Brugman, H.H., H.C. Dickey and J.C. Goater. 1969. J. Animal Sci. 29:153 (abstr).
Calhoun, M.C. and M. Shelton. 1969. J. Animal Sci. 29:154 (abstr).
Calhoun, M.C. and M. Shelton. 1970. J. Animal Sci. 31:237 (abstr).
Campion, D.R., R.A. Field, M.L. Riley and G.M. Smith. 1976. J. Animal Sci. 43:1218.
Chase, L.E. et al. 1970. J. Animal Sci. 31:173.
Church, D.C., C.W. Fox and T.P. Davidson. 1966. Proc. West. Sec. Amer. Soc. Animal Sci. 17:241.
Church, D.C. and W.H. Kennick. 1977. Oregon Agr. Expt. Sta. Tech. Bul. 140.
Church, D.C., J.A.B. McArthur and C.W. Fox. 1961. J. Animal Sci. 20:644.
Church, D.C. et al. 1967. Oregon Agr. Expt. Sta. Sp. Rpt. 241.
Clapperton, J.L. 1969. Proc. Nutr. Soc. 28:57a (abstr).
Colovos, N.F. et al. 1970. J. Animal Sci. 30:819.
Craddock, B.F., R.A. Field and M.L. Riley. 1973. Proc. West. Sec. Amer. Soc. Animal Sci. 24:120.
Crookshank, H.R. 1966. J. Animal Sci. 25:1005.
Crouse, J.D. et al. 1978. J. Animal Sci. 47:1207.
Croyle, R.C., T.A. Long and T.V. Hershberger. 1975. J. Animal Sci. 40:1144.
Dinius, D.A., A.D. Peterson, T.A. Long and B.R. Baumgardt. 1970. J. Animal Sci. 30:309.
Doyle, J.J. and C.V. Hulet. 1978. Proc. West. Sec. Amer. Soc. Animal Sci. 29:72.
Dunn, B.H., R.J. Emerick and L.B. Embry. 1979. J. Animal Sci. 48:764.
Elvidge, D.G. and I.E. Coop. 1975. N. Zeal. J. Expt. Agr. 2:397.
Ely, D.G. et al. 1979. J. Animal Sci. 48:32.
Eng, K.S. and J. McNeil. 1975. Feedstuffs 47(51):20.
Farlin, S.D., U.S. Garrigus and E.E. Hatfield. 1968. J. Animal Sci. 27:785.
Field, R.A. et al. 1978. J. Animal Sci. 47:858.
Garrett, W.N., H.G. Walker, G.O. Kohler and M.R. Hart. 1979. J. Animal Sci. 48:92.
Goatcher, W.D. and D.C. Church. 1969. Proc. West. Sec. Amer. Soc. Animal Sci. 20:151.
Hagsten, I., T.W. Perry and J.B. Outhouse. 1975. J. Animal Sci. 40:329.
Hartman, R.H., D.L. Staheli, R.G. Holleman and L.H. Horn. 1959. J. Animal Sci. 18:1114.
Hartnell, G.F. and L.D. Satter. 1978. J. Animal Sci. 47:935.
Hoard, D.W., L.B. Embry, H.R. King and R.J. Emerick. 1968. J. Animal Sci. 27:557.
Hoar, D.W., R.J. Emerick and L.B. Embry. 1969. J. Animal Sci. 29:647.
Horton, G.M.J. 1979. Proc. West. Sec. Amer. Soc. Animal Sci. 30:241.
Horton, G.M.J. and P.H.G. Stockdale. 1979. Proc. West. Sec. Amer. Soc. Animal Sci. 30:251.
Huntington, G.B., R.J. Emerick and L.B. Embry. 1977. J. Animal Sci. 47:119.
Huston, J.E. and M. Shelton. 1971. J. Animal Sci. 32:334.
Jagusch, K.T. 1975. N. Zeal. J. Agr. Res. 18:9.
Javed, A.H. and E. Donefer. 1970. J. Animal Sci. 31:245 (abstr).
Jensen, Rue. 1974. Diseases of Sheep. Lea & Febiger, Philadelphia, PA.
Johnson, R.R. and K.E. McClure. 1972. J. Animal Sci. 34:501.
Jones, J.R. and D.E. Hogue. 1960. J. Animal Sci. 19:1049.
Jordan, R.A., H.G. Croom and H. Hanke. 1958. J. Animal Sci. 17:819.
Jordan, R.M. and H.E. Hanke. 1958. J. Animal Sci. 17:1162.
Jordan, R.M. and H.E. Hanke. 1970. J. Animal Sci. 31:593.
Joyner, A.E., L.J. Brown, T.J. Fogg and R.T. Rossi. 1979. J. Animal Sci. 48:1065.

Karr, M.R. et al. 1965. J. Animal Sci. 24:469.
Kirton, A.H., B. Crane, D.J. Paterson and N.T. Clare. 1975. N. Zeal. J. Agr. Res. 18:267.
Kromann, R.P., E.E. Ray and A.B. Nelson. 1966. Proc. West. Sec. Amer. Soc. Animal Sci. 17:391.
L'Estrange, J.L. and T.A. Mulvihill. 1975. J. Agr. Sci. 84:281.
Loggins, P.E., C.B. Ammerman, J.E. Moore and C.F. Simpson. 1968. J. Animal Sci. 27:745.
Loggins, P.E., K.R. Fick, W.G. Hillis and C.B. Ammerman. 1971. J. Animal Sci. 32:384 (abstr).
Martin, F.H., D.E. Ullrey, H.W. Newland and E.R. Miller. 1968. J. Nutr. 96:269.
McManus, W.R., M.L. Bigham and G.B. Edwards. 1972. Aust. J. Agr. Res. 23:331.
McManus, W.R., J.A. Reynolds and E.M. Roberts. 1973. Aust. J. Agr. Res. 24:413.
Menzies, C.S. 1968. In: Proc. Symposium on Sheep Nutrition and Feeding. Sheep Industry Development Program, Denver, Colo.
Menzies, C.W., D. Richardson and R.F. Cox. 1960. Kansas Agr. Expt. Sta. Cir. 378.
Merion, H., N.S. Raun and E. Gonzalez. 1965. J. Animal Sci. 24:897 (abstr).
Meyer, J.H. and A.O. Nelson. 1961. Calif. Agr. 15:14.
Moir, R.J., M. Sommers and A.C. Bray. 1968. Sulphur Inst. J. 3:15.
Neale, P.E. 1958. New Mexico Agr. Expt. Sta. Bul. 429.
Nishimuta, J.F., L.B. sherrod and R.D. Furr. 1969. Proc. West. Sec. Amer. Soc. Animal Sci. 20:259.
Noble, R.L., K. Urban, F. Harper and G. Waller. 1961. Oklahoma Misc. Pub. MP-64:79.
NRC. 1975. Recommended Nutrient Allowances for Sheep. Nat. Acad. Sci., Washington, D.C.
Ørskov, E.R., W.R. Duncan and C.A. Carnie. 1975. Animal Prod. 21:51.
Ørskov, E.R., C. Fraser and E.L. Corse. 1971. Animal Prod. 13:397.
Ørskov, E.R., C. Fraser, I. McDonald and R.I. Smart. 1974. Brit. J. Nutr. 31:89.
Ørskov, E.R. and D.A. Grubb. 1977. Animal Feed Sci. Tech. 2:307.
Ørskov, E.R., I. McDonald, C. Fraser and E.L. Corse. 1971. J. Agr. Sci. 77:351.
Ørskov, E.R., I. McDonald, D.A. Grubb and K. Pennie. 1976. J. Agr. Sci. 86:411.
Ørskov, E.R. et al. 1972. Animal Prod. 15:183.
Park, R.J., A.L. Ford and D. Ratcliff. 1978. J. Food Sci. 43:1363.
Park, R.J., R.A. Spurway and J.L. Wheeler. 1972. J. Agr. Sci. 78:53.
Perry, T.W., W.M. Beeson, M.H. Kennington and C. Harper. 1959. J. Animal Sci. 18:1264.
Perry, T.W. et al. 1976. J. Animal Sci. 42:1104.
Pierson, R.E. 1967. J. Amer. Vet. Med. Assoc. 150:298.
Pierson, R.E. 1970. J. Amer. Vet. Med. Assoc. 157:1504.
Preston, R.L. 1966. J. Nutr. 90:157.
Preston, R.L. 1967. J. Animal Sci. 26:1483 (abstr).
Preston, R.L. and W. Burroughs. 1958. J. Animal Sci. 17:140.
Preston, R.L., D.D. Schankenberg and W.H. Pfander. 1965. J. Nutr. 86:281.
Preston, R.L., M.S. Zuber and W.H. Pfander. 1964. J. Animal Sci. 23:1182.
Prigge, E.C., R.R. Johnson, F.N. Owens and D.E. Williams. 1976. J. Animal Sci. 43:705.
Ranhotra, G.S. and R.M. Jordan. 1966. J. Animal Sci. 25:630.
Rea, J.C., C.V. Ross and W.H. Pfander. 1965. Missouri Agr. Expt. Sta. Bul. 827.
Reid, R.L., K. Powell, J.A. Balasko and C.C. McCormick. 1978. J. Animal Sci. 46:1493.
Reynolds, R.J., D.A. Dinius, C.K. Lyon and G.O. Kohler. 1978. J. Animal Sci. 46:732.
Robinson, J.J. and T.J. Forbes. 1970. Animal Prod. 12:95.
Rounds, W., T. Klopfenstein, J. Waller and T. Messersmith. 1976. J. Animal Sci. 43:478.
Schaadt, H., R.R. Johnson and K.E. McClure. 1966. J. Animal Sci. 25:73.
Schultz, T.A., E. Schultz and C.F. Chicco. 1972. J. Animal Sci. 35:865.
Scott, G.E. 1976. The Sheepman's Production Handbook. 2nd ed. Sheep Industry Development Program, Denver, CO.
Shell, L.A., F.D. Dryden, A. Mata-Hernandez and W.H. Hale. 1978. J. Animal Sci. 46:1332.
Shelton, M., J.E. Huston and M.C. Calhoun. 1969. J. Animal Sci. 28:147 (abstr).
Shelton, M. and M.C. Calhoun. 1970. Proc. West. Sec. Amer. Soc. Animal Sci. 21:207.
Sherrod, L.B. and R.C. Albin. 1973. Proc. West. Sec. Amer. Soc. Animal Sci. 24:330.
Shiehzadeh, S.A. and L.H. Harbers. 1974. J. Animal Sci. 38:206.
Steeds, W.G., T.J. Devlin and K.M. Wittenberg. 1979. Can. J. Animal Sci. 59:35.
Theurer, B., W. Woods and G.E. Poley. 1968. J. Animal Sci. 27:1059.

Thomson, D.J. and S.B. Cammell. 1979. Brit. J. Nutr. 41:297.
Ullrey, D.E. et al. 1968. J. Animal Sci. 27:1772 (abstr).
Vesely, J.A. and R. Hironaka. 1975. Can. J. Animal Sci. 56:51.
Vetter, R.L. and G.S. Ternus. 1971. Animal Nutr. Health 26(4):8.
Wainman et al. 1972. J. Agr. Sci. 78:441; 79:435.
Weir, W.C. et al. 1959. J. Animal Sci. 18:805.
Weston, R.H. 1971. Aust. J. Agr. Res. 22:307.
Weston, R.H. 1974. Aust. J. Agr. Res. 25:349.
Wheeler, J.L., R.J. Park, R.A. Spurway and A.L. Ford. 1974. J. Agr. Sci. 83:569.
Wilkins, R.J. 1970. J. Brit. Grassland Soc. 25:125.
Woods, W. and R.W. Rhodes. 1962. J. Animal Sci. 21:479.
Woolfitt, W.C., W.E. Howell and J.M. Bell. 1964. Can. J. Animal Sci. 44:179.
Zorrilla, R. and Z.H. Merino. 1972. Nutr. Abstr. Rev. 42:339.

Chapter 14—Feeding and Nutrition of Goats

by C. Devendra

INTRODUCTION

Goats are valuable animals and constitute an important component of the livestock resources of the world. The total world populations of goats was estimated to be 410 million in 1977 (FAO, 1977). Table 14-1 reflects the distribution by region of this total world goat population.

In terms of regional distribution the largest populations are found in Asia and Africa, which together accounted for about 89.4% of the total. About 82% of the goats are found in the tropics. The temperate regions are relatively unimportant with respect to goat production. In the tropics the bulk of the population is found in the less developed countries (LDC's; Fig. 14-1,2) and demonstrate the value and importance of the species to the peoples of the area. Within Asia, India alone accounted for as many as 70 million goats (FAO, 1977).

Goats are widely distributed within the tropics and are found in the highland ecozones of East Africa, Near East, and the Himalayas, arid and semiarid reigons of North and East Africa including the Sahel, to the subhumid to humid regions of South East Asia, East and West Africa, and the Near East. In many instances in the LDC's, goats are more numerically important than sheep and this situation is found, for example, in Nigeria, Haiti, Dominan Republic, Venezuela, Malaysia, and Indonesia. An analysis of goat population changes between 1961-65 to 1977 indicates that the rate of growth over this period was about 9%, equivalent to an annual rate of increase of +0.5%.

Table 14-1 also includes the ratio of human:sheep:goat, and the data suggest that in Africa and Asia, in particular, goats were important animals. By comparison, sheep were far more important than goats in the temperate regions, especially in Europe and the USA.

Goats are presently making a valuable contribution in terms of meat, milk, fiber, and skins, as well as variety of miscellaneous contributions (Table 14-2). Meat, the most important of the products (Fig. 14-3), is produces mainly in Asia and Africa; milk in Asia, Europe and Africa, and goat skins again from Asia and Africa. It is significant to note that meat production is of importance in the tropics, and by comparison, milk production in temperate regions (Fig. 14-4). In many countries such as Ghana and Nigeria, the West Indies, Pakistan, Nepal, and parts of India and South East Asia, there is even a preference for goat meat. This is partly due to the fact that there are no religious taboos against eating of the meat or milk from goats. Fresh skins are

Figure 14-1. Red Sokoto goats and crossbred goats in Nigeria. Photo by C. Devendra.

Figure 14-2. Jamnapari goats in India. Photo by C. Devendra.

Table 14-1. The distribution of the world goat population by region.[a]

Region	Total goat population, 10^3	Distribution, %	Ratio Human:Sheep:Goat
Asia	231,453	56.4	1:0.12:0.10
Africa	131,126	32.0	1:0.38:0.31
South America	18,230	4.4	1:0.45:0.08
North and Central America	12,333	3.0	1:0.06:0.04
Europe	11,506	2.8	1:0.26:0.02
USSR	5,539	1.4	1:0.54:0.02
Oceania	155	0.0	1:9.01:0.01
World	410,342	100.0	1:0.25:0.10

[a]FAO (1977)

Table 14-2. Productivity and importance of goats.[a]

Region	Meat		Milk		Fresh skins	
	Prod.*	Dist.**	Prod.*	Dist.**	Prod.*	Dist.**
Asia	1,075	62.5	3,027	46.2	206.6	62.6
Africa	425	24.7	1,271	19.4	83.6	25.4
Europe	92	5.4	1,491	22.8	12.4	3.8
South America	64	3.7	126	1.9	14.2	4.3
USSR	40	2.3	400	6.1	6.6	2.0
North & Central America	23	1.3	233	3.6	5.8	1.8
Oceania	1	0.1	---	---	0.2	0.1
World	1,720	100.0	6,548	100.0	329.4	100.0

[a]FAO (1977)
*Prod. = production in metric tons x 10^3; **Dist. = distribution as % of total.

very important by-products and in countries such as India and Pakistan their export is a major source of revenue earned.

In recent years there has been enlightened thinking about the value of goats as livestock. This view is not confined to the tropics alone, and has also been the subject of considerable interest in the temperate regions, notably in Australia, Europe and also the USA. The view stems from the realization that goats' milk has certain special characteristics, that these animals are, therefore, of potential importance for milk production in tropical and temperate regions, and, more particularly, the fact that they possess many attributes that can be exploited to advantage. For example, in the humid tropics the importance of goats as milk producers in comparison to buffaloes and cattle has recently been discussed (Devendra, 1979b). It is also being increasingly realized in the LDC's that these animals are extremely common among smallholders or peasant farmers to whom they serve a variety of important functions. For all these reasons, increased productivity in the future in the form of more animal protein supplies from the species is being stressed. Therefore, attention to the factors of production is essential, and efficient feeding and nutrition represents an important aspect of this approach.

FEEDING HABITS AND BEHAVIOR

Goats are very inquisitive animals, much more so than other domestic ruminants.

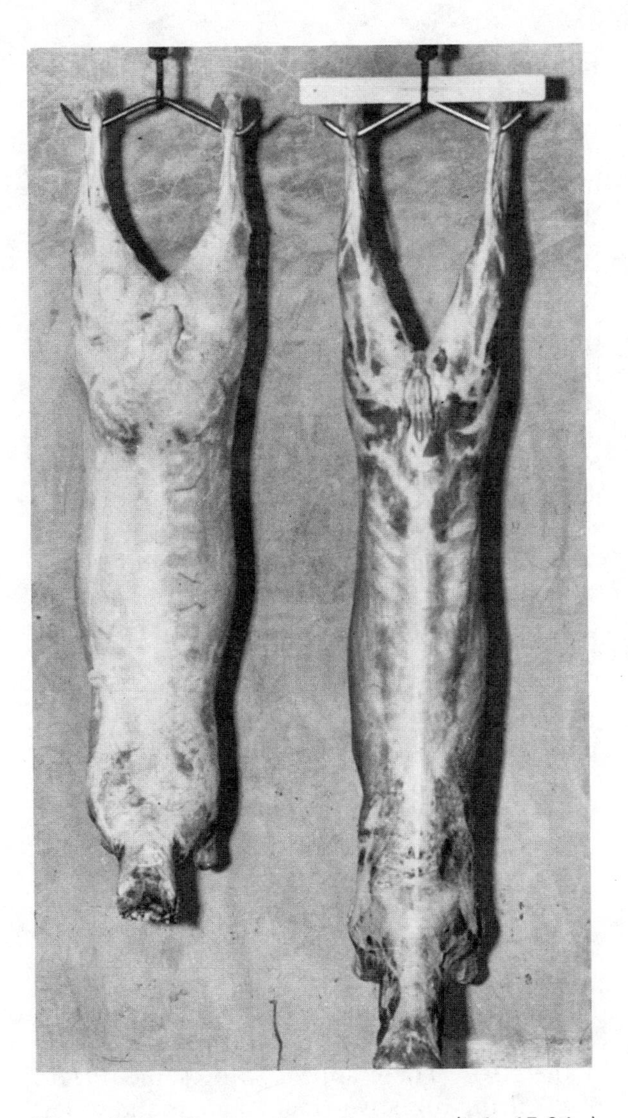

Figure 14-3. Carcass of yearling sheep (left, 15.0 kg) compared to one of a Saanen goat (right, 15.4 kg) in New Zealand. Note that the goat carcass is leggier and carries very much less fat cover. Courtesy of A.H. Kirton, Ruakura Animal Research Station, Hamilton.

This inquisitive feeding behavior is consistent with their ability to walk long distances in search of feed. The wide distribution of goats, from the temperate, semiarid to super humid environments is partially due to their habit of feeding on a wide variety of feedstuffs mainly tree leaves, browse, forbs and grasses. They utilize feeds that would not normally be eaten by cattle or sheep. In the USA, 90% of the total Angora goat population is found in the Edwards Plateau of Texas where there is a variety of browse and complex range vegetation which suits these goats.

Figure 14-4. Angora goats grazing in the bush range country of Texas. USDA photo.

Goats can distinguish between bitter, sweet, salty, and sour tastes, and show a higher tolerance for bitter taste than cattle (Bell, 1959; Goatcher and Church, 1970). Because of a higher tolerance for bitter tastes as well as a preference for a wide variety of chemical compounds (see Ch. 11, Vol. 2), goats consume a wider range of plant species than either sheep or cattle.

While goats can and will accept a wide variety of feed, contrary to popular opinion they have fastidious feeding habits. Feeds that are acceptable to one goat may not be acceptable to another, and goats usually refuse anything which has been soiled by other animals. They relish variety and do not thrive when kept on a single type of feed for any length of time. Rather, they prefer to select from many varieties of feeds, such as combination of grasses and browse or tree leaves. Grazing habits of goats vary not only with the ecological environment but also with the season of the year in the same locality. Such variations are as characteristic of domesticated and highly productive goats as of the wild mountain types (French, 1970). Goats tend to nibble at the shoots and leaves of growing plants and reject the stems. Even for the same plants, goats will consume them at definite stages and reject them at other times.

VALUE OF BROWSE

Given a choice, goats prefer browse plants, as studies in East Africa (Wilson, 1957), Texas (McMahan, 1964) and in Australia (Wilson et al, 1975, 1976) have shown. For example, McMahan (1964) found that over 50% of goats' diet consisted of browse and mast in all pastures and seasons.

Seasonal variations and intensity of stocking appear to influence the nature of the intake as is evident in studies in Spain (Palazon, 1953) and Texas (Malachek, 1970). Palazon (1953) stated that goats in Spain ate brush in the dry season and grasses, legumes and forbs in the wet season. In Texas, Malachek (1970) reported from studies on goats fitted with esophageal fistulas, that the preferences for feed were seasonal, depending on forage availability and growth. Goats showed distinct preferences for grass from June to October even though browse was available, and then a changed preference for browse, especially in winter and early spring. The botanical composition of browse was also different on two different stocking rates.

In Mexico, observations of 1,728 goat bites in a mixed brush-grass-forb community revealed that 83% of the bites were on browse and 17% on grass (Carrera, 1971). Here, goats ate browse almost exclusively in the arid zones of Mexico.

More recently the utilization of a woodland climax suggests that there is preference for some shrubs and tree leaves and not others. A study on Belah *(Casuarina cristata)* - Rosewood *(Heterodendrum oleifolium)* woodland community in western New South Wales, Australia, showed that Rosewood and mature trees of this species were stripped of foliage to a height of 2 m, whereas the weed shrub Turpentine *(Eremophila strutii)* was not browsed at any time (Wilson et al, 1976). However, goats can also use both pasture and fodder grasses and other roughages such as cereal straws efficiently, especially if browse plants are not readily available. Huss et al (1970) and Zertuche (1970) noted in Mexico that goats preferred browse even when exposed to an abundance of grass.

One aspect of the browsing behavior of goats is that there is a definite need to have control over them. This includes control over numbers and also the grazing management. This need is justified by the fact that, if left uncontrolled, goats cause serious damage to the environment, leading to removal of soil cover and soil erosion. For the same reason, foresters justly blame goats for their browsing propensities and the damage they can do to trees if they are carelessly herded in large flocks or left to wander at will. A case in point is Pakistan where, because of inadequate control and lack of adequate grazing management, there was considerable damage to the environment leading to legislation for the extermination of goats which existed up to 1964.

On the other hand, and in the light of current knowledge, goats are valuable animals for the control of brush and waste vegetation (Fig. 14-4). They are, therefore, being used to effectively clear scrubland in Texas and Northern Australia, in situations where it is unlikely that cattle or sheep could survive. This value is even more apparent in parts of Africa where goats have the added advantage of being trypano-tolerant and can be used in large areas of trypano-infested bushland where cattle and sheep are not able to survive. In this context goats are expected to make a significant impact on the future of meat production in Nigeria (Devendra and Burns, 1970).

THE GASTRO-INTESTINAL TRACT OF THE GOAT

It is perhaps useful to keep in perspective the nature of the gastro-intestinal (GI) tract of the goat. There appear to be some differences in the digestive tract of goats and other ruminants (see Vol. I for details). In the goat, the capacity of the four compartments of the ruminant stomach is as follows: rumen 28 ℓ, reticulum 2.3 ℓ, omasum 1.2 ℓ and abomasum 4.0 ℓ. Spedding (1975) has reported that the length of the tract is approximately similar to sheep (22-43 m), but the weight of stomach + intestines as percentage of empty body weight is smaller in goats (5.3 kg) as compared to sheep (6.2 kg). The weight of empty GI tract (including mesenteric fat) of adult Kambing Katjang goats in Malaysia weighing approximately 25 kg was found to be 16.2% of live weight (Devendra, 1966). The latter appears to be somewhat lower than the corresponding figure for sheep weighing approximately 27 kg (excluding mesenteric fat) (Devendra and Wan Zahari Mohamed, 1977). The higher value for goats is due to their tendency to promote a relatively higher deposition of mesenteric fat characteristic of this species.

Table 14-3. Some parameters of digestive efficiency between desert goats and sheep on three types of feed.[a]

Parameter	Goats	Sheep
I.[b] High quality roughage		
Dry matter digestibility, %	61.0	62.2
Crude fiber digestibility, %	64.3	65.8
Total VFA conc.,[c] meq/dℓ	7.91	6.59
N retention, g	3.37	1.02
Fermentation rate in vitro, ml	13.6	12.1
II.[b] Medium quality roughage		
Dry matter digestibility, %	58.9	57.3
Crude fiber digestibility, %	74.4	71.6
Total VFA conc., meq/dℓ	8.81	7.29
N retention, g	4.54	1.03
Fermentation rate in vitro, ml[d]	9.7*	7.4
III.[b] Low quality roughage		
Dry matter digestibility, %	54.5	51.7
Crude fiber digestibility, %	59.6**	54.2
Total VFA conc.,[c] meq/dℓ	5.24*	4.95
N retention, g	-0.27	-1.68
Fermentation rate in vitro, ml[d]	6.5**	4.8

[a]Adapted from El Hag (1976)
[b]I Berseem hay *(Medicago sativa)*
II Lokh grass *(Dianthium annulatum)*
III Hummra *(Dactyloctenium aegyptium, Sheenfeldia gracilis, Ergostic pilosa, Aristida juniculate* and *Aristida* sp.)
[c]3 hours after feeding. [d]For one hour; $*P < .04$, $**P < .01$

Digestive Efficiency

One of the features that appear to be characteristic about the feeding and nutrition of goats is their digestive efficiency. The situation is perhaps more relevant in the tropics where the roughage feeds are abundant and are often coarse. Devendra (1978b) has made a critical review of this situation, and the evidence suggests that goats are more efficient than sheep and cattle in digesting fiber, especially when it is coarse. This is illustrated in Table 14-3.

The implication of these differences, if biologically inherent, is that on fibrous feeds the ME intake is higher and, in marginal areas, goats may well be making more efficient use of the available feeds than either sheep or cattle. In particular, any real differences in digestive efficiency also means that separate nutrient requirement standards will have to be used for goats.

NUTRIENT REQUIREMENT OF GOATS

Dry Matter Intake

The dry matter intake of goats is an important consideration since it indicates their capacity in terms of voluntary food intake to utilize feed. Mackenzie (1967) considers that consumption of 5-7% of live weight is suitable for dairy goats in a temperate environment. He considers that for very high producing goats their intake can be as much as 8.5%. Majumdar (1960) reported a value of 3.1% for Jamnapari goats (dairy) showing a positive balance on major nutrients. In Malaysia, Devendra (1967) reported values of 2.2, 2.6 and 2.7% for Kambing Katjang goats (meat) on three planes of nutrition above maintenance. By comparison, the same breed of goats on maintenance consumed 1.6% of dry matter as percent of live weight (Devendra, 1967). In these studies there was a highly significant correlation between live weight and dry matter intake ($r = 0.94$, $P < .01$). More recently, Devendra (1978) has reported values of 2.8, 2.8, 2.4, 2.4 and 2.2% for Guinea grass *(Panicum maximum)* fed ad libitum at 16-19, 21-28, 28-35, 35-42 and 42-49 days stage of growth, respectively, for the same breed of goats.

For meat and fiber production goats seldom exceed intakes above 3% body weight. On the other hand dairy goats have higher intakes. In a recent review on this aspect for dairy goats, Devendra (1979b) concluded that there were differences between indigenous goats in the tropics, exotic breeds introduced in the tropics, and goats in temperate regions. The dry matter intakes for these three groups were 3.3, 3.6 and 5.0% of body weight, respectively, with corresponding values of 79.6, 88.0 and 110.3 g/kg$^{0.75}$. The relatively lower intake for tropical dairy breeds and also imported exotic breeds into the tropics is due to a combination of body size and high ambient temperatures which tend to depress appetite. This is an adaptive mechanism to reduce heat production. Imported temperate dairy goats in the tropics seldom appear to have intakes above 5% of live weight. In the temperate regions the feed intakes are highest and about 6% of live weight.

In cognizance of these conclusions, approximate dry matter intakes of goats in both tropical and temperate situations are shown in Table 14-4. It has been calculated, based on experimental evidence, that the intake of goats at maintenance was 1.6% of live weight. Production values of 3 and 6% have been used for the tropical and temperate regions, respectively. The table has been constructed for the live weight range of 10 to 60 kg. Most breeds of goats in both tropical and temperate situations will fall into this category as Fig. 14-5 demonstrates.

It should be noted that the dry matter intake will vary with the energy concentration of the diet. In Table 14-4 the ME concentration is about 2.0 Mcal/kg DM which is considered typical of an average diet. For milk production the ME concentrations will have to be increased to about 2.5 Mcal/kg DM.

Energy Requirements for Maintenance

French (1944) was the first to recommend requirements for energy of browsing goats, which under outdoor conditions were

Table 14-4. Dry matter intake of goats.

Body weight, kg	Tropics		Temperate	
	kg	% BW	kg	% BW
Maintenance				
10	0.2	1.6	0.2	1.6
15	0.2	1.6	0.2	1.6
20	0.3	1.6	0.3	1.6
25	0.4	1.6	0.4	1.6
30	0.5	1.6	0.5	1.6
35	0.6	1.6	0.6	1.6
40	0.6	1.6	0.6	1.6
45	0.7	1.5	0.7	1.5
50	0.8	1.5	0.8	1.5
55	0.8	1.5	0.8	1.5
60	0.9	1.5	0.9	1.5
Production				
10	0.3	3	0.6	6
15	0.5	3	0.9	6
20	0.6	3	1.2	6
25	0.8	3	1.5	6
30	0.9	3	1.8	6
35	1.1	3	2.1	6
40	1.2	3	2.4	6
45	1.4	3	2.7	6
50	1.5	3	3.0	6
55	1.7	3	3.3	6
60	1.8	3	3.6	6

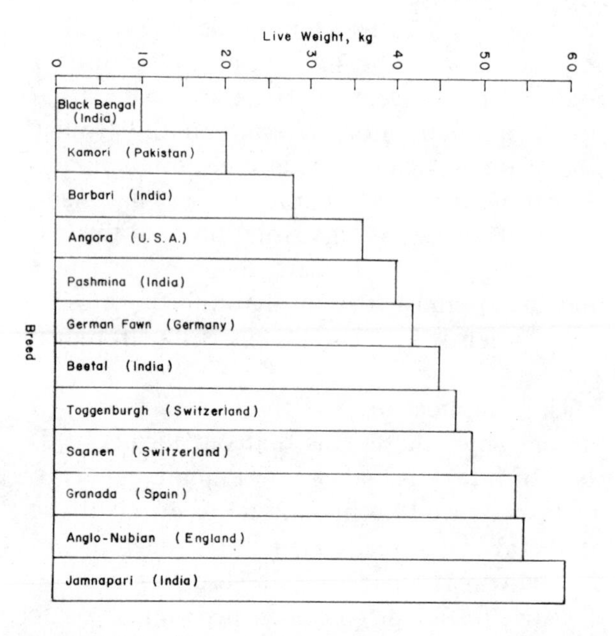

Figure 14-5. Mature body weights of does of some major breeds of goats.

estimated to have a requirement of 0.59 starch equivalents (SE)/day/kg$^{0.75}$, equivalent to 1.80 Mcal ME/day/kg$^{0.75}$. Webster and Wilson (1960) and Mackenzie (1967) have also made alternative recommendations.

The first report to emerge from well controlled experiments was that of Devendra (1967) under pen-fed conditions in Malaysia. The results, based on maintenance of live weight and digestibility trials were: 0.42±0.01 kg digestible organic matter (DOM); 0.47± 0.03 kg total digestible nutrients (TDN); or 0.41±0.01 kg SE/day/45.5 kg$^{0.73}$ of body weight, equivalent to 90.8 Kcal ME/day/ kg$^{0.75}$. A somewhat lower value of 0.6-0.7 ffu (fattening feed units) was reported by Opstvedt (1967) for the maintenance requirements of lactating goats. Using a conversion factor of 1650 Kcal ME/ffu, the equivalent requirement is 57.0 Kcal ME/day/kg$^{0.75}$.

More recently, Akinsoyinu (1974) has reported a value of 92.2 Kcal ME/day/kg$^{0.75}$ in Nigeria. Table 14-5 presents a summary of the available data including both recommendations and experimental values. The conversion of SE, used by English workers, to TDN and ME can only be approximate, since the ratios vary with type of feed used. The ratio between TDN and SE is approximately 1.43 to 1 for roughage and 1.08 to 1 for concentrates (Leroy, 1949). It is clear from Table 14-4 that the recommendations are inaccurate. Of the four experimental values, that of Opstvedt (1967) appears to be low; if it is excluded, the mean value is 94.3 Kcal ME/ day/kg$^{0.75}$ which is comparable to the value of 97.6 Kcal ME/day/kg$^{0.75}$ recommended for sheep (NRC, 1975). The energy requirements for sheep are, therefore, comparable to those of goats (Devendra and Burns, 1970).

Using the mean value of 94.3 Kcal ME/ day/kg$^{0.75}$, the daily maintenance requirements of goats for the live weight range 10 to 60 kg are given in Appendix Table 20. For convenience of application, the energy requirements are presented in terms of SE, TDN, DE and ME. The table presents requirements for goats in stages of production: confined (pen-fed), semi-intensive, and intensive. Under range conditions, French (1944) has suggested a practical maintenance requirement of browsing goats, and in comparison to the pen-fed requirement (Devendra, 1967), the value is about 44% higher. Accordingly, for the semi-intensive and intensive situations increments of 20 and 40% have been used in the calculations in this table. These later values are suggestions, pending the availability of more critical data on the energy requirements for growth.

Protein Requirements for Maintenance

The first digestible protein (DP) requirements for maintenance of goats were determined by Majumdar (1960) from balance experiments using Jamnapari goats in India. The author reported a value of 1.61 g DCP/ kg$^{0.75}$. From a review of the literature, Devendra and Burns (1970) recommend a value of 1.68 g DP/kg$^{0.75}$. Akinsoyinu (1974) has reported a value of 0.59 g DP/kg$^{0.75}$ for West African Dwarf goats in Nigeria. Singh and Mudgal (1978) used castrated Beetal bucks to determine endogenous urinary nitrogen and metabolic fecal nitrogen components, from which they derived a requirement of 116.9 g DP/100 kg live weight, equivalent to 2.57 g DP/kg$^{0.75}$. More recently, Devendra

Table 14-5. Energy requirements for maintenance.

Type of goat	Location	ME/day, Kcal BW$^{0.75}$	Reference
I. Recommendations			
Tropical, dry, meat	East Africa	1805	French (1944)
Tropical, meat	Tropics	2147	Webster and Wilson (1966)
Temperate, milk	England	2221	Mackenzie (1967)
II. Experimental			
Tropical, humid, meat	Malaysia	90.8	Devendra (1967a)
Temperate, milk	Norway	57.0	Opstvedt (1967)
Temperate, milk	USA	100.0	Flatt et al (1972)
Tropical, dry, meat	Nigeria	92.2	Akinsoyinu (1974)

(1979c) fed low-N diets to uncastrated male Kambing Katjang goats and reported EUN and MFN values of 0.13 g/kg$^{0.75}$ and 0.22 g/100 g DMI, respectively, from which the author derived a requirement of 1.41 g DP/kg$^{0.75}$ using the factorial method. Of these 5 values, that of 0.59 g DP/kg$^{0.75}$ (Akinsoyinu, 1974) appears to be low. The mean of the other four is 1.82 g DCP/kg$^{0.75}$. Appendix Table 21 gives the DP recommendations for maintenance of goats of various body weights.

The variability in the DP values reported is due to the techniques used to determine protein requirements. If dietary energy supply is adequate, determination of DP requirements when feeding N-free or low-N diets is an accurate measure of maintenance. Diets with a variable or rich supply of N tend to increase the maintenance requirement. This point is demonstrated in the subsequent study of Majumdar (1960) from which he reported a relatively high value of 1.12 g DP/kg liveweight.

Energy and Protein Requirements for Milk Production

The energy requirements for milk production in goats have been considered to be similar to that of the cow. Using the concepts of Mollgard, Kalaisakki (1959) estimated the NK_f-equivalent for milk production to be 950 NK_f/100 Kcal milk. Mackenzie (1967) has recommended a value of 326 g SE/ℓ of milk, and Opstvedt (1969) reported a value of 0.4 ffu/kg 4% fat-corrected milk (FCM), equivalent to 600 Kcal/kg 4% FCM.

The importance of adequate nutrients for milk production has been demonstrated recently in the studies of Sachdeva et al (1974). Using 124, 100, and 75% of Morrison(s (1956) standards for sheep, the effect of varying treatment planes of nutrition was studied with Barbari and Jamnapari goats. The lactation milk yields for the high-energy group were 101, 130, 110, 107 and 109 ℓ for Barbari for 5 lactations. The highest yields were recorded for the second lactation and the lowest for the low-energy groups. There were no appreciable differences in milk composition.

The requirements for lactation depend upon the composition of the milk and the amount produced/day. Using the ARC (1965) method to derive energy standards factorially, and also using a figure of 70% efficiency of ME utilization for milk production established for the dairy cow, a table of requirements (Appendix Table 22) has been calculated for milk varying in fat content from 3.5 to 5.5%. The ME concentration of diets for lactation should be in the range of 2.3 to 2.5 Mcal/kg DM.

The table also gives the mineral and vitamin requirements. Goats in milk have a high requirement of NaCl, and salt licks and/or mixed licks should be available. Equally important is the provision of clean drinking water in adequate amounts.

Energy and Protein Requirements for Growth

The rate of growth and the mature weight of goats vary widely in different parts of the world. This is mainly because of breed differences and level of nutritional management. More particularly, and especially for several indigenous goat breeds in the tropics, the rate of growth is also a manifestation of the extent of selection. Often there has been inadequate selection, and in many instances the indigenous breeds may reach little more than half the weight attained by goat breeds that have undergone improvement by selective breeding, good feeding and management.

After birth, the most rapid gains are recorded during the first 4-6 mo. of age. High birth weights are important, since they are correlated to both weaning and mature weights. Older data suggest that female Toggenburg kids reached 50% of their mature weight by 4-5 mo. of age. By comparison, male and female kids of the Saanen breed reached 50% of their mature weight in about 8 mo. (Spector, 1956); Fig. 14-6 illustrates the growth curve.

The significance of plane of nutrition on growth is well illustrated by Wilson (1958) with East African dwarf goats. These studies demonstrated that kids given a low-high level of nutrition did about as well as those given a high-high level. A high-low treatment depressed growth, particularly of the females.

The weight gain of kids after the point of inflection of the growth curve occurs at a slow rate and apparently is difficult to alter by high energy or protein supplementation. Lindahl (1954) found that the growth rate of female kids beyond 4 mo. age was not significantly different from female kids receiving good quality roughage. Also, he was not successful in increasing the growth rate or rate of

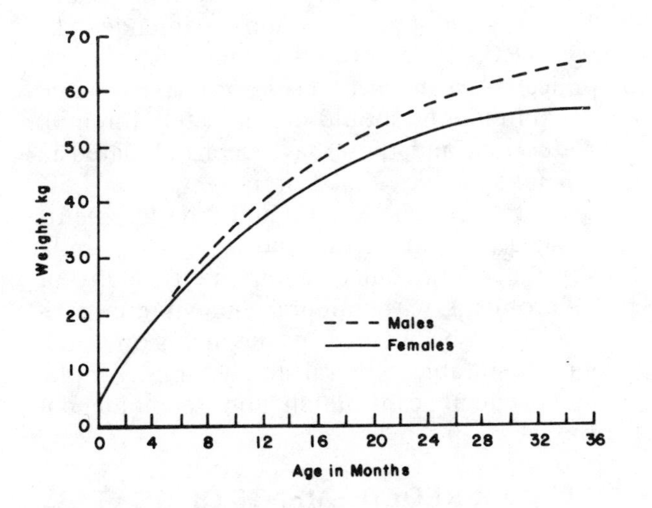

Figure 14-6. Growth curve of Saanen male and female kids. From Spector (1956).

fattening of Toggenburg male kids and mature does by dietary means.

French (1944) assumed that the requirements for growth of goats were similar to lambs and he recommended requirements over the live weight range 9 to 50 kg. This assumption is questionable. For example, he suggested that the SE requirements for 1 kg live weight increase for goats weighing 9, 27 and 50 kg were 0.45, 0.68 and 0.80 kg. Devendra (1967) has reported that over the weight range of 18 to 26.4 kg, the cost of live weight gain in Kambing Katjang goats was 2.7 g DOM/g gain, equivalent to 10.4 Kcal ME/g gain. By comparison, Akinsoyinu (1974) has reported a value of 5.14 Kcal ME/g, which is approximately half that of the former. The difference clearly indicates the need for much more information on this aspect. For present purposes a table of requirements (Appendix Table 23) has been calculated using the mean value of 8.44 Kcal ME/g from the two experimental values reported.

This table has been calculated for the live weight range of 10 to 60 kg and at three rates of growth (50, 100 and 150 g/day). Additionally, the total ME requirements have been calculated from a diet with an ME concentration of 2.3 Mcal/kg DM. In view of the extreme paucity of information on protein requirements for growth, the digestible protein (DP) requirements have been calculated on the basis of 1 Mcal DE = 20 g DP.

Energy and Protein Requirements for Reproduction

There is no doubt that a sufficient intake of energy and protein has a significant effect on reproductive performance. Sexual maturity, conception and pregnancy are enhanced, and, finally, the kids are born with a relatively high birth weight and a strong constitution.

Evidence for this is seen in extensive study on Barbari and Jamnapari does by Schadeva et al (1973). In the absence of any specific recommendations for goats, three levels of energy and three levels of protein based on 125, 100, and 75% of Morrison's standard were used. The effect of plane of nutrition was more apparent after the first kidding. In the low-energy groups, irrespective of protein levels, the does did not kid in the fifth kidding. Based on 4 complete kiddings for Barbari and 3 for the Jamnapari breeds, there were more twin births: 47.5% in each of the high-high and medium-medium groups for the Barbari, and 45.3 and 40.4% in the high-medium and high-high groups, respectively, in the Jamnapari breeds. In addition, both breeds had 2.0 and 1.5 kids/doe/year in the high-energy cum medium-protein groups compared with 1.0 and 0.4 in the low-energy groups. Equally important was the finding that the high and medium-energy groups had shorter intervals between kidding compared to the low-energy treatments. The results demonstrate that a higher energy level together with adequate protein significantly affect the reproductive performance of goats.

In Germany Kalaissakis (1958, 1959), used German Fawn does and also demonstrated the need for high nutrient requirements of does carrying multiple fetuses during late pregnancy. Four does were kept in metabolism cages from about the 3rd mo. of pregnancy until 14 to 19 weeks of lactation. A ration providing 410 SU and 125 g crude protein/day was considered to be adequate for the does. Two does received this level while the other two does received about 80% of the recommended level. One doe gave birth to a single kid weighing 4.1 kg, two does had twins with total birth weights of 5.5 and 7.4 kg, and the 4th had quadruplets weighing 7.6 kg. Calculations of energy balances during the last 34 to 57 days of pregnancy showed that only the doe producing the single kid was in positive balance. The other three does were underfed 27-37% in terms of energy. The doe producing the single kid was essentially maintained in N balance but the other three does lost from 50-145 g of N.

In Nigeria Akinsoyinu (1974) has reported an energy value of 179.3 Kcal ME/day/$kg^{0.75}$ for West African dwarf goats during pregnancy, which appears to be the only experimental value available. Using this value, a table of energy and protein requirements has been constructed (Appendix Table 24). In view of the need for higher energy during pregnancy, an ME concentration of 2.5 Mcal ME/kg DM has been used. The DP requirements have been calculated on the basis of 1 Mcal DE = 20 g DP. In order to ascribe a realistic requirement due to single or twin fetuses, a 20% allowance has been used, based on published recommendations for ewes with single or twin fetuses.

Vitamin and Mineral Requirements for Growth and Production

Very limited information exists on the vitamin and mineral requirements for goats. Published data are limited to isolated studies on carotene metabolism (Majumdar and Gupta, 1960) or to the experimental production of Zn deficiency (Miller et al, 1964). Typical goat diets should have adequate carotene to prevent vitamin A deficiency. It is possible that the storage capacity of vitamin A in goats is less than sheep due to less total fat depots, especially in the viscera. Access to good quality green feeds will enhance carotene supply to goats. Under normal grazing conditions a deficiency of vitamin D is unlikely.

There is no evidence of vitamin E deficiency in goats and, since vitamin K is plentiful in feedstuffs, a deficiency is unlikely. Of the B vitamins, only vitamin B_{12} (cobalamin) is likely to be deficient, and a deficiency can be avoided by the presence of cobalt. In general, B vitamins probably should be included in the diets of young kids.

There is evidence that deficiency of dietary Na significantly affected daily feed intake and daily milk yield, and this was also reflected in reduced levels of the element in the colostrum and milk (Schellner, 1972/1973). The author also showed that Mn and Zn deficiency reduced the level of both elements in colostrum, milk and the hair. Clearly, it is important that these and other minerals are provided in sufficient quantities in the diet.

Until more definite information is available, it is suggested that the recommended values (NRC, 1975) for vitamins and minerals applicable to sheep be used for goats. Ca and P requirements should be adjusted for milk production, and these have been tabulated in Appendix Table 22. For milk with a 4.5% butter fat content, the Ca and P requirements are suggested to be 0.9 and 0.7 g/kg of milk over the maintenance requirement. Many of the problems with mineral and vitamin deficiencies can be avoided by the inclusion in the diet of suitable mineral and vitamin supplements which can match the recommendations.

WATER REQUIREMENTS OF GOATS

Goats are efficient animals in the use of water. They have a low rate of water turnover/unit of body weight (Maloiy and Taylor, 1971); however, ample quantities of clean water are essential for high milk production by lactating goats and for maximum growth and mohair production. However, goats are obviously adapted to water shortages. Rao and Mullick (1965) showed that the intake of water and its ratio to the intake of DM increased with increasing temperature, while the water excreted in the feces and urine decreased. At temperatures of 37°C water turnover rates are 185, 188, 197, and 347 ml/$kg^{0.82}$/25 h, respectively, for camels, goats, sheep and cattle (French, 1970).

Goats found in semi-arid and arid regions have the ability to withstand water deprivation and to conserve and utilize it efficiently. In these desert conditions food resources are meager and water holes are widely spaced. The Bedouin or Hejaz goat has been shown to adapt to these conditions remarkably well and also to have a low caloric demand. In normal Bedouin husbandry these goats are watered once every 2-4 days, and it has been shown that when watered after continuous grazing after this period, the goats consumed as much as 30-45% of their dehydrated body weight (Shkolnik et al, 1975).

Pen-fed goats in Malaysia drank 680 ml of water/day of which 80% was drunk between 0700 and 1900 h (Devendra, 1967). In winter in temperate areas goats can obtain sufficient water from their feeds to eliminate the need for drinking water, especially if the feed contains over 60% moisture. In hot environments (38°C ±), goats pant at half the rate of sheep, do not sweat, and lose less water in their feces and urine. When the water intake is low and animals go for days without drinking water, the excretion of urine is reduced. Goats are not quite as efficient as camels in reduction of the kidney glomerular filtration rate, but are far more efficient than sheep or cattle (French, 1970). For further information on water metabolism the reader is referred to Ch. 2, Vol. 2.

PRACTICAL RATIONS FOR DAIRY GOATS

Annual Nutrient Requirements

The first step in formulating rations for goats is to calculate the nutrient requirements for a specific function (meat, milk or mohair). The required amounts of energy, protein, Ca and P for maintenance and production are found in the tables of nutrient requirements (Appendix Tables 20-24), and these are then applied to determine the needs of goats in a particular situation. The requirements will vary in relation to size and age of the goats. The calculations should begin with the roughage source, since this is the least expensive source of nutrients on the farm. The difference between the nutrients provided by a particular roughage source and the total nutrient required will have to be fulfilled by a concentrate mixture. Care is also taken to insure that, in addition to energy and protein needs, mineral and vitamins are present or added as needed to meet dietary recommendations.

Having considered the nutrient requirements of goats in various situations, it is perhaps appropriate to consider the annual dry matter, energy and protein requirements. Table 14-6 presents the calculations using the data from Appendix Tables 20-24. For early lactation the nutrient allocation is assumed to be about 65% of the full requirement in late lactation. The DP requirements for lactation have been calculated on the basis of an average daily milk yield of 2 kg.

General Recommendations

This section is intended to be useful to nutritionists in the efficient nutritional management of goats under both temperate and tropical conditions. Practical feeding programs for goats should be based on the type and quality of forage available, since the quality of forage determines the amount of and the quality of concentrates needed to supplement the diet (Fig. 14-7). Concentrate feeding is important for milk production as shown by Lindahl (1956) who found that comparable milk production was obtained when concentrate mixture containing 14 or 16% crude protein was fed to lactating does along with alfalfa hay containing about 15% crude protein. However, milk production was significantly higher when the 16% crude protein concentrate was fed in comparison with the 14% protein concentrate, along with mixed hay containing 11% crude protein. By comparison, meat goats probably do not need concentrates and can be grown on pasture. Variety in feeds is greatly relished by goats,

Table 14-6. Annual dry matter, energy and protein requirements of a 40 kg doe.

Components	Maintenance, 10 weeks	Gestation Early, 15 weeks	Gestation Late, 5 weeks	Lactation, 22 weeks	Total requirements
Dry matter					
kg/day	0.6	1.0	1.4	1.5	
kg/period	420	105	49	231	805
Metabolizable energy					
Mcal/day	1.50	2.26	3.42	3.75	
Mcal/period	105.0	237.3	119.7	577.5	1039.5
Digestible protein					
g/day	31.7	45.1	68.4	118.0[a]	
kg/period	2.2	4.7	2.4	18.2	275.0

[a]Based on a daily milk yield of 2 kg.

Figure 14-7. Dairy goats grazing under lush pasture conditions in Maryland. USDA photo.

Figure 14-8. Saanen goats eating ensilage on a New Zealand goat dairy. Courtesy of E.L. Collins, Auckland, N.Z.

and it is very common in the tropics to feed various tree leaves and also crop residues (Devendra, 1978b). In France, a 4-year study on dairy goats on the use of dried forages (grass, corn or alfalfa) indicated that the best results were obtained when the diet included 2 or 3 forages (Kaszas et al, 1974).

Good quality legume hays, such as alfalfa, alsike clover, red clover, ladino clover, soybean hay, vetch, and birds foot trefoil, or early-cut mixed hay containing legumes and grasses should be used, if possible. Concentrate mixtures containing 14% crude protein are recommended in conjunction with good pasture and high quality legume hay. However, a concentrate mixture containing 16-18% crude protein is required when grass hays form the basic ration.

There must be a balance between the amount of roughage and concentrates fed. An excess of either is unsuitable for milk production, and practical diets must aim for a sufficient intake of dry matter and, therefore, of energy so as to insure high milk production. A 60% level of good quality alfalfa hay diet has been shown to give the highest yield of milk (Simiane and Fehr, 1976).

It has been reported that feeding good quality forages before and after parturition favorably affects the onset of lactation (Morand-Fehr and Sauvant, 1978). The authors suggest that the supply of concentrate intake of the goats should not be too rapid. In the middle of lactation and especially at the end of lactation, maintenance of milk production at a high level requires a slightly higher supply of concentrates than that necessary to meet the requirements of energy. Good quality hay is eaten more readily than poor quality hay (Martin et al, 1976). It was found in the same experiment that the consumption of poor hay increased the intake of salt.

Silages, roots, cabbage and other high moisture feeds are relished by goats and add variety to the diet (Fig. 14-8). Good pasture usually can take the place of half of the concentrate or grain requirement.

Non-protein N sources like urea or biuret are satisfactory for goats. Protein from oilseeds like peanut meal or soybean meal, and also proteins of animal origin like fish meal are generally expensive. The inclusion of urea in the diet to meet part of the crude protein requirements tends to reduce the cost of feeding (Fujihara and Tasaki, 1975; Devendra, 1979b; Harmeyer and Martens, 1979). In Nigeria it has been demonstrated that feeding urea to West African dwarf goats promoted faster growth rates and was also associated with better utilization of the N content in urea as compared to comparable rations with peanut meal (Mba et al, 1974).

Suitable concentrate mixtures containing approximately 14, 16 and 18% crude protein are given in Table 14-7.

Table 14-7. Concentrate mixtures for dairy goats.

	Approximate crude protein content		
Ingredient	14%	16%	18%
	..... % of mixture		
Corn, gr.	37.0	35.0	32.0
Oats, gr.	37.0	35.0	32.0
Wheat bran	16.0	14.0	15.0
Soybean meal	9.0	15.0	20.0
Iodized salt	1.0	1.0	1.0

Lactating Does

Suggested ration for lactating does are:

Clover or alfalfa hay	1.4 kg (3 lb)/day
+ Concentrate (14% protein)	0.45-0.90 kg (1.2 lb) or more depending on quantity of milk produced
Mixed or grass hay	1.4 kg (3 lb)/day
+ Concentrate (16-18% protein)	0.45-0.90 kg or more
Legume or good mixed hay	0.9-1.4 kg (2-3 lb)/day
+ Silage or roots	0.7-0.9 kg (1.5-2 lb)
+ Concentrate (16% protein)	0.45-0.90 kg or more

The concentrate should be fed at the rate of 0.3-0.5 kg for each kg of milk produced. However, if the doe has access to good pasture, the concentrate allowance can be cut in half. The following procedures have been suggested as an aid in keeping high producing does on feed (Hofmeyr, 1969): (1) if the doe receives concentrates before kidding, reduce the amount to 0.2 kg (0.4 lg)/day during the last week before parturition; (2) feed from 0.2 to 0.45 kg (0.5 lb) of concentrate/day during the first 2 weeks after kidding; (3) after 2 weeks gradually increase the concentrate level to that suggested by the milk yield. If the does leaves some concentrate, reduce the amount such that she consumes everything and there is no waste. Does should be fed the concentrate on an individual basis. This can be done most easily by feeding the concentrate at milking time, allowing equal parts for each milking.

Pregnant Dry Does

Pregnant dry does should be fed so as to rebuild any lost body reserves, to provide for the developing fetuses, and to gain some reserve fat before kidding. Free access to good pasture and roughage plus concentrate at a level of 0.2-0.7 kg (0.5-1.5 lb)/day (depending on the condition of the doe) is usually recommended for pregnant dry does. Feeding of good quality forages affects the onset of lactation, and it has been suggested that the supply of concentrates before parturition should not exceed 8 g DM/kg live weight (Morand-Fehr and Sauvant, 1978).

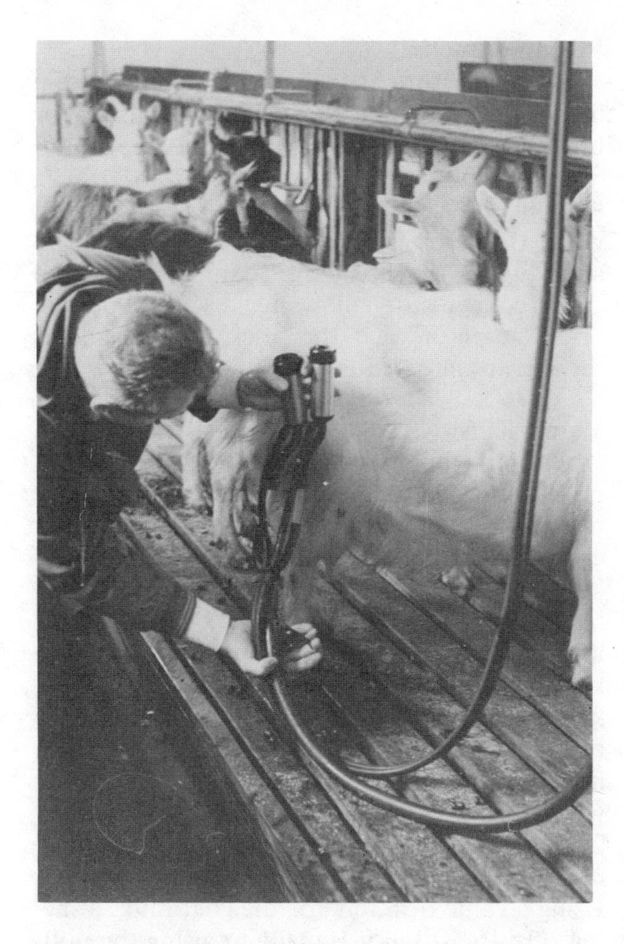

Figure 14-9. Milking Saanen boats in Norway. Courtesy of T. Skejvdal, AAS-NLH, Norway.

Yearling Does

Yearling does should be fed enough for maintenance and growth but not enough to fatten them. Browse, good pastures, high quality hay, and a place to exercise are desirable. Concentrates may have to be fed at the rate of 0.12-0.7 kg (0.5-1.5 lb)/day to obtain the desired growth rate, depending on the quality of the roughage available.

Breeding Bucks

Good pasture, alone, will maintain the bucks in good health when not used for breeding. If pasture is not available, feed good quality hay and concentrate up to 0.7 kg (1.5 lb)/day. Do not feed excessive grain when the buck is inactive. Two weeks before and during the breeding season increase the daily concentrate allowance by 0.45-0.9 kg (1-2 lb) or more if the buck is large and is serving numerous does.

Young Kids

Due to the relatively high value of goat milk, most kids are fed on cows' milk or milk replacer. Lindahl (1954) compared the growth rate and the cost of raising wether kid kids on goats' milk and a milk replacer. Kids receiving goats' milk averaged 0.17 kg (0.37 lb) gain/day for a 65-day period while the kids receiving the milk replacer averaged 0.12 kg (0.26 lb) gain/day. Although the kids receiving the milk replacer gained 0.05 kg less/day than those receiving goats' milk, they were thrifty at weaning. They also gained 0.01 kg/day faster than those receiving goats' milk for the first 91 days following weaning. Using a value of 53¢/ℓ for the goat milk, a saving of $46.50/kid was made by using the milk replacer. It might be noted that improved milk replacers, intended for lambs, are now available (see Ch. 12). These might be expected to improve performance of young kids as compared to those used by Lindahl.

Most kids are either fed on a bottle or from a pan, rather than on commercial nursing machines. If it is intended to rear the kid on a bottle, it is best not to let the kid nurse its dam. Milk the colostrum from the doe and feed it at body temperature from a mipple bottle. Continue this for 3-4 days and then shift to milk or milk replacer. The following suggestions have been given for feeding kids (Anon., 1969): warm the milk or milk substitute to about 40°C (103-105°F); wash and sanitize the bottle and nipple or pan after each feeding; feed the kid 0.7 to 0.9 ℓ of milk or milk substitute; feed the kid 3-5X/day, particularly the first 2 weeks; provide access to good calf starter when the kids are 3-4 weeks of age. Increase the amount as the kids can take it without digestive upsets; offer the kids fine second cutting hay and calf starter at 3-4 weeks of age; discontinue the feeding of milk at 3-4 months of age or as soon as the kid is eating hay and concentrate.

It has been shown in studies in Japan that the development of the rumen and maintenance of rumen function were considered to be normal when the kids were given basal diets containing about 5% crude fiber (Haryu and Kameoka, 1974).

FEEDING OF ANGORA GOATS

Angora goats are (Fig. 14-10) valuable for fiber production. About 85% of the total world production of mohair comes from the USA and Turkey. However, only limited information exists concerning the nutrition of Angora goats. Gray (1959) reports that a wide variety of supplemental feeds may be used, ranging from shelled corn to 20% protein range cubes. The amount of supplement needed varies from about 0.1 kg to 0.45 kg (0.2-1 lb)/head/day, depending on range conditions. Some ranchers prefer to feed roughages such as alfalfa, sorghum, peanut, sudan, or Johnson grass hay; 0.45 kg/head/day is usually considered to be minimal. Hand-feeding is generally not practical under range conditions. In such cases grain mixtures with added salt to control intake can be fed in self-feeders. A popular mixture is 3 parts of ground grain, 1 part cottonseed meal and 1 part salt, when fed near a water source.

While both energy and protein are important for fiber production, the response to protein is significant. Young and growing kids

Figure 14-10. An angora goat typical of those in Texas. Courtey of J.E. Huston, San Angelo, TX.

have been shown to respond consistently to increased fiber production with dietary protein levels up to 20% (Huston and Shelton, 1967; Stewart et al, 1971). Concentrate feeding beyond the 50% level in the diet has been shown by the latter authors to have no advantage.

METABOLIC DISORDERS OF GOATS

A survey was conducted in England during 1965 to determine the cause of death of dairy goats (Payne, 1966). A total of 207 replies were received giving details of the health of 975 goats. Of these, 69 or 7.1% of the 975 goats died during the survey (Table 14-8). The metabolic disorders, abdominal disorders, enterotoxemia, ketosis (pregnancy toxemia) and milk fever accounted for 33.3% of all deaths. Enterotoxemia usually occurs with high producing dairy goats and the onset of the disorder is very rapid–animals that are normal in the morning may be dead before night. The disorder undoubtedly is often diagnosed as lead poisoning. Apparently, the organisms responsible for enterotoxemia *(Clostridium perfringens)* are always present in the intestinal tract. High level of grain feeding favor the rapid growth of the organism and production of large amounts of toxin.

Table 14-8. Causes of death in goats.[a]

Cause of death	Deaths	
	Number	%
Exterotoxemia	13	18.8
Poisons	7	10.1
Cardiovascular	6	8.7
Parturition	6	8.7
Injuries	5	7.3
Old age	5	7.3
Pneumonia	5	7.3
Abdominal disorders	4	5.8
Ketosis	4	5.8
Mastitis	3	4.4
Unknown	3	4.4
Lameness	2	2.9
Milk fever	2	2.9
General infection	2	2.9
Endoparasites	1	1.4
Kidney disease	1	1.4
Total	69	100.0

[a]From Payne (1966)

However, other factors must be present since animals on a constant feeding regimen can develop enterotoxemia. Immunization of animals with *Clostridium perfringens* type D toxoid is indicated where enterotoxemia is a problem.

By comparison in humid tropical countries, mortality can be very high. In Bangladesh the mortality rates were 82.2 and 47.8%, respectively, out of a total population of 214 kids born and 115 adult does (Table 14-9). Among the kids born respiratory disorders, gastro-intestinal parasitism, and contagious ecthyma caused 43.8, 25.0 and 18.3% of the total mortality. With adults respiratory disorders and gastro-intestinal parasitism accounted for 36.2 and 43.5% of the total mortality.

Table 14-9. Causes of mortality of goats in Bangladesh.[a]

Cause	Kids,[b] %	Adults,[c] %
Respiratory disorders	43.8	36.2
Gastro-intestinal parasitism	25.0	43.5
Contagious ecthyma	18.8	7.3
Digestive disturbances	2.3	1.8
Miscellaneous (undiagnosed)	10.1	11.2

[a]From Abdur Rahman et al (1976); [b]Total mortality was 82.2% out of a total of 214 kids born; [c]Total mortality was 47.8% out of a total adult population of 115 does.

Disorders of the digestive tract include bloat and acute indigestion. Acute indigestion most often occurs when animals not accustomed to a grain diet are allowed to consume large quantities of grain (see Ch. 17, Vol. 1). Changes in feeding of grain and concentrates should be done on a gradual basis. Bloat results from inability of animals to eructate during periods of rapid gas formation (Ch. 17, Vol. 1). Cessation of eructation can result from mechanical causes such as blockage of the esophagus by pieces of apples, roots, etc. Removal of the blocking agent is successful at times. There is some evidence that bloat can result from psychological factors. Adrenalin can inhibit eructation. Sudden outbreaks of bloat have been reported following management changes in dairies without alteration in feeding patterns. Gentle treatment of animals of high temperament cannot be overstressed.

Acute cases of bloat have followed the feeding of large amounts of dried alfalfa leaves; this practice should be avoided. It is obvious that a number of interacting factors must play a role in the etiology of frothy bloat. Substances that alter surface tension and act as froth breakers can be used as preventative or as therapeutic agents. Management practices such as providing adequate amounts of coarse roughage when animals are grazing on lush legumes are indicated.

Pregnancy toxemia (ketosis) is often a problem with high producing does which have been carrying multiple fetuses (see Ch. 12, Vol. 2). Adequate nutrition during the last 3-4 weeks of pregnancy is essential. Underfeeding at the end of pregnancy can cause toxemia although it has been reported that an excess of concentrates appears to be associated with an increase in stillbirths and subsequent metritis (Fehr et al, 1976). Practices effective with ewes include increasing the feeding of concentrates and provision of high quality forage. This helps maintain the nutrient intake at a time when nutrient need is increased while, at the same time, the capacity of the stomach is reduced by the volume of the fetus and reproductory organs. Treatment during early stages of the disease is essential if it is to be effective. Glycerol or propylene glycol drenches are often effective treatments. Glycerol given as a drench twice daily (200 ml) or propylene glycol (56 g) given 3-4 times/day are usually recommended. The most successful treatment is intravenous glucose plus oral glycerol and insulin given intramuscularly; this treatment should only be given by a competent veterinarian or skilled animal husbandryman.

Milk fever may appear soon after kidding. Early symptoms are loss of appetite followed by restlessness, excitement, and trembling of the muscles. Later symptoms are incoordination and coma. Treatment includes intravenous injection of a mixture of minerals and glucose (see Ch. 12, Vol. 2). European data indicate that milk fever is not a big problem in goats (Payne, 1966).

CONCLUSIONS

It is patently clear that there is an extreme paucity of knowledge specific to nutrient requirements of goats in different phases of production, including both pen-fed and extenextensive situations. This conclusion is explained essentially by the fact that research on the species has been limited in temperate regions due to the general lack of importance of goats in these parts. Goats are relatively more important in underdeveloped countries, areas which also generally do not do much research on animal production. Therefore, a major task in the future is acceleration of research into the nutrition of goats; substantial financial and other rèsource use is necessary for this purpose.

Where goats are numerous in the tropics, one of the most important factors affecting their productivity is adequate nutrition, in particular the supply of energy and protein. There is no doubt that the present level of productivity is low, due to a combination of underfeeding, diseases and poor husbandy. For maximum productivity from goats, more information is necessary, but even more so, application of existing information is required. Effective nutritional management represents in this contex an important means of insuring maximum food and fiber production from goats.

References Cited

ARC. 1965. The Nutrient Requirements of Farm Livestock. No. 2 Ruminants. Agr. Res. Council, London.

Akinsoyinu, A.O. 1974. Ph.D. Thesis, Univ. Ibadan, Nigeria.

Anonymous. 1968. Dairy goats. Breeding/Feeding/Management. Amer. Dairy Goat Assoc., Spindale, N.C.

Bell, F.R. 1959. J. Agr. Sci. 52:125.

Carrera, C. 1971. I.T.E.S.M., Monterray, Mexico, p. 168.

Devendra, C. 1966. Malays. Agr. J. 45:345.

Devendra, C. 1967. Malays. Agr. J. 46:80, 98, 191.

Devendra, C. 1970. 1st National Seminar on Goat Production, Barquisemeto, Venezuela, 12-14th Nov., 1970 (Mimeo., 11 pp.).

Devendra, C. 1978a. MARDI Res. Bul. 5:91; World Rev. Animal Prod. 14:9.

Devendra, C. 1978b. In: An Introduction to Animal Husbandry in the Tropics, 3rd ed. Longmans Green and Co. Lts., London.
Devendra, C. 1979a. J. Animal Sci. (In press).
Devendra, C. 1979b. Int. Symp. on Dairy Goats, 26-27th June, Logan, Utah (Mimeo., 56 pp.).
Devendra, C. 1979c. The protein requirements for maintenance of indigenous Kambing Katjang goats of Malaysia. MARDI Res. Bul. (In press).
Devendra, C. and Burns, M. 1970. Goat production in the Tropics. Tech. Commun. No. 19, Comm. Bur. Animal Nutr. Genet., Commonwealth Agr. Bureau.
Devendra, C. and Wan Zahari Mohamed. 1977. Malays. Agr. J. 51:191.
El Hag, G.A. 1976. World Rev. Animal Prod. 12:3.
FAO. 1977. Production Yearbook, Vol. 31. FAO, Rome.
Fehr, P.M., J. Hervieu and J. Delage. 1976. Nutr. Abstr. Rev. 46:921.
Flatt, W.P. et al. 1972. Handbuck der Tierernahrung, Vol. 2. Verlag Paul Parey, Hamburg and Berlin.
French, M.H. 1944. E. Afric. Agr. J. 10:66.
French, M.H. 1970. Observations on the Goat. FAO Agr. Studies No. 80. FAO, Rome.
Fujihara, T. and I. Tasaki. 1975. J. Agr. Sci. 85:185.
Goatcher, W.D. and D.C. Church. 1970. J. Anmal Sci. 31:373.
Gray, J.A. 1959. Texas Agr. Expt. Sta. Bul. B-926.
Harmeyer, J. and H. Martens. 1979. Int. Symp. on Dairy Goats, 26-27th June, Logan, Utah (Mimeo., 43 pp.).
Haryu, T. and K. Kameoka. 1974. Bul. Nat. Inst. Animal Industry, No. 26.
Hofmeyr, H.S. 1969. S. Africa Dept. of Agr. Tech. Bul. 387.
Huss, D.L. et al. 1970. I.T.E.S.M., Monterray, Mexico.
Huston, J.E. and M. Shelton. 1968. Texas Agr. Expt. Sta. PR 2523.
Kalaissakis, P. 1958. Ztschr. Tierphysiol. Tierreranhrung Futtermittelk 13:355.
Kalaissakis, P. 1959. Ztschr. Tierphysiol. Tierreranhrung Futtermittelk 14:204.
Kaszas, L., H. Miossec, R. Disset and M. Simiane. 1976. Nutr. Abstr. Rev. 46:921.
Lerroy, A.M. 1949. Norms for energetic feeding. General Reports. Ve Congre's International de Zootechnic. Paris, Nov. 3-10, 1949.
Lindahl, I.L. 1954. ARS, USDA, APH 155.
Lindahl, I.L. 1956. Goat Feeding Investigations. Annual Report, Animal and Poultry Husbandry Research Branch, ARS, USDA.
Lindahl, I.L. 1968. Amer. Dairy Goat Assoc. Handbook 14:83
Mackenzie, D. 1967. Goat Husbandry. 2nd ed., Faber and Faber Pub. Co., London.
Majumdar, B.N. 1960. J. Agr. Sci. 54:329, 335.
Majumdar, B.N. and B.N. Gupta. 1960. Indian J. Med. Res. 48:388.
Martin, D., P.M. Fehr, J. Hervieux and A. Cucci. 1976. Nutr. Abstr. Rev. 46:921.
Mayone, B. 1949. Ve Congress International de Zootechnic., Paris, Nov. 3-10, 1949.
Macquot, G. and M. Bejambes. 1960. Dairy Sci. Abstr. 22:1.
Malachek, J.C. 1970. Thesis, Texas A & M Univ.
Maloiy, G.M.O. and C.R. Taylor. 1971. J. Agr. Sci. 77:203.
Mba, A.U. et al. 1974. Nigerian Soc. Animal Prod. 1:209.
McMahan, C.A. 1964. J. Wildl. Mgmt. 28:798.
Miller, W.J., W.J. Pitts, C.M. Clifton and S.C. Schmittle. 1964. J. Dairy Sci. 47:556.
Morand-Fehr, P. and D. Sauvant. 1978. Livestock Prod. Sci. 5:203.
Morrison, F.B. 1956. Feeds and Feeding, 22nd ed., Morrison Pub. Co.
NRC. 1975. Nutrient Requirements for Sheep. Nat. Acad. Sci., Washington, D.C.
Opstvedt, J. 1967. Tech. Bul. No. 134. Inst. Animal Nutr. Agr. Coll. Norway.
Opstvedt, J. 1969. European Study Comm. Animal Nutr., Rpt. No. 2.
Palazon, J. 1953. Ganado Cabrio. Salvat Editores S.A., Madrid, Spain.
Payne, J.M. 1966. Vet. Rec. 78:31.
Rao, M.V.N. and D.N. Mullick. 1965. Indian J. Vet. Sci. 35:288.
Rogers, A.L. 1956. Amer. Dairy Goat Assoc. Handbook 11:93.
Sachdeva, K.K. et al. 1973. J. Agr. Sci. 80:375.
Sachdeva, K.K. et al. 1974. Milchwissensachaft 29:471.
Shkolnik, A. et al. 1975. Vulcani Centre, Special Publ. 37:79.

Schellner, G. 1972/1973. Tierreranhrung Feutterung 8:246.
Simiane, M. and P.M. Fehr. 1976. Nutr. Abstr. Rev. 46:920.
Singh, N. and V.D. Mudgal. 1978. 20th Int. Dairy Congr., Paris.
Spector, W.S. 1956. Handbook of Biological Data. W.B. Saunders Co.
Spedding, C.R.W. 1975. The Biology of Agricultural Systems. Academic Press, New York.
Stewart, J.R.M., M. Shelton and H.G. Haby. 1971. Texas Agr. Expt. Sta. PR 2933.
Suzuki, S. 1968. Abstr., 64th Annual Meeting, Amer. Dairy Goat Assoc., Bethseda, Maryland.
USDA. 1970. Agricultural Statistics (1970). U.S. Government Printing Office, Washington, D.C.
Varma, A. and P.C. Sawhney. 1969. J. Nutr. Dietet. 6:301.
Webster, C.C. and P.N. Wilson. 1966. Agriculture in the Tropics. Longmans, Green & Co. Ltd., London.
Williamson, G. and W.J.A. Payne. 1960. An Introduction to Animal Husbandry in the Tropics. Longmand, Green & Co. Ltd., London.
Wilson, A.D. et al. 1975. Aust. J. Exptl. Agr. Animal Hus. 15:45.
Wilson, A.D., W.E. Mulham and J.H. Leigh. 1976. Aust. Rangel. J. 1:7.
Wilson, P.N. 1957. E. Afr. J. 28:501.
Wilson, P.N. 1958. J. Agr. Sci. 50:198.
Zertuche, T.M. 1970. I.T.E.S.M., Monterray, Mexico.

Chapter 15—Nutrition of Ruminants in the Tropics

by the late T.H. Stobbs and D.J. Minson

INTRODUCTION

Over half the cattle in the world, 30% of the sheep and most of the buffalo and goats are found in the tropics (Table 15-1). Despite the presence of this high proportion of domestic ruminants in these areas, only 34% of the world's beef and 21% of the world's milk is produced in the topics and the human consumption of protein is low. The daily consumption of animal protein in the tropics varies from 9 to 23 g compared with 42 to 69 g in developed countries (Jasorowski, 1976). Therefore, large increases in animal production are required to cater to the domestic demands of the rapidly increasing population of these regions.

Table 15-1. Number and distribution (millions) of cattle, buffalo, sheep and goats inthe tropics.[a]

Region	Cattle	Buffalo	Sheep	Goats
Africa	133.5	2.3	97.9	105.4
Central & S. America	216.0	0.2	85.5	35.2
Asia	280.4	118.8	147.3	161.2
Oceania	11.6	---	13.4	0.2
Total numbers in tropics	641.5	121.3	337.1	302.0
Proportion in tropics, %	53	92	32	75

[a]From FAO (1975)

The most important need in the tropics is to increase the productivity of animals rather than to increase their numbers. However, in Africa there are vast areas (about 7 million km^2) of high potential grazing land which are capable of carrying a high cattle population once trypanosomiasis-carrying tsetse flies are controlled. Large areas in South America are also capable of development for grazing ruminants.

Social constraints against increased production are evidenced in Africa where cattle are often kept for prestige, rather than for production. This is especially true in India which has about 17% of the cattle and buffalo population of the world (230 million), yet maintains a strict religious taboo against cattle slaughter. In this chapter only technical problems associated with increasing livestock production in the tropics are considered.

Improved nutrition is the main requirement for increasing both reproductive efficiency (Lamond, 1970) and animal production (Stobbs, 1976a) provided that healthy well-adapted stock are available. Poor performance of stock is not entirely due to poor nutrition for direct climatic stress, ticks, tick-borne diseases, reproductive diseases, and poor herd management all contribute to the extremely low productivity of ruminants in many tropical environments. It is therefore axiomatic that livestock development programs should be based on both improved nutrition and better herd management.

High yields of beef and milk can be achieved by feeding grain, but production costs are usually high and the grain is often required for direct human consumption. However, large areas of the moist tropics (equatorial, tropical monsoonal and humid subtropical regions) can be used to produce meat and milk from improved pastures. Pasture development is considered to be a practical proposition on about half the 38×10^8 ha of the tropics without using land required for producing crops for direct human consumption (Minson et al, 1978). In this chapter emphasis will be given to increasing livestock production from improved tropical pastures together with some consideration of supplementation with agricultural by-products which are often wasted.

NUTRITIONAL VALUE OF TROPICAL PASTURE FEEDS

In this chapter the nutritional value of tropical pastures will be considered by comparing and contrasting with temperate climate feeds. This approach has been adopted since most readers understand temperate pastures and a comparative approach should make possible a clear identification of the additional factors that must be taken into account when considering the nutritional value of tropical pastures.

Energy

The milk production and growth rate of cattle grazing tropical pastures is generally lower than that of cattle grazing temperate pastures. This lower production is caused by the smaller quantity of net energy (NE) absorbed each day. Three factors control the intake of NE; the daily intake of food energy (I), the proportion of the feed digested (D) and the efficiency of utilization of the products of digestion (E). Thus, NE = I x D x E.

These three factors will depend on both the chemical and physical composition of the pasture and therefore depend on the pasture species, stage of maturity, soil nutrient status, part of the plant being eatern, as well as the climate of the region. For convenience the intake and digestibility of pasture energy will be considered separately. Unfortunately, very little is known about E since very few tropical pasture species have been fed in calorimetric studies.

Intake of Energy

The gross energy (GE) content of tropical grasses and legumes cut at different stages of maturity is relatively constant, varying from 4.11 to 4.48 Kcal/g of dry matter with a mean of 4.33 Kcal/g. This value is only slightly lower than that of temperate pastures which vary between 4.30 and 4.59 Kcal/g. The slightly lower GE value is mainly caused by the lower crude protein (CP) content of tropical pastures. Since the GE of tropical feeds is relatively constant, voluntary feed intake is usually expressed as the quantity of dry matter or GE eaten/unit of metabolic weight (body weight$^{0.75}$; Crampton et al, 1960). When expressed in this way the daily intake can vary from 30 g/kgW$^{0.75}$ for mature tropical pasture to 140 g/kgW$^{0.75}$ for immature temperate alfalfa. In energy terms this is a range of 129 to 602 Kcal/kgW$^{0.75}$.

Tropical grasses are usually eaten in smaller quantities than temperate grasses grown for the same time period. The difference occurs even when the tropical grasses are grown under irrigation in very fertile soils and is caused by a higher fiber content, lower dry matter digestibility and hence greater proportion of indigestible residues which reduce appetite (Table 15-2).

Table 15-2. Comparative nutritional value of monthly regrowths of tropical and temperate grasses when fed to sheep.

Parameter	Tropical grasses	Temperate grasses
Voluntary intake, g/day/kgW$^{0.75}$	56	71
Crude fiber, %	30	24
DM digestibility, %	62	71
Indigestible DM, %	38	29
Intake of indigestible DM, g/day/kgW$^{0.75}$	21	21

As herbage matures there is an increase in the proportion of fiber, a decrease in the digestibility, and a fall in the voluntary feed intake. However, the intake of tropical grasses is about 20% higher than that of temperate grasses with similar digestibilities due to structural differences between the two groups of grasses. At 60% digestibility the tropical grasses are young and leafy compared with the temperate grasses which are mature and stemmy.

There is a widespread belief that leaves are eaten in greater quantities than stem because of their lower fiber content and higher dry matter digestibility. Recent studies with tropical grasses have shown that leaf is eaten in much greater quantities than stem of similar fiber level and dry matter digestibility (Table 15-3). The higher intake of the leaf fraction is associated with the shorter time the leaf fraction was retained in the reticulo-rumen. The cause of this shorter rentention time is not yet clear. There were no differences between leaf and stem fractions in the level of pepsin-soluble dry matter or rate of digestion in vitro, two well recognized indicators of intake. The most likely reason for the differences in retention time was the longer time required by the indigestible part of the stem fraction to reach a particle size sufficiently small to leave the reticulo-rumen. This idea is supported by the larger quantity of energy required to grind samples of stem in a hammer mill.

The higher intake of leaf is not a phenomenon restricted to tropical grasses. Differences, although smaller, have also been found with temperate rye grass. With the tropical legume *Lablab purpureum,* the intake of leaf was twice that of the stem even though there was no difference in digestibility.

The limitation of high fiber in tropical pastures may be reduced by grinding and pelleting. Grinding changes the physical

Table 15-3. Voluntary intake and other nutritional parameters of separated leaf and stem of five tropical grasses and one temperate grass.[a]

	Tropical grasses			Temperate grasses		
	Leaf	Stem	Diff.	Leaf	Stem	Diff.
Voluntary intake, g/d/kgW$^{0.75}$	58	40	-18	74	62	-12
DM digestibility, %	53	56	+3	67	65	-2
Rumen contents, g	6200	5900	-300	---	---	---
Rumen DM, %	11.5	11.5	0	---	---	---
Retention time of DM, h	24	32	+8	---	---	---
Surface area, sq. cm/g	130	40	-90	190	90	-100
Grinding energy, j/g	240	410	+170	130	260	+130
Density, g/cc	0.08	0.24	+0.16	0.08	0.12	+0.04
ADF, %	35.1	39.4	+4.3	29.0	32.2	+3.2
Pepsin-soluble DM, %	23.0	23.4	+0.4	37	32	-5

[a]From Laredo and Minson (1973, 1975a,b)

characteristics of the fiber allowing it to leave the reticulo-rumen after a shorter time while pelleting increases density of the ground feed. Grinding and pelleting have increased mean voluntary intake of dry matter by 49% with a variety of tropical grasses; the advantage varying from 7 to 136%. Small increases were associated with feeds of low protein content. Although grinding and pelleting increase the quantity of dry matter eaten, this advantage is partly offset by a reduction in dry matter digestibility (DMD). This drop occurred in all studies and was caused by a reduction in the time the feed is subjected to microbial digestion within the reticulo-rumen. The average reduction in DMD following pelleting is about 6% with a range of 3-12%.

Although the fiber level in pasture is the principal factor controlling the quantity of food eaten, this applies only if there are no deficiencies of protein, vitamins, and minerals. If there is a deficiency of any of these nutrients, then the intake of food will be limited by these deficiencies and not by the level of fiber in the diet. When the CP content of mature grass falls below 6-8% of the dry matter, then the appetite drops and intake of grass is reduced. With a protein deficiency the intake of pasture is no longer controlled by the quantity of food in the rumen and the intake is virtually unaffected by grinding and pelleting (Minson, 1967). For high producing animals the level at which protein becomes limiting is probably >8% CP.

The adverse effects of protein deficiency on intake may be overcome by applying fertilizer N or by growing tropical grass in association with a tropical legume. Applying fertilizer to a mature stand of *Digitaria decumbens* has increased the CP content from 4.1% in the unfertilized grass to 9.9%, resulting in an increased consumption of beef cattle from 4.3 to 7.7 kg/day and changed a live weight loss of 0.22 kg/day to a gain of 0.69 kg/day (Chapman and Kretschmer, 1964). Fertilizer N will also increase CP content of immature grass containing >8% CP. In these cases food intake is not usually limited by a protein deficiency and food intake is not increased.

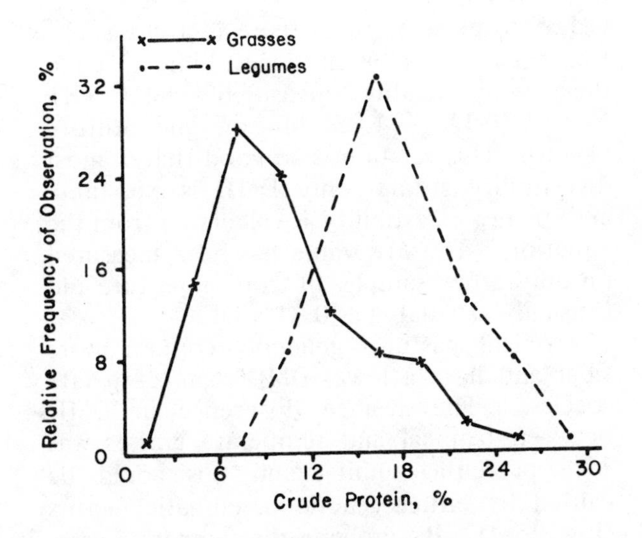

Figure 15-1. Frequency distribution of crude protein % in a wide range of tropical grasses and legumes cut at many stages of growth. From Minson (1976).

Legumes have a higher protein content than grasses (Fig. 15-1) and can act as a protein supplement to the grass. Addition of as little as 10% of legume to a grass diet containing 3.6% protein has led to a spectacular 50% increase in the consumption of sheep (Minson and Milford, 1967). Field studies have also shown that the presence of legumes in the pasture will increase the CP level of the diet in two ways, through the legume and through the grass.

Both N and S are required for protein synthesis and when feeds are S deficient, appetite is low. Low levels (0.08%) of S have been found in young regrowths of *Digitaria decumbens* and feed intake was increased 28% by feeding a S supplement. S deficiencies in pasture feed may also be overcome by applying S-containing fertilizer which can increase feed intake as much as 44% (Rees et al, 1974).

Voluntary food intake will be depressed by mineral deficiencies. The time taken for this to occur depends on both the level of the element in the diet and the reserves of the animal. For instance, animals have good reserves of Na and deficiency symptoms will develop only where the animal is on the low-Na feed for a period of time and is unable to obtain sufficient Na from the drinking water. Where animals have become Na-deficient and are eating pastures low in Na, then feeding Na supplements will increase food intake, growth rate and milk production.

Digestibility

For tropical pastures there is a close relation between the energy digestibility (Y) and the more readily measured DMD (X). $Y = 0.981X - 1.08$, Minson and Milford (1966). This relation is so good that in most digestibility studies only DMD is measured, and energy digestibility is calculated from the equation. The ME value has been measured on only a few samples of tropical pasture and is usually calculated as 0.81 x DE.

Tropical pastures generally contain more fiber and have a lower DMD than temperate species. The average difference in DMD between tropical and temperate grasses was 12.8 percentage units, and this could be caused by either genetic or climatic factors (Fig. 15-2). By growing the temperate grass *Lolium perenne* under irrigation in the subtropics it was shown that DMD was depressed by high temperatures and when grown at the

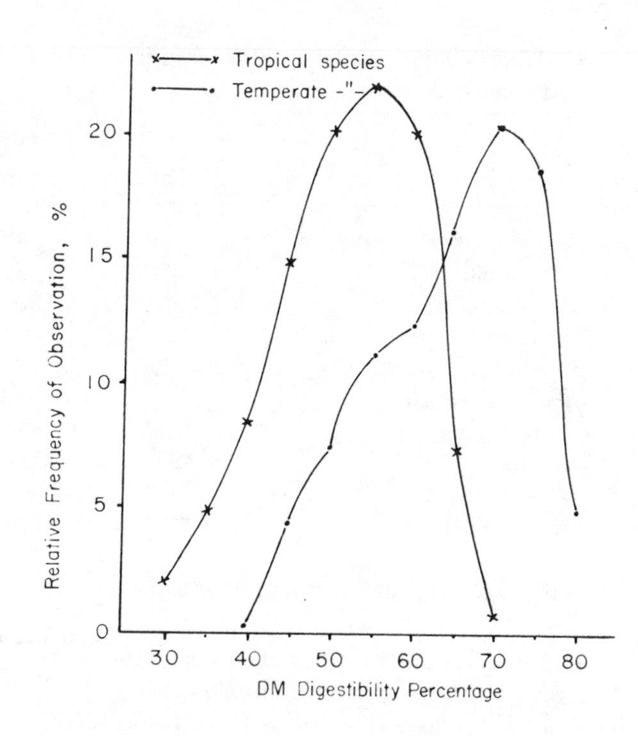

Figure 15-2. Frequency distribution of dry matter digestibility of a wide range of tropical and temperate grasses cut at many stages of growth. From Minson and McLeod (1970).

same high temperature there was no difference in digestibility between temperate and tropical grasses (Minson and McLeod, 1970). Conversely, the occasional high digestibility values for tropical grasses reported in the literature are usually caused by the tropical pastures being grown at low temperatures. Although temperature has a profound effect on digestibility, it is important to recognize that there are also genetic differences in the plants which affect digestibility between tropical grass species and varieties of up to 8 percentage units even when grown at the same temperature.

As grasses mature there is a decrease in the proportion of cell contents (neutral detergent solubles, NDS) and a rise in the level of cell wall components (neutral detergent fiber, NDF; acid detergent fiber; ADF; and lignin). The increase in fiber level and the higher lignification of this fiber lead to a decrease in the digestibility of the dry matter and organic matter and a fall in the ME value of the grass (Table 15-4). There is also a decrease in the voluntary food intake and a change from liveweight gain on the first cut to a liveweight loss with the mature herbage.

Table 15-4. The effect of stage of growth on the composition and nutritional value of six tropical grasses fed to sheep.[a]

Item	Days regrowth		
	28	70	98
Yield, kg DM/ha	3,600	10,000	11,900
Leaf, %	45	25	20
Feed composition			
Crude protein, %	12	7	6
Phosphorus, %	0.34	0.22	0.20
Ash, %	16	11	10
NDF, %	60	69	68
ADF, %	30	38	40
Lignin, %	2.2	3.8	4.7
Digestible DM, %	64	53	49
Digestible OM, %	65	55	50
ME, Mcal/kg	1.97	1.75	1.60
Intake, g/kg$W^{0.75}$/day	55	49	42
Weight change, g/day	30	-4	-66

[a]From Minson (1972) and unpublished data

Protein

The CP content of tropical pastures depends largely on the pasture species, level of N supply from the soil and fertilizer and the maturity of the herbage.

Pasture Species. The most important species difference in protein content is between grasses and legumes. This difference is illustrated in Fig. 15-1 which shows the frequency distribution of the CP content of a very large number of samples of tropical grasses and legumes cut at different stages of growth and grown in many countries. The mean protein content of the grasses was 9.5% compared with 17.5% for the legumes. Of special practical importance is the very large proportion (22%) of grass samples which contained <6% CP.

Leaf blade generally contains more protein than leaf sheath or stem (Table 15-5) so stemmier grasses tend to have lower levels of protein than leafy species. Thus, the major aim in many plant breeding projects is to increase the leaf percentage of tropical grasses.

In comparing tropical and temperate pasture species, it is interesting to note that tropical grasses generally have less protein than temperate grasses and for a wide range of samples the mean protein levels were 9.5% and 13.2%, respectively. For legumes there is no difference between tropical and temperate species in their mean protein content (Minson, 1976).

Table 15-5. Mean crude protein content of separated leaf and stem fractions of five tropical grasses.[a]

Regrowth, days	Crude protein, %	
	Leaf	Stem
52	12.4	9.2
75	8.6	6.6
88	7.5	5.6

[a]From Laredo and Minson (1973)

Plant Maturity. As grasses mature there is a decrease in the protein percentage (Table 15-5). This fall is caused by both a decrease in the protein level of the leaf and stem fraction and an increase in the proportion of stem which contains less protein than the leaf fraction. The minimum protein level will be determined by the level of soil fertility, species grown, and the suitability of the environment to allow pasture to grow beyond the soil N supply. Thus, low protein levels in mature pastures are generally associated with poor soil fertility, high rainfall, and high pasture yield.

Fertilizers. N fertilizers will increase the CP content of grasses if cut relatively soon after the N is applied. N fertilizer stimulates growth and stem development so the final protein content of the grass may be similar to or even lower than that of the unfertilized grass if the grass is allowed to grow for long periods. This problem of dilution may be overcome by applying the fertilizer within a month of harvesting. Under these conditions quite large increases in protein content can be achieved (Table 15-6).

Soil N may also be increased by growing legumes and this will increase the protein level in tropical grass. This is the most important practical method of overcoming the low inherent level of protein in tropical grasses although it requires a knowledge of the management factors necessary for legume survival.

Table 15-6. Rise in crude protein content achieved by applying fertilizer nitrogen.[a]

Species	Unfertilized	Fertilized
Chloris gayana	10.7	14.1
Pennisetum clandestinum	10.6	14.7
Digitaria decumbens	8.1	13.2
Digitaria decumbens (mature)	4.2	8.0

[a]From Minson (1967, 1973)

Protein Availability. The simplest measure of the availability of CP to the animal is the quantity apparently digested/100 g of feed DM. For tropical grasses and legumes a good relation (r = +0.98) has been reported by Milford and Minson (1965) between digestible crude protein % (DCP) and the CP % of the DM. DCP = 0.90 CP - 3.25. This equation applies to both leaf and stem fraction of tropical grasses and to fertilized and unfertilized grass. It is also very similar to equations derived for temperate forages, illustrating that the same nutritional principles apply to tropical and temperate forages.

Protein Utilization. Part of the protein in tropical forages is degraded in the reticulo-rumen, and some of this is excreted as urinary urea. This loss is partly offset by the synthesis of microbial protein in the reticulo-rumen from ammonia derived from the feed protein and saliva. At the present time there appear to be no data on the extent of these two processes and whether deamination of the protein in tropical grasses leads to a deficiency of amino acids for meat production. It has been shown recently that cows grazing *Chloris gayana* pastures containing 20% CP responded to the feeding of formaldehyde treated casein but not to untreated casein, indicating that the cows were absorbing insufficient amino acids from the small intestine (Table 15-7). More information on protein breakdown and resynthesis is obviously required for ruminants fed tropical forages.

Minerals

Animal production in many areas of the tropics is severely limited by mineral deficiencies, imbalances and toxicities (McDowell, 1976). Geographical locations in the tropics of confirmed or suspected mineral deficiencies and toxicities are shown in Fig. 15-3.

Major Limiting Minerals. At least 15 mineral elements are nutritionally essential for ruminants (see Vol. 2). The most widespread deficiencies in the tropics appears to be P, Co, Cu, Na and I. In some areas toxic concentrations of F, Se and to a lesser extent Cu, Mn and Mo appear to limit animal production. The main mineral deficiencies encountered in tropical environments will now be considered.

Phosphorus. Many tropical soils are deficient in P. In addition large areas of the humid tropics have acid soils containing large concentrations of Fe and Al which accentuate a P deficiency by forming insoluble phosphate complexes. Some species such as *Stylosanthes* varieties can grow in soils which are low in available P and these plants usually contain a low concentration of P. In the early stages of growth the uptake of P is usually sufficient to match the relatively slow production of DM. As photosynthetic area increases, DM production outstrips the mineral supply and there is a decrease in the concentration of P. Thus, the P content of most tropical species falls from about 0.30% P during the early stages of growth to <0.15% P when mature. It is assumed that deficiencies will occur in animals grazing this mature herbage. However, plasma P concentrations are often normal due to the mobilization of bone P. Responses to

Table 15-7. Response of Jersey cows grazing N fertilized Rhodes grass pastures to supplements of casein and formaldehyde-treated casein.[a]

Characteristics	Control	Casein	Formal-casein
Milk yield, kg/cow/day	12.3	12.7 (3)[+]	14.7 (20)[+]
Butterfat yield, kg	0.63	0.66(5)	0.71(13)
Solids-not-fat yield, kg	1.11	1.15(4)	1.36(23)
Protein yield, kg	0.41	0.42(2)	0.52(27)

[a]From Stobbs et al (1977); [+]( ) percentage increase

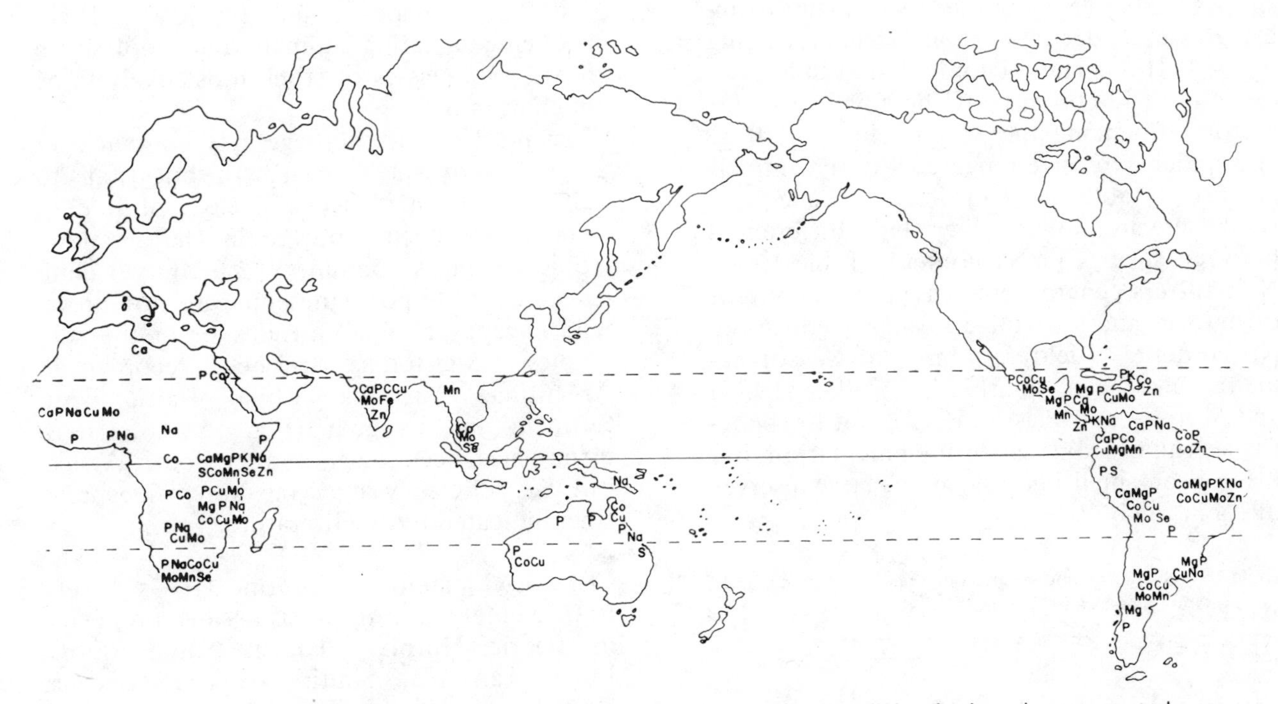

Figure 15-3. Geographical locations in the tropics where mineral deficiencies have been reported.

feeding P supplements occur only when the amount of P absorbed from feed and released from the bone reserves is insufficient to maintain an adequate level of plasma P.

The most important economic result of P deficiency in the tropics is the inhibition of estrus. In P-deficient areas of South Africa, Brazil and Panama, supplementation with bone meal and other P supplements has dramatically increased fertility in grazing cattle with calving percentages increasing by 20-50% (McDowell, 1976). In Senegal and Brazil there are reports of deaths of cattle caused indirectly by P deficiency. Cattle in these areas develop a craving for bones and may become infected with *Chlostridium botulinum.*

Cobalt. Co deficiency occurs in at least 20 countries (Fig. 15-3) and is caused by animals consuming herbage which contains <0.05 ppm of Co on a DM basis. Leaves of apparently healthy pastures of *Panicum maximum* and *Puerana phaseoloides* have been found to contain 0.009 ppm of Co when growing on Co-deficient soils in Malaysia. Co-deficient areas are to be found on acid soils derived from older, coarse, sandy geological formations.

Deficient animals quickly respond to an injection of vitamin B_{12}, to Co supplied in salt mixtures or as a drench, and to application of Co-containing fertilizers. Slow-release Co oxide bullets have also been used. These are placed in the rumen together with a large grub screw which gradually wears away the bullet. Recently it has been found that some mineral incrustation on cobalt pellets is so hard that it will only be removed if the grub screw is case-hardened before dosing.

Copper and Molybdenum. Cu deficiency can occur when animals graze Cu deficient pastures, but most Cu deficiencies occur when apparently adequate amounts of Cu (6-16 ppm) in herbage are made unavailable by excess of Mo (>3 ppm) or sulfate. Conversely, chronic Cu toxicity can occur where intake of Mo and sulfate is low. It is likely that there are large areas of the tropics where subclinical Cu deficiencies occur. Both Cu deficiency and Mo toxicity are corrected by providing additional Cu in the animals' diet.

Sodium. Na deficiency is widespread in high-producing grazing ruminants in the tropics (see Fig. 15-3), deficiencies being reported in South America, Africa, and Australia.

There is a very large variation in the Na content of both tropical and temperate forage species (Table 15-8) with levels of many tropical grasses and most tropical legumes being below 0.10%. The Na content, and to a lesser extent the K content, of the soil is a major factor influencing the Na content of herbage and under extensive range conditions animals are usually able to select diets with adequate Na. There are, however, very large differences between species in Na content (Table 15-8). N fertilizers generally increase the Na content of herbage and K fertilizers usually reduce the Na content. Field responses to Na supplements have been reported by Walker (1957) and Murphy and Plasto (1973), but the benefit accruing from salt supplementation has often been small because of the large reserves of Na in the body.

Table 15-8. Sodium concentration in a range of tropical and temperate grasses and legumes.

Species	Mean, %	Range
Tropical grasses		
Panicum coloratum	0.81	0.31-1.65[a]
Panicum maximum	0.28	0.11-0.80
Setaria anceps cv. Kazungula	1.59	1.42-1.80[b]
Setaria ancepts cv. Nandi	0.06	0.05-0.10
Tropical legumes		
Centrosema pubescens	0.04	0.04-0.05[c]
Leucaena leucocephala	0.02	0.01-0.03
Macroptilium atropurpureum	0.03	0.02-0.04
Temperate grasses		
Dactylis glomerata	0.69	0.12-1.01[d]
Festuca pratensis	0.07	0.05-0.18
Temperate legumes		
Trifolium repens	0.28	0.26-0.32[e]
Trifolium pratense	0.06	0.05-0.11

[a]Minson (1975); [b]Hacker (1974); [c]Minson (1977); [d]Griffith et al (1965); [e]Davies et al (1966).

Iodine. I deficiency is widespread in sheep and cattle grazing in the tropics and goiter or other I deficiency symptoms have been reported in most areas where human goiter is evident. I deficiency may be induced by the consumption of plants containing goitrogens, such as *Cynodon plectostachyus.* Most deficiencies can be overcome by feeding appropriate I sources in a mineral mixture.

Other Minerals. Herbage with a low S concentration has been reported in many tropical areas and improved animal production following S supplementation has been reported in Australia, Ecuador, and Uganda. High-producing lactating animals and wool sheep are the two classes of stock most likely to be deficient in S.

Tropical pasture herbage may contain over 2% K at early stages of growth, but it falls to 0.3% K with maturity and cases of K deficiency have been reported in Haiti. Many tropical forages contain <0.2% Mg, yet clinical cases of hypomagnesemia are not prevalent except at high altitudes or in the subtropics. Mg tetany has been recorded in Argentina, Australia, Chile, Haiti, Peru, Surinam and Uruguay (Fig. 15-3). Se deficiency has been diagnosed in South Africa, but Se deficiency may be more widespread than indicated by the literature.

Toxic Elements. Fluorine (F) is a very toxic element, and fluorosis has been reported in Algeria, Morocco, Tunisia, Saudi Arabia, India, Tanzania, South Africa, Argentina, Guyana and Mexico (Fig. 15-3). The F content of herbage is seldom more than 1 to 2 ppm but intake can be increased by consumption of contaminated herbage, high-F mineral supplements, or drinking water high in F (>15 ppm). Animals on a low plane of nutrition during an extended dry season are particularly susceptible to F toxicity.

Selenium toxicity (selenosis) is found in Australia, Columbia, Iran, Mexico, South Africa, and Venezuela. Toxicity is most often associated with the consumption of Se-accumulator plants which may contain from 100 to >9,000 ppm. Control at present is limited to supplementing with non-seleniferous food and pasture management aimed at reducing accumulator plants.

Correcting Mineral Deficiencies

A mineral deficiency in a forage diet may be eliminated in two different ways. The level of the deficient mineral in the forage may be increased by the application of the appropriate fertilizer or the animals may be supplied with the mineral in the form of a lick, drench, injection, or slow-release pellets.

The method adopted will usually depend on the type of deficiency and local economic factors, but, in the case of the fertilizer, additional benefits may be achieved. These arise from the fertilizer changing the growth form of the plants leading to changes in the intake and digestibility of the forage over and above the direct effect achieved by feeding a mineral supplement. With S-deficient Pangola grass applying fertilizer S increased intake by 44% compared with only 28% when the sheep were drenched with sodium sulfate (Rees et al, 1974).

Toxins, Taints and Odors

As a group tropical pasture species are relatively free of toxins and other undesirable compounds. However, a few contain toxic factors which can result in death, depression of production, or can cause adverse taints in milk or meat.

Oxalic Acid. Most tropical grasses contain some oxalic acid but three genera, *Setaria, Cenchrus,* and *Pennisetum,* contain substantially higher levels (Table 15-9). Levels of oxalate vary between cultivars and levels of soil fertility. For example, *Kazungula setaria* consistently contains a higher level of oxalate than *Nandi setaria.* Oxalate levels in both cultivars are increased by the addition of N and K fertilizers.

Table 15-9. Range of soluble oxalate in tropical pasture grasses.[a]

Grasses	Oxalate, %
Setaria anceps	2.8-5.6
Cenchrus ciliaris	1.4-3.4
Pennisetum purpureum	1.9-2.6
Pennisetum typhoides	1.8-2.7
Pennisetum clandestinum	0.7-1.8
Panicum maximum	0.2-2.3
Brachiaria spp.	0.2-0.9
Digitaria spp.	0.4-0.7
Chloris gayana	0.07-0.10

[a]From Garcia-Rivera and Morris (1955) and Jones and Ford (1972)

A few cattle deaths through oxalate poisoning have occurred following acute signs of hypocalcemia accompanied by deposition of calcium oxalate crystals (Jones et al, 1970). Absorption of oxalate is rapid in monogastric animals and in horses *Oesteodystrophia fibrosa* has been reported when grazing *Cenchrus* and *Setaria* (Walthall and McKenzie, 1976). Depressions in milkfat percentages have been recorded when cows graze *Pennisetum typhoides* and *Kazungula setaria* pastures.

Bloat. The only tropical legumes reported to cause bloat are *Lablab purpureum* (Hamilton and Ruth, 1968) and *Trifolium semipilosum* (Stobbs, 1976b). This occurred only when the legume was grazed at a young, rapidly growing stage.

Estrogenic Activity. Small quantities of the estrogens, genistein, formononetin and biochanin A, have been found in the tropical legume *Stylosanthes humilis.* The levels in this and other tropical legumes are too small to produce signs of estrogenic activity (Little, 1976).

Indospicine. *Indigofera spicata* is a vigorous and potentially useful tropical legume but contains the hepatotoxic amino acid, indospicine, which interferes with metabolism (Hegarty and Pound, 1970). A number of other Indigofera species also contain indospicine and it is generally recommended that they should be fed with care and not constitute more than 50% of the diet of cattle and should not be fed to pigs or poultry.

Hydrocyanic Acid (HCN). Sorghum *(Sorghum vulgare),* Sudan grasses *(Sorghum halepense),* and to a lesser extent *Cynodon plectostachyus* and Tanner grasses *(Brachiaria radicans),* contain the cyanogenetic glucoside dhurrin which can be reduced to HCN. Deaths have occurred when hungry animals graze young herbage or material which has had a check to growth. Deaths occur only when HCN levels exceed 0.02% of dry matter. One gram of HCN requires 1.2 g of S for detoxification so HCN can induce S deficiencies in animals grazing tropical herbage containing marginal levels of S (Stobbs and Wheeler, 1977).

Mimosine. The shrub legume *Leucaena leucocephala* contains 1-7% mimosine, a depilatory agent which shows promise as a chemical defleecing agent. The mimosine is broken down by bacteria in the rumen to 3,4-dihydroxypyridine (DHP), an active goitrogen which may cause poor growth, enlarged thryoid, and dead calves (Hegarty et al, 1977). When cattle receive less than 30% *Leucaena* in the diet or when it is fed as the sole constituent for periods under 2 mo., there is little risk of toxic effects. Feeding an I supplement fails to overcome the goiter caused by DHP.

Tannins. Various types of tannins and other polyphenolic substances are to be found in tropical forages (McLeod, 1974). Tannins inhibit breakdown of plant material by enzymes such as proteolytic enzymes in the rumen. This protection may be valuable in enabling a higher proportion of plant protein to pass into the small intestine. The low palatability of some herbage plants such as *Lespedeza cureata* and *Imperata cylindrica* have been attributed to their high tannin contents.

Other Miscellaneous Toxins. A number of *Crotoleria* species contain pyrrolizidine alkaloids (mainly monocrotaline, retusine and retusamine) which can result in death. Photosensitization of animals has been reported when ruminants graze *Brachiaria decumbens* and *Cassia* spp. Occasionally, toxic effects have been recorded when cattle graze *Pennisetum cladestinum,* but the cause is unknown.

Taints and Odors in Milk and Meat. Cows grazing certain tropical legumes sometimes product milk with objectionable odors and flavor, the strongest occurring with *Lablab purpureus* and *Leucaena leucocephala* (Stobbs and Fraser, 1971). Fortunately, these taints are lost on pasteurization. Objectionable odor and flavor have been reported from lambs grazing *Glycine wightii* prior to slaughter and occasionally when grazing *Macrotyloma axillare* (Park and Minson, 1972).

NUTRITIONAL PROBLEMS OF THE GRAZING ANIMAL

Although the basic principles of nutrition apply in the field, there are many factors which influence animal productivity, over and above those found under indoor conditions (McDonald, 1968; Stobbs, 1975). Therefore, the true feeding value of selectively grazed pasture can only be measured under grazing conditions. In this section factors which modify animal production under tropical grazing conditions will be considered under four major headings: selective grazing, increased nutrients required, greater climatic stress, and influence of pasture-born parasites.

Selective Grazing

Grazing animals are generally offered a range of pasture species, each with varying proportions of leaf and stem. Given the opportunity, all grazing animals are selective in the diets they choose, and therefore the amount and presentation of desired herbage greatly influence their eating behavior and production. Since the grazing animal moves in a horizontal plane and selectes in a vertical plane, sward heterogeneity (variation in structure and nutrients within the sward) influences the diet selected.

Sward Heterogeneity. Tropical pasture swards often contain high proportions of stem and dead material but these components are consumed only in small quantities when excess feed is offered. Leaf is the major fraction which is preferred and eaten by grazing animals. The quantity and accessibility of leaf are major factors influencing the quantity and quality of feed consumed by grazing ruminants.

Tropical pastures vary greatly in composition from the top to the bottom of the sward. This is illustrated in Table 15-10 for 6-week regrowths of *Kazungula setaria.* It is not uncommon for the top leaves of mature tropical pasture swards to be 10% more digestible than leaves at the base of the sward.

The density of many tropical pastures is low (Table 15-11) with a low leaf weight/unit of height (18 to 170 kg DM/ha/cm). This may influence the ease with which leaf can be selected from the sward. Sward bulk density is highest in the lower layers but swards which are continually being cut are generally less dense in the base than grazed swards, so extrapolation of results from cutting experiments to a grazing situation can be misleading.

Selection Within Single Species Swards. The nutritive value of the diet selected from a single species pasture depends on the pasture species being grazed, stage of maturity, and grazing intensity. As a pasture is grazed down animals are forced to eat herbage which contains more stem, and diets become more fibrous and lower in digestibility, N, and some minerals. This is illustrated in Fig. 15-4 which

Table 15-10. Variation in nutrients with a *Setaria anceps* cv. Kazungula sward.[a]

Sward sample	N, % of DM		In vitro DMD		P, % of DM		Ca, % of DM	
	Leaf	Stem	Leaf	Stem	Leaf	Stem	Leaf	Stem
Top	1.60	0.64	59.1	58.5	0.28	0.29	0.27	0.12
Middle	1.10	0.42	57.1	54.4	0.21	0.25	0.29	0.12
Bottom	0.67	0.21	55.6	49.2	0.21	0.22	0.41	0.12

[a]From Stobbs (1973, 1975)

Table 15-11. Range of sward densities of tropical and temperate pastures.

Parameter	Tropical	Temperate
Density sward, kg DM/ha/cm	27-215	160-410
Leaf, kg leaf DM/ha/cm	18-170	130-340

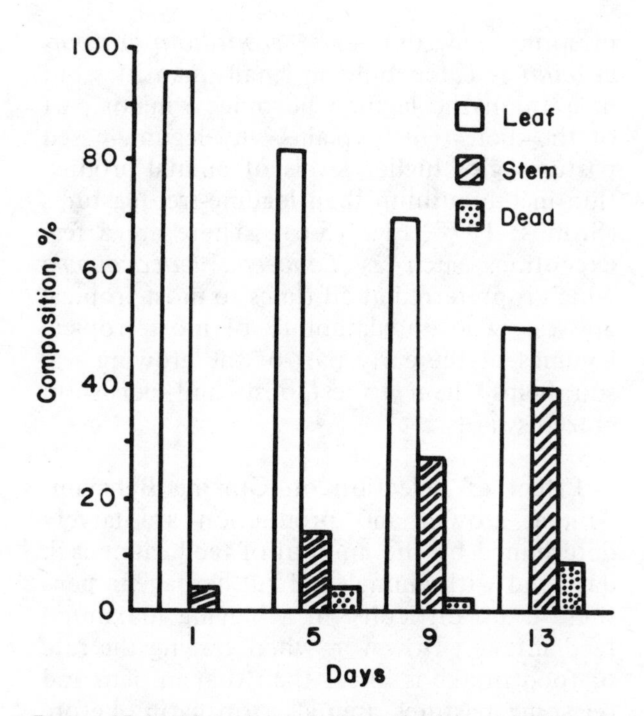

Figure 15-4. Mean botanical composition of diets of cattle grazing autumn-grown *Kazungula setaria* pastures at different stages of defoliation. From Chacon and Stobbs (1976).

shows the botanical composition of diets selected by cattle grazing *Setaria anceps* with progressive defoliation.

Seed heads may be selectively removed from pastures, as in the case of white panic millet (*Echinochloa crusgalli* var *edulis;* Stobbs, 1975). Alternatively, fallen seed can be harvested from the ground, by sheep and goats, and in some environments this provides the animal with good feed in the dry season of the year. For example, ingested *Stylosanthes humilis* seed plus pod has been shown to contain a high content of N (5.5%), P (0.36%), S (0.30%) and Ca (0.88%). It is relatively digestible (DMD of 58.6%) compared with mature leaf and stem (Playne, 1969).

Fouling of pastures with excreta and treading with hooves can deter animals from eating certain areas within a pasture, particularly during hours of darkness, and thus reduce the effective grazing area. Herbage contaminated with urine is not usually rejected by grazing animals for long periods, but herbage grown in the vicinity of feces is avoided and the area lost to grazing depending upon climate, stocking rate, length of the grazing period, and the presence of invertebrate organisms such as dung beetles (Simpson and Stobbs, 1979).

The opportunity to select the more nutritious leaf is greater at low intensities of grazing and with tropical pastures a large excess of herbage is necessary to ensure maximum consumption and production (Fig. 15-5). This contrasts with more homogeneous temperate pastures in which there is less selective grazing and no increase in milk production is achieved by offering cows more than 20 kg DM cow^{-1} day^{-1}.

High animal performance will be achieved from tropical pastures only if maximum selection is allowed, but this leads to underutilization of pasture and low production per unit area because of the low carrying capacity. A management practice which allows high producing cows to selectively graze the sward first (leaders) and offer the remainder of the sward to other animals (followers) with a lower nutrient requirement has been used to achieve higher utilization of the pasture. Milk yield of leader cows grazing 3-week regrowths

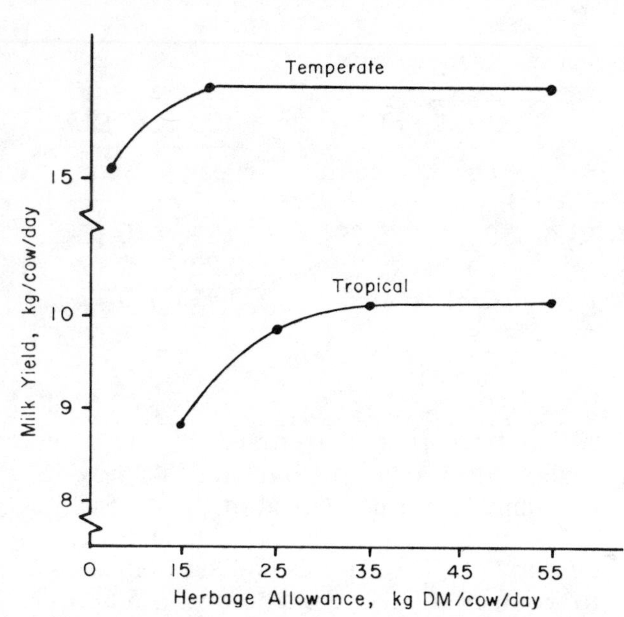

Figure 15-5. Milk yields of cows grazing tropical pastures and temperate grass pastures at various herbage allowances (Stobbs, 1977).

of both *Panicum maximum* and *Chloris gayana* was 38% higher than that of followers. These differences are higher than those found with temperate pastures (Table 15-12).

Table 15-12. Response in milk yield to leader grazing on tropical and temperate pastures.

Pasture	Increase, %*
Tropical	
Rhodes	38[a]
Gatton panic	38
Temperate	
Ryegrass	8[b]
Ryegrass	5[c]
Ryegrass	22[d]
Ryegrass	16

*% of leaders over followers.

[a]Stobbs (1978); [b]Bryant et al (1961); [c]Castle (quoted by Archibald et al, 1975); [d]Archibald et al (1975).

Selection Within Mixed Swards. The problem of selection is even more complex with mixed swards. The species or plant part selected by grazing animals is dependent upón many factors including the species combinations, stage of growth, and previous grazing history of the animals (Arnold, 1964). Young regrowths of tropical grasses are generally preferred to tropical legumes, but with mature pasture the legume is often preferred. Thus,

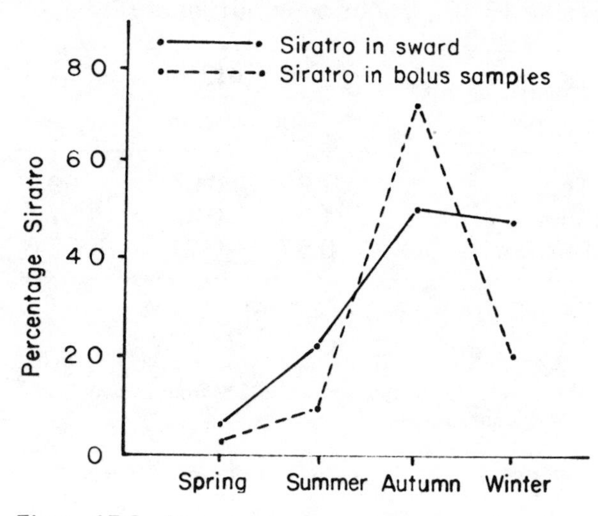

Figure 15-6. Mean percentage of *Siratro* offered and selected by cows grazing *Nandi setaria-Siratro* swards Stobbs, 1977).

in spring and summer *Macroptilium atropurpureum* is eaten only in small quantities but in autumn the legume becomes a major part of the diet. This explains why legume-based pastures give higher levels of animal production in the autumn than legume-free pastures (Stobbs, 1977; Fig. 15-6). There are a few exceptions such as *Leucaena leucocephala* which is preferred at all times to most tropical grasses. The unpalatability of most tropical legumes in the early part of the growing season helps them to establish and persist in grazed swards.

Effect of Selection on Grazing Behavior. Animal growth and production are largely determined by the amount of feed eaten each day, and with animals fed cut herbage in pens there is no difficulty in achieving maximum feed intake. However, when grazing the rate of food intake is lower than that in pens and on some pastures animals stop eating before they have reached maximum rumen fill (Chacon and Stobbs, 1976). The intake (I) of herbage depends on the time spent grazing (T), the rate of biting (R), and the size of each bite eaten (S); $I = T \times R \times S$.

The ease with which animals can satisfy their appetite depends on the yield and spatial distribution of leaf within the sward. On leafy temperate pastures cattle generally have no difficulty satisfying their feed requirements, taking in large quantities of herbage with each sweep of the tongue and spending only short

Table 15-13. Mean grazing time, rate of biting and size of bite taken by Jersey cows grazing pastures of varying age and quality.[a]

Item	Pasture			
	Immature		Mature	
	Temperate*	Tropical	Tropical grass**	Tropical legume
Mean grazing time, min/24 h	464	561	677	719
Mean bites/24 h				
Grazing	13,900	22,300	29,900	--
Rumination	14,900	22,200	31,800	--
Total	28,800	44,500	61,700	--
Mean bite size, mg OM/bite	430	340	170	130

[a]From Stobbs (1974); *<3 wk-old regrowths; **> wk-old regrowths; + calculated from 3-4 animals each grazing 3-6 pasture replicates.

periods grazing (Table 15-13). In contrast cattle tend to graze tropical pastures for long periods each day. This is particularly the case where pasture yields are low or when leaf is inaccessible because of the presence of stem or inflorescence makes it difficult for animals to harvest leaf. Cows grazing under these unfavorable conditions take small bites and attempt to compensate by increasing the time spent grazing and rate of biting (Stobbs, 1973). Cattle have been found to graze up to 750 min/24 h and take up to 36,000 grazing bites/24 h. Where swards are very mature and leaf is inaccessible, then the time spent grazing and the number of bites appear to be insufficient for cattle to fill their rumens to maximum capacity. Feeding behavior studies indicate that animal production should be improved by: using leafy pasture plants which are easier for cattle to harvest; fertilizing pastures to produce dense leafy swards; and grazing management practices which maintain swards in a dense leafy condition.

Domestic ruminants modify their grazing behavior patterns in an attempt to overcome the low rate of herbage intake associated with the selection of leaf. For example, when the quantity and quality of the pasture is good, most grazing occurs during daylight hours. However, under adverse pasture conditions longer grazing times are achieved by extending grazing into the night.

The proportion of night grazing will also depend upon climate. With cattle under high altitude and cooler tropical conditions, only 9% of the grazing might be at night, but in hot, humid environments cows have been found to consume up to 77% of their food at night. For dairy cattle grazing tropical pastures it is quite usual for more than 50% of grazing to occur between evening and morning milking (Stobbs, 1970). Therefore, it is essential that cattle grazing tropical pastures should be allowed to graze at night, particularly when the feed is poor. The advantages of better night grazing are illustrated in Fig. 15-7, which shows responses in milk production to increasing areas of night grazing on commercial farms in southeast Queensland (Rees et al, 1972).

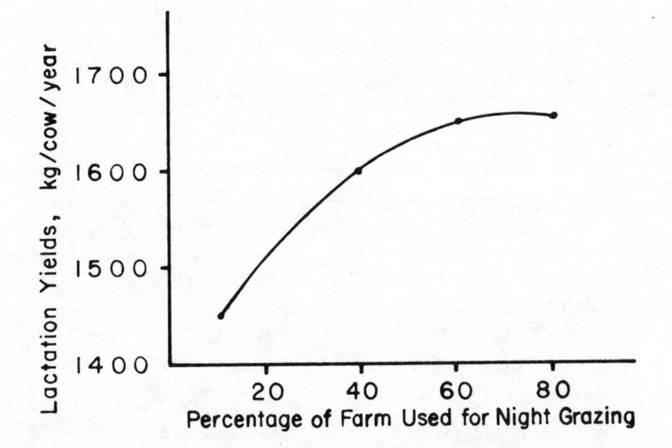

Figure 15-7. Milk production with various proportions of the farms available for night grazing. From Rees et al (1972).

Increased Nutrient Requirements of Grazing Animals

On good quality dense pastures cattle travel about 2-3 km in each 24 h period. The distance traveled on most tropical pastures is usually higher (3-4 km/24 h) and depends on pasture yield, sward heterogeneity plus any additional traveling required between shade, water, and yards. In arid and semiarid areas walking activity is centralized around water holes. Under extremely hot conditions (max temperatures of 36-43°C) in shadeless arid country, Yeats and Schmidt (1974) have observed cattle 11 km from water. They report an extreme case where cattle were 32 km from a water hole to which they returned after 4 days. Sheep are better adapted to semiarid conditions than cattle due to the lower loss of water in the feces.

The longer distances traveled together with the long times spent grazing and ruminating probably lead to high maintenance requirements of stock in the tropics. No direct measurements have been made of this higher requirement for cattle, but an estimate may be obtained from studies of the energy expenditure of sheep and cattle involved in different activities while kept in respiration chambers (Table 15-14). Care is obviously necessary when extrapolating these results to outdoor conditions. Additional energy rates for a 400 kg steer on tropical pasture compared to good temperature pasture could consist of an additional 5 h grazing, a further 4 h longer rumination and another 2 km longer walking if maintenance requirements are to be satisfied. The additional energy cost would be approximately 1,350 Kcal/day and the energy cost for maintenance would increase from 11,000 Kcal to 12,350 Kcal, an increment of about 12%. Increases of up to 40% seem possible under less favorable pasture conditions in hilly country with widely spaced watering points.

Table 15-14. Energy costs of various activities.[a]

Activity	Mean energy cost in Kcal/kg BW Sheep	Cattle
Walking on level surface, Kcal/horizontal km	0.6	0.5
Walking on sloping surface, Kcal/vertical km	6.0	6.2
Grazing or eating, Kcal/h	0.54	0.45
Ruminating, Kcal/h	0.24	0.23
Standing, Kcal/h	0.18	0.17

[a]From Osuji (1974) and Ribeiro et al (1977)

Climatic Effects

Great varieties of climates are encountered in tropical areas, varying from hot-wet conditions of the desert areas. These climatic effects manifest themselves in two ways—directly on the animal or indirectly through the effect on the quantity and quality of forage, as discussed earlier. The effects of heat, humidity, solar radiation, and daylight/darkness ratios have a marked effect on food intake, water balance, and grazing behavior.

Food Intake

When the environmental temperature rises above 21°C for European cattle or 35°C for tropical breeds, voluntary intake and production decline. This decline is caused by the difficulty in dissipating heat associated with the fermentation and metabolism of the food. The rate of heat loss depends on both temperature and water vapor deficit, so the adverse effect of temperature on food intake is much greater in humid environments. With continued heat stress rumen contractions become smaller in amplitude and slightly less frequent, and food is retained in the rumen for a longer time. Associated with these changes is a reduction in the concentration of volatile fatty acids in the rumen. A lower proportion of acetic acid in rumen fluid has usually been reported from heat-stressed animals. The waste heat production from all feeds is a relatively constant % of GE intake. Where heat stress is a problem there is an obvious advantage in feeding a highly digestible feed.

Cows kept at high temperatures produce milk with a low content of solids-not-fat and protein due to a reduced energy intake and direct effects of temperature. A lower fat and protein percentage has been found in milk from Jersey cows kept at 30°C compared with cows kept at 15°C and receiving the same quantity of feed.

Water Intake

In a hot environment an increased evaporative loss of water provides an important mechanism for regulating body temperature and animals' requirements for water intake increase accordingly. The effect of ambient temperatures on water intake is greater with European breeds of cattle than with tropical breeds although acclimatized cattle require less water than unacclimatized cattle. Increased intensity of radiation has also been shown to increase the water consumption of a range of cattle breeds (see Ch. 2, Vol. 2).

Supplying insufficient water will lead to reduced feed intake and lower animal production. When cattle in hot environments are totally deprived of water, their urine excretion, fecal water output, total body evaporation, and urinary Na output are reduced and urinary output of K is increased. In contrast, cattle receiving unlimited water in hot environments overcompensate for increased water loss and they have a dilution of blood and urine and an increase in urine volume (McDowell et al, 1969).

Pastures grown in the humid tropics contain large quantities of water, especially at a young stage of growth when water may constitute 90% of the fresh weight. It is often suggested that a high water content may limit the quantity of dry matter eaten. However, several studies of tropical grasses have failed to show any increase in voluntary intake of dry matter when the water was removed by low-temperature drying. All these studies have been with pasture containing <80% water, so it is still possible that very high levels of water might depress feed intake.

Grazing Behavior

The grazing ruminant modifies feeding behavior to alleviate climatic stress. In hot humid environments *Bos taurus* cattle stop grazing during the hottest time of the day and grazing is confined entirely to early morning, late afternoon and during the night. *Bos indicus* cattle are more tolerant of climatic stress and spend more time grazing during the hotter periods of the day. Under extensive grazing conditions with few watering points there is a limit to the distance animals are prepared to walk each day. Areas immediately around water points become seriously overgrazed whereas more distant areas (2 km or more for cattle) are only lightly grazed.

Parasites and Pathogens

Grazing conditions encountered in many tropical environments result in a large range of ecto- and endoparasites which influence both feeding behavior and production. Insects and ticks increase stress in two ways: firstly, as vectors of disease and secondly by physical irritation and blood sucking. In some potentially productive tropical areas disease-carrying parasites can prevent animals being kept. For example N'dama cattle, which have some resistence to trypanosomiasis, are the only animals that can be kept in tsetse-infested areas. In other areas special husbandry practices have been devised to protect animals. In the Lake Shore areas of Uganda, cattle are kept in smokey, darkened buildings during the mid-day period when *Stomoxys* flies are most ferocious.

Ticks are widespread in the tropics but *Bos indicus* cattle and their crosses are more resistant than *Bos taurus* animals (Turner, 1969). Exposure to moderate levels of tick infestation caused a reduction in liveweight gain of Hereford-Shorthorn cattle of 34 kg, Africander cattle of 15 kg, but had negligible effects on Brahman crosses.

Many hot, wet, humid environments provide an ideal environment for the build-up of internal parasites which affect growth, particularly of young animals. The effect of anthelmintic treatments varies between breeds. Brahman crosses show less response to drenching than Hereford-Shorthorn animals.

SEASONAL PROBLEMS OF PASTURE PRODUCTION

Many tropical pastures have a high annual yield of dry matter, but annual animal production is seriously limited by the seasonal nature of this production. The main factor limiting pasture growth is the lack of soil moisture for long periods of the year. With the start of the rainy season there is a flush of high quality forage with a growth rate exceeding the appetite of the limited number of animals normally available for grazing. The pasture grows more rapidly than it can be consumed, leading to a rapid decline in quality of the herbage which is partly offset by selective grazing. When pasture growth ceases at the end of the rainy season, the pasture consists of a large bulk of mature feed from which some of the leafier parts have already been removed. In some countries this dry season may last 8 mo. Since pasture growth virtually ceases during the drier months, continuity of feed supply throughout the year is achieved by understocking during the periods of rapid pasture growth. Thus, ruminants have to graze for the greater part of the year on what is essentially standing hay. This results in the familiar pattern of animal growth illustrated in Fig. 15-8, with rapid weight gains in the early wet season, slower gains until some time after the end of the summer rains; then losses in the dry season and early rainy season. The most rapid losses in weight are normally before and at the commencement of the rains. Under this system calving rates are low and the stock are rarely ready for slaughter until 4 years old.

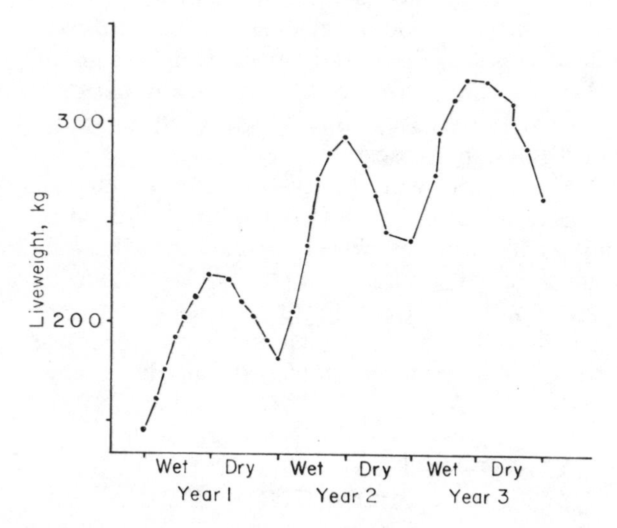

Figure 15-8. Seasonal live weight changes in cattle grazing native pastures in northern Australia. From Norman (1965).

Poor liveweight gain in the dry season is usually caused by the low protein content of the diet, but it can also be due to a lack of pasture DM or difficulty in harvesting the forage. In the more favorable of these drier environments (600-750 mm of rainfall), the protein content of native pastures may be increased by oversowing with legumes such as *Stylosanthes humilis, Stylosanthes hamata* and *Centrosema pascuorum* together with superphosphate. These legumes are eaten mainly at the end of the wet season, thus extending the period of animal growth into the dry season. Improved reproductive performance has also been reported following the introduction of legumes into pastures.

Lack of protein in the diet of grazing ruminants can also be overcome by feeding small quantities of high protein supplements such as cottonseed meal, blood and bone meal or NPN in the form of urea. While urea may be beneficial in preventing large weight losses, it has rarely promoted gains (Loosli and McDonald, 1968) and the improvement achieved in the dry season is often offset by a lower rate of liveweight gain in the wet season (Table 15-15).

Table 15-15. Growth of weaners during molasses-urea supplementation and after-supplementation (mean of 3 years).[a]

Treatment	Average daily gain, kg/head/day	
	During supplementation	After supplementation
Native pasture	-0.16	0.64
Native pasture plus urea & molasses	0.08	0.62

[a]From Winks et al (1976)

Conservation

In the wetter tropical regions (>750 mm) the seasonality of production of forage may be alleviated by ways which will now be considered. One solution to the problem of seasonal variation in grass growth is to store excess forage produced during seasons of flush growth for use during months when growth is slow or absent. This reserve of feed can be stored as standing feed, stored hay or silage. Hay is usually difficult to make in the humid tropics because of the thick stems of many tropical grasses, and artificial drying is usually uneconomic. Examples can be found of fairly good quality hay being produced from tropical grass pastures, but there are many failures which are not reported. Irrigated alfalfa makes excellent hay in drier areas and late crops of *Lablab purpureus* grown under dry land conditions are reported to produce good hay.

In areas where topography permits mechanization, silage can be made from tropical pastures (Catchpoole and Henzell, 1971) or specially grown forage crops such as corn. At best, silage produced from tropical pastures provides low to medium quality roughage feed. It is satisfactory only when fed with concentrates. The densities of these silages can be low, and air tight silos are required if excessive wastage is to be avoided, especially with wilted material. Silage has been made from a number of tropical pastures without additives, e.g. *Setaria sphacelata* cv. Nandi, *Desmodium intortum, Lotononis bainesii, Cynodon dactylon.* Additives such as molasses, sodium metabisulphite, and formaldehyde can improve preservation of silage. Some pastures from the humid tropics, such as *Pennisetum purpureum,* have high moisture contents and wilting to 30% DM assists the ensiling process but mold growth can be a problem. The regrowth of many tropical pastures can be severely impaired by cutting.

Integration of Different Pastures

Pastures exhibiting different growth patterns can be integrated into the whole farm situation. This is illustrated by the feed-year program developed for dairy production in the subtropical environment of Wollongbar in Australia (Hudson et al, 1965) where varying patterns of growth of four species resulted in even production throughout the year. However, unreliability of the temperate legumes *(Trifolium subterraneum* and *Vicia sativa),* a decline in productivity of *Glycine wightii,* and unreliable spring rain reduced the effectiveness of the system. Nevertheless, it may be possible to find other combinations of species which give uniform herbage production throughout the year in areas with a favorable distribution of rainfall. Where water is available irrigated pastures can provide valuable feed, but they are expensive and have a high labor requirement. However, very small quantities of irrigated pasture can have a large effect on annual animal production. In some areas swamp margins are available for grazing in the dry season but the advantages of this green feed are often outweighed by the presence of liver flukes.

Undergrazing and Deferred Grazing

Undergrazing during periods of rapid pasture growth is usually the only economically viable management practice. This generally results in poor quality feed being available for the remainder of the year. The problem can sometimes be alleviated by careful timing of fertilizer and grazing management practices.

N fertilizer applied in the spring and autumn can help spread production throughout the year and improve herbage quality. Up to 100 kg N ha^{-1} can be applied on grass legume pastures for one year without loss of legume but higher or repeated applications will reduce the legume content of the pasture and the productivity of legumes (Cowan and Stobbs, 1976). Higher stocking rates are usually necessary if the fertilizer costs are to be recovered.

Some pasture species maintain their high feeding value for long periods and may be left ungrazed during the rainy season for use in the autumn and winter. Pastures such as *Lablab purpureus, Leucaena leucocephala* and *Pennisetum purpureum* have all proved valuable for this purpose.

Energy Supplements

When the energy requirements cannot be met from the pasture some form of supplementary feed may be economically justified. Grain can be fed but in some parts of the tropics various by-products are available as energy sources. The extent to which these are used varies according to local economic conditions and in some cases the by-product may become the major component in the diet. A comprehensive list of tropical by-products, together with their chemical composition, is given by Gohl (1975), and only the more important of these tropical feeds are discussed in this chapter.

Sugarcane By-Products. Molasses is a valuable source of energy and minerals (see Ch. 2) which is usually exported at a low price. It is a palatable feed which is used an an energy source (0.2-0.5 kg/head/day) with urea. There are reports that molasses may comprize up to 80% of the ME for beef cattle if some true protein such as fish meal is provided (Preston et al, 1967), but it is more normal to feed no more than 25% molasses in the diet.

Response to molasses feeding is 76% of the response to grains of similar digestibility. Substantial increases in milk production have been recorded from cows grazing tropical pastures when supplemented with molasses (0.4-0.7 kg milk/kg of molasses). Molasses is not always of uniform feeding value so, when feeding at high levels, careful analysis is recommended.

Bagasse is the main by-product of sugar production and consists of the fiber of the sugar cane. It is a low quality feed but, when mixed with protein and a source of energy such as molasses, moderate levels of bagasse (up to 30%) can promote satisfactory growth of stock. It is possible to increase the digestibility of bagasse (as well as straws and wood pulp) by treating with NaOH, steam or anhydrous ammonia, but this is rarely economic. Sugarcane leaves are not usually utilized but can provide low to medium quality feed for stock.

Cassava and Bananas. Both cassava (Coursey and Halliday, 1974) and bananas (Ruiz et al, 1973) can provide a valuable source of energy in the dry season when pasture is in short supply. Fresh peeled cassava roots normally contain 60-70% water and the DM is almost completely starch with only very small percentages of fat and protein. High moisture content may result in damage due to molding, and it is usual to dry the roots into chips or pellets. Cassava root contains a cyanogenetic glycoside, linimarin, which is readily hydrolyzed by enzymes naturally present in cassava roots to form toxic HCN. Work is currently being undertaken at two international research centers, CIAT in Columbia and IITA in Nigeria, to develop high-yielding new varieties with a low content of cyanogenetic glycoside and tubers which can be easily harvested.

Citrus Pulp and Spent Grains. By-products from citrus and pineapple processing and the spent grains from fermentation processes are often wasted due to their high water content (>80%), rapid deterioration and the cost of transporting high-moisture feeds. Dehydration procedures are generally not economical in the tropics, but these products have a high sugar content and can be ensiled satisfactorily.

POTENTIAL FOR INCREASING ANIMAL PRODUCTION

Output per Animal

The potential for increasing animal production depends on the potential for higher output/animal and higher production/ha. Beef cattle grazing improved tropical pastures are capable of producing good gains (0.9-1.2 kg/steer/day) over relatively short periods during the early grazing season, but some of this gain may be attributable to compensatory growth. However, animal performance from a large number of long-term experiments with improved pastures in the tropics showed that annual gains averaged only 0.35 kg/steer/day (Stobbs, 1976). Obviously, the genetic potential of stock can limit performance but even when data were obtained from animals with high potential, average daily gains rarely exceeded 0.60 kg/day. Although this is a great improvement in levels of production, it is below the genetic potential of the animals as illustrated by the higher growth rates achieved in feedlots (McDowell and Hernandez-Urdaneta, 1975).

Lactating animals require more nutrients than fattening cattle to achieve maximum production. Even the best improved tropical pasture swards are incapable of supplying the nutrients required for maximum milk production (Table 15-16). For smaller breeds, such as the Jersey, maximum milk yields over extended periods on unsupplemented tropical pastures rarely exceed 8-9 ℓ/cow/day, and larger breeds, such as the Holstein, yields of

Table 15-16. Maximum beef and milk production achieved from tropical and temperate feeds of varying digestibilities.[a]

Diet	DMD, %	Maximum milk production, kg/cow+	Maximum beef production, kg gain/day
Tropical pastures			
Immature	60-65	1,800-2,400	0.7-0.9
Semimature	50-55	1,000-1,400	0.4-0.5
Immature temperate pastures	70-80	3,300-3,800	0.9-1.2
Concentrate ration	80-85	4,400-4,900	1.2-1.4

[a]From Stobbs and Thompson (1975); +Jersey cows

Table 15-17. Estimates of beef production from natural grasslands and sown grasslands in temperate and tropical environments with various inputs, kg gain/ha/year.[a]

	Monsoonal tropics, 5-6 m dry	Humid tropics, long growing season	Temperate
Unimproved grazing	10-80	60-100	100-400
Oversown with legumes and fertilized with phosphate	120-170	250-450	200-500
Grass/legume mixtures with phosphate	200-300	200-300	400-1,200
N fertilized grass	300-500	400-1,200	700-1,400

[a]From Stobbs (1976a)

Table 15-18. Expected levels of milk production/ha with various inputs.[a]

	Milk production, ℓ/ha	
	Tropical	Temperate
Unfertilized	1,000- 2,000	3,000- 6,000
Grass-legume	3,000- 8,000	6,000-14,000
N fertilized grass, +P, S, K	4,500- 9,500	8,000-17,000
N fertilized and irrigated, +P, S, K	15,000-20,000	10,000-18,000*

[a]From Stobbs (1976c); *few data

12-14 ℓ/cow/day. In many situations the pastures mature rapidly, and have a low digestibility and protein content resulting in lactation yields of <1,000 ℓ.

Production per Hectare

Animal production/ha is dependent upon both production/animal and carrying capacity. Low rainfall areas (250-400 mm) with a prolonged dry season are generally not suitable for domestic ruminants. Talbot et al (1965) have shown that substantial yields of lean meat can be obtained from mixed game ranching in these areas. Expected levels of milk and beef production/ha from tropical and temperate pastures in higher rainfall areas are shown in Tables 15-17 and 15-18.

Despite the lower lactation yields, production/ha from tropical pastures compares favorably with production from temperate pastures because of the higher pasture yield and carrying capacity. Although production levels are closely related to carrying capacity, care is needed in interpreting the results of some short-term experiments because high stocking rates can often lead to pasture deterioration. This is particularly the case with legume-based swards. Improved pastures appear to offer a practical way of increasing ruminant production in the tropics using land not required for arable crops. However, the level of pasture improvement adopted will obviously depend on local economic conditions and the availability of by-products.

References Cited

Archibald, K.A.E., R.C. Campling and W. Holmes. 1975. Animal Prod. 21:147.
Arnold, G.W. 1964. Proc. Aust. Soc. Animal Prod. 5:258.
Bryant, H.T., R.E. Blaser, R.E. Hammes and W.W. Hardison. 1961. J. Dairy Sci. 44:1733.
Catchpoole, V.R. and E.F. Henzell. 1971. Herb. Abstr. 41:213.
Chacon, E. and T.H. Stobbs. 1976. Aust. J. Agr. Res. 27:709.
Chapman, H.L. and A.E. Kretschmer. 1964. Proc. Soil & Crop Sci. Soc. Florida 24:176.
Coursey, D.G. and D. Halliday. 1974. Outlook on Agr. 8:10.
Cowan, R.T. and T.H. Stobbs. 1976. Aust. J. Exp. Agr. Animal Hus. 16:829.
Crampton, E.W., E. Donefer and L.E. Lloyd. 1960. In: Proc. 8th Int. Grassl. Cong., Reading, England.
Davies, E.W., A.G. Griffith and A. Ellington. 1966. J. Agr. Sci. 66:351.

FAO. 1975. Production Yearbook. FAO, Rome.
Garcia-Rivera, J. and H.P. Morris. 1955. Science 122:1089.
Gohl, B. 1975. Tropical feeds. Feeds Information Summaries and Nutritive Values. FAO, Rome.
Griffith, A.G., D.I.H. Jones and R.J.K. Walker. 1965. J. Sci. Fd. Agr. 16:94.
Hacker, J.B. 1974. Trop. Grassl. 8:145.
Hamilton, R.I. and G. Ruth. 1968. Trop. Grassl. 2:135.
Hegarty, M.P., R.D. Court, G.S. Christie and C.P. Lee. 1976. Aust. Vet. J. 52:490.
Hegarty, M.P. and A.W. Pound. 1970. Aust. J. Biol. Sci. 23:831.
Hudson, W.T.C. et al. 1965. In: Proc. 9th Int. Grassl. Cong., Sao Paulo.
Jasiorowski, H.A. 1976. In: Beef Cattle Production in Developing Countries. Centre for Vet. Med., Univ. of Edinburgh.
Jones, R.J. and C.W. Ford. 1972. Trop. Grassl. 6:201.
Jones, R.J., A.A. Seawright and D.A. Little. 1970. J. Aust. Inst. Agr. Sci. 36:41.
Lamond, D.R. 1970. Animal Breed. Abstr. 38:50.
Laredo, M.A. and D.J. Minson. 1973. Aust. J. Agr. Res. 24:875.
Laredo, M.A. and D.J. Minson. 1975. Aust. J. Exp. Agr. Animal Hus. 15:203; Brit. J. Nutr. 33:159.
Little, D.A. 1976. Aust. J. Agr. Res. 27:681.
Loosli, J.K. and I.W. McDonald. 1968. Nonprotein Nitrogen in the Nutrition of Ruminants. Agricultural Studies No. 75, FAO, Rome.
McDonald, I.W. 1968. Nutr. Abstr. Rev. 38:381.
McDowell, L.R. 1976. In: Beef Cattle Production in Developing Countries. Centre for Vet. Med., Univ. of Edinburgh.
McDowell, R.E. et al. 1969. J. Dairy Sci. 52:188.
McDowell, R.E. and A. Hernandez-Urdaneta. 1975. J. Animal Sci. 41:1228.
McLeod, M.N. 1974. Nutr. Abstr. Rev. 44:803.
Milford, R. and D.J. Minson. 1965. J. Brit. Grassl. Soc. 20:177.
Minson, D.J. 1967. Brit. J. Nutr. 21:587.
Minson, D.J. 1972. Aust. J. Exp. Agr. Animal Hus. 12:21.
Minson, D.J. 1973. Aust. J. Exp. Agr. Animal Hus. 13:153.
Minson, D.J. 1975. Forage Res. 1:1.
Minson, D.J. 1976. In: From Plant to Animal Protein. Armidale Univ.
Minson, D.J. 1977. In: Tropical Forage Legumes. FAO, Rome.
Minson, D.J. and M.N. McLeod. 1970. In: Proc. 11th Int. Grassl. Cong., Surfers Paradise.
Minson, D.J. and R. Milford. 1966. Aust. J. Agr. Res. 17:411.
Minson, D.J. and R. Milford. 1967. Aust. J. Exp. Agr. Animal Hus. 7:546.
Minson, D.J., T.H. Stobbs and L. t'Mannetje. 1978. Proc. Aust. Soc. Animal Prod. 12:205.
Murphy, G.M. and A.W. Plasto. 1973. Aust. J. Exp. Agr. Animal Hus. 13:369.
Norman, M.J.T. 1965. Aust. J. Exp. Agr. Animal Hus. 5:227.
Osuji, P.O. 1974. J. Range Mange. 27:437.
Park, R.J. and D.J. Minson. 1972. J. Agr. Sci. 79:473.
Preston, T.R., M.B. Willis and A. Elias. 1967. Rev. Cubana Cienc. Agr. 1:33.
Rees, M.C., D.J. Minson and J.D. Kerr. 1972. Aust. J. Exp. Agr. Animal Hus. 12:553.
Rees, M.C., D.J. Minson and F.W. Smith. 1974. J. Agr. Sci. 82:419.
Ribeiro, J.M. de C., J.M. Brockway and A.J.F. Webster. 1977. Animal Prod. 25:107.
Ruiz, M.E., K. Vohnout, M. Isidor and C. Jimeneg. 1973. Proc. 4th Conf. Assoc. Latin-Amer. Animal Prod., Guadalajara.
Siebert, B.D. and W.V. Macfarlane. 1969. Aust. J. Agr. Res. 20:613.
Simpson, J.R. and T.H. Stobbs. 1979. (in press) In: Grazing Animals, Vol. 16, World Animal Sci. Series, Elsevier Scientific Pub. Co.
Stobbs, T.H. 1970. Trop. Grassl. 4:237.
Stobbs, T.H. 1973. Aust. J. Agr. Res. 24:809, 821.
Stobbs, T.H. 1974. Proc. Aust. Soc. Animal Prod. 10:299.
Stobbs, T.H. 1975. Trop. Grassl. 9:151; Aust. J. Exp. Agr. Animal Hus. 15:211; Aust. J. Agr. Res. 26:997.
Stobbs, T.H. 1976a. In: Beef Production in Developing Countries. Centre for Vet. Med., Univ. of Edinburgh.
Stobbs, T.H. 1976b. Proc. Aust. Soc. Animal Prod. 11:477.

Stobbs, T.H. 1976c. In: Proc. Int. Sem. Animal Prod., Acupulco, Banco de Mexico.
Stobbs, T.H. 1977. Aust. J. Exp. Agr. Animal Hus. 17:892; Trop. Grassl. 11:87.
Stobbs, T.H. 1978. Aust. J. Exp. Agr. Animal Hus. 18:5.
Stobbs, T.H. and J.S. Fraser. 1971. Aust. J. Dairy Tech. 26:100.
Stobbs, T.H., D.J. Minson and M.N. McLeod. 1977. J. Agr. Sci. 89:137.
Stobbs, T.H. and P.A.C. Thompson. 1975. World Animal Rev. 13:27.
Stobbs, T.H. and J.L. Wheeler. 1977. Trop. Agr. 54:229.
Talbot, L.M. et al. 1965. Tech. Commun. No. 16, Comm. Bur. Animal Breed and Genet.
Turner, H.G. 1969. Aust. Vet. J. 48:162.
Walker, C.A. 1957. J. Agr. Sci. 49:394.
Walthall, J.C. and R.A. McKenzie. 1976. Aust. Vet. J. 52:11.
Winks, L., A.R. Laing, G.S. Wright and J. Stokoe. 1976. J. Aust. Inst. Agr. Sci. 42:246.
Yeates, N.T.M. and P.J. Schmidt. 1974. Beef Cattle Production. Butterworths, London.

Chapter 16—The Nutrition of Wild Ruminants

by R.E. Dean

The amount of information on the nutrition of wild ruminants, especially deer, has grown tremendously in the past 20 years. Hopefully, this trend will continue in the future, especially as the importance of wild ruminants continues to increase. The number of people hunting game animals for meat and trophies is increasing each year. Millions of people also photograph and observe wildlife; although the value of these non-consumptive uses cannot be determined accurately, it appears to be increasing steadily. Consequently, the importance of wild animals to the people of this country will continue to increase in the future. This means that research on the nutrition of these animals will have to continue if herds are to be managed to provide optimum opportunities to meet the demands of our growing population.

Wild ruminants in their native state are difficult to study from a nutritional standpoint. It is especially difficult to obtain quantitative nutritional information on free-ranging animals. Thus, most of the specific information on nutrient requirements comes from captive animals. Researchers have barely scratched the surface in the understanding of nutrient needs of wild ruminants. Energy, protein, Ca, and P are the only nutrients that have received much study, and these primarily with reference to white-tailed deer. Other species have not been studied in any depth. Most of the nutritional data regarding free-ranging animals concerns diets including the frequency and abundance of plants found in the diet. The levels of the various nutrients in important forage plants have also been well documented. However, until quantitative measures of nutrient intake of free-ranging animals can be made, the nutritional requirements obtained from the study of captive animals or diets and nutritional data on forage plants have limited practical application in terms of managing game ranges to furnish needed nutrients.

The current status of known nutritional requirements of wild ruminants based on studies with captive animals and the availability of nutrients in forage plants are discussed in this chapter. There is also some discussion of nutritional maladies and winter feeding of big game animals.

NUTRIENT REQUIREMENTS OF WILD RUMINANTS

The nutrient requirements and general nutritional processes of wild ruminants are not nearly as well understood as those for domestic ruminants. Some physiological processes involving nutrition appear to be different for the wild species which have evolved under regimens where tremendous differences exist in forage availability between seasons of the year. In general, forage should be considered satisfactory if it keeps the wild animals alive and allows an adequate level of reproduction. This is an important concept for free-ranging animals and differs from that of domestic ruminants where nutrient needs are evaluated primarily on a daily basis. Thus, it appears that the nutritional requirements of wild ruminants should be evaluated on seasonal rather than a daily basis, a situation which coincides with nutrient availability patterns of native ranges. The amount of various nutrients required during one particular time of the year usually depends on the nutrient availability during the other seasons. For example, nutritional needs for a winter range

Figure 16-1. A black-tailed deer in Oregon in good body condition. Summer forage is usually of high quality and abundant. This allows animals to deposit large amounts of body reserves which are necessary for survival during the winter stress period.

Figure 16-2. Elk are shown wintering on a ridge in deep snow country where the wind has blown the tops bare in places. This is the only available forage and consists of low growing plants, mainly weathered grasses.

depend largely on what the animals were able to obtain and store from summer and fall ranges (Fig. 16-1).

The actual amount of nutrients required from the winter range is less for animals in excellent body condition than it is for animals coming to the same winter range in poor body condition. Nagy et al (1974) showed that mule deer under captive conditions could survive complete starvation for as long as 64 days and recover. This suggests that deer can survive long periods with little or nothing to eat provided they have an opportunity to store ample body reserves (Fig. 16-2). Holter and Hayes (1977) studied the effects of restricted energy intake on fawns during the fall. Energy restriction of 40% resulted in a 76% reduction in fat deposition. Fawns had a tremendous ability to store fat during the fall when consuming diets high in productive energy.

If the objective of the wildlife manager is to provide forage that allows animals to live and reproduce satisfactorily, then making recommendations on requirements for a specific age or sex of animal can be difficult without knowing the entire nutrient availability facing the animal on a year-long basis. The amount actually needed by the animal at any time depends, in part, on what has been available in the past and what will be available in the future. This should be kept in mind when attempting to apply nutritional data that were obtained from captive animals to free-ranging wildlife.

Energy Requirements

More is known about the energy needs of captive wild ruminants than any other dietary requirement. Energy requirements are commonly expressed as dry matter (DM) intake (lb or kg) or, more specifically, as calories; both are used in practice. DM intake is important as a rough estimator of energy needs and dietary bulk. After forage production has been determined, DM intake is used in the calculation of range carrying capacity. When forage is adequate in terms of providing the needed nutrients, specific nutrient requirements are of little value in determining carrying capacity because animals eat in excess of their needs.

Dry matter intake by ruminants in North America can vary greatly depending on season of year and behavior patterns. This has

Table 16-1. Dry matter intake levels for various wild ruminant species.

Species	Level	Comments	Source
White-tailed deer	92-98 $g/kg^{0.75}$ 54-61 $g/kg^{0.75}$	Intake from September through November Intake during February through April	Holter et al (1977)
White-tailed deer	3.5 lb/day	Adult does, winter, pelleted ration	Verme (1969)
White-tailed deer	3 lb/day. 4.5-6.5 lb/day	Ad lib intake rate, adult deer, winter Ad lib intake rate, adult deer, summer	McEwen et al (1957)
Black-tailed deer	2.85 lb/100 lb BW	Captive deer, winter	Brown (1961)
Mule deer	31 g/kg/day 32 g/kg/day 21 g/kg/day 17 g/kg/day	Fawns, summer Fawns, winter Yearling and adults, summer Yearling and adults, winter	Wallmo et al (1977)
Mule deer	2.5-3 lb/day	Wild deer, ad lib, winter, pelleted ration	Dean (1973)
Bison	3.5-4.5 lb/100 lb BW	Estimate for free ranging animals	Pedan (1972)
Elk	22.7 g/kg BW	Free ranging elk, late gestation requirements	Thorne et al (1976)
Elk	4.47 lb/100 lb BW 3.51 lb/100 lb BW	Short term ad lib intake, pelleted alfalfa Short term ad lib intake, baled alfalfa	Thorne and Dean (1973)
Moose	10 kg/day 6.7 kg/day 14.5-18.6 kg/day	Cow moose, Canada Calf moose, Canada Fresh weight of balsam fir, captive bull, Canada	Crete and Bedard (1975)
Moose	1.7-3.4 kg/day 17.6 kg/day	Alaska, winter Alaska, summer	LeResche and Davis (1973)

been demonstrated repeatedly with captive deer (Tables 16-1,2) and may occur with other species. In general, DM intake by adult deer is greatest during the summer and early fall months and is lowest during the winter. This pattern of intake is voluntary and probably has evolutionary significance since it coincides with annual forage availability, i.e., high in the summer and low in the winter. The degree of voluntary variations in intake with other wild ruminants remains largely unknown. It is very likely that seasonal fluctuations exist with most species which evolved in areas where forage supplies were seasonal. Studies with bighorn sheep demonstrated that they voluntarily consumed a maximum during October and the least during February (Chappel and Hudsen, 1978). However, work with Rocky Mountain elk indicates that intake during the winter months can be high (Thorne and Dean, 1973). Unlike deer, captive pregnant cow elk which are properly fed can gain weight throughout the winter months. It is very likely that sexually active males of all wild ruminant species voluntarily consume less forage during breeding seasons. This has been shown with white-tailed deer (Short, 1969).

Actual DM intake rates are shown for several species in Table 16-1. Most of these data were gathered under captive conditions with white-tailed deer. DM intake rates are influenced by many factors, and variation is expected between studies and different environments. Young, growing animals will consume a greater percentage of DM per unit of body weight than adults, and smaller species such as deer and antelope will consume more per unit of body weight than larger animals such as elk or moose. As DM digestibility increases so will DM intake, especially with moderate quality forage.

Forage availability is always a factor affecting intake with free-ranging animals. Although intake figures for wild ruminants are far from being complete, it is likely that most of the larger species will consume 1.5-2.2% of their body weight daily during the winter and 2.5-4.0% during periods of plentiful vegetation. In general, smaller species, such as deer, antelope, bighorn sheep, and mountain goats, will consume 2.0-3.5% of their weight in the winter and 4.0-6.5% during the summer and fall.

More specific measures of energy intake and requirements have been made for several wildlife species under captive conditions. However, many of these have limited value at this time because the utilization of energy of most game forage plants by game animals has not been determined. DE, ME and NE

Table 16-2. Energy requirements and levels of intake for wild ruminants.

Species	Level	Comments	Source
White-tailed deer	52.2 Kcal/kg	FMR, summer, adults	Silver et al (1969)
	143.6 Kcal/kg$^{0.75}$	FMR, summer, adults	
	33.8 Kcal/kg	FMR, winter, adults	
	97.1 Kcal/kg$^{0.75}$	FMR, winter, adults	
	53.2 Kcal/kg	FRM, summer, fawns	
	130 Kcal/kg$^{0.75}$	FMR, summer, fawns	
	38.2 Kcal/kg	FMR, winter, fawns	
	90.2 Kcal/kg$^{0.75}$	FMR, winter, fawns	
White-tailed deer	173 Kcal/kg$^{0.75}$	ME required for fawns during periods of growth	Thompson et al (1973)
White-tailed deer	298.35-309.48 Kcal	DE, ad lib intake of milk by 3 kg fawns	Robbins and Moen (1975)
White-tailed deer	160 Kcal/kg$^{0.75}$	DE requirement for maintenance of does	Ullrey et al (1969)
White-tailed deer	131 Kcal/kg$^{0.75}$	ME requirement for maintenance of does	Ullrey et al (1970)
White-tailed deer	97 Kcal/kg$^{0.75}$	NE requirement for fawns, December through April	Holter et al (1977)
	153 Kcal/kg$^{0.75}$	ME requirement for fawns, December through April	
	125 Kcal/kg$^{0.75}$	NE requirement for maintenance, yearlings May through Oct.	
	162 Kcal/kg$^{0.75}$	ME requirement for maintenance, yearlings May through Oct.	
White-tailed deer	5000-6000 Kcal/day	Bucks, April through October	Short (1969)
	3500-4000 Kcal/day	Bucks, winter	
Mule deer	158 Kcal/kg$^{0.75}$	ME, maintenance requirement	Baker et al (1979)
Black-tailed deer	193.8 Kcal/kg$^{0.8}$	DE, ad lib intake, female fawns, summer-fall	Nordan et al (1970)
	118.7 Kcal/kg$^{0.97}$	DE, ad lib intake, male fawns, summer-fall	
	149 Kcal/kg$^{0.75}$	DE, birth to 22.5 kg body weight, resting heat prod.	
	80 Kcal/kg$^{0.94}$	DE, birth to 22.5 kg body weight, resting heat prod.	
Black-tailed deer	2475 Kcal/day	DE, weighing 50 lb and gaining 1% of BW daily	Cowan et al (1957)
Bighorn sheep	1760 Kcal/day	DE, rams weighing 62-75 lbs	
Antelope	92 Kcal/kg$^{0.75}$	Fawns, fasting heat production, fall	Wesley et al (1970)
Reindeer	125 Kcal/kg$^{0.75}$	Resting metabolism	White and Yousef (1978)
Roe deer	117.0 Kcal/kg$^{0.75}$	FHP, spring molt	Weiner (1978)
	99.4 Kcal/kg$^{0.75}$	FHP, summer	
	91.3 Kcal/kg$^{0.75}$	FHP, winter	

requirements of wild ruminants can be determined with captive animals. Before these determinations have much practical importance the ability of each forage plant to furnish energy must be learned in relation to each animal species. Energy values for many important forage plants have been determined for domestic ruminants but not for wild ruminants.

There are great differences in stomach capacity between species which greatly affect the rate of passage and digestibility by the animal. For example, the forestomach comprises 6-8% of the body weight of deer, 8-10% of elk and 12-14% of cattle (see Vol. 1). Differences of these magnitudes coupled with large variations in nutrient levels and availability in the various forage plants make it difficult to project values from one species or situation to another. Thus, caution should be exercised if these values are used for free-ranging ruminants.

Table 16-2 is a compilation of some of the more detailed work on energy intake and requirements of various wildlife species. These data show the variation that can exist between species, study conditions, and animals. There are a few studies that have been conducted which indicate how various factors influence energy metabolism and partitioning. Holter et al (1975) studied the effects of temperature on energy requirements of white-tailed deer. Mean values for energy (ME) expenditures were 146 and 152 Kcal/$kg^{0.75}$ with and without wind chill, respectively. Temperature had the greatest effect on deer during the summer. The level of energy in the diet may affect the age that puberty was attained as, in one study white-tailed deer fawns matured faster sexually if fed 3.06 Mcal of DE/kg than if fed 2.50 (Abler et al, 1976). Wesley et al (1973) studied the effects of age and temperature on heat production by antelope. Total heat production was 152 Kcal/$kg^{0.75}$ for 2 mo. old fawns and 122 Kcal/$kg^{0.75}$ for 18 mo. old fawns. Total heat production for mature antelope was 114 Kcal/$kg^{0.75}$ at -1°C and 163 at -23°C. White and Yousef (1978) stated that the energy expenditure of reindeer increased 22 and 74% when the animals climbed slopes of 5 and 9%, respectively, in comparison to horizontal walking. Work with bighorn sheep showed that the act of eating increased the metabolic rate 31 and 33.4% for rams and ewes, respectively, which appeared to be related to the time spent eating (Chappel and Hudson, 1978). The energy expenditure was 0.43 and 0.46 Kcal/kg/h for ewes and rams, respectively, while eating alfalfa-brome hay. Other work showed that 2-year old bighorn sheep used 18.5% more energy while standing than while lying; wind during cold weather also greatly increased their metabolic rate. For example, at -2°C increasing wind speed from 0 to 7 m/s increased the metabolic rate from 108 Kcal/$kg^{0.75}$ to 165 for ewes and from 125 to 160 Kcal/$kg^{0.75}$ with rams (Chappel and Hudson, 1979).

Protein Requirements

The protein requirements for white-tailed deer have been studied in several experiments, but protein needs for other wild ruminant species have not received much attention. Murphy and Coates (1966) fed white-tailed does diets ad libitum where the protein levels ranged from 7.4, 11.4, and 13.0%. Deer on both of the lower levels lost body weight during the study while the deer on the 13% level showed no weight change. The number of fawns born to deer on the lower levels was less than for those on the 13% level (1.85 fawns/doe). The group receiving 7% lost half of their fawns following birth, and the 11% group lost nearly a third of their fawns and the does of the 13% group successfully raised all of their fawns.

Ullrey et al (1967) found a sex difference in the protein requirements for maximum growth with white-tailed deer fawns. Their work showed that 12.7% protein produced maximum growth with female fawns while 20% was needed for maximum growth with male fawns. They also showed that as the protein content of the diet increased, so did the DM intake. In other work reports suggest that white-tailed does reached puberty as quickly with 9.6% protein as those receiving 18% (Abler et al, 1976) and that 19 g of digestible crude protein per $kg^{0.75}$ was required for white-tailed fawns for optimum growth (Smith et al, 1975). Smith et al also indicated that maximum body protein deposition occurred when the deer received 3 g of digestible N/$kg^{0.75}$. When fed 2 g, growth was not reduced greatly. Holter et al (1977) stated that yearling deer being fed a concentrate feed ad libitum could meet their needs with 11% crude protein. Low levels of crude protein (4-5%) resulted in stunted, runty deer with no antlers while greatest skeletal growth and antler development occurred with 17% crude protein (McEwen et al, 1957). Thompson et al (1973) indicated that 16.8% crude protein (DM basis) was in excess of actual needs for growth and maintenance.

Information regarding the protein requirements for other wild species is somewhat sparse. Thorne (1969) indicated that 14.5% crude protein was at least adequate for adult Rocky Mountain elk. McEwan and Whitehead (1970) indicated that caribou and reindeer needed 2.9 g/$kg^{0.75}$ digestible protein to maintain an even N balance.

Water Requirements

The dietary requirements for water vary tremendously between animal species (see Ch. 2, Vol. 2). For example, some rodents stay in positive water balance while consuming foods that are air dry, while other species must consume large amounts of free water in addition to the water supplied in their food. The water needs of most ruminant species are correlated to the amount of DM consumed, i.e., water consumption increases as DM consumption increases.

The amount of free water consumed by ruminants depends on the amount of moisture in the forage and the physiological mechanisms for conserving water within the body. Consumption of free water is not required by wild ruminants when snow is available and probably not when lush green forage is being consumed or when forage is covered with dew. The following reports indicate some levels of water intake by wild species.

Knox et al (1969) found that mule deer consumed 140 g/kg$^{0.75}$ of water for every 65 g/kg$^{0.75}$ of DM. Elder (1954) observed that mule deer in Arizona drank from 3.8-10.4 liters of water daily in July and August. Longhurst et al (1970) stated that black-tailed deer had a relatively larger water body pool than did domestic sheep and that the turnover rates were slower for deer than sheep. The water flux in the deer was 0.053 and 0.104 ℓ/kg/day in the winter and summer, respectively. Wesley et al (1970) reported that antelope fawns 108-182 days of age consumed free water at levels of 84-107 ml/kg/day and 52-64 ml/kg/day for females and males, respectively. Beale and Smith (1970) reported that antelope in Utah had water intakes that varied with the succulence of the forage. During extremely dry periods of the year when moisture comprised about 39% of the diet, the animals would consume as much as 3 ℓ of free water daily. When forage was succulent and contained at least 75% moisture, antelope did not consume free water even though it was available. Macfarlane et al (1966) said 60-70% water in forage was adequate for domestic sheep.

Cameron and Luick (1972) found that season of the year affected the water flux of reindeer in Alaska. These authors concluded that the factors which affected the variability were climate, forage quantity and quality, pregnancy and lactation. With these animals the water flux varied from 188 ml/day/kg in June to 29 in December, although the flux the following December was 57 ml/day/kg. Hoppe (1977) reported on the water intake of African species that were adapted to different environments. The dikdik from semi-arid areas consumed 278 ml/day while consuming 3.8% of its body weight in DM, and the suni from dense mountain coastal forests consumed 401 ml of water while eating 3.5% of its body weight in DM. Water consumption for the dikdik increased 52, 170, and 162% during gestation and lactation (the 4th week prepartum and 1st and 2nd week of lactation, respectively). Hoppe (1977) estimates that suni can live without free water if they consume a diet of 72-82% water. The dikdik can survive on a diet containing 65% water.

Mineral Requirements

The best available information on wild ruminant mineral requirements is for Ca and P requirements of white-tailed deer. McEwen et al (1957) found that the greatest skeletal growth occurred when a diet containing 0.65% Ca and 0.56% P was fed to fawns. Levels as low as 0.3% Ca caused no apparent stress during the winter months. When the deer received levels as low as 0.09% Ca and 0.25% P, they failed to survive. Magruder et al (1957) found that 0.25-0.30% P would allow for optimum growth in white-tailed deer. Work in Michigan (Ullrey et al, 1973) suggested that white-tailed deer needed 0.4% Ca for optimum gains, bone strength, and antler composition when fed with 0.25-0.27% P. In other work Ullrey et al (1975) reported that the P requirement was no greater than 0.26% (could be higher for males during antler growth) when fed with 0.46-0.51% Ca.

For the most part required levels for the other minerals remain unknown. Some researchers have begun looking for possible deficiencies in wild ruminants and some have possibly been identified even though required amounts have not been determined.

Weeks and Kirkpatrick (1976) postulated that white-tailed deer in Indiana were not consuming adequate levels of Na. They felt that the desire to consume Na in the spring of the year was the result of Na deficiency brought on by the high consumption of K (high in green forage) and water. Franzmann et al (1975) presented evidence indicating that Mg, Cu, and Mn could be inadequate in the diets of some moose in Alsaka. This was based on the analysis of hair samples taken from winter-killed moose and compared to normal values from healthy moose and domestic ruminants. Se deficiency had been observed in the Rocky Mountain goat in British Columbia (Hebert and Cowan, 1971) and Hyvarinen et al (1977) felt that reindeer in Finland may be suffering from Ca, Mg, and Cu deficiencies as indicated by low serum levels. While these papers do not shed light on the mineral requirements of wild animals, they do indicate that mineral needs may not be met in many situations. This is to be expected since grazing domestic ruminants often develop mineral deficiencies.

Until the mineral requirements of wild ruminants have been determined, the best estimates of their needs can be obtained from domestic species. However, mineral requirements can be highly specific for certain species (i.e. Cu), so caution must be used. When working with the mineral needs of wild ruminants, it must be remembered that there is a range of intake that allows normal physiological functions and that below this deficiency symptoms result in some animals, and above this toxicity symptoms appear in some individual animals. Also, several minerals are known to interact with other minerals or vitamins and cause deficiencies in spite of absolute levels. So when mineral deficiencies result, not only must the level of the mineral be studied but also the levels of the other nutrients (see Ch. 4,5 of Vol. 2).

Summary of Nutrient Requirements

Energy. Most wild ruminants should have near unlimited access to high quality feed during late gestation and early lactation (Fig. 16-3). This is necessary to assure good reproduction, especially if the preceding winter was nutritionally stressing to the animals. If the animals are faced with a nutritional stress period, such as winter, and are to survive, then energy intake must be high sometime prior to entering the period. Specific information on maintenance requirements for deer based on a daily basis indicates that about 160 Kcal/$kg^{0.75}$ are needed by adult does and 160-175 Kcal/$kg^{0.75}$ of ME are needed by immature deer.

Figure 16-3. Elk on summer range in nursery area. Good nutrition is especially important in late gestation and early lactation for reproducing females.

Protein. Smaller ruminants, such as deer and antelope, probably require in excess of 15% crude protein for maximum growth and 12-15% for reproduction. Large ruminants, such as elk, probably require 10-15% for adequate reproduction.

Water. Water requirements are highly variable depending on the species, moisture in forage, weather, and many other factors. Free water intake can vary from virtually nothing to needing water daily.

Minerals. Approximately 0.45-0.50% Ca and 0.25% P are required by growing deer. Other mineral needs remain largely undetermined.

NUTRITIVE CHARACTERISTICS OF FORAGE PLANTS

Characteristics of plants that affect their nutritive value have been examined by numerous researchers. Sufficient data are available to predict the general nutritive value (energy, protein, and vitamin A) of most plants with minimal examination. Training in plant identification and knowledge of diets make it possible to make fairly accurate assessments of range adequacy with respect to general nutrient availability. The following are plant characteristics relating to forage quality that can be used to make assessments of range adequacy. Specific or detailed data require further information on nutrient content or availability.

Energy

Most actively growing plants will have a relatively high digestibility and, if consumed in proper quantities, will provide sufficient energy. Figure 16-4 shows the general seasonal trends of available energy in forage plants. As the phenology of the plant advances, the amount of available (digestible) energy decreases. The important point in the figure is that young, rapidly growing plants are more digestible than slow growing or mature plants. This is true of browse, forbs,

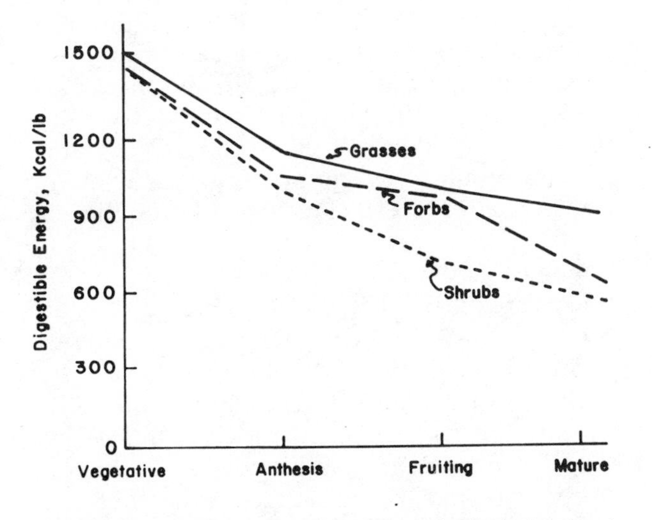

Figure 16-4. The effects of phenological development on digestible energy in plant tissues. Taken from Cook (1972).

Figure 16-5. Mountain sheep and elk are primarily grass eaters. During the winter months short growth forms of weathered grasses and sedges are all that are available. Forage of this type is low in most nutrients but does supply energy which together with stored body energy allows animals to survive long winter periods.

and grasses. The relative importance of each of these classes of forage plants could change depending on the species of animal using them. The DE levels found in grass species are higher than for forbs and shrubs throughout the year. Blair et al (1977) showed large differences in DM digestibility with white-tailed deer in Texas for grasses and forbs between summer and winter. The digestibility of grasses decreased about 30% from summer to winter and forbs dropped about 42%. Browse leaf and twig digestibility decreased about 12% as did residual twigs. Exceptions were pine needles which increased slightly from summer to winter and mushrooms which improved 10% in digestibility. Ward (1971) studied the in vitro DM digestibility of plant species eaten by elk during the winter months in Wyoming. The values were all <50%, which indicates the energy was not readily available at this time from these plants. Oldemeyer et al (1977) found the in vitro DM digestibility of browse grazed by moose in Alaska to be higher (37-50%) in the summer than in the winter (29-45%), although none of the values are very high and Snider and Asplund (1974) report similar seasonal differences for white-tailed deer forages in the South.

Protein

Actively growing plants should provide adequate protein for wild ruminants, but the amount of available protein decreases rapidly with advancing maturity of the plants (Hagen, 1953; Bissel and Strong, 1955; Lay, 1969; Short, 1969; Dietz, 1972). Figure 16-6 shows that all growth forms are high in protein when the plant tissues are young and growing rapidly, but shrubs appear to retain more digestible protein in their tissues than forbs and grasses with advancing maturity. The amount of protein in grasses decreases rather rapidly and is of questionable value as a source of protein after its growth has slowed

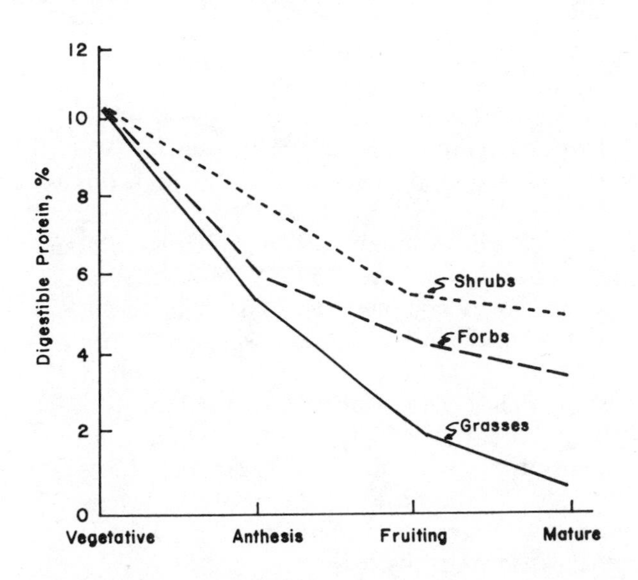

Figure 16-6. The effects of phenological development on digestible protein in plant tissues. Taken from Cook (1972).

Figure 16-7. Antelope in Colorado are shown selecting green forage in the fall. By the late summer and early fall many plants have stopped their active growth. Forage in this phenological state is marginal in providing protein and phosphorus.

(Fig. 16-7). For example, the crude protein levels in short grass prairie grasses decreased from a range of 12.8-19.6% in early June to 3.4-7.9% in late August (Jefferies and Rice, 1969). Urness et al (1975a) studied mule deer forage in Arizona and found the crude protein to drop from 25% in May to 10% in September. Grasses in Texas were shown to decrease about 6% from summer to winter with forbs dropping about 7% (Blair et al, 1977). These workers indicated that browse leaves and twig tips decreased about 5% and residual twigs dropped only 3%. Pine needles remained about the same while mushrooms and fruits increased about 3% from summer to winter.

Short (1971) examined forage plants consumed by white-tailed deer in Texas and found that the crude protein levels of forbs dropped from 14.8-7.9% in the winter. Mixed grasses dropped from 13.3-7.2% and mixed browse dropped from 16.2-6.1%. Pine needles did not show much seasonal change. Harshberger and McGinnes (1971) found that sourwood leaves decreased from 12.6 in the summer to 4.9% in the winter. Work in Wyoming indicated that several grass species contained very high crude protein levels in the bloom stage but fell off to <5.8% when mature (Hamilton and Gilbert, 1972). Mautz et al (1976) found the crude protein levels in browse in New Hampshire to be <10% during the winter months.

Calcium

Ca is generally adequate in most forage plants. Based on existing data, it would seem unlikely that a Ca deficiency would develop with wild ruminants in most areas, one exception being those depending largely on weathered grasses. The greatest likelihood of a Ca problem would result in a wide Ca:P ratio which could precipitate a deficiency. Chatterton et al (1971) showed that Ca levels in the leaves, new stems, and old stems remained above 0.60% in desert saltbush for the entire year. Dietz (1972) stated that Ca levels in shrubs in the western states tended to be high enough that Ca deficiencies would be unlikely. He indicated that spring values of various browse species varied from 0.61-1.28% for snowberry and serviceberry, respectively. Winter values ranged from 0.75 for snowberry to 1.55% for serviceberry with the other species having intermediate values. Browse eaten by black-tailed deer in Washington appears to be adequate in Ca with all winter values exceeding 0.46%, although western hemlock was 0.36% (Brown, 1961). Browse and forbs eaten by white-tailed deer in Texas had high levels of Ca year round; grasses were as low as 0.4% which may be marginal for optimum growth of deer (Short, 1971). Several browse species in Colorado have adequate levels of Ca (Dietz et al, 1962). Other studies on a variety of browse indicate adequate levels of Ca (Harshberger and McGinnes, 1971; Oldemeyere et al, 1977; Urness et al, 1975).

The results of most of these studies show that levels stay relatively uniform through the year or increase during the winter months. Contrary to the studies mentioned, Demarchi (1968) found that the grass species important to bighorn sheep in British Columbia during the winter months were very low in Ca. During March, only bluebunch wheatgrass had Ca levels >0.2%. Four other grass species were <0.1%. Forbs were adequate in Ca although Ca:P ratios were 10:1 and 31:1 for yarrow and *wyeth eriogonum.*

Phosphorus

P levels tend to be associated with the phenological development of the plant (Fig. 16-8). Generally, those plants that are actively growing should be adequate in P, but as the growth rate slows and the plant matures, P levels decrease. Since P levels decrease with plant development and Ca levels remain stable or increase, P deficiencies

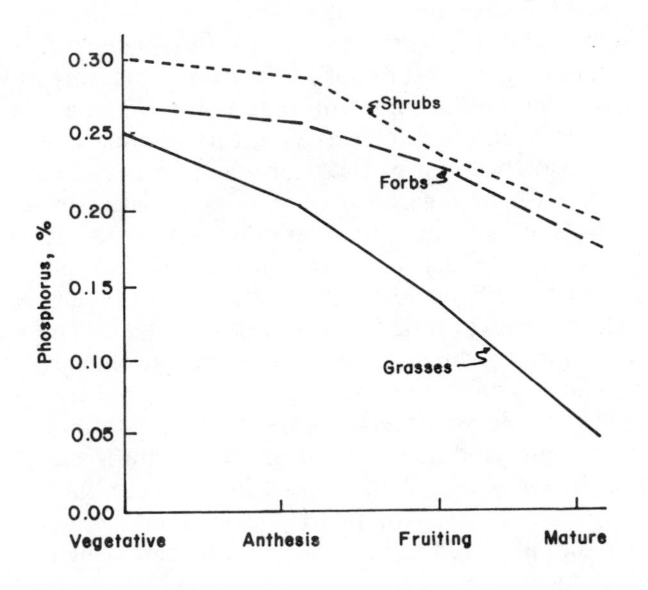

Figure 16-8. The effects of phenological development on phosphorus levels in plant tissues. Taken from Cook (1972).

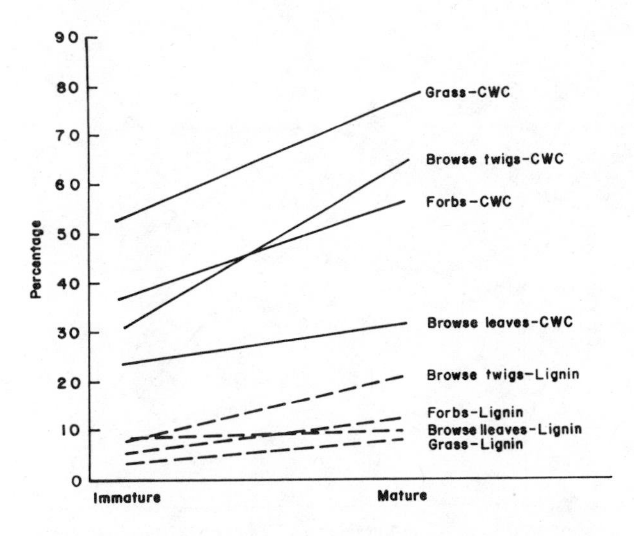

Figure 16-9. This shows the effects of plant development on fibrous components (cell wall contents–CWC and lignin) in plant tissues. Based on figures taken from Short et al (1974).

could possibly result from an improper Ca:P ratio. Dietz (1972) stated that P may be deficient in the winter months with browse species and Cook (1972) indicated that P was adequate for domestic animals up to flowering for grasses and until near maturity with both forbs and browse. Figure 16-8 shows that grass levels of P tend to be generally lower than that of browse and forbs. The level of P in grasses also drops more quickly with the development of the plant. Blair et al (1977) reported that P levels dropped in grasses, forbs, browse, pine needles, mushrooms, and fruits. The greatest variation was with the forbs which dropped from 0.25% to 0.07%. Browse had the best winter and summer values, being 0.15 and 0.28%, respectively. Other reports showing a season decline in P levels include those of Urness et al (1975) and Lay (1969).

Fiber

The fiber content in the plant increases with advancing plant maturity which causes a reduction in digestibility. A plant cell is comprised of the cell contents which are highly digestible and the less digestible cell wall, made of hemicellulose, cellulose, lignin, and minerals. The digestibility of plant material depends largely on the amount of lignin present and, to some extent, on the amount of other cell wall constituents, which increase with maturity. Figure 16-9 illustrates the general seasonal trends of the fiber components in plant material. Generally, both cell wall constituents and lignin increase throughout the year. Blair et al (1977) showed that the cell wall constituents increased in grasses from 70.0 in the summer to 84.6% during the winter. Forbs increased from 43.8 to 74.8% from summer to spring. The increase in browse was from 35.9 to 39.0% for leaves and terminal twigs and from 65.2 to 70.8% for residual twigs. Short (1971), studying white-tailed deer forage in Texas, found that the ADF levels in mixed browse increased from 27 in the summer to 46% in the winter. Levels in forbs increased from 29 to 37%, mixed grasses from 38 to 44%, and pine needles from 36 to 37%. Other reports showing seasonal changes in fiber include one on Alaskan moose browse (Oldemeyer et al, 1977) and for Wyoming forage (Hamilton and Gilbert, 1972).

Carotene

The carotene content of a plant is closely associated with the growth of the plant. Young, actively growing plants with good green color should be adequate in carotene. Figure 16-10 shows the relative levels of carotene in the various classes of forage. Cook (1972) indicates that browse should be adequate in carotene throughout the year while

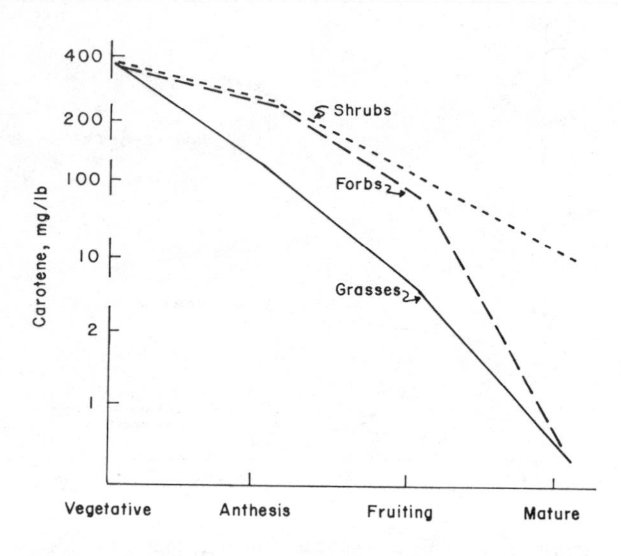

Figure 16-10. The effects of phenological development on the amount of carotene (mg/lb) in plant tissues. Taken from Cook (1972).

both grasses and forbs become deficient as they approach maturity.

Summary of Nutritive Characteristics of Forage Plants

Green, actively growing plants are usually high in DE, digestible protein, P and carotene. As plants mature the amount of fiber increases and the levels of available energy, protein, and P decrease. Carotene is usually adequate in green plant tissues. Most forage plants have adequate Ca levels throughout the year, weathered grasses being the exception. Grasses are usually high in K.

In general, the color of the plant indicates good quality forage. Grasses, forbs, and browse that are green in color are normally high in DM digestibility, crude protein, P, and carotene. This is because plants that are green in color are actively growing and the structural fiber components are at a minimum level.

ADEQUACY OF FORAGE PLANTS FOR WILD RUMINANTS

The determination of the adequacy of various forage plants via feeding trials, digestion studies, etc., is largely limited to deer with some work on elk. Chemical analysis of forage plants also provides good nutritional information on adequacy, although it can only be used to provide general guidelines until it has been tested with animals. Since the winter months are viewed as being the critical period for most wild ruminants, most of the studies determining plant adequacy have been conducted during this period. This is also the time of the year when deer tend to lose weight regardless of how nutritious their diets might be. In examining study results involving deer weight changes, note that weight will be lost during the winter regardless of the type or amount of feed. The important thing is the relative weight changes when several plant species are compared in a study. It is sometimes difficult to compare studies conducted at different locations and time of year. Weather, season of year, individual variation in plant species, and animal handling techniques can all affect weight changes.

There has been considerable research done on the adequacy of forage plants (Table 16-3). Robbins and Moen (1975) examined white-tailed deer browse in New York and reported paper birch, nannyberry, staghorn sumac, and witch-hazel all had relatively high in vitro DM digestibility (>59%), cell solubles (>47%) and low lignin (<12%). They examined 14 other browse species and found the DM digestibility of these ranged from 42-57%. The cell solubles in these plants ranged from 35-47% and the lignin values varied from 12.5-18.6%. Ullrey et al (1972) studied the adequacy of northern white cedar and large-toothed aspen in Michigan for white-tailed deer. Neither plant kept the deer in a positive energy or N balance, although the weight and N losses were considerably less for deer consuming cedar. By using values presented in the paper to calculate weight losses, it appears that the deer on the cedar lost about 17% of their body weight in 4 weeks and those on aspen lost about 25%. Ullrey et al (1975), in subsequent studies with cedar and aspen, found that if the deer received protein and energy supplements with browse, the intake of the browse increased and weight losses were greatly reduced. Mautz et al (1976) studied seven other northern browse species. The browse species were fed together in various combinations to white-tailed deer and none was able to keep the animals in positive energy or N balances.

Several researchers have examined dietary adequacy of forage plants for white-tailed deer in many parts of the South. Snider and

Table 16-3. Dry matter digestibilities of various forage species by wild cervids.

Species	Plants	DMD	Comments
White-tailed deer	Paper birch nannyberry staghorn sumac whitch-hazel	>59	New York, winter browse[a]
	14 other species	42-57	
White-tailed deer	Browse leaves and twig tips	56	Texas, winter[b]
	Forbs	27	
	Grasses	20	
Mule deer	Alfalfa	40-52	Wyoming-Colorado[c]
	Big sagebrush	42-50	
	Mountain mahogany	43	
	Bitterbrush	41-61	
	Juniper	45-55	
	Rabbitbrush	52-59	
	Willow	57-66	
Elk	Grasses	39-51	Wyoming, winter[d]
	Bitterbrush	23-33	
	Big sagebrush	49-58	

[a]Robins and Moen (1975); [b]Blair et al (1977); [c]Dean et al (1975) and Dietz et al (1962); [d]Ward (1971)

Asplund (1974) studied forage plants in Missouri and found whiteheath aster, pokeweed, and spiderweed to have very high in vitro DM digestibilities (>65%) during the summer. In the winter, corn was found to be 70% digestible and coralberry was 53%. A total of 41 plant species was studied and the remaining species were <41% digestible. Snider and Asplund found that, as a group, forbs had the highest DM disappearance rate (54%). Fruits had values that averaged 50%, leaves of woody species were 46% and twigs dropped to 28%.

Segelquist et al (1972) studied forage samples obtained from the rumen of white-tailed deer in Arkansas during the winter. They found that green forbs and grass during the winter had crude protein levels >15%. Browse at this time was <8%. Sedges had crude protein values of 9.6% and browse ranged from 6.5 to 8.0%. Blair et al (1977) studied deer browse in Texas in considerable detail. Their figures show that browse leaves and twig tips were more digestible (in vitro) than grasses, forbs, residual twigs, and pine needles and twigs on the year-long basis. The browse leaves and twig tips had digestion rates ranggin from 55.6% in the winter to 67.2% in the spring. The range of digestibilities of forbs and grasses were 68.2 to 26.7% and 49.4 to 19.6%, from spring to winter, respectively. The residual twigs of browse stayed fairly uniform from summer to winter, varying from 26.6 to 28.4%, while the spring values were 40.0. Although not available in great quantities, mushrooms and fruits were the most digestible of any class of forage. These ranged from a low of 56.6% in the spring to 76.3% in the fall.

Short (1976) studied the nutritive value of acorns from different oak species in Texas. He found that, in general, acorns were low in protein (4.4 to 7.0%) and P (0.07 to 0.14%). The crude fat levels of white oaks were <6.7% while the black oaks had fat values >14%. The amount of cell wall constituents was 35% for black oaks and 45% for white oaks. Burns and Viers (1973) studied various species of mast and fruits in Louisiana. Carya species (pecan and hickory) had the highest gross energy values, averaging 7.319 Kcal/g. Red oaks averaged 5.008 and white oaks 4.140 Kcal/g. Various fruits (Yaupon, huckleberry, mulberry, honeysuckle, sumac, crabapple, pyracanthia, and dogwood) varied from 5.361 Kcal/g for dogwood to 4.246 Kcal/g for crabapple. The high GE levels of the Carya species and some oaks are probably due to high levels of fats or oils. However, GE values, by themselves, are not a good indicator of available energy for animals.

Rodgers and Box (1967) studied the protein value of several grass species in Texas. They found that the crude protein content corresponded directly with the periods of high rainfall. Seldom did any of the grass species exceed 10% in crude protein.

Studies with forage plants in the western states and Canada indicate the same general patterns as found in the east and south, i.e., plants in the spring, summer, and fall provide the animals with good nutrition, but during the winter the plants are unable to meet the needs of the animals for maintenance nutrition. In Utah curl leaf mahogany, bitterbrush, and Utah juniper were unable to keep weight on mule deer during winter studies; deer consuming bitterbrush were nearly able to maintain their weight (Smith, 1952). Smith (1950) fed mule deer sagebrush for an average of 26 days in March and April. They lost an average of about 7 lb during this period, which was about 8% of their body weight. Other work by Smith (1957) on the nutritive value of mule deer winter browse indicated that chokecherry had the highest crude protein levels (9.9%) and Gambel oak had the lowest (5.4%). Birchleaf mahogany, cliffrose, and juniper were intermediate between these two. Urness et al (1975b) made chemical and digestion analysis of many Arizona mule deer forages. Forbs had an average DM digestibility of nearly 70% in May. In September the digestibility for the forbs tested had dropped to 53.6%. Grass species remained rather constant in digestibility throughout the period of May through September. Most of the digestibilities were between 49 and 59%. Browse digestibilities were highest in May (55 to 59%) and decreased as the year progressed (41 to 50% in August). Pine stayed about the same during the entire period, fluctuating between 39 and 43%.

There has been a limited amount of work done with other wild ruminant species concerning the nutritive value of forages. Ward (1971) studied elk browse in Wyoming. In vitro DM digestibilities indicated that grasses during the winter months were more digestible than were browse species. Cheatgrass had a digestibility of 50.7%, bluebunch wheatgrass varied from 38.7 to 48.3%, Indian ricegrass fluctuated from 41.3 to 44.0%, and needle and thread grass ranged from 41.3 to 45.3%. With the exception of big sage, which had digestibilities ranging from 48.7 to 58.0%, browse species were usually <40%. Antelope bitterbrush was the lowest, varying from 23.3 to 33.3%.

Thorne and Butler (1976) presented data on the adequacy of alfalfa for captive Rocky Mountain elk. In general, about 10 lb of baled alfalfa were required to keep weight on pregnant cow elk during winter months. When the alfalfa was pelleted, then 8 to 9 lb would keep the elk from losing weight. Calf elk gained nearly 12% of their initial weight when consuming 5.5 lb of baled hay.

In summary, these studies indicate that summer forage is, in general, adequate for rapid growth and nutrient storage for deer. Winter forage is unable to keep weight on deer and nutrient levels are usually below those calculated to be optimum. These values on nutritive content of plants illustrate the importance of nutrient storage during periods of abundant food supply to allow animals to survive the winter period. Commonly, the length and severity of the winter period is as important as the quality of winter forage. Wild ruminants seem to be able to recycle nutrients within their bodies allowing them to make some use of poor quality food. This, coupled with ample body reserves allow them to survive long periods of inadequate nutrition.

Mineral Licks

The importance of mineral licks in providing nutrients to animals is questionable. Most studies on licks have been concerned with use patterns by animals and the levels of the various elements found in the licks. Whether or not the animals are able to obtain nutrients that are vital from the licks has not been the subject of study.

Observations on mule deer and Rocky Mountain elk indicate that the use on natural licks is seasonal (Dalke et al, 1965; Carbyn, 1975). These studies found that most of the use was just prior to and after parturition. The numbers of elk using licks may vary greatly between years (Carbyn, 1975) and elk will not paw through the snow to reach a lick (Dalke et al, 1965).

Elk in Wyoming readily come to blocks of salt placed on mountain ranges for domestic livestock, and feedground elk consume trace mineralized salt when placed before them during the winter months. Intake is greatest just after the salt is put out. After a couple of weeks the salt intake is greatly reduced. This indicates the elk may be salt hungry initially, but can quickly satisfy their desire for salt. Na dietary deficiency symptoms have not been observed with feedground elk when salt has not been fed.

The minerals and their concentrations vary greatly between mineral licks. Henshaw and Ayeni (1971) studied licks in Africa in considerable detail. They examined mineral concentrations at the licks proper, from areas immediately next to licks, and from areas more distant from the licks. Their data indicate that "salt lick" may not always be the most descriptive term. They examined four licks and found that the Na level in the lick proper was less than in the immediate surrounding area in 1 of 4 licks examined and about equal in the others. It was only slightly greater in two licks than that in the remote areas. In the other two licks, the concentration of Na was decidedly greater than in the more distant areas. Mg was the only mineral studied that was consistently in greater concentration in the licks than in nearby areas. N, P, and K were present in all licks and were possibly the result of excreta of the animals. On the average, Ca and K were present in the licks in greater concentrations than in areas next to the licks. In general, all minerals were in greater concentration in or near the licks than in the more distant areas. Cowan and Brink (1949) also felt that Na may not be the element that the animals are seeking. However, Stockstad et al (1953) and Dalke et al (1965) felt that salt was the element attracting the animals. Dalke et al (1965) reported Mg, K, Cl, SO_4, Na, F, and P to be present in licks in Idaho. Calef and Lortie (1975) found that Na, Cl, SO_4, and NO_3 were present in large amounts; Mg, P, and Ca were present in lesser amounts. Herbert and Cowan (1971b) examined licks used by mountain goats in British Columbia. They found the licks to contain 1,050-85,000 ppm Ca, 0-25 ppm P, and 115-5,500 ppm Na.

In summary, the contribution that mineral licks make to the nutritional well-being of wild ruminants is unknown. Animals are certainly attracted to natural licks and at a time when their nutritional needs are very high. Whether or not the minerals consumed at these licks are essential to the animals is speculation at this time.

NUTRITIONAL MALADIES IN WILD RUMINANTS

Very little is known about nutritional maladies in wild ruminants. Unless a problem is very widespread and affecting many animals, it commonly goes unnoticed. Consequently, information on nutritional maladies of wild species is rather limited.

Rumenitis

Rumenitis has been found in white-tailed deer in Saskatchewan (Wobeser and Runge, 1975). The cause in this case was felt to be the consumption of grain (wheat and barley). The deer had been subjected to an abrupt ration change from roughage to grain. The high grain intake resulted in rumenitis and rumen overload in several deer. Woolf and Kradel (1977) also reported rumenitis in deer maintained on high carbohydrate diets. When the deer were taken off the diets, the recovery appeared to be complete and it was believed that rumenitis was not a factor in herd mortality. Church and Hines (1978) observed

Figure 16-11. An old area of rumenitis from the caudodorsal blind sac of Roosevelt elk. Areas have been denuded of papillae which are beginning to grow back. Healed ulcers are present in the left center and top. From Church and Hines (1978).

Table 16-4. Pathological findings in the rumen and liver of elk fed pelleted and baled alfalfa for 120 days.[a]

Ration	Elk	Parakeratosis*		Rumen stain*	Hepatic granulomas
		Gross	Microscopic		
Group D	W-58	0	0	3+	1+
10 lb baled alfalfa	W-55	0	0	0	0
	W-52	0	0	3+	0
Group G	B-61[1]	0	ND	0	0
8 lb baled alfalfa	W-19	0	ND	0	0
	R-144	0	ND	0	0
Group F	R-10	0	1+	0	0
6 lb pelleted alfalfa	R-140	3+	2+	0	0
+ 2 lb baled native	P-229	3+	0	3+	1+
Group E	B-62[2]	0	0	0	3+
8 lb pelleted alfalfa	B-57[2]	1+	1+	0	3+
	B-60	1+	2+	0	0
	B-58[3]	2+	3+	0	3+
	B-54	1+	0	0	0
	B-52	0	ND	0	1+
Group H	R-143	3+	2+	0	3+
8 lb pelleted alfalfa	W-21	1+	ND	0	0
	B-75	0	2+	0	1+
Group I	B-36	1+	2+	0	0
8 lb pelleted alfalfa					

*Comparatively, 2+ was mildest and 3+ the most extensive. None were severe.

[1]Extensive adhesions between reticulum-diaphragm and rumen-liver.

[2]Fundic portion of the abomassum inflammed.

[3]Old splenic abscesses present.

[a]From Thorne and Butler (1976)

rumenitis in the rumens of all elk samples they examined (Fig. 16-11). Some of the cases were old while others were new. These animals had not been subjected to high concentrate feeds. Alfalfa and browse were found in the rumen. Thorne and Butler (1976) observed rumen parakaratosis in the rumens of elk that had been fed pelleted alfalfa for 90-120 days (Table 16-4). The condition was not severe and appeared to be just beginning. If the consumption of the pellets had continued, it could have become detrimental to the animals. The consumption of pelleted feeds over long periods of time has long been known to cause parakaratosis in domestic ruminants.

Essential Oils

Much speculation exists as to the possible effects of essential oils on rumen function. These oils are found in high concentrations in may species such as sagebrush, Douglas fir, and juniper. Some in vitro studies show that these oils inhibit or reduce rumen microbial activity. Nagy et al (1964) showed that the essential oils in sagebrush reduced cellulose digestion on bacterial cultures; VFA production was also reduced. Oh et al (1968) studied the effects of essential oils on VFA and gas production from 8 California plant species. They grouped the essential oils into 4 categories. One of the groups (oxygenated monoterpenes) inhibited VFA and gas production. Other essential oil fractions actually promoted VFA and gas production. Smith et al (1963) found that essential oils did not interfere with the in vitro digestion of alfalfa, except at very high concentrations. They did find inhibitions in vitro when essential oils were used where pure cellulose was the substrate.

In vitro studies do not necessarily give the true picture of the actual result of essential oils in vivo. Most in vitro systems are closed and the essential oils cannot escape or be removed, thus having a direct and constant effect. However, in vivo these could be absorbed from the GI tract or inactivated. Deer and antelope commonly consume large amounts of plants containing essential oils without apparent ill effects. It appears that a properly functioning rumen can adequately handle large amounts of essential oils from some sage species under natural conditions or that the animals avoid individual plants that have especially high levels. At any rate, essential oils have not been shown to cause problems to free-ranging animals consuming plants characteristically high in these oils.

Mineral Toxicities and Deficiencies

Perhaps one of the greatest potential nutritional dangers to wild ruminants is mineral imbalances. Wild ruminants probably require no less than 14 different minerals in their diets. When present in excessive or deficient amounts, these can result in nutritional problems ranging from death to slight deficiency symptoms that may or may not have much effect on important physiological processes. Nutritional mineral imbalances are common in domestic ruminants and are probably present to some degree with wild ruminants, even though they go relatively unnoticed. One of the more common mineral problems involves Se. Se is an element that tends to accumulate in certain plants in certain areas. Deer, antelope, and moose have been observed exhibiting typical Se toxicity symptoms in Wyoming (see Vol. 2, Ch. 5 for symptoms). Hebert and Cowan (1971) reported that Se deficiencies had been observed in Rocky Mountain goats in British Columbia and Brady et al (1978) found that white-tailed deer were susceptible to Se depletion when fed diets of 0.04 ppm Se. However, this level did not result in preweaning mortality of offspring. It was felt that vitamin E was important for the survival of young since diets without vitamin E resulted in increased mortality. Even if Se was present in the diet, white muscle disease would result and mortality would increase if vitamin E was deficient. In summary, they concluded that the requirement for Se by white-tailed deer was low and that the main expression of vitamin E deficiency was increased juvenile mortality.

Molybdenosis is a potential problem for wild ruminants in coal mining spoils with surface mining in the Northern Great Plains. Erdman et al (1978) found that Cu to Mo ratios in most vegetation samples ranged from 0.44:1 to 5:1. They felt that ratios of 5:1 or less could cause molybdenosis. Toxicity levels have not been defined for wild ruminants but these workers felt it could possibly be a problem. Nagy et al (1975) fed mule deer very high levels of Mo (2,500 mg/day or more) which resulted in reduced food intake and diarrhea in some cases. When the Mo was removed from the diet, intake returned to normal. They stated that mule deer appeared to be tolerant of Mo, perhaps as much as 10X that of domestic ruminants.

Fluoride toxicity was reported in black-tailed deer near an industrial fluoride source in northwestern Washington (Newman and Yu, 1976). Marked dental disfigurement and abnormal tooth wear patterns were observed.

Ammerman et al (1977) reported on several contaminating elements and their potential toxicity for domestic animals. These same elements are perhaps of some concern to wild ruminants in certain cases. Accidental introductions of these elements may occur if animals are feeding on municipal garbage or industrial contamination. Lead contamination may occur around smelters or from automobile exhause on herbage near highways. Other minerals that are discussed include arsenic, cadmium, aluminum, tin, and mercury.

Starvation-Exposure

Several wild ruminant species have evolved in the colder climates of the northern hemisphere under the influence of severe winter conditions. As mentioned earlier, voluntary food intake commonly decreases during the winter months, which coincides with forage availability patterns. Ozoga and Gysel (1972) found that white-tailed deer seek dense cover during cold weather and that most of their activity was during mid to late afternoon. Both of these adaptive features allow the deer to conserve needed body reserves. It is also likely that the basic metabolic rate of wild ruminants decreases during the winter months. Seal et al (1972) suggested that deer enter a state of hypothyroidism and decreased metabolic rate which allows further conservation of energy reserves. Hair coats in the winter months are obviously of better insulating quality than summer coasts, and this allows for more efficient conservation of energy. Animals can store a trenmendous amount of energy as fat and protein in body tissues if summer and fall forage are adequate. In addition to these adaptive traits, ruminants seem to have a digestive system capable of retaining nutrients within the body. For example, saliva glands, the rumen, and kidneys are able to recycle certain nutrients within the body, thus allowing their reuse. This process would be very important in keeping the rumen environment favorable for the microorganisms which are essential in providing nutriment from relatively poor quality forage. In spite of these adaptations, winter mortality can be high in some herds at various times.

Winter starvation is common in wild ruminant populations in the northern USA and Canada. In addition to the sick and weak animals that commonly die during the winter, otherwise healthy individuals also die if the winter period is unusually long or these animals entered the winter with low body reserves.

Although starvation is the term commonly used, death from exposure which is precipitated by inadequate nutrition is probably more descriptive. In general, inadequate food supplies and/or body reserves can make animals very susceptible to additional stress such as cold temperature and/or wind. Death by exposure can vary from situations where strong, healthy individuals simply freeze to death (rare but it does happen) to situations where a weak animal dies during the night as temperatures cool. There seems to be certain situations where winter mortality via starvation-exposure commonly occurs. One is when the winter period is simply too long and the animal cannot obtain enough nutriment from body reserves or forage to survive. The second type is when temperatures drop suddenly and kill animals that would otherwise have survived the winter. This happens most commonly in the late winter or early spring when many animals are in a weak and susceptible condition. Another type can occur in some areas where the animals are suddenly attracted to new spring growth. Animals that are particularly weak following the winter are very susceptible at this time. They are attracted to new growth which contains mostly water and, consequently, they actually consume very little DM. This is also a time when the rumen microorganisms must adjust from a diet high in fiber to one high in digestible carbohydrates. Body reserves must be great enough to keep the animals alive until the new growth becomes plantiful and the digestive system is adjusted. Weak animals often die during this period.

Complete data are lacking on the starvation process in wild ruminants. Hershberger and Cowan (1972) starved deer 16 to 30 days and found rumen fluid, protozoa number and VFA concentrations decreased. In vitro studies showed cellulose and alfalfa DMD and VFA formation dropped while lactic acid formation from starch digestion increased. Other work by DeCalesta et al (1975) indicated that while bacteria numbers decline with starvation, a viable population still remains following as long as 47 days of complete starvation. They concluded that the digestive function of rumen bacteria was not significantly impaired by starvation. Other workers have shown starvation to decrease rumen motility in ruminants, which would decrease rumen efficiency (see Vol. 1). These studies, plus field observations, indicate that while major changes may occur in the digestive system of starving ruminants, ruminants possess the ability to recover following periods of prolonged malnourishment.

It appears on the surface that energy may be the most critical nutrient. The reasoning for this being that saliva glands, the rumen and kidneys seem to be able to recycle N and many of the minerals; the liver can store large amounts of vitamin A; and the rumen microorganism can synthesize the B vitamins. As the animals enter into an energy deficient state, fat and protein reserves are gradually used. The first fat reserves to be completely used will be the subcutaneous stores. The omental stores then disappear. As the animal continues to utilize its reserves, the amount of fat on the heart and kidneys becomes sparse. The amount of fat in the marrow of the bones also decreases as the body reserves are utilized. As this happens the color of the marrow in the femur changes from a white color to one of red with the marrow being gelatinous in consistency (Fig. 16-12).

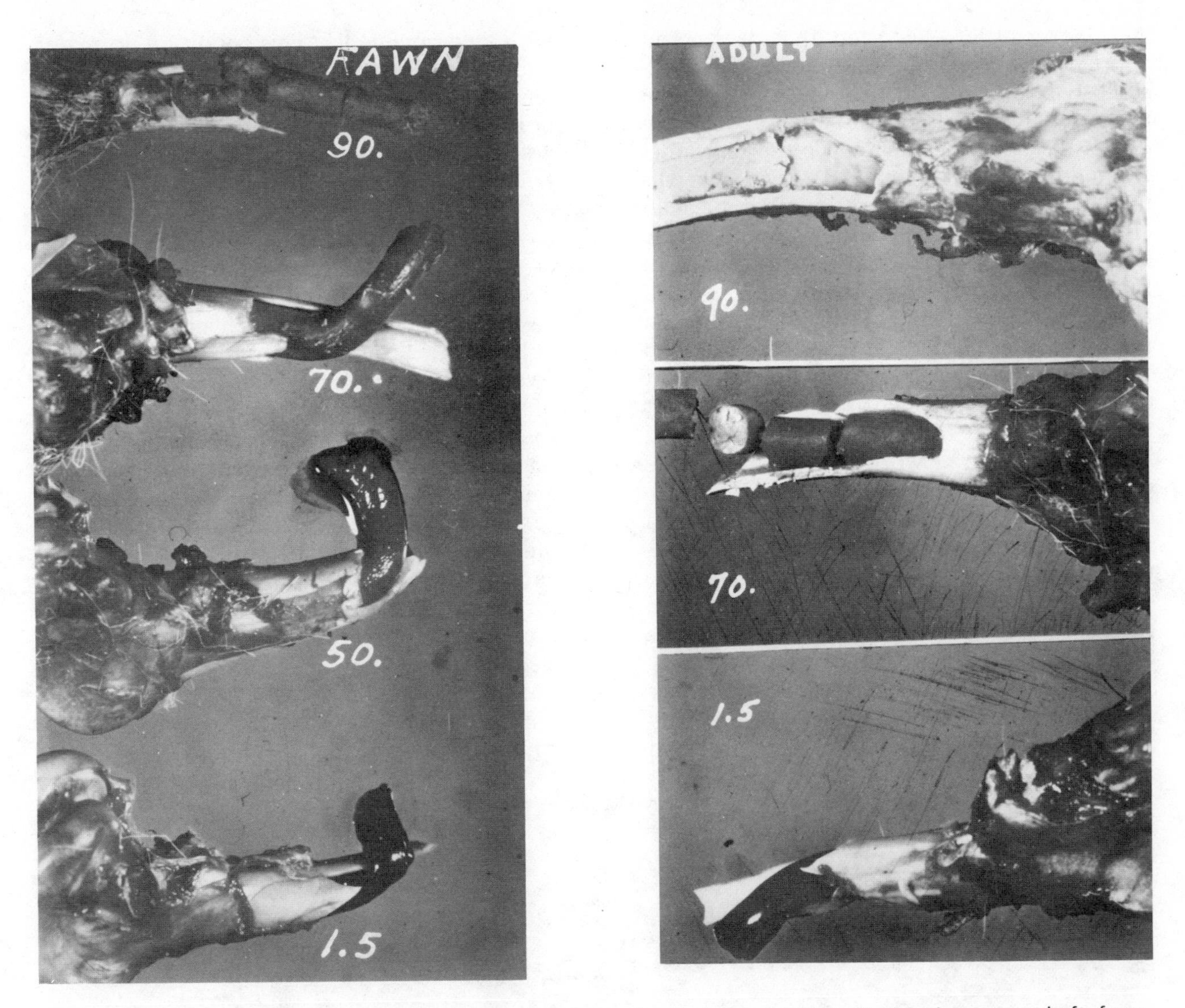

Figure 16-12. Bone marrow of white-tailed deer. Numbers refer to the approximate % of fat in marrow. Left, fawn. 90. Marrow firm, brittle, crumbly. Color usually whitish but may vary to reddish or brown. 70. Firm but slightly more flexible; usually mottled with red on cross section. 50. Wet, rubbery, but still some body to marrow. Color usually red but may be yellowish in some animals. 1.5. Wet, homogeneous and jelly-like. Right, adult. 90. Marrow firm, brittle, crumbly. 70. Firm but slightly more flexible, slightly moist, breaks when bent sharply. 1.5. Wet. homogenous and jelly-like; usually red but may be yellowish in some animals. Courtesy of Wildlife Pathology Laboratory, East Lansing, MI.

Both the amount of fat in the bone marrow and the amount of fat on the kidneys have been used as indices of the level of energy reserves in deer. When using the bone marrow, caution must be used because in younger animals erythropoiesis may still be occurring in the femur. If so, the marrow would be red in color and may have nothing to do with the energy levels of the animal. Protein tissues also provide the body with energy and are catabolized during starvation. Animals that have died of starvation will commonly have no visible fat on the heart and kidneys, the femur marrow will be red and gelatinous, and the animal will be emaciated (Dean, 1973).

Miscellaneous Maladies

There are nutritional maladies that have occurred in isolated cases which indicate that various conditions have been observed, but their significance to the general well-being of wild ruminants is unknown. For example, Call and James (1976) reported that deer could abort if they consumed pine needles. The possibility exists that the consumption of pine needles, which are known to cause abortion in cattle and sheep, could reduce the reproductive ability of wild animals. However, it seems unlikely that a significant number of wild animals would voluntarily consume toxic levels of poisonous plants that were native to the area since the animals and plants evolved together. The chance that an introduced poisonous plant could cause a major problem seems more likely. Cases where this might occur have not been identified but should not be overlooked. Dean and Winward (1974) found that tansy ragwort, which is poisonous to cattle and horses, and which is an introduced plant to black-tailed deer ranges in western Oregon, did not produce toxic symptoms when force fed to deer.

Necrotic stomatitis has been observed with elk consuming plant material that punctures the lining of the mouth. Grasses with long awns, woody parts of browse, and stemmy alfalfa have all been associated with this condition in elk. Frothy bloat has been observed with feedground elk. This condition normally occurs early in the feeding season when the animals are fed high quality alfalfa. Nitrate poisoning is also suspected to occur with feedground elk consuming oat hay, although positive evidence has not been gathered.

Vitamin deficiencies with wild ruminants have not been reported to any degree. Youatt et al (1976) examined the livers of road killed deer for vitamin A and found young deer to have less (81 mg/g of fresh liver) than adults (117 mg/g). They also found that vitamin A levels were lowest in late winter. Only 2-3% of the deer examined had levels low enough to be indicative of incipient deficiency (based on deficiency levels observed with domestic animals). Anderson et al (1976) showed basically the same trends in carotene and vitamin A levels. Apparently, the liver can store enough vitamin A during periods when the forage is green to satisfy body needs during the time when the vegetation may be deficient (winter). Young animals could be a possible exception.

Schoonveld et al (1974) fed mule deer alfalfa diets very high in fiber. Their data

Figure 16-13. Moose are shown here at a feedground. They can be kept in a herd and away from private property by feeding good quality alfalfa.

showed that the omasum and abomasum were subject to distention because of impaction. However, deer would not normally be fed diets this high in fiber. It does indicate that fiber levels are critical to small ruminants, such as deer, which have small digestive tracts and which depend on rapid passage of ingesta in order to obtain adequate nutrients.

ARTIFICIALLY FEEDING BIG GAME

Whether or not to give big game animals supplemental food has long been a subject of controversy. Much of the old literature indicated that supplemental feeding, especially of deer, did more harm than good. Recent research and experience have shown that wild ruminants can be fed safely in emergencies as well as during annual feeding operations if done properly.

Prior to initiating a feeding program, the advantages should be weighed against the disadvantages. The merits of a feeding operation change depending on the area, situation, etc., but the major advantages are as follows: (1) feeding programs make it possible to maintain a population of animals at a level that would otherwise be impossible. Examples of when this might be advantageous are when winter range is lost via encroachment of man and when migration routes are blocked by man. In general, this situation occurs when the animals have been deprived of a place to live and/or food. (2) Supplemental feeding can sometimes keep animals in a herd and away from private property, thereby reducing damage otherwise done by the animals (Fig. 16-13). (3) Emergency feeding programs have the potential to reduce large die-offs during unusually severe winters (Fig. 16-14).

There are two primary disadvantages of supplemental feeding. First, feeding operations are very costly under most situations. Figures from Dean (1977) showed that it cost an average of $43.53 to feed an elk during the winter in northwestern Wyoming in 1977 on established feedgrounds (Fig. 16-15). Costs in 1979 have risen to nearly $50 per elk fed. During the winter of 1977-78, it costs in excess of $150.00 per elk to feed them during an emergency situation where food had to be hauled great distances over snow. The second major problem with supplemental feeding is that the chance of disease transmission between animals is greatly increased when the animals are bunched together during the time of feeding (Fig. 16-16).

Feeding for Damage Prevention

The most extensive feeding programs in the USA are done in Wyoming. Each year about 21,000 elk and 500 moose are fed on a regular basis. Most of the animals are fed to keep them from doing damage to private property, primarily hay stacks (Fig. 16-17). Many factors dictate how successful such a feeding program will be. These include snow depth, nearness of escape cover, amount of human activity in the area, palatability of feed being fed, and the distance to potential damage areas. Feeding to prevent damage to standing crops, especially if the crops are green and/or highly palatable, is difficult, if not impossible. On the other hand, animals such as elk and moose can be fed to keep them away from private property during the winter months in many situations. Elk are relatively easy to feed because they readily stay in large herds and are easily attracted to hay.

The selection of a feeding site is very important. It should be located between the summer-fall range and the potential damage area and in a location that the animals will pass by or can be driven to while moving from the fall range to wintering areas. Having escape cover nearby and being in an area with minimal human activity are also important factors. These animals should be fed very palatable feed. Last cutting alfalfa is used in Wyoming to hold elk and moose. It is futile to feed a poor quality feed. In damage prevention situations it is advisable to feed all they will eat, which will be in excess of actual nutritional needs. Should they get hungry they will likely leave the feeding area in search of food. This is especially true early in the feeding season. It is best to be more concerned with appetite of the animal rather than its nutritional requirements in these situations.

Preventing damage by mule deer and antelope by means of supplemental feeding is not as practical as with elk. If snow conditions do not allow the animals much mobility, then they can be fed and kept in a particular area. However, deer and antelope prefer to eat some browse each day if available and move around freely if snow conditions allow. In this situation it is difficult to keep them in a certain area. Other damage prevention techniques must be used in such cases.

Figure 16-14. The emergency feeding of deer is shown here. Deer, elk and antelope can be saved from starvation with properly conducted feeding programs. Other wild ruminant species also could probably be fed successfully in emergency situations.

Figure 16-15. The cost of feeding is the main disadvantage of feeding programs. Hay purchases and personnel are the two items that are the most costly.

Figure 16-16. Elk can be fed in large herds (several thousand). This is advantageous in feeding programs designed to prevent damage, yet is condusive to rapid spread of disease.

Figure 16-17. Wildlife can cause much damage to private property. Elk and moose are relatively easy to attract to a feedground, thus keeping them away from the damage area. Deer and antelope are more difficult to feed in this manner.

Feeding to Sustain a Population on a Regular Annual Basis

This type of feeding can usually be designed to fit the nutritional needs of the animals unless damage prevention is also involved. The feeding levels should be designed to meet the needs of the type of animal in the herd requiring the greatest nutrients (based on the percentage of body weight). Early in the winter this is usually the young of the year which are 7-10 mo. old. As the winter progresses and the pregnant females near term, attention should be given to them and their nutritional needs. Nutrition during mid and late pregnancy affects calf weight and survival in elk (Thorne et al, 1976) and presumably all species. Table 16-5 shows the influence of weight loss of pregnant cow elk on calf birth weight and chances for survival. Calves weighing 16 kg or more at birth had a 90% chance of living to 4 weeks of age under captive conditions. Those weighing less than 11.4 kg had less than a 50% chance of surviving.

Elk feeding in Wyoming usually begins in December and the elk are given 6-7 lb/day of supplemental feed to go with the native feed they get off the feedground. As snow depths increase and available forage decreases, the amount fed is increased to 9-10 lb/day. By March the amount is increased to ad libitum feeding, or about 10-12 lb/elk. The gradual increase of the feeding level is designed to coincide with the increasing needs of the pregnant females.

The type of feed can be varied depending on the feedstuffs available, costs, storage facilities, etc. Elk do well on baled, cubed and pelleted hay (Table 16-6, 7). Straight alfalfa, alfalfa-grass mixtures or grass hay are all acceptable. However, the better the quality, the less that is needed. Alfalfa hay

Table 16-5. Effects of weight changes in cow elk during the last 4 months of gestation on abortions, calf birth weights and survival rates under captive conditions.[a]

Group	No. elk cows	% weight change[b]	Abortions	No. calves: Born	No. calves: Dead at 4 weeks	Birth wt, kg: Av	Birth wt, kg: Range	4-week wt, kg: Av	4-week wt, kg: Range
A	7	-17.1	2	3	2	11.8	8.04-16.1	44.4	ND[c]
B	9	-20.6	1	6	4	11.8	7.2 -17.3	44.3	ND
C	6	+ 0.8		6		18.4	16.4-21.2	43.3	41.4-44.1
D	6	- 6.6		6		15.5	13.2-18.2	34.1	30.0-38.7
E	6	+ 7.1		6		17.8	16.4-19.9	40.5	33.6-44.5
F	6	- 6.8		6	1	14.8	11.5-18.7	31.6	30.4-36.8
G	6	-17.2		3	1	12.8	10.9-13.7	23.2	16.8-29.5

[a]From Dean et al (1976)

[b]Percent weight difference during gestation is based upon the beginning weight and weight after parturition and includes parturition loss.

[c]No data.

Table 16-6. Weight changes in elk fed various amounts of baled and pelleted alfalfa.[a]

Group	Date	Number		Weight, lb Start		Weight, lb End		Change, %	
		Cows	Calves	Cows	Calves	Cows	Calves	Cows	Calves
Baled alfalfa									
A 10 lb	1/17-4/24 (1973)	10	3	460	256	459	258	+0.2	+0.8
D 8 lb	1/24-5/21 (1974)	9	4	461	240	421	246	-9.5*	+2.4†
Pelleted alfalfa									
B 8 lb	1/17-4/24 (1973)	10	3	407	243	425	264	+4.2*	+7.2†
E 8 lb	1/24-5/21 (1974)	9	4	460	233	442	267	-4.1	+12.7*
F 8 lb	1/24-5/21 (1974)	9	4	459	227	440	282	-4.3	+19.5†
Baled native + pelleted alfalfa									
C 2+6 lb	1/17-4/24 (1973)	10	3	463	238	482	251	+3.9*	+5.2

[a]From Thorne and Butler (1976)

*Weight change significant at P = .99; †Weight change significant at P = .95

Table 16-7. Weight changes in elk fed various amounts of baled and cubed alfalfa.[a]

Animal and ration	Date	Number in group	Weight, lb Beginning	Weight, lb Ending	Percent weight change
Cow elk					
8 lb baled	1/12-4/26	6	508	488	-3.9
8 lb cubed	1/12-4/26	6	507	512	+1.1
6.5 lb cubed	1/12-4/26	6	495	463	-6.5*
Calf elk					
5.5 lb baled	1/12-4/27	8	215	243	+11.7
5.5 lb cubed	1/12-4/27	8	221	250	+11.8

[a]From Thorne and Butler (1976)

*Change significant at P = .95

provides elk with more nutriment than does grass hay. In general, captive elk can maintain weight on 8-9 lb of good quality alfalfa, whereas around 10 lb of grass hay are required.

Emergency Feeding

Emergency feeding of big game animals can easily be unsuccessful and costly. However, there are times when it seems to be the lesser of evils, and emergency feeding programs are initiated. Success of such feeding programs depends, in part, on the species of animal being fed. For example, elk can be fed under emergency situations, and success can be expected. Feeding deer and antelope is a different matter. These species are much smaller than elk and seem to be more delicate to feed.

Feeding deer and antelope under emergency situations is not always successful and, in fact, many wildlife people feel it is not possible. In general, they feel that for some reason deer are unable to digest artificial feedstuffs or that these feeds cause digestive upsets which result in the death of the animal. Research studies designed specifically to study the digestive process during deer starvation and subsequent refeeding with a variety

Table 16-8. Daily intake rates of various feedstuffs (complete concentrate pelleted ration, fresh green grass and soybean meal, or soybean meal and whey) by deer following abrupt ration changes from poor quality roughages.[a]

	Ration					
			SBM		SBM-Whey	
Days	Concentrate	Green grass[b]	Limited	Ad lib	Limited	Ad lib
1-5	1.92	0.77	0.64	0.67	0.72	1.06
6-10	2.24	1.43	0.74	1.04	0.72	1.04
11-15	2.66	1.25	0.80	1.19	0.92	1.72
16-20	2.18	1.42	0.71	1.44	0.92	1.82
21-30	2.59	1.46				

[a]From Dean (1973); consumption in lb.

[b]Dry weight

of feeds have not shown any reason that starved deer could not be fed. Work in Oregon by Dean (1973) showed that deer could be switched from high fiber diets to high concentrate diets without significant mortality. Table 16-8 shows intake rates of deer following a change from poor quality roughage diets to several concentrate rations. Intake rates indicate that consumption is slow following the change in diets and continues to increase for up to 20 days. In this study 16 deer were fed soybean meal (SBM) and whey in various combinations and levels. Diarrhea was the only abnormality observed. One deer died in the limited SBM-whey group. These data showed that starved deer adjusted to high fiber diets could survive changes to high concentrate diets without excessive mortality under captive conditions. Nagy et al (1974) showed that mule deer could be starved for as long as 64 days and still recover and DeCalesta et al (1977) found that deer could survive refeeding after being starved to lose about 16% of their body weight.

Figure 16-18. Dead deer are a common sight around hay stacks during some winters. Hay is often blamed for the deaths but the deer usually die in spite of the hay (especially if it is poor quality) or they are too near death before coming to the hay to survive.

Although this work indicates that deer can consume artificial food successfully during periods of starvation or following starvation, the studies do not explain why so many deer feeding programs are unsuccessful (Fig. 16-18). The causes are unknown at this point but probably are related to several factors. (1) The deer would have probably been consuming a diet very high in fiber prior to emergency feeding. If they were fed a high quality ration, additional time would be required before the rumen microorganisms could adjust. Since some of the deer would likely be in poor condition at the time, the additional time needed for the adaptation could result in deaths, primarily because of starvation. Also, any mild digestive upset that might be caused by switching diets would be particularly hard on weak individuals. (2) At the beginning of an emergency feeding program many deer will probably already have died of starvation, and more will be very near death. These deer will probably die regardless of what or how they are fed. They have gone beyond a point of recovery. They will die after the feeding begins and it will appear on the surface that the feeding program killed them. (3) If the deer are being fed something other than a complete ration, malnutrition may continue. While they will be getting some nutriment, they will not be getting enough to be completely sufficient and additional deaths may result. Therefore, when feeding deer under emergency situations, it is to be expected some will die after feeding begins but also it must be realized that if they are being fed properly, feeding is saving deer that would otherwise die.

Feed the best diet available, i.e., formulate a ration designed for starving deer if possible. There are many ration formulations being used throughout the country that would be adequate to feed deer during emergency situations. Any university or experimental station having captive deer probably has a basal ration that would be acceptable. Table 16-9 shows an example of ingredients used in a ration that has been used successfully with mule and black-tailed deer in Oregon.

Research in Oregon with black-tailed deer showed that deer preferred pelleted grains over the rolled and whole forms (Table 16-10). This work also indicated that soybean meal was the most palatable high protein source and that corn and wheat were the most acceptable high energy sources. If feeding a formulated, pelleted ration is not possible, then good quality alfalfa is probably the best single feed for deer. It should be noted that alfalfa is likely not a complete ration and that the deer will continue to lose weight while consuming it, but at a much lesser rate than if eating poor quality diets.

Animals being fed during emergency conditions should be fed at near ad libitum rates. This would amount to 10-12 lb/day for elk

Table 16-9. Ration formulated for feeding captive deer. The ration should be in ¼ inch pellets for deer.[a]

Ingredient	Percentage
Ground oats	15
Ground wheat	9
Ground corn	30
Cottonseed meal	20
Soybean meal	10
Alfalfa meal	5
Molasses	10
Tricalcium phosphate	0.5
Salt, trace mineralized[b]	0.5
Vitamin A	2,500,000 IU

[a]From Dean (1973)

[b]One oz contains 70.9 mg of Zn, 56.7 mg of Mn, 35.4 g of Fe, 7.09 mg of Cu, 1.42 mg of I and 1.42 mg of Co.

Table 16-10. Intake (lb/day/deer) by black-tailed deer of concentrate feeds when offered in various combinations.*

	Daily Consumption[a]					
	Trial 1	Trial 2	Trial 3	Trial 4	Trial 5	Trial 6
Alfalfa[c]	$0.44 \pm .20^b$	$0.20 \pm .10^b$	$0.10 \pm .02^b$	$0.16 \pm .10^b$	$0.93 \pm .02^b$	$0.08 \pm .05^b$
Barley	—	—	$0.20 \pm .02^a$	—	—	$0.26 \pm .19^a$
Beet pulp	$0.00 \pm .00^a$	—	—	—	$0.08 \pm .03^a$	—
Corn	—	—	$1.60 \pm .45^b$	—	—	$0.71 \pm .24^b$
Cottonseed meal	—	$0.00 \pm .00^a$	—	—	$0.06 \pm .05^a$	—
Linseed meal	$0.00 \pm .00^a$	—	—	—	$0.24 \pm .05^a$	—
Oats	—	—	$0.32 \pm .05^a$	$2.62 \pm .36^a$	—	$0.16 \pm .25^c$
Peas	—	—	—	$0.00 \pm .99^b$	—	—
Soybean meal	$1.90 \pm .25^b$	—	—	—	—	$0.94 \pm .15^d$
Wheat	—	$1.54 \pm .74^b$	—	—	—	$0.83 \pm .29^e$

*From Dean and Church (1972)

[a]Mean ± the standard deviation

[b]Means within a trial that are followed by different superscript letter differ significantly ($P < 0.01$)

[c]Consumption of alfalfa not compared with kinds of concentrates

and 2.5-3.5 lb/day for deer. If the diet is being changed from one of roughage (browse) to one with concentrates, then the ration should be relatively high in roughage until the animal has time to adjust. This can be accomplished by feeding a limited amount of the pelleted ration with ad libitum levels of baled hay and gradually increasing the amount of pellets fed. Alternatively, several rations can be formulated where the level of roughage in the formulation is varied accordingly.

In contrast to deer, elk are relatively easy to feed under emergency situations. They are larger in size and seem to be able to recover more quickly when given supplemental feed. This is probably because they are able to digest diets higher in fiber, and good quality roughage would serve as a good emergency feed. Their digestive systems would not have to undergo major changes. Also, the larger body size would have some advantages in terms of being more resistent to cold weather. Alfalfa hay or alfalfa-grass hay are adequate for elk in emergency situations. The hay should be fed ad libitum, at least until the weaker animals have gained strength.

References Cited

Abler, W.A., D.E. Buckland, R.L. Kirkpatrick and P.F. Scanlon. 1976. J. Wildl. Manage. 40:442.

Anderson, A.E., D.E. Medin and D.C. Bowden. 1972. Comp. Biochem. and Physiol. 41B:745.

Ammerman, C.B., S.M. Miller, K.R. Fick and S.L. Hansard II. 1977. J. Animal Sci. 44:485.

Baker, D.L. et al. 1979. J. Wildl. Manage. 43:162.

Beale, D.M. and A.D. Smith. 1970. J. Wildl. Manage. 34:570.

Bissell, H.D. and H. Strong. 1955. Cal. Fish and Game 41:145.

Blair, R.M., H.L. Short and E.A. Epps. 1977. J. Wildl. Manage. 41:667.

Brady, P.S. et al. 1978. J. Nutr. 108:1439.

Brown, E.R. 1961. Wash. State Game Dept. Biol. Bull. 13.

Burns, T.A. and C.E. Viers. 1973. J. Wildl. Manage. 37:585.

Calef, G.W. and G.M. Lortie. 1975. J. Mammology 56:240.

Call, J.W. and L.F. James. 1976. J. Amer. Vet. Med. Assoc. 169:1301.

Cameron, R.D. and J.R. Luick. 1972. Can. J. Zool. 50:107.

Carbyn, L.N. 1975. Can. J. Zool. 53:378.

Chappel, R.W. and R.J. Hudson. 1978. Can. J. Zool. 56:2388; Acta Theriol. 23:359.

Chappel, R.W. and R.J. Hudson. 1979. J. Wildl. Manage. 43:261.

Chatterton, N.J. et al. 1971. J. Range Manage. 24:37.

Church, D.C. and W.H. Hines. 1978. J. Wildl. Manage. 42:654.

Cook, C.W. 1972. Interm. For. Range Exp. Sta. USDA Tech. Rep. INT-1.

Cowan, I.M. and V.C. Brink. 1949. J. Mammology 30:379.

Cowan, I.M., A.J. Wood and W.D. Kitts. 1957. N. Amer. Wildl. Conf. 22:179.

Crete, M. and J. Bedard. 1975. J. Wildl. Manage. 39:368.

Dalke, P.D. et al. 1965. J. Wildl. Manage. 29:319.

Dean, R.E. 1973. Ph.D. Thesis. Oregon State University, Corvallis.

Dean, R.E. 1977. Wyoming Wildl. 41:24.

Dean, R.E. and D.C. Church. 1972. Feedstuffs 44(51):37.

Dean, R.E. and A.H. Winward. 1974. J. Wildl. Dis. 10:166.

Dean, R.E. et al. 1975. J. Wildl. Manage. 39:601.

DeCalesta, D.S., J.G. Nagy and J.A. Bailey. 1975. J. Wildl. Manage. 38:815.

DeCalesta, D.S., J.G. Nagy and J.A. Bailey. 1977. J. Wildl. Manage. 41:81.

Demarchi, R.A. 1968. J. Range Manage. 21:385.

Dietz, D.R. 1972. Wildland Shrub—Their Biology and Utilization. USDA Intermount. For. Range Exp. Sta., Tech. INT-1.

Dietz, D.R., R.H. Udall and L.E. Yeager. 1962. Colo. Game and Fish Dept. Tech. Publ. 14.

Elder, J.B. 1954. J. Wildl. Manage. 18:540.

Erdman, J.A., R.J. Ebens and A.A. Case. 1978. J. Range Manage. 31:34.

Franzmann, A.W., A. Flynn and P.D. Arneson. 1975. J. Wildl. Manage. 39:374.

Hagen, H.L. 1953. Cal. Fish and Game. 39:163.

Hamilton, J.W. and C.S. Gilbert. 1972. Univer. Wyoming, Agr. Exp. Sta. Res. J. 55.
Harshbarger, T.J. and B.S. McGinnes. 1971. J. Wildl. Manage. 35:668.
Henshaw, J. and J. Ayeni. 1971. E. Afr. Wildl. J. 9:73.
Hebert, D.M. and I.M. Cowan. 1971. J. Wildl. Manage. 35:752; Can. J. Zool. 49:605.
Hershberger, T.V. and R.L. Cowan. 1972. J. Animal Sci. 35:266.
Holter, J.B. and H.H. Hayes. 1977. J. Wildl. Manage. 41:506.
Holter, J.B., W.E. Urban and H.H. Hayes. 1977. J. Animal Sci. 45:365.
Holter, J.B. et al. 1975. Can. J. Zool. 53:679.
Hoppe, P.P. 1977. E. Afr. Wildl. J. 15:41.
Hyvarinen, H. et al. 1977. Can. J. Zool. 55:648.
Jefferies, N.W. and R.W. Rice. 1969. J. Range Mange. 22:192.
Knox, K.L., J.G. Nagy and R.D. Brown. 1969. J. Wildl. Manage. 33:389.
Lay, D.W. 1969. White-tailed Deer in the Southern Forest Habitat. Proc. Symp. S. Forest Exp. Sta.
LeResche, R.E. and J.L. Davis. 1973. J. Wildl. Manage. 37:279.
Longhurst, W.M., N.F. Baker, G.E. Connolly and R.A. Fisk. 1970. Amer. J. Vet. Res. 31:673.
Macfarlane, W.V., C.S.H. Dolling and B. Howard. 1966. Aust. J. Agr. Res. 17:491.
Magruder, N.D., C.E. French, L.C. McEwen and R.W. Swift. 1957. Penn. Agr. Expt. Sta. Bul. 628.
Mautz, W.W. et al. 1976. J. Wildl. Manage. 40:630.
McEwan, E.H. and P.E. Whitehead. 1970. Can. J. Zool. 48:905.
McEwen, L.C. et al. 1957. N. Amer. Wildl. Conf. 22:119.
Murphy, D.A. and J.A. Coates. 1966. N. Amer. Wildl. and Nat. Resource Conf. 31:129.
Nagy, J.G., H.W. Steinoff and G.M. Ward. 1964. J. Wildl. Manage. 28:785.
Nagy, J.G., W. Chappell and G.M. Ward. 1975. J. Animal Sci. 43:412.
Nagy, J.G. et al. 1974. Colo. Game and Fish, P-R Proj. W-38-R, Plan 14, Job 6.
Newman, J.R. and M. Yu. 1976. J. Wildl. Dis. 12:39.
Nordan, H.C., I.M. Cowan and A.J. Wood. 1970. Can. J. Zool. 48:275.
Oh, H.K., M.B. Jones and W.M. Longhurst. 1968. Appl. Microbiol. 116:39.
Oldemeyer, J.L. et al. 1977. J. Wildl. Manage. 41:533.
Ozoga, J.J. and L.W. Gysel. 1972. J. Wildl. Manage. 36:892.
Pedan, D.G. 1972. Ph.D. Thesis. Colorado State University, Ft. Collins.
Robbins, C.T. and A.N. Moen. 1975. J. Wildl. Manage. 39:337, 355.
Rodgers, J.D. and T. W. Box. 1967. J. Range Manage. 20:177.
Schoonveld, G.G., J.G. Nagy and J.A. Bailey. 1974. J. Wildl. Manage. 38:823.
Seal, U.S. et al. 1972. J. Wildl. Manage. 36:1041.
Segelquist, C.A., H.L. Short, F.D. Ward and R.G. Leonard. 1972. J. Wildl. Manage. 36:174.
Short, H.L. 1969. White-tailed Deer in the Southern Upland Forest. Proc. Symp. Southern For. Expt. Sta.
Short, H.L. 1971. J. Wildl. Manage. 35:698.
Short, H.L. 1976. J. Wildl. Manage. 40:479.
Short, H.L., R.M. Blair and C.A. Segelquist. 1974. J. Wildl. Manage. 38:197.
Silver, H., N.F. Colovos, J.B. Holter and H.H. Hayes. 1969. J. Wildl. Manage. 33:490.
Smith, A.D. 1950. J. Wildl. Manage. 14:285.
Smith, A.D. 1952. J. Wildl. Manage. 16:309.
Smith, A.D. 1957. J. Range Manage. 10:162.
Smith, G.E., D.C. Church, J.E. Oldfield and W.C. Lightfoot. 1963. Proc. West. Sec., ASAS 17:373.
Smith, S.H., J.B. Holter, H.H. Hayes and H. Silver. 1975. J. Wildl. Manage. 39:582.
Snider, C.C. and J.M. Asplund. 1974. J. Wildl. Manage. 38:20.
Stockstad, D.S., M.S. Morris and E.C. Lory. 1953. N. Amer. Wildl. Conf. 18:247.
Thompson, C.B. et al. 1973. J. Wildl. Manage. 37:301.
Thorne, E.T. 1969. Wyoming Game and Fish. P-R Job Prog. Rep. FW-3-R-16, Work Plan 2, Job 1W.
Thorne, T. and G. Butler. 1976. Syoming Game and Fish Dept. Tech. Rep. No. 6.
Thorne, E.T. and R.E. Dean. 1973. Wyoming Game and Fish. P-R Prog. Rep. FW-3-R-20, Plan 2, Job 6W.
Thorne, E.T., R.E. Dean and W.G. Hepworth. 1976. J. Wildl. Manage. 40:330.
Ullrey, D.E. et al. 1967. J. Wildl. Manage. 31:679.
Ullrey, D.E. et al. 1969. J. Wildl. Manage. 33:482.
Ullrey, D.E. et al. 1970. J. Wildl. Manage. 34:863.

Ullrey, D.E. et al. 1971. J. Wildl. Manage. 35:57, 732.
Ullrey, D.E. et al. 1972. J. Wildl. Manage. 36:885.
Ullrey, D.E. et al. 1973. J. Wildl. Manage. 37:187.
Ullrey, D.E. et al. 1975. J. Wildl. Manage. 39:590, 699.
Urness, P.J., D.J. Neff and J.R. Vahle. 1975a. J. Wildl. Manage. 39:670.
Urness, P.J., D.J. Neff and R.K. Watkins. 1975b. USDA For. Ser. Res. Note RM-304.
Verme, L.J. 1969. J. Wildl. Manage. 33:881.
Wallmo, O.C. et al. 1977. J. Range Manage. 30:122.
Ward, A.L. 1971. J. Wildl. Manage. 35:681.
Weeks, H.P. and C.M. Kirkpatrick. 1976. J. Wildl. Manage. 40:610.
Weiner, J. 1977. Acta Theriol. 22:3.
Wesley, D.E., K.L. Knox and J.G. Nagy. 1970. J. Wildl. Manage. 34:908.
Wesley, D.E., K.L. Knox and J.G. Nagy. 1973. J. Wildl. Manage. 37:563.
White, R.G. and M.K. Yousef. 1978. Can. J. Zool. 56:215.
Wobeser, G. and W. Runge. 1975. J. Wildl. Manage. 39:596.
Woolf, A. and D. Kradel. 1977. J. Wildl. Dis. 13:281.
Youatt, W.G., D.E. Ullrey and W.T. Magee. 1976. J. Wildl. Manage. 40:172.

Chapter 17—The Nutrition of Captive Wild Ruminants

by D.E. Ullrey

The large number of species of wild ruminants and the limited information on nutrient needs of individual species dictates that this subject be treated in general terms. Relevant areas of discussion include digestive tract anatomy and natural dietary habits as they relate to nutrient requirements, management considerations, data on qualitative and quantitative nutrient requirements, and diets used in captivity. Proposed diets and feeding systems for captive wild ruminants are presented.

DIGESTIVE TRACT ANATOMY AND NATURAL DIETARY HABITS

Ruminants have the most differentiated, specialized and complex stomach among all the mammals (see Vol. 1 for further detail). Of the 142 species distinguished by Simpson (1945) only about twelve—sheep, goats, cattle, llama, alpaca, dromedary camel, bactrian camel, water buffalo, yak, banteng, gaur and reindeer—have been domesticated. Wild or domestic, ruminants have the collective advantage of a rumen inhabited by microbial symbionts which ferment cellulose and hemicellulose and synthesize amino acids, vitamin K and B-vitamins. These phenomena are essential features of the adaptation of wild ruminants to their natural environment and are of major significance in managing these species in captivity. Despite this common feature, natural dietary habits of ruminants in the wild differ appreciably, and Hofmann (1973) has divided East African game ruminants into (1) selectors of juicy, "concentrate" herbage, (2) bulk and roughage eaters, and (3) intermediate feeders (seasonally and regionally adaptable). This classification system is applicable to other wild ruminants, as well, and consideration of digestive tract anatomy and dietary habits is potentially useful in developing diets for captive wild ruminants. Average body weights and capacities of ruminoreticula of species studied by Hofmann are presented in Table 17-1.

Concentrate Eaters

Concentrate eaters have a ruminoreticulum with a small capacity in relation to that of the peritoneal cavity and as compared to that of the other two feeding types. Hungate (1959) reported that the four stomach compartments of concentrate eaters comprise 8-10% of body weight when full and about 2% when empty. In contrast, the stomachs of rougage eaters comprise 14-15% of body weight full and 3-3.5% empty. Passage of ingesta is rapid, fermentation rate is high and the rumination pattern is irregular. Absorptive surfaces of the forestomach are large due to papillation throughout the rumen (including dorsal surfaces) and to dense papillation of the dorsal quarter to third of the reticulum.

Hofmann (1973) has divided East African concentrate eaters into two subdivisions. One includes small bush and forest antelope that tend to select fruit and dicotyledon foliage. Representative species are Gunther's dik-dik *(Madoqua guentheri)*, Kirk's dik-dik *(Madoqua kirki)*, suni *(Nesotragus moschatus)*, klipspringer *(Oreotragus oreotragus)*, gray duiker *(Sylvicapra grimmia)*, Harvey's duiker *(Cephalophus harveyi)*, bushbuck *(Tragelaphus scriptus)*, sitatunga *(Tragelaphus spekei)*, blue

Figure 17-1. Dorcas gazelle *(Gazella dorcas)* are very selective in their dietary habits and when fed alfalfa hay, mainly eat the leaves.

Table 17-1. Body weight and capacity of ruminoreticula of East African wild ruminants.[a]

Species	Body weight, kg	Ruminoreticulum, ℓ
Concentrate Selectors		
Gunther's dik-dik	4.1	0.8
Kirk's dik-dik	5.2	0.9
Suni	6.2	0.9
Klipspringer	11.4	2.6
Gray duiker	14.0	3.2
Harvey's duiker	16.0	5.0
Bushbuck, male	55.5	7.6
female	43.5	
Giraffe	750	105
Lesser kudu, male	98	13.2
female	83	
Greater kudu, male	257	58-71
female	170	27
Gerenuk, male	46	6.2
female	40	
Bulk and Roughage Eaters		
Buffalo, male	751	107
female	447	
Oribi	16	4.0
Bohor reedbuck	45	9.7
Uganda kob, male	96	9.8
female	62	
Waterbuck, male	226	41
female	175	
Wildebeest, male	201	40
female	163	
Kongoni	116-160	23
Hartebeest	152-196	40
Topi, male	130	31
female	108	
Mountain reedbuck, male	26	7.3
female	21	
Oryx (fringe-eared and beisa), male	203	35.5
female	160	
Intermediate Feeders		
Impala, male	63	11.9
female	48	
Thomson's gazelle, male	22	5.8
female	20	
Eland antelope, male	510	
female	420	53
Grant's gazelle, male	64	12.8
female	46	
Steenbok	10.5	2.5

[a]From Hofmann (1973)

duiker *(Cephalophus monticola)* and bongo *(Taurotragus euryceros).*

The second subdivision includes larger ruminants inhabiting the tree and bush savannah and the hill regions of semi-arid country. They tend to select tree and shrub foliage and include giraffe *(Giraffa camelopardalis),* lesser kudu *(Tragelaphus imberbis),* greater kudu *(Tragelaphus strepsiceros)* and gerenuk *(Litocranius walleri).*

Bulk and Roughage Eaters

The bulk and rougage eaters include species which Hofmann (1973) considers to have the most advanced stomach forms among East African herbivores. Forestomach structure delays passage of the preferred fibrous food and increases time of exposure to fermentative processes. The relative capacity of the ruminoreticulum is large in relation to the capacity of the peritoneal cavity and to that of the other two feeding types. The ruminoreticulum of roughage eaters is filled amost to capacity, while concentrate selectors rarely have more than a 50-60% fill.

Hofmann (1973) has separated this group into three subdivisions based on specialization in feeding and drinking habits and finer differentiation of stomach structures. The first subdivision includes fresh grass grazers which require surface water at regular, relatively short intervals and may migrate to ensure that fresh grass and water are available. They have a phenotype adapted for grazing and include African buffalo *(Syncerus caffer),* oribi *(Ourebia ourebi),* Bohor reedbuck *(Redunca redunca),* Uganda kob *(Adenota kob),* water buck *(Kobus ellipsiprymnus)* and wildebeest *(Connochaetus taurinus).*

Figure 17-2. Plains bison *(Bison bison bison),* a roughage grazer, eating sudan grass hay fed on a concrete pad.

The second subdivision includes the roughage grazers, mainly larger plains animals and the mountain reedbuck. They feed entirely or predominantly on grasses, appear less dependent upon water than those in the former subdivision, and show little tendency to migrate unless drought is severe. This subdivision includes hartebeest *(Alcephalus buselaphus),* topi *(Damaliscus lunatus),* mountain reedbuck *(Redunca fulvorufula),* roan antelope *(Hippotragus equinus)* and sable antelope *(Hippotragus niger).*

The third subdivision includes dry region grazers which are well adapted to arid, occasionally hot climates and are represented by fringe-eared oryx *(Oryx gazella callotis)* and beisa oryx *(Oryx gazella gallarum).*

Intermediate Feeders

The intermediate feeders include species which are more adaptable to changing habitats and vegetation than most of the concentrate selectors or grazers previously discussed. Forestomach structures show significant variations which can be related to food plants eatern by preference or necessity. Adaptations to gradual changes in vegetation during rainy or dry seasons are particularly apparent in ruminal papillation, with concentrate selection increasing papillation and grass selection decreasing it.

Intermediate feeders have been subdivided into those which prefer grasses and those which prefer forbs and shrub or tree foliage. Both groups have stomachs of the simpler type, resembling those of the concentrate selectors but with variations in capacity, structures which delay passage, and degree of papillation and omasal development. The first subdivision includes two species which are well adapted to heat and water restrictions. These are the impala *(Aepyceros melampus)* and Thomson's gazelle *(Gazella thomsoni).*

The second subdivision includes eland antelope *(Taurotragus oryx),* Grant's gazelle *(Gazella granti)* and steenbok *(Raphicerus campestris).* The first two species adapt well to extremely dry, hot climates although they are widely distribution and may live in cooler, well watered areas. The bulk of food consumed by this subdivision is foliage of dicotyledons, but some grass is eaten, and the lush green grass that appears after rains or fire is especially favored.

Microstructure of the Ruminoreticulum

Detailed studies (Hofmann, 1973) of the microstructure of the ruminoreticulum demonstrate variations between species which correlate with food habits. Roughage grazers not only have few or rudimentary papillae in the dorsal ruminal sac, but the epithelium has several layers of cornified cells with very narrow intercellular spaces. These cornified cells undoubtedly offer protection to underlying structures and probably interfere with absorption. Houpt (1970) has shown that disruption of the cornified layer increased the rate of urea nitrogen transport about 50 times. Changes regularly occur in the microstructure of the small dorsal papillae of intermediate feeders during the various seasons. With the advance of the dry season, impala and Thomson's gazelle supplement their diet with dicotyledon foliage although dry grass remains the principal food. The dorsal ruminal papillae of that period are superficially keratinized and have a thick, well defined cell barrier with a poorly developed subepithelial vascular system. Two to three weeks after lush green grass becomes available, surface vesiculation begins and there is a dramatic increase in subepithelial vascular contact surface by penetration of capillaries into the epithelium via slender *papillae occultae.* While the barrier layer remains thick for some time, gradual attenuation of its

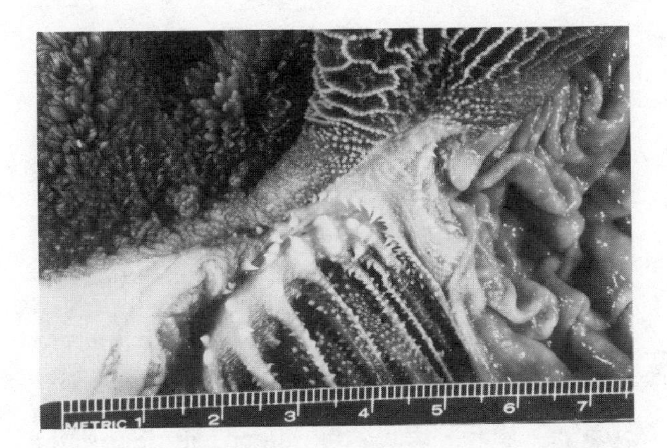

Figure 17-3. Internal structure of the compound stomach of the aoudad *(Ammotragus lervia).* Clockwise from upper left—rumen, reticulum, abomasum and omasum.

cells is accompanied by increased distal transformation into balloon cells which ultimately burst (Henrikson, 1970), providing multiple delicate processes with a glycocalyx for temporary attachment of bacteria or protozoa. Five to six weeks after the onset of rain, there is little microstructural difference between absorptive papillae of the ruminal atrium and those of the dorsal wall. Habitual concentrate selectors such as dik-diks, which tend to eat rapidly fermenting food throughout the year, also show maximum development of superficial balloon cells on dorsal ruminal papillae during the rainy season.

MANAGEMENT CONSIDERATIONS IN CAPTIVITY

While these differences in gastrointestinal anatomy and dietary habits have obvious adaptive advantages in a natural ecological setting, feeding captive wild ruminants thousands of miles from their native environment dictates dependence on dietary items far different from those eaten in the wild. Fortunately, most wild species, just like domestic ones, require specific nutrients not specific feedstuffs. Thus, the adequate nutrition of captive wild ruminants assumes knowledge of qualitative nutrient requirements and their provision in sufficient amounts and in an acceptable form. Few studies have been conducted with wild ruminants to define nutrient requirements, and most diets for captive animals are empirically formulated. Some consider natural dietary habits. Others are based on domestic animal models. Most are pragmatic formulations, derived from more or less satisfactory experience, and are infrequently altered due to fear of failure.

In any case, feeding wild ruminants in captivity involves compromise. Animals of different ages and physiological states may be housed together. Social groupings may or may not correspond to those typical for the species. Groups frequently include one or more adult males, several adult females (non-gravid, gravid, lactating), weaned but immature young and nursing infants. Fairly often, animals of different species are exhibited together to simulate, as nearly as possible, animal distribution in a native environment. Since the normal variety of food plants usually cannot be provided, alternate means must be devised to properly nourish this divergent

Figure 17-4. Addax *(Addax nasomaculatus)*, a desert-dweller, consuming a pelleted diet fed in a concrete trough hidden by rocks.

animal population. Traditionally, hay and concentrates have been fed in a manner very similar to management of domestic ruminants. This can be done very successfully if the animals in a single exhibit are of the same species, age, sex and physiological state. Hay may be provided free choice, and concentrates may be fed once or twice a day. Consideration must be given to hay quality and the amount actually eaten as compared to that which is wasted. Adequate feeder space must be available so that timid animals will not be crowded out by those that are socially dominant. This is a particular problem when the time that food is available is limited. If the exhibit contains animals of different age, sex and species, the proper nutrition of each individual becomes much more difficult. Feed troughs can be placed at different locations or at different heights consistent with differences in anatomy or behavior. Creep feeders may be necessary to ensure that weaned youngsters get their fair share. Some feeding systems use pellets that contain roughage and concentrates which have been ground and mixed together. These may be fed ad libitum as the sole diet or along with hay to promote rumination and to reduce the boredom associated with close confinement and consequent vices such as fence chewing. Experience has shown that a number of different feeding systems will produce satisfactory results. However, two relevant scientific questions remain. Which nutrients and what concentrations should be provided in the diets of captive wild ruminants?

QUALITATIVE NUTRIENT REQUIREMENTS

Although few experimental data exist, qualitative nutrient requirements are probably identical to those of domestic ruminants. Thus, adult wild ruminants need the items shown in Table 17-2 in their diet. In addition, some indigestible fiber must be present to support normal function of the digestive tract. Since nursing young are basically monogastrics without functional ruminoreticula, they require these listed nutrients plus vitamin K and the usual B-vitamins. Essential amino acid needs of these preruminants probably include arginine, histidine, isoleucine, leucine, lysine, methionine, phenylalanine, threonine, tryptophan and valine.

Table 17-2. Probable qualitative nutrient requirements of adult wild ruminants.

Water	Cobalt
Energy	Manganese
Nitrogen	Selenium
Essential fatty acids (?)	Chromium
Calcium	Fluorine
Phosphorus	Nickel
Magnesium	Silicon
Sodium	Vanadium
Chlorine	Tin
Potassium	Arsenic
Sulfur	Molybdenum
Iron	Vitamin A
Copper	Vitamin D
Iodine	Vitamin E

QUANTITATIVE NUTRIENT REQUIREMENTS

Water

Quantitative nutrient requirements have been established in only a few instances. Water requirements vary with climatic conditions, type of food, physiological state (growth, maintenance, lactation), amount of activity and evolutionary adaptation to regions with limited water. Voluntary water intake of red deer *(Cervus elaphus)* was 50% greater than that of sheep in Scotland (Blaxter et al, 1974) and may relate to the origins of sheep in seasonally arid areas of Eurasia as compared to red deer which are natives of temperate woodland. Captive

Table 17-3. Minimum water requirements of East African ruminants in liters/100 kg/day (liters/100 $W_{kg}^{0.75}$/day).[a]

Species	Metabolic	Preformed	Drinking	Total
		Ambient temperature 22°C		
Eland	0.4	0.2	3.2	3.7 (12.7)
Buffalo	0.4	0.2	2.8	3.4 (12.7)
Beisa oryx	0.4	0.2	1.3	1.9 (5.7)
Wildebeest	0.5	0.2	2.2	3.0 (10.5)
Grant's gazelle	0.6	0.2	1.2	2.1 (4.6)
Thomson's gazelle	0.7	0.2	1.2	2.2 (4.4)
		Ambient temperature 22°C (12 h) 40°C (12 h)		
Eland	0.4	0.2	4.9	5.5 (18.7)
Buffalo	0.3	0.1	4.2	4.6 (17.1)
Beisa oryx	0.6	0.2	2.1	3.0 (9.0)
Wildebeest	0.5	0.2	4.1	4.8 (16.4)
Grant's gazelle	0.6	0.2	3.0	3.9 (8.6)
Thomson's gazelle	0.5	0.1	2.1	2.7 (5.4)

[a]From Taylor (1968)

pregnant white-tailed deer *(Odocoileus virginianus)* typically consume 2-3 times as much water as dry matter when ambient temperatures are in the thermal neutral zone. Taylor (1968) paired East African species of similar size, with one able to tolerate arid regions and the other dependent on water–eland and buffalo, beisa oryx and wildebeest, Grant's gazelle and Thomson's gazelle. He determined minimum water requirements by gradually reducing drinking water from ad libitum intakes to a point where the animals were just able to maintain their weight at about 85% of initial levels, assuming that this procedure maximally stimulated mechanisms of water conservation. Water balance was determined for 14 days at an ambient temperature of 22°C, followed by a like period with alternating 12 h periods of 22 and 40°C temperatures. Metabolic water, preformed water in food, drinking water and total water requirements determined in this study are shown in Table 17-3. Beisa oryx, Grant's gazelle and Thomson's gazelle were found to be specialized for conserving water and maintained constant weight on total water inputs of approximately 2% of their body weight/day. Eland did not appear to be particularly efficient in conserving water, but may gain independence from surface water by behavioral and other physiological means. For example, eland select succulent food, avoid the midday sun, form dry feces and have a narrow thermal neutral zone.

The low humidity in arid regions is associated with large diurnal variations in ambient temperature, which each night drops below the eland's lower critical temperature. The consequent increase in metabolism increases food intake (and preformed water intake) and production of metabolic water, while the heat produced is dissipated into the cool night air. Few data exist on water requirements of other species, but this deficiency may be of minor consequence since water should be provided ad libitum for captive species.

Dry Matter Consumption

The amount of food consumed each day is a function of appetite. Food intake is physiologically regulated over both the long and short term, otherwise starvation or obesity would be more common.

Physical limitations of the digestive tract very likely limit the intake of coarse foods. Difficult to digest foods are retained for longer periods of time in the ruminoreticulum than are easily digested foods, restricting the amount of food consumed per unit of time. In domestic ruminants the voluntary intake of forage dry matter (DM) has been shown to increase as the apparent digestibility of energy in the forage DM increased, providing forage DM digestibility didn't exceed 70% (Blaxter et al, 1966). When forages alone, or forages plus concentrates, were used to produce a range of DM digestibility from 52 to 80%, voluntary DM intakes of lactating dairy cows increased as DM digestibility increased from 52 to 67% and decreased as dry matter digestibility increased from 67 to 80% (Conrad et al, 1964). On diets with low digestibility, intake was a function of body weight, passage rate of digesta and DM digestibility. On high-digestibility diets, intake was related to metabolic body size ($W_{kg}^{0.75}$), level of milk production, and DM digestibility. Montgomery and Baumgardt (1965) have shown that in the lower ranges of dietary nutritive value, physical factors may be most important in limiting DM intake, and digestible energy intakes may never reach need. In the upper ranges, chemostatic or thermostatic mechanisms may regulate intake such that energy consumption corresponds to need, while DM intake declines with increasing nutritive value (see Ch. 11, Vol. 2).

White-tailed deer (and presumably other wild ruminants) consume DM in conformity with this model, at least in winter when voluntary food intake of fawns is less than in the previous fall or in the following summer (Ammann et al, 1973). When digestible energy (DE) density of an artificial diet was increased from 1.9 to 3.5 Kcal/g of dry diet, DM intake in grams per $W_{kg}^{0.75}$ increased to a dietary DE density of 2.2 Kcal/g and then declined (Fig. 17-5). A DE density of 2.2 Kcal/g is equivalent to a DM digestibility of about 50%. When fed diets of less than 50% digestibility, physical limitations of the digestive tract would limit DM intake to less than maintenance requirements and the deer would lose weight. While environmental factors, species and animal individuality, and other characteristics of the food may alter this 50% digestibility limit, it is apparent that diets for captive concentrate selectors, such as white-tailed deer, should be more than 50% digestible, particularly for growth and lactation. When diet digestibility ranges from 50 to 70%, DM intakes for maintenance of large roughage eaters typically range from 1.5 to 2.5% of body weight. For maintenance of small concentrate selectors, DM intakes typically range from 3-4% of body weight. Ruminants of intermediate size and dietary habits

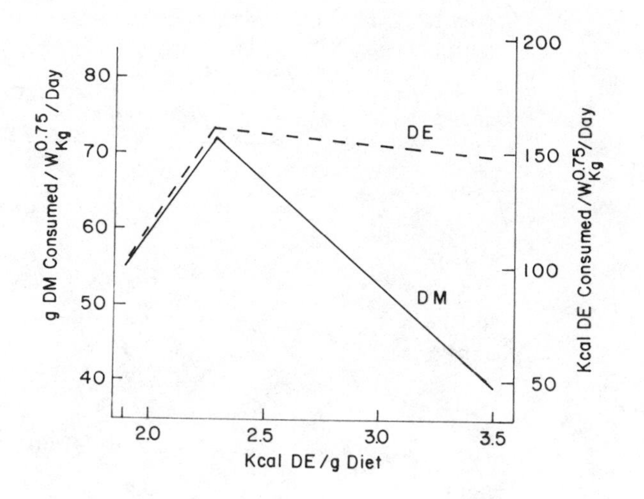

Figure 17-5. Influence of dietary digestibile energy (DE) concentration upon dry matter (DM) and DE consumption/$W_{kg}^{0.75}$/day by white-tailed deer. From Amman (1973).

consume intermediate amounts of DM for maintenance.

Energy

Because ruminants are never in a post-prandial state, they do not conform to those conditions prescribed for measurements of basal metabolism. As a result, the term fasting heat production is frequently used for that minimum energy expenditure measured under practical conditions. Complicating these measurements in wild ruminants is the excitement induced by confinement in a strange environment. Consequently, few reported values are as low as the interspecific mean of 70 Kcal/$W_{kg}^{0.75}$. The value which comes closest is 79 Kcal/$W_{kg}^{0.75}$ determined on red deer in Scotland by Blaxter et al (1974). Maintenance requirements for adult red deer in terms of metabolizable energy (ME) were about 150 Kcal/$W_{kg}^{0.75}$.

Studies of energy utilization by eland and wildebeest have been conducted by Rogerson (1968) in Kenya. Mean estimates of fasting heat production for these two species were 103 and 96 Kcal/$W_{kg}^{0.75}$, respectively. Maintenance ME requirements were estimated to be 20-30% higher than for cattle, or about 160-190 Kcal/$W_{kg}^{0.75}$.

Silver et al (1969) reported that fasting heat production of adult white-tailed deer in the winter was 97 Kcal/$W_{kg}^{0.75}$ with appreciably higher values in the summer. The daily digestible energy (DE) requirements for maintenance

Figure 17-6. A pair of situtunga (African swamp antelope) out in the snow. These thin-skinned antelope can tolerate moderately cold weather if adequate energy is provided. Courtesy of L. LaFrance.

Figure 17-7. A yearling addax calf suckling a female. Sufficient energy must be provided captive animals to prevent excessive weight loss during lactation. Courtesy of L. LaFrance.

or pregnant does in a Michigan winter were found to be 155-160 Kcal/$W_{kg}^{0.75}$ (Ullrey et al, 1969, 1970) with ME requirements of 131 Kcal/$W_{kg}^{0.75}$. Croyle (1969) found that male and female fawns required 168 and 155 Kcal DE/$W_{kg}^{0.75}$, respectively, for maintenance in a temperate environment. Thompson et al (1973) found that fawns required 199 Kcal DE/$W_{kg}^{0.75}$ daily for growth during their first summer and 144 Kcal DE/$W_{kg}^{0.75}$ for maintenance during their first winter.

Protein

Although estimates of protein requirements have been made from protein concentrations of foods eaten in the wild, it is not clear that such values provide an appropriate basis for such estimates. In any case there are large seasonal variations in protein levels in plant food items. Thus, dietary protein concentrations are far from constant over a years time. Controlled studies to define quantitative need are very limited. Protein requirements for growth of white-tailed deer fawns after weaning have been estimated (Ullrey et al, 1967) to be between 14 and 22% (dry basis). The lower value seemed adequate for females, while the requirement for males was higher. Holter et al (1977) have suggested that 11% protein (dry basis) is adequate for yearling deer. Protein requirements for weaned aoudads *(Ammotragus lervia)* were estimated to be no more than 14% on a dry basis (Brady and Ullrey, 1975).

Minerals

Calcium requirements (dry basis) to support growth, skeletal and antler development of weaned white-tailed deer fawns are about 0.45% (Ullrey et al, 1973). Phosphorus requirements (dry basis) are about 0.28% (Ullrey et al, 1975). Requirements for selenium probably do not exceed 0.2 ppm in the presence of 80 IU vitamin E/kg. Iodine requirements do not exceed 0.25 ppm, and cobalt requirements do not exceed 0.1 ppm.

Other Nutrients

Quantitative data are not available.

DIETS USED IN CAPTIVITY

The diets used for captive wild ruminants are nearly as diverse as the numbers of species themselves. This diversity has little relation to anatomical or physiological differences but seems to have evolved for each zoo or wild animal park from some dimly recorded precedent. In few instances are there adequate records on reproduction, health or mortality to permit objective judgement of dietary adequacy. One study (Blakely, 1966) of 73 American and 20 foreign zoos reported that 48 different feeds in a near infinite variety were being fed to hoofed stock. About half of the 174 species or subspecies were wild ruminants. Eighty-seven percent of all animals were fed hay, 63% received grain, 55% were fed commercial formulas, 52% were fed vegetables and 19% were fed fruits. Periodically, diet formulas have been published in the International Zoo Yearbook. Recently, diets have been published for a host of domestic and wild mammals, including new world camelids, giraffids, North American elk, caribou, pronghorn, gazelle, antelope and deer (Rechcigl, 1977).

New World camelids depend largely on grasses for food in their natural environment. When confined, they can eat any forage eaten by other ruminants. Forages such as alfalfa hay, oat hay or barley hay are accepted readily. Concentrate mixtures such as those used for domestic ruminants are also eaten readily. Since camelids are unable to lick, mineral supplements must be incorporated into the diet or provided in granular form rather than as mineral licks. Approximate adult weights of New World camelids are as follows (kg): alpaca, 75; llama, 120; guanaco, 100; vicuna, 40. Daily intakes of 2 kg DM/100 kg body weight, or 65 g/$W_{kg}^{0.75}$, have been estimated from digestion trials to be adequate for maintenance. When confined with no grazing, overgrowth of the lower incisors may be a problem requiring the attention of a veterinary dentist.

Giraffe feeding in captivity seems to be a relatively simple matter. Most are fed a good quality forage, such as alfalfa hay, plus a concentrate mixture. Some also receive fruits and vegetables. When available, freshly cut browse may be helpful in increasing contentment in confined quarters and in reducing vices such as persistent licking of objects in the exhibit. Okapi are fed in a similar manner with greater attention to fruits and vegetables. These succulent foods may be helpful in preventing excessively dry feces and anal prolapse that is sometimes seen when only dry feeds are offered. DM intakes of about 2% of body weight appear adequate for adult maintenance for both species.

Figure 17-8. Okapi *(Okapia johnstoni)* consuming browse placed in the exhibit daily. This helps relieve boredom and may be more important for behavioral reasons than as a source of nutrients.

North American elk *(Cervus elaphus)* depend very heavily on grass in their native environment, but may browse extensively on certain ranges or during certain times of the year, especially in winter and in dry summer periods (see Ch. 16). Alfalfa hay was found to be a satisfactory supplement food for elk on inadequate winter range. Based on their large ruminoreticulum volume, rumen flora and fauna, and ability to digest cellulose, elk may be more akin to cattle than North American deer. DM intakes of about 2% of body weight appear adequate for adult maintenance.

Caribou *(Rangifer tarandus)* are a circumpolar species with catholic tastes. They prefer green vascular plants and mushrooms, but are opportunistic and eat what is available.

Reindeer lichens *(Cladonina alpestris* and *C. rangiferina),* dwarf birch *(Betula nana),* sedge *(Carex aquatilis),* cottongrass *(Eriophorum vaginatum)* and horsetail *(Equisetum fluviatile)* are heavily used in both Eurasia and North America. Caribou are fastidious feeders that strip the leaves from woody stems and pick only new sprouts and finer stem tips when food is abundant. In winter, most types of food will be eaten out of necessity. Lichens may be the primary food in most areas because of their wide distribution and relative abundance, but they are certainly not required and are frequently less nutritious than sedges and grasses. Captive caribou and domestic reindeer have been fed a variety of diets including alfalfa hay. timothy hay, sugar beets, lichens and a variety of grains and commercial feeds. Successful diets should be fairly high in digestibility. Adult females weight about 70-85 kg and adult males about 115 kg. ME requirements for maintenance of acults have been estimated to be in the range of 130-180 $Kcal/W_{kg}^{0.75}$. Daily DM intakes which will supply these amounts of ME approximate 2-3% of body weight.

Pronghorn *(Antilocapra americana)* habitat is very diverse and over 200 plant species are consumed in some areas. When available, big sagebrush *(Artemisia tridentata)* and silver sagebrush *(A. cana)* are most important. Where sagebrush does not occur, other shrubs are used–snowberry *(Symphoricarpos* sp.) in Montana and snakeweed *(Gutierrezia* sp.) in Colorado and Wyoming. Forbs are also used in large amounts, particularly in summer and fall. Grasses are not consumed in large quantities except during early spring or at times of rapid growth following rain. Adult pronghorn have been maintained in captivity on a diet of second cutting leafy alfalfa hay and the pelleted concentrate shown in Table 17-4. Adult males weigh about 40-65 kg while adult females weigh about 30-50 kg. DM intakes of 2-3% of body weight are sufficient for maintenance.

Gazelles are found in eastern and northern Africa, southwest Asia (including parts of the USSR), India, Tibet, Sinkiang and Mongolia. Although taxonomists do not entirely agree, 12 species have been recognized.

Diets of gazelle vary from one part of their range to another. Mountain gazelle are primarily browsers in the desert where they feed on trees and shrubs. In agricultural areas, heavy use is made of crops such as wheat and barley. Thomson's gazelle are primarily grazers while goitered gazelle are primarily browsers. Most species are intermediate feeders, consuming edible plants that are available depending on climate and season. The majority are very selective in their dietary habits, and diets for captive gazelle should accommodate this behavior. Digestible nutrient concentrations need to be relatively high. High quality leafy alfalfa hay is a good source of nutrients and helps prevent restlessness. Many gazelle will eat only the leaves. Concentrate mixtures to be fed with hay should provide 12% or more protein.

Table 17-4. Pelleted pronghorn concentrate.[a]

Ingredient	%
Barley	10
Corn	30
Sorghum grain	5
Oats	7.5
Wheat middlings	6.5
Beet pulp	2.5
Brewer's grain, dried	35.0
Dicalcium phosphate	1.0
Cane molasses	2.5
Vitamin A, D and E premix	+
Trace mineral premix	+
	100
Analysis (DM basis)	
Dry matter	90.6
Crude protein	19.3
Ether extract	2.8
Ash	7.9
Cell wall const.	52.8
Acid detergent fiber	13.6
Lignin	3.6
DM digestibility, %	70.7
DE, Kcal/g	3.45
ME, Kcal/g	3.05

[a]From Carpenter and Baker (1976)

PROPOSED DIETS AND FEEDING SYSTEMS

Complete Pelleted Diet

Two systems which have proven successful in rearing captive wild ruminants are proposed. The first involves feeding a complete pelleted diet containing both roughage and concentrates. Additional hay is not needed for nutrients but may be useful for behavioral or economic reasons. Because only a short time is required to consume sufficient pellets to meet nutrient needs, some animals become restless and develop vices such as persistent licking or chewing of non-food items in their environment. Provisions of hay or browse may increase contentment and minimize this problem. Composition of such a pellet is shown in Table 17-5. It was developed originally for white-tailed deer but has since been used successfully for more than 20 species of wild ruminants. The nutrient composition of this diet is based on the determined or presumed needs of the most demanding periods in the life cycle, i.e., immediate post-weaning when the young are still growing rapidly but don't have access to mother's milk, and lactation when nutrient demands are also particularly high. There may be some nutrient waste when this diet is consumed by animals in other phases of the life cycle when nutrient needs are not so great. However, this waste is compensated for by the reproductive success which results in the simplicity of the system, which has both direct and indirect economic benefits. One can usually obtain a lower unit price for a large order of a single diet than for smaller orders of several diets. In addition, problems of diet identification, storage and errors in feeding are minimized. Ad libitum feeding makes the diet available 24 h/day so that all animals have maximum opportunity to meet their nutrient needs. This is particularly beneficial to small or timid animals that do not compete effectively with dominant pen-mates when feed is available for only a limited time. Species or animals that become overweight on an ad libitum regimen must be fed controlled amounts of this diet. The problem is greatest with older animals which are non-productive, closely confined and members of quiet species such as eland. Reproduction and lactation performance on this diet has been determined in a controlled study. White-tailed does (48) were exposed to bucks (6 does to 1 buck) from November 1 to January 31. At least 94% of the does conceived and 90% gave birth to live fawns (1.74 fawns/doe). Of the live fawns born, 95% lived to at least 90 days of age. All but 9 of these were weighed near the time of weaning at an average age of 120 days. Their average daily gain from birth was 0.18±0.004 kg. Alternative formulas have been developed for monogastric herbivores or for mixtures of ruminants and monogastric herbivores in the same exhibit. These are shown in Table 17-6.

Table 17-5. Complete pelleted[a] wild ruminant diet.

Ingredient	%
Corn cob product[b]	35
Corn grain	18.7
Soybean meal, 48% CP	23.95
Alfalfa meal, dehyd., 17% CP	5
Cane molasses	5
Wheat	10
Soybean oil	1
Trace mineral salt	0.5
Limestone, grd	0.4
Vit A, D, E and Se premix[c]	0.25
Calcium propionate	0.2
	100
Analysis, DM basis	
Dry matter	90
Crude protein	18
Ether extract	2
Cell wall const.	40
Soluble carbohy.	35
Ash	5
Calcium	0.45
Phosphorus	0.32
DE, Kcal/g	3.1

[a]Diameter 5 mm. May constitute the total diet or may be fed with hay.
[b]Consists of bracts and pith (soft parenchyma without vascular bundles). Cell wall constituents 81.2%, acid detergent fiber 37.5% and lignin 6.5%.
[c]Supplies per kg of diet: 3300 IU vitamin A, 220 IU vitamin D, 88 IU vitamin E and 0.2 mg Se from sodium selenite.

Roughage and Concentrates

A second system for rearing captive wild ruminants involves feeding roughage and concentrates separately. Appropriate combinations can meet the nutrient needs of most, and perhaps all, species. A variety of concentrates can be developed using grains that are locally available plus supplements providing protein, minerals and vitamins. An example of a supplement formula which can be mixed with ground grains and the mixture pelleted is shown in Table 17-7. Alternatively, the supplement could be pelleted and then mixed with rolled grains. Example concentrate formulas plus nutrient levels provided when

Table 17-6. Complete pelleted[a] wild monogastric herbivore or ruminant diets.

Ingredient	1	2
Corn cob product[b]	31.55	---
Corn grain	20	36.7
Soybean meal, 44% CP	25	11
Alfalfa meal, dehyd, 17% CP	5	5
Alfalfa meal, sun-cured, 15% CP	---	27
Cane molasses	5	7.5
Wheat	10	10
Soybean meal	1	1
Trace mineral salt	0.5	0.5
Limestone, grd	0.7	0.15
Mono-dical phos, 18% Ca, 21% P	0.8	0.7
Vitamin and Se premix[c]	0.25	0.25
Calcium propionate	0.2	0.2
	100	100
Analysis, DM basis		
Dry matter	92	92
Crude protein	18	18
Ether extract	2	2
Cell wall const.	38	38
Soluble carbohy.	37	37
Ash	5	5
Calcium	0.7	0.7
Phosphorus	0.5	0.5
DE, Kcal/g	3.1	3.1

[a]Diameter 5 mm. May constitute the total diet or may be fed with hay.
[b]Bracts and pith (soft parenchyma without vascular bundles).
[c]Supplies per kg of diet: 3300 IU vitamin A, 220 IU vitamin D, 88 IU vitamin E, 1.1 mg riboflavin, 5.5 mg niacin 3.3 mg pantothenic acid, 12.3 μg vitamin B_{12} and 0.2 mg Se from sodium selenite.

Table 17-7. Supplement for grains for captive wild ruminants.[a]

Ingredient	%
Alfalfa meal, dehyd, 17% CP	10.6
Wheat middlings	6
Soybean meal, 48% CP	65
Mono-dical phosphate, 18% Ca, 21% P	8
Limestone	3
Salt	5
Trace mineral premix[b]	0.25
Selenium premix[c]	1
Vitamin premix[d]	1
Calcium propionate	0.15
	100
Calculated analysis, DM basis	
Crude protein	38
Calcium	3.1
Phosphorus	2.3

[a]May be pelleted, 4 mm diameter.
[b]Supplies per kg supplement: 0.8 mg cobalt, 10 mg copper, 4 mg iodine, 100 mg iron, 80 mg manganese, 140 mg zinc.
[c]Supplies per kg supplement: 0.8 mg Se from sodium selenite.
[d]Supplies per kg supplement: 20,000 IU vitamin A, 2000 IU vitamin D_3, 400 IU vitamin E, 16 mg riboflavin, 160 mg niacin, 80 mg pantothenic acid, 120 μg vitamin B_{12}, 1.2 g choline.

fed in equal parts with hay are shown in Table 17-8. Other combinations sould be developed on the basis of probable need.

Raising Orphans

It is sometimes necessary to raise a young wild ruminant without benefit of its mother. The mother may die during parturition, may reject its offspring or may not produce sufficient milk. Even if one could formulate an artificial milk which chemically duplicates the nutrient composition of milk from that species, it is difficult to mimic the natural nursing pattern in frequency and amount. There are also other types of social interaction between mother and offspring that influence factors as mundane as elimination of urine and feces by the youngster, and as complex as the learned behavior that permits members of the newest generation to successfully enter the social structure of the species. Nevertheless, a number of persons have succeeded, and the following protocol has proved particularly useful in rearing white-tailed deer fawns. It should be useful for many other species as well.

Upon arrival fawns were identified with ear tags and were injected intramuscularly with 1 mg of Se and 50 mg D-α-tocopheryle acetate (1 ml of BO-Se®, Burns-Biotec Laboratories Div., Chromalloy Pharmaceutical Inc., Oakland, CA 94621), 125,000 IU vitamin A and 18,750 IU vitamin D_3 (0.25 ml Injacom®, Roche Chemical Div., Hoffman LaRoche Inc., Nutley, NJ 97110), and 150,000 IU penicillin G benzathine and 150,000 IU penicillin G procain (1 ml Flocillin®, Veterinary Prod., Bristol Lab., Syracuse, NY 13201). They were also treated orally with 200 mg mebendazole by placing Telmin® paste (Pitman-Moore, Washington Crossing, NJ 08560) on the tongue. This latter worming treatment was repeated once weekly for 4 weeks.

The milk feeding schedule is shown in Table 17-9. At first the milk was warmed to 38°C and placed in a bottle fitted with a

Table 17-8. Concentrate formulas for captive wild ruminants.

Ingredients	1	2	3	4
Corn, rolled or cracked	50	55	65	70
Oats, rolled or crimped	20	20	20	20
Cane molasses	5	--	5	--
Wild ruminant supplement (Table 17-7)	25	25	10	50
	100	100	100	100
Calculated analysis, dry basis				
Crude protein	17	17	13	13
Calcium	0.9	0.8	0.4	0.35
Phosphorus	0.8	0.8	0.5	0.5
Calculated analysis (dry basis) when fed in equal parts with hay				
Alfalfa hay, mid bloom				
Crude protein	17	17	15	15
Calcium	1	1	0.9	0.9
Phosphorus	0.5	0.5	0.4	0.4
Alfalfa-bromegrass hay				
Crude protein	16	17	14	14
Calcium	0.9	0.9	0.6	0.6
Phosphorus	0.5	0.5	0.35	0.35
Bromegrass hay, mid bloom				
Crude protein	14	14	12	12
Calcium	0.6	0.6	0.4	0.4
Phosphorus	0.5	0.5	0.35	0.35

Table 17-9. Milk feeding schedule for artificial rearing of white-tailed deer fawns.

Age, days	Frequency	Amt/feeding, ml	Type
1-2	3 h intervals	30- 60	Bovine colostrum
3-4	4 h intervals	85	Bovine colostrum
5	4X daily	100-200	Bovine colostrum, 50%; milk replacer, 50%[a]
6	4X daily	100-200	Bovine colostrum, 25%; milk replacer, 75%
7	4X daily	100-200	Milk replacer
8-20	3X daily	250-350	Milk replacer
21-34	2X daily	850	Milk replacer
35-70	2X daily	1000	Milk replacer

[a]Lamb Milk Replacer®, Land O'Lakes, Inc., Agricultural Services, Fort Dodge, IA 50501. Min. crude protein, 24%; min. crude fat, 30%; max. crude fiber, 0.35%; 44,000 IU vitamin A/kg; 11,000 IU vitamin D_3/kg; 44 IU vitamin E/kg. One part solids diluted with 2½ parts water. Undiluted evaporated bovine milk also works well.

rubber nipple designed for lambs. The fawns were restrained and the nipple forced into their mouths. The longer the fawns had been with the does, the more difficult it was to acclimate them to a rubber nipple. Once all fawns were suckling on their own, a nursing rack was used to hold the bottles at a height appropriate for the fawns. The milk was then fed at room temperature. Nursing bottles and nipples were thoroughly cleaned between each feeding. Normal patterns of urination and defecation were established by gently rubbing the perineal region with a warm, moist cloth at the time of feeding. This was necessary only for a 1-2 week period. In cases of diarrhea, 5 ml of Kaopectate® (The Upjohn Co., Kalamazoo, MI 49007) were added to the milk at feeding. If the diarrhea persisted after 2 days, the percentage of milk solids to water was changed (90% water, 10%

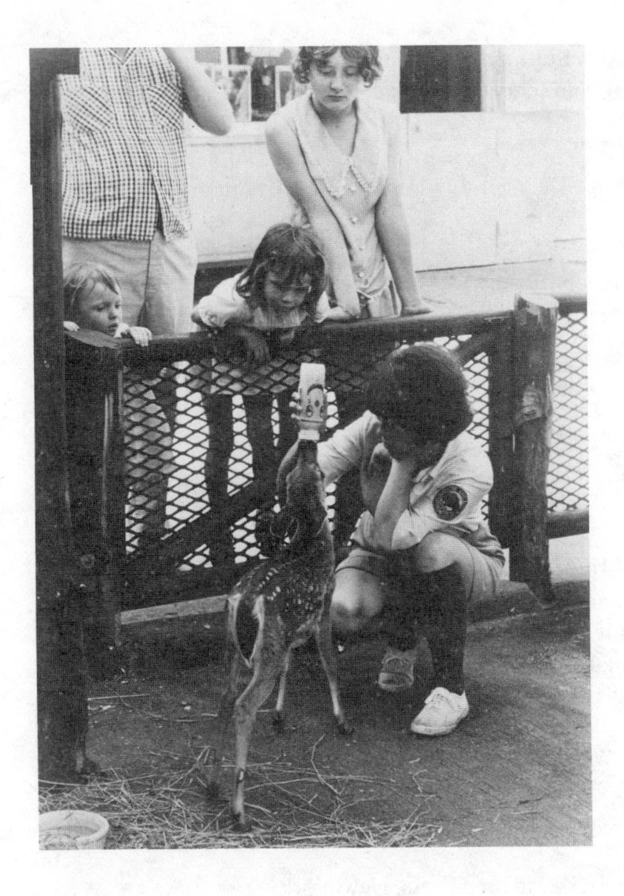

Figure 17-9. Bottle rearing of young ruminants in children's zoos is a common practice. Courtesy of L. LaFrance.

milk solids) and the Kaopectate® treatment continued until improvement was noted. Water was available ad libitum from the beginning. Parasite-free soil was placed in the pen at 1 week, and alfalfa hay, large-toothed aspen and dogwood browse, and the complete pelleted wild ruminant diet (Table 17-5) were provided at 2-3 weeks. Rumination was first observed at 50-60 days of age. Weaning was accomplished between 1½ and 3 mo. of age by gradually reducing the amount of milk provided and by forcing dependence on solid food items.

NUTRITIONAL PROBLEMS

Nutritional problems among captive wild ruminants are not well documented in the scientific literature. Boever (1978) cited only a case of parturient paresis in an eland in his discussion of artiodactyla in Fowler's Zoo and Wild Animal Medicine. This animal was unsuccessfully treated with Ca gluconate. It has been suggested that peracute mortality in giraffe has a nutritional component (Fowler, 1978), although a specific etiological agent was not defined. White-muscle disease has been seen in captive white-tailed deer when dietary Se and vitamin E concentrations were inadequate (Brady et al, 1978). However, the muscle lesions were more common in animals that had been stressed in capture, similar to the capture myopathy described by Harthoorn and Young (1974). Nitrite toxicosis has been described in klipspringer and Speke's gazelle fed hydroponic grasses heavily fertilized with nitrate. Yearling reindeer subsisting largely on lichens during a Norwegian winter have shown a marked depletion of liver vitamin A reserves. The consequence was barren cows, abortion, and weak calves at birth (Skjenneborg and Slagsvold, 1968). Although not captive, buffalo, elk and mule deer inhabiting geothermal areas of Yellowstone Park show evidence of fluoride toxicosis. When forage is scarce during the winter, these animals cluster around the warm springs, eating vegetation which grows there and drinking the high fluoride water (Shupe et al, 1979).

Many nutritional problems are the consequence of careless husbandry. Sudden dietary changes may lead to digestive upsets, rumen atony, diarrhea and other sequelae. Large intakes of concentrates may result in lactic acidosis and, coupled with close confinement, to founder. If young wild ruminants do not have access to a creep feeder or cannot compete successfully for food, inanition or starvation may result. Excess energy consumption in older animals may lead to obesity and impaired reproductive efficiency.

Literature Cited

Aman, A.P., R.L. Cowan, C.L. Motherhead and B.R. Baumgardt. 1973. J. Wildl. Mgmt. 37:195.
Blakely, R.L. 1966. Zoological Hoofed Animal Study. Sedgwick Co. Zoo, Wichita, KS.
Blaxter, K.L., F.W. Wainman and J.L. Davidson. 1966. Animal Prod. 8:75.

Blaxter, K.L. et al. 1974. Farming the Red Deer. Her Majesty's Stationary Office, Edinburgh.
Boever, W.J. 1978. In: Zoo and Wild Animal Medicine. W.B. Saunders Co., Philadelphia.
Brady, P.S. and D.E. Ullrey. 1975. J. Zoo Animal Med. 6:24.
Brady, P.S. et al. 1978. J. Nutr. 108:1439.
Carpenter, L.H. and D.L. Baker. 1976. Middle Park Cooperative Deer Study—Deer Habitat Evaluation. P-R Rep. W-38-R-29. Colorado Game and Fish Dept., Denver.
Conrad, H.R., A.D. Pratt and J.W. Hibbs. 1964. J. Dairy Sci. 47:54.
Croyle, R.C. 1969. M.S. Thesis, Pennsylvania State Univ., University Park, PA.
Fowler, M.E. 1978. J. Amer. Vet. Med. Assoc. 197:1088.
Harthoorn, A.M. and E. Young. 1974. Vet. Rec. 95:337.
Hendrickson, R.C. 1970. Z. Zellf. 109:15.
Hofmann, R.R. 1973. The Ruminant Stomach. East African Literature Bureau, Nairobi, Kenya.
Holter, J.B., W.E. Urban and H.H. Hayes. 1977. J. Animal Sci. 45:365.
Houpt, T.R. 1970. In: Physiology of Digestion and Metabolism in the Ruminant. Oriel Press, Newcastle upon Tyne, England.
Hungate, R.E. 1959. J. Amer. Assoc. Adv. Sci. 130:1192.
Montgomery, M.J. and B.R. Baumgardt. 1965. J. Dairy Sci. 48:569.
Rechcigl, M. (ed.). 1977. CRC Handbook Series in Nutrition and Feed. Section G: Diets, Culture Media, Food Supplements. Vol. 1, Diets for Mammals. CRC Press, Inc., Cleveland, OH.
Rogerson, A. 1968. In: Comparative Nutrition of Wild Animals. Academic Press, New York.
Shupe, J.L., H.B. Peterson and A.E. Olson. 1979. In: Animals as Monitors of Environmental Pollutants. Nat. Acad. Sci., Washington, D.C.
Silver, H., N.F. Colvos, J.B. Holter and H.H. Hayes. 1969. J. Wildl. Mgmt. 33:490.
Simpson, G.G. 1945. Bull. Amer. Mus. Nat. Hist. 88:1.
Skjennebert, S. and L. Slagsvold. 1968. (Reindeer Husbandry and Its Ecological Principles). Universitetsforlaget, Oslo, Norway.
Taylor, C.R. 1968. In: Comparative Nutrition of Wild Animals. Academic Press, New York.
Ullrey, D.E. et al. 1967. J. Wildl. Mgmt. 31:679.
Ullrey, D.E. et al. 1969. J. Wildl. Mgmt. 33:482.
Ullrey, D.E. et al. 1970. J. Wildl. Mgmt. 34:863.
Ullrey, D.E. et al. 1973. J. Wildl. Mgmt. 37:187.
Ullrey, D.E. et al. 1975. J. Wildl. Mgmt. 39:590.

Chapter 18—Therapeutic Nutrition or Veterinary Dietetics

by J.H. Ternouth

When the reader has progressed to this point, he will be aware of the considerable amount of research effort expended in quantitating the nutrient requirements of normal healthy ruminant animals. There has been much less research undertaken on animals suffering from disease, especially diseases with an etiology involving nutrition, except for certain relatively discrete types of metabolic diseases, nutrient deficiencies and digestive tract disorders.

Therapeutic nutrition can be defined from the Shorter Oxford Dictionary as "the science and art of healing by supplying food". Veterinary dietetics may be defined from the same source as "the medical treatment of domestic animals by the regulation of the kind and type of food to be eaten". More practically, Swenson (1962) has stated that therapeutic nutrition "entails providing the essential nutrients deficient in the animals' body, either orally or parenterally, to restore the normal composition of body fluids and tissues". The reference to restoring the normal composition of body fluids and tissues immediately suggests that a subclinical or clinical disorder has been diagnosed, the quantitative physiology of which is known and proper therapy instituted. The therapy may be partly or wholly nutritional.

The average nutritionist is concerned with improving and expanding the information on healthy animals and is not particularly interested or trained in diseases of animals, or their treatment. Most veterinarians are more concerned with pharmacological or surgical treatment of animals and few have a sound working knowledge of nutrition. At a research level, it is often difficult to repeatably induce diseases, and the duration of these diseases is short. The result of this meager research output and the gulf of interest between the nutritionist and the veterinarian is that only limited quantitative information is available to provide a sound basis for recommending adequate nutritional therapeutic regimens.

It is unfortunately true that the lack of popularity of veterinary dietetics as an integral part of therapeutic medicine is due to the inability of commercial companies to make large profits from the sale of therapeutic diets. This may be contrasted with the popularity of chemical therapeutics.

This author concludes that therapeutic nutrition or veterinary dietetics includes: (1) the correction of nutritional deficiencies or excesses (malnutrition); (2) the replacement of body fluids or nutrients to ameliorate or correct a disease or disorder which does not have a nutritional cause; and (3) the feeding of nutrients and drugs (including anti-infectious agents) to aid in the inhibition of specific diseases.

General discussions of the subject may be found in papers by Anon. (1967), Rader (1969), Whitehair (1969, 1970), Mather (1976), Morris (1977) and Blaxter (1979).

THE EFFECT OF NUTRITION UPON INFECTIOUS DISEASES

Food restrictions due to wars and famines commonly result in the increased appearance and frequency of appearance of human and animal diseases. Certain diseases may appear in a more malignant form during a period of malnutrition. Well-fed animals are generally more resistant to bacterial (Whitehair, 1969, 1970; Newberne, 1973) and parasitic diseases (Hungerford, 1975; Blood et al, 1979). However, there are several exceptions to this generality. A number of diseases caused by *Clostridial* spp bacteria cause death among the best fed animals; enterotoxemia *(Clostridium perfringens)* commonly occurs in the lambs and calves which are the fattest in the group. Well-fed animals are more susceptible to some viral infections (Whitehair, 1969, 1970); the incidence of foot-and-mouth disease is greatest in well-fed cattle and occurs in a milder form during periods of food shortage. Haemonchosis causes death more frequently when the nutrition is high and the environmental conditions result in massive infestations (Blood et al, 1979). Anaplasmosis infection has been observed to cause considerable weight-loss in well-fed cattle but not in cattle losing weight (Wilson and Trueman, 1978). In calves and lambs, overfeeding may predispose these young animals to some enteric infections (Roy and Ternouth, 1972).

Garber (1956) and Sprunt and Flanigan (1960) have proposed that there are two environmental influences in any particular ecological niche which affects the equilibrium between host and the microorganism. These are: (a) the nutrient supply to the microorganism; (b) the potential antagonism to the microorganism of the host.

Only when the nutrient supply is high and antagonism low does the microorganism become virulent and the host susceptible to the disease. Whitehair (1969) has listed tissue integrity, antibody production, detoxification ability to the liver and kidneys and maintenance of the reticulo-endothelial system as antagonistic factors which are of considerable importance in limiting the virulence of diseases.

Energy and Nitrogen Deficiency

Energy deficiency, fasting or starvation have been shown to decrease the resistance of mice to several species of bacteria *(Staphylococcus aureus, Klebsiella pneumoniae* and *Shigella flexmeii).* Acute energy deficiency, commonly observed as ketosis of ruminant animals, has been discussed elsewhere (Vol. 2, Ch. 12) and is normally of too short a duration of time for microbial invasion to be involved. In ruminant animals the effects of moderate prolonged energy deficiency are insidious (Blood et al, 1979) and commonly result in: (1) reducing the size of the neonates of energy deficient dams with an increased risk of neonate mortality; (2) inadequate colostrum production, especially from heifers; (3) retarding the growth and delaying the onset of puberty of young animals; and (4) reducing the milk yield and delaying the return to estrus of lactating animals.

These are not clear signs of energy deficiency and the energy deficiency may be secondary to some other deficiency (e.g. protein deficiency). Severe energy deficiency and starvation, commonly observed in drought situations, results in emaciation, unthriftyness and weakness, and if uncorrected, eventually in death.

Much experimental and demographic evidence may be quoted showing that protein deficiencies increases the susceptibility of man and monogastric animals to bacterial, rickettsial and viral diseases (Newberne, 1973). After weaning, ruminant animals have no obligatory need for protein, and thus only when diets deficient in N are fed are deficiency signs likely to occur. The formation of amino acids from simple nitrogenous substances has been dealt with elsewhere (Vol. 2, Ch. 3). Appetite is reduced when low N diets are fed (Baile and Forbes, 1974) so that N and energy deficiencies are commonly confounded in ruminant animals. Impaired immunoglobulin synthesis resulting from N-caloric deficiency is the most widely attributed cause of increased disease susceptibility (Hudson et al, 1974).

Mineral and Vitamin Deficiencies

Specific mineral and vitamin deificiencies may have a marked effect upon the susceptibility of animals to infection as well as upon animal production. Co deficiency is known to cause diarrhea in cattle and sheep and a high incidence of infectious diseases are known to be a sequelae of Fe deficiency (Blood et al, 1979). In cattle, the relationship between P deficiency, pica and botulism is well known. Many mineral and vitamin deficiencies result in significant reductions in food (energy and protein) intake (Baile and Forbes, 1974) so that concomitant energy-protein deficiencies occur.

Nutrient Excesses

Excesses or imbalances of nutrients in diets may also increase the susceptibility of animals to certain infectious diseases. The occurrence of such diseases is not in contrast with the proposals of Garber (1956) mentioned earlier. Because the quantities of nutrient available to the microorganism are large, they directly or indirectly reduce the antagonism of the host to the microorganism. Calves under hygienic research conditions may be fed liquid milk diets ad libitum without diarrhea occurring (Ternouth et al, 1978). However, under field conditions, the feeding of similar large quantities of milk may result in nutritional fermentative or putrefactive diarrhea (Roy, 1969; Ternouth and Roy, 1973). Strains of *Escherichia coli* may be secondary colonizers of the diarrheic intestine.

Excesses of grains in diets may cause non-infectious acute indigestion, bloat and parakeratosis of the ruminal wall. Hepatic foci of *Sphaerophorus necrophorus* are common sequelae in such grain-fed cattle and lambs, the bacteria reaching the liver from the primary lesion in the ruminal wall via the ruminal veins.

EFFECT OF STRESSES AND DISEASES ON NUTRIENT INTAKE AND METABOLISM

The effects of stresses and diseases on the nutrition of any animal is dependent upon the type of tissue affected and on the severity and duration of the stress or disease. Some diseases and stresses have an effect upon a single tissue, which may in itself be diagnostically significant, but most influence a number of tissues or have secondary effects which are evident in other tissues.

Food Intake

Voluntary food intake is used by many veterinarians and animal scientists as a sign of health. Most forms of stress and disease result in some reduction in food intake while an improvement of food intake after a period of inappetance generally indicates that the animal is recovering from the disease or stress.

In ruminant animals fed dry food, there is a common tendency for them to refuse concentrates (grains) and eat only hay (roughage) when moderately diseased or stressed and to become anorexic when the disease is more severe. However, the eating of less concentrates and more hay does not invariably occur; chronic ruminal bloat, due to a high concentrate diet, results in some cattle refusing to eat hay but continuing to eat concentrates.

There are many reasons for reduced food intake associated with disease or stress. Just as the control of food intake is a multifunctional control mechanism (Baile and Forbes, 1974; see also Vol. 2, Ch. 11), so the disease and stress processes may modify the regulatory mechanisms in a variety of ways. Fever (hyperthermia) which occurs as part of many acute diseases may modify the thermoregulatory control mechanisms. Changes of environmental temperature, or body insulation are known to affect food intake, the changes being made in an apparent attempt to maintain energy balance. It appears that toxins produced by infectious agents and circulating metabolites released by damaged tissue cells, may reduce food intake either directly or possibly by the resultant hyperthermia. Dehydration is known to reduce food intake in monogastric and ruminant animals. Dry matter intake is reduced when milk-fed calves have diarrhea. Diarrhea, by causing massive losses of nutrients in the feces (Blaxter and Wood, 1953), results in a complex series of changes in the bloodstream (Fisher and Martinez, 1976) so that dehydration, change in osmolality, ion concentration, acid-base balance, pH or increased circulating metabolites or toxins may be the cause of the reduced food intake (Vol. 2, Ch. 17).

Artificial infection with large numbers of nematode larve have resulted in marked depressions in food intake of sheep (Bawden, 1969; Roseby, 1973). Similarly, Berry and Dargie (1978) have reported depressions in food intake when sheep were artificially infected with metacercariae of *Fasciola hepatica.*

The reason for a reduced food intake is easy to understand when the animal lacks teeth or areas of the gastro-intestinal tract are inflamed, e.g. gingivitis or pharyngitis. Lesions in the oral cavity occur in a number of viral diseases including rinderpest, foot-and-mouth disease, vesicular stomatitis, bovine malignant catarrh in cattle and blue tongue in sheep.

In young animals reduced food intake has an immediate and severe effect upon the animal as their nutrient reserves are limited. In older animals the effect of a reduced food intake will depend upon the size of available body reserves of the animal and the amount of consumed nutrient(s) in most limited supply. When the stress or disease is chronic, then the reduced food intake will result in the appearance of one or a number of nutrient deficiency signs, unless some form of therapeutic nutrition is instituted. If the animal is anorexic, some means (e.g. feeding by stomach tube or by parenteral administration) must be found for supplying the essential nutrient requirements, to allow the animal to recover, and to allow any therapeutic drugs to have an optimal effect upon the disease. Whitehair (1969) points out that adequate nutrition will promote the effectiveness of other therapeutic agents. If the animal is consuming some food, increasing the concentration of nutrients in the diet may be a sufficiently large dietary change to insure that the maintenance nutrient requirements of the animal are met.

The normal order in which nutrients may become deficient in the animal during a period of inappetance is energy, protein, vitamins and minerals. It has been shown that blood proteins decline rapidly and there is ample evidence in man and laboratory animals to indicate that a protein deficiency results in reduced ability to form antibodies (Hudson et al, 1974). Vercoe (1970) found that protein oxidation accounted for about 25% of the fasting heat production (96 h fast) in steers and Butterfield (1966) found that the ratio of loss of dissectible muscle to fat in fasting steers was 1.2:1. In diseased or fasting animals there is almost invariably a *negative* N balance.

In young animals with less nutrient reserve in body tissues, fluids are the first limiting "nutrient", particularly in conditions such as diarrhea where water loss is very high (Vol. 2, Ch. 17). Likewise, blood electrolytes may be rapidly depleted and some, such as Na, are stored to a very limited extent in the tissues. Energy reserves are depleted rapidly also in the young animal, particularly liver glycogen. Furthermore, most young have very limited reserves of fat to draw upon. Vitamin reserves, even the fat-soluble vitamins, are low in the young, although the hepatic reserves of the fat-soluble vitamins may be sufficient for some months in older animals. The water-soluble vitamins are rapidly depleted, particularly in the pre-ruminant animal or in an animal with impaired rumen function.

Gastro-Intestinal Motility

There is an appreciable amount of information relating the effect of some stresses, particularly non-pathogenic ones, to rumen function (Vol. 1, Ch. 17; Vol. 2, Ch. 16 & 17). However, there is much less information relating the incidence of infectious diseases to rumen function. Abnormal ruminal function (secondary indigestion) is associated with a wide variety of causes; disturbed function may develop as a result of bacterial or parasitic disease—such as respiratory tract infection, anthrax, malignant catarrhal fever, coccidiosis, and severe infestations of stomach worms or liver flukes (Jones, 1965; Blood et al, 1979). Secondary indigestion may be associated also with febrile response to inflammatory involvement of a non-gastro-intestinal organ (acute mastitis, endometritis); with dislocations of the abomasum or intestinal tract; with milk fever or grass tetany, ketosis, a variety of plant toxins, and especially with peritonitis and pericarditis arising from traumatic reticulitis.

Motility is the result of the interaction of the stimulation of sympathetic and parasympathetic parts of the autonomic nervous system with the intestinal intrinsic nervous plexuses and musculature. Autonomic imbalance may result in hypermotility or hypomotility. Hypermotility is observed to accompany inflammatory conditions found in the gastro-intestinal tract. The effects of various excitatory and inhibitory influences on the motility of the rumeno-reticulum have been tabulated by Blood et al (1979). Gastro-intestinal nematodes, trematodes and cystodes, bacteria, protozoa, fungi and viruses, either by causing an inflammatory response in the gastro-intestinal wall or by the production of toxins, cause diarrhea, the normal manifestation of gastro-intestinal hypermotility. Parasitic infections have not been observed to cause any change in the rate of passage of water-soluble markers (Roseby, 1977), but have reduced the rate of passage of stained particles (Bawden, 1970). When diarrhea was produced artificially in calves using *Salmonella bovis-morbifans,* antibiotics and a water-soluble marker passed through the large intestine much faster than in healthy calves (Mylrea, 1968). Non-infectious stresses, toxins, laxative foods, cathartics and other drugs may have similar effects. When the integrity of the gastro-intestinal wall is damaged by any type of infectious agent, body fluids and electrolytes (or in severe cases whole blood) pass into the intestine (Mylrea, 1968), greatly increasing the quantity of abnormally watery feces produced by the animal (Blaxter and Wood, 1953; Fisher and Martinez, 1975). When the losses include significant amounts of protein and occur from the upper portions of the gastro-intestinal tract, partial digestion of the proteins by endogenous and microbial enzymes may occur resulting in a putrefactive (foul smelling, gaseous) diarrhea.

Digestion and Absorption

The degree of digestion and absorption occurring at any particular site in the gastro-intestinal tract is dependent upon the suitability of the microenvironment for digestion and absorption and the length of time the nutrients remain in that environment. Thus, hypermotility, by reducing gastro-intestinal transit time, results in reduced food digestibility and absorption.

There are over 100 species of microbes present in significant numbers in the rumenoreticulum (Vol. 1, Ch. 11). The author is unaware of any known direct effects of any infectious condition upon microbial digestion of food. Ruminal acidosis results in a massive change in the ruminal microflora and the disappearance of protozoa which must influence ruminal digestion. The synthesis of cellulase by ruminal microorganisms and its activity upon plant cellulose is suppressed at low pH levels. Rumenitis, hyperkeratosis, parakeratosis and ruminal stasis may all be expected to reduce the absorption across the rumenoreticular wall.

In calves, secretion of abomasal acid and pepsin and pancreatic fluid and enzymes has been reduced by a variety of non-infectious stresses (Tagari and Roy, 1969; Ternouth et al, 1974a; Ternouth and Roy, 1978; Williams et al, 1976). Reduction in the quantities of pancreatic enzymes has reduced the apparent digestibility of milk diets in calves (Ternouth et al, 1974b) and in pigs (Anderson and Ash, 1971; Rerat et al, 1976). Infections of the liver (including ketosis and *Fascioliasis*) which reduce bile production, or infections of the abomasum and small intestine which reduce gastric and enteric enzymes, acid and bicarbonate secretion must also depress digestion and hence absorption of nutrients. Coop et al (1972) has found a depression in the lactase and maltase activity of the intestinal mucosa of sheep following artificial *Nematodirus battus* infection.

In ruminant animals there is massive secretion of minerals and nitrogenous compounds into the gastro-intestinal tract. Na and P are secreted in large quantities in saliva (Vol. 1, Ch. 5), Cl by the abomasum and Na and bicarbonate by the liver and pancreas. These minerals are secreted in isotonic aqueous solutions to aid the process of digestion and any failure of the re-absorption portion of the cycle results in severe body losses (Vol. 2, Ch. 17). Thus, many of the losses associated with diarrhea are due to failure to reabsorb endogenous secretions rather than to absorb dietary nutrients *per se.* The osteoporosis observed in lambs artificially infected with *Trichostrongylus columeriformis* larvae may be the result of chronic P losses (Sykes et al, 1975).

Non-Fecal Losses of Nutrients

Water and some nutrients may be lost by other non-fecal routes in stress and disease situations. In the healthy animal there is an obligatory loss of fluids associated with respiration and a loss in the urine. When animals become dehydrated, urinary fluid losses are reduced to minimal levels. Polyuria (increased urinary flow) occurs when there is nephrosis or nephritis (damage to the renal glomeruli or tubular epithelium). The polyuria is accompanied by a variable loss of blood constituents including protein and electrolytes, depending on the severity of the renal damage.

Hyperthermia, fevers and heat stress increase the respiration and sweating rates resulting in increased water and, in the case of sweating, electrolyte loss.

Tissue Metabolism

Hepatitis, hepatosis and diseases such as ketosis, which result in derangement of liver function, interfere with the storage of vitamins and trace minerals in the liver. Utilization of glycogen and fatty infiltration of the hepatic cells occurs in ketosis. Following periods of high Cu intake, the ingestion of hepatotoxins and acute starvation have resulted in Cu poisoning due to the release of Cu stored in liver cells. Diseases affecting bone may result in bone resorption. The stress associated with the change of body weight also influences bone resorption (Siemon and Moodie, 1973).

Nematode infection has recently been shown to depress the efficiency of utilization of digested energy in cattle (Jordan et al, 1977) and sheep (Reveron et al, 1974), although the observed effects were partly due to reduced food intake and hence a reduced quantum of energy for weight gain.

Acid-Base Balance

The subject of acid-base balance is a complicated one and complete discussion of it is outside the scope of this work. The author would refer the reader to the presentations by Houpt (1970), Michell (1970, 1979), Ardington (1977) and Blood et al (1979).

A derangement tending to reduce pH is defined as acidosis and an actual reduction in blood pH is defined as acidemia. Corresponding increases in pH are defined as alkalosis and alkalemia. When alterations occur in blood and tissue fluid bicarbonate, the condition—either acidosis or alkalosis—is referred to as "metabolic" and when the alteration primarily affects blood carbonic acid and CO_2, it is referred to as "respiratory".

Respiratory alkalosis is caused by hyperventilation which may be associated with such things as fear, high fever, heat stress, some central nervous system lesions and a few other miscellaneous factors. Metabolis alkalosis is, apparently, not much of a problem in ruminant species, although it has been observed in calves which respond by increasing urinary excretion of bicarbonate. Metabolic alkalosis may result from either ingestion or production of excess base or by loss of hydrogen ions (pooled in the abomasum) or by therapeutic administration of bicarbonate, overdosage with mercurial diuretics, and overdosage or hypersecretion of adrenal corticoids or ACTH. In cases of urea toxicity, alkalosis results from absorption of excess ruminal ammonia which cannot be removed rapidly enough by the liver, but there is a partially compensated lacticacidemia and hyperkalemia (Lloyd, 1970). Increased ruminal or blood ammonia levels are associated with a reduction of parotid salivary secretion (Obara and Shimbayashi, 1979). When jugular blood ammonia levels reached 0.28 mmol/ℓ, parotid salivary secretion was inhibited.

Respiratory acidosis is caused mainly be hypoventilation of the lungs and is, apparently, of little consequence in most diseases. Metabolic acidosis, on the other hand, is a very severe problem in many situations for ruminants. Metabolism of compounds such as the S-containing amino acids to S acids which must be excreted via the kidney, cause part of the problem. In addition, relatively large amounts of benzoic acid are consumed in forage. Also, the large amount of volatile fatty acids formed in the rumen and gut, and metabolism of P-containing proteins all add to the acid burden that the body must handle. Ingestion of chemicals such as ammonium chloride, an acid used to prevent and treat urinary calculi, may add to the problem as well as the ingestion of mineral acids that are added to silages in some areas.

In addition to these dietary factors, acidosis (acute indigestion, heavy grain feeding) may develop as a result of excessive lactic acid production in the rumen, particularly in animals not adapted to readily available carbohydrates. Acidosis may also be due to excessive production or incomplete metabolism of acids in extra-ruminal tissues (ketosis, pregnancy toxemia), as a result of restricted acid excretion (renal failure or adrenal insufficiency), or due to loss of bicarbonate as a result of diarrhea or following obstructions which cause alkaline secretions to accumulate in the gut.

The urine of ruminants consuming large amounts of forage is alkaline, largely because of the excess intake and need to excrete cations, primarily K, which must be balanced electrically with a base, primarily bicarbonate. In animals given high levels of grain or other sources of readily available carbohydrate, the urine usually will be acid, partly because of the reduced intake of K, which competes with H for kidney buffers, and partly because of the much larger amounts of H which must be excreted. In the young ruminant on a milk diet, the urine is acid, due, apparently, to the low intake of Na and K and to the acid-forming nature of milk or milk replacers.

Information published by Lebeda et al (1970) indicates that young calves are not any less resistant than older animals to acid or base loads. Their data also indicate that cattle are more resistant to bases than acids, but their systems recover more quickly when loaded with a variety of acids.

In the young calf suffering from diarrhea and, usually, acidosis, the tissues are apt to be dehydrated, large amounts of bicarbonate are lost accomplanied by large amounts of Na and K. Kidney function may or may not be affected, although Dalton (1967) has shown that calves have the ability to concentrate urine much more than many non-ruminant species. This would be beneficial for excretion when the calf is dehydrated due to high fecal excretion or reduced intake of water.

EVALUATION OF THE NUTRIENT FLUX AND STATUS

Before any form of therapy can be instituted, a diagnosis of the cause of the disease or stress must be made. This diagnosis should include estimation of etiology, pathogenesis and prognosis of the disease or stress under circumstances in which therapy is and is not instituted. It is beyond the scope of this chapter to enter into a discussion of differential diagnosis. Readers requiring this type of information are referred to the standard veterinary medicin texts including Hungerford (1975) and Blood et al (1979). Diagnoses are commonly made upon a clinical examination of the patient or herd. A clinical examination normally includes reviewing the history of the appearance of the disease, the present and previous management of the animals, an examination of the environment as well as a visual and physical examination of the patient. This clinical examination may then indicate the need for further pathological, microbiological, hematological or other laboratory type examinations. The difficulty with these laboratory examinations is that they commonly involve a time delay which results in the veterinarian having to make a tentative diagnosis and to institute immediate therapy based upon that tentative diagnosis.

As this chapter is concerned with therapeutic nutrition, the diagnosis and proposed dietary treatment result from an evaluation of the animal's food requirements, the modifications necessary to the normal dietary requirements to ameoliorate the effect of the disease or stress and the animal's present nutrient flux and status. In adult animals the food offered would commonly be intended to supply the maintenance nutrient requirements.

Visual Appraisal

It is possible to make a general evaluation of an animal's condition from visual observations. If such an examination is to be meaningful, it is axiomatic that the person making the evaluation be cognizant of normal body size, condition and conformation of that species, breed and type of animal. For example, a normal Jersey cow may look emaciated compared to an Angus bull.

An experienced herdsman or caretaker will perceive general skeletal and growth aberrations as well as poor conditioning of animals. Skeletal malformations in ruminants are generally due to altered metabolism of, or imbalances or deficiencies of, minerals. Rickets is most evident as bowing of the legs and enlargement of the carpal and tarsal joints. Fluorosis, usually due to ingestion of minerals or forages high in fluorine, causes lameness and exostoses of the long bones.

Young bovine animals suffering from generalized chronic starvation often have concomitant stomach worm infections and will usually have rough hair coats and appear thin or emaciated. The size of their heads, length of tails and volume of their abdomens or bellies will often seem exaggerated. Older animals suffering from generalized starvation, after attaining full growth via a normal plane of nutrition, may appear weak, emaciated and thin. If starvation is prolonged, animals will appear depressed and with their eyes sunken in their orbits. This condition is common in beef herds in overgrazed lands or during wintering on ranges or corn stalk fields. Starvation in sheep may be difficult to evaluate by visual observations, due to the wool covering, and physical palpation is desirable.

A protein deficiency in cattle or sheep may be due to a dietary deficiency *per se* or may be due to infections of blood-sucking internal and external parasites, e.g. gastro-intestinal worms or sucking lice or ticks. Hypoproteinemia (subnormal blood serum protein) is the usual result. A common symptom of this condition is a generalized edema, most evident as a watery swelling of the throat and neck area and best known as "bottle jaw". Sheep heavily infected with *Haemonchus contortus* often may appear only depressed and anemic on superficial examination.

The appearance of the skin and hair coats of cattle is important as an indicator of general systemic health. The hair of healthy cattle will generally appear smooth and bright, often with lick marks. During acute febrile diseases the hair will often stand up from the skin and even appear sweaty. In chronic or wasting disease conditions, the hair of cattle will usually lack luster and the wool covering of sheep will appear dry with occasional missing patches; the skin will seem to adhere to underlying tissues, a condition known as hidebound. On the other hand, during dehydration the skin is loose but will fail to return to a normal position when pulled into a fold.

Rapid or difficult breathing is usually an indication of disease of the lungs or air passages. However, rapid breathing may also occur in toxemias, anemia or alkalosis. The state of exercise or excitement should always be considered when observing animals. Animals which have excess saliva around the mouth may be suffering from infections of the mouth or tongue. Some animals that appear starved may be unable to chew feed.

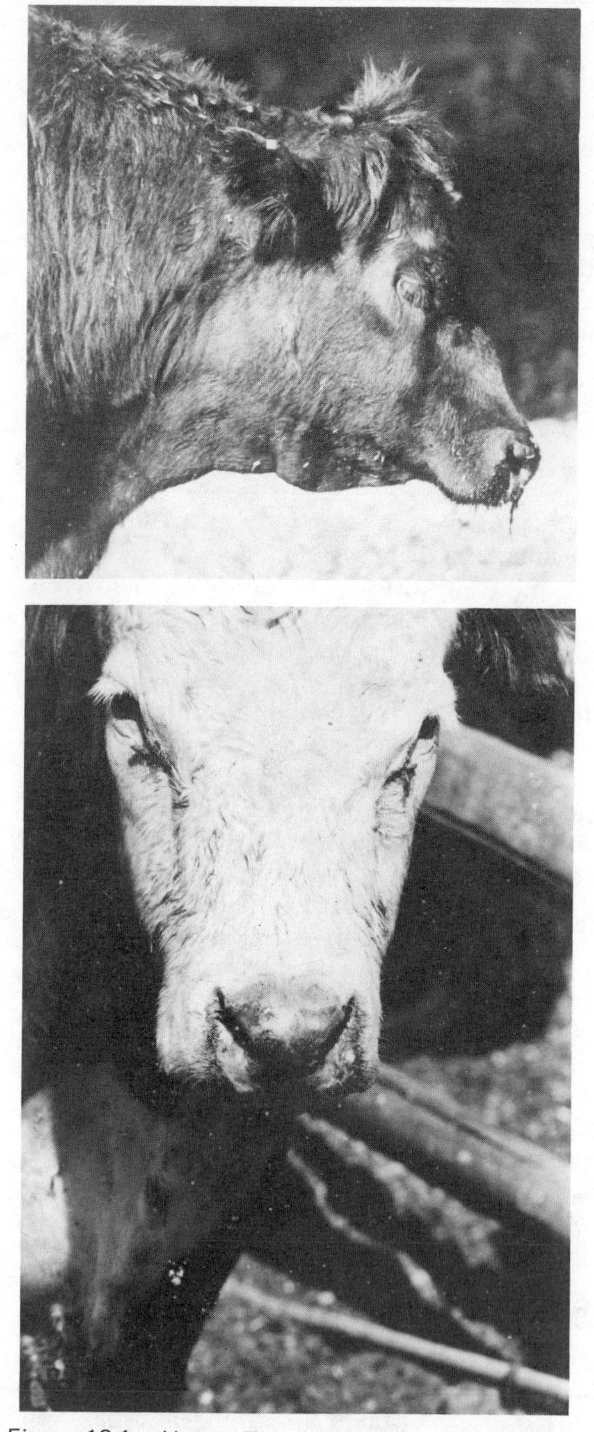

Figure 18-1. Upper. The "typical" symptoms of an animal with a respiratory ailment–half closed eyes, head forward and down to allow easier breathing, a slight nasal discharge and a depressed look. In most cases a pen rider would pull this animal in a hurry. Lower. Here, the symptoms are less obvious–a slight discharge from each eye, plus a dry, dirty-looking nose that usually means a feverish animal. An animal such as this will usually respond quickly to proper treatment. Courtesy of Bill Flemming, Beef Magazine.

Figure 18-2. This steer hasn't started to show eye and nose discharges, yet he is obviously sick. Note the gaunt, "stove-up" appearance, dirty, matted coat, lowered head and a generally depressed appearance. Courtesy of Bill Flemming, Beef Magazine.

Bloat, excluding the common frothy type induced by ingestion of certain leguminous forages, is often an indicator of an animal's condition. Bloat may be due to ruminal stasis resulting from excess ingestion of carbohydrate feedstuffs or of some nitrogenous feedstuffs. Chronic bloating may be due to lesions in the esophagus, or may develop following acute indigestion or after oral medication with antibacterial drugs.

Animals should be observed for signs of diarrhea which, if prolonged, results in serious dehydration and correction of this dehydration is of utmost importance to the survival of the animal. Diarrhea of various types is observed as a common symptom of many infectious diseases, such as Johne's disease, bovine virus mucosal-diarrhea disease, salmonellosis, coccidiosis and colibacillosis. Diarrhea is a common sequelae of gastro-intestinal parasite infections. In addition, diarrhea is a common symptom of various forms of indigestion or toxemias.

The general posture and movement of animals should be observed. Those that appear stiff or tucked-up may be suffering from diseases of the digestive tract, such as traumatic reticulitis, peritonitis, rumenitis or severe enteritis. These signs are common to several nutritional, metabolic, toxicological or infectious disease conditions.

Physical Examination

In veterinary diagnostic situations, the physical examination of the animal follows automatically after the initial visual examination. It commonly involves inspecting in more detail the signs of abnormality initially observed from a distance; for instance, an abdominal swelling may be palpated to check if it is hard and unyielding or soft and malleable, if it is in the subcutaneous tissue or deeper in the abdominal tissues, if it has well defined edges or merges into the surrounding tissue.

During the examination, rectal body temperature, pulse rate and rhythm, and respiration rate and rhythm are checked as indicators of pathological processes. Examination of skin texture and elasticity, as well as palpation, percussion, ballottement, fluid percussion and auscultation of the underlying tissues and organs are other important techniques. Virtually all these techniques are useful in diagnosing gastro-intestinal abnormalities. Rectal examinations, again by palpation and ballottement, are also useful for aiding in diagnosis of abnormalities present in the more caudal portions of the abdominal cavity. More detailed discussion of the useage of these examination techniques are beyond the scope of this chapter and readers are referred to texts on general veterinary medicine.

Quantitative Examination

The nutrients required by an animal may be obtained directly from the food or from some body reserve. Considerable quantities of some nutrients are stored in the body, e.g. energy as fat, Ca and P as bone, Na and water in the rumen, vitamin A in the liver. Other nutrients are not stored in any quantity, e.g. Mg, K and the B group vitamins.

The quantities of nutrient in these reserves may be estimated by various techniques. When a veterinarian judges the "condition" of an animal, he is judging not only the health of the animal but intrinsically its fat reserves or energy status and its lack of dehydration. The quantities of some nutrient reserves may be judged by the concentration in the blood because there is a very close relationship between the body reserves and blood concentration. The quantity in the blood may represent the total body reserve. The use of such blood tests has been widely used in human and animal medicine. Other techniques are available for judging the status of the animal in terms of some other nutrient reserves; e.g. X and γ ray absorption for Ca status of bones (Siemon et al, 1974), bone biopsy and neutron activation for P (Little, 1972; Little and McMeniman, 1973; Whineray et al, 1980), liver biopsy cores for vitamin A, Cu and Fe (Sherman et al, 1959; Andrews et al, 1960; Marston, 1970), Na/K ratios in saliva for Na (Murphy and Plasto, 1973; Rai et al, 1973).

These tests are used to verify a tentative clinical diagnosis. Hypoglycemia (energy deficiency) and depleted bone P have been associated with infertility (McClure, 1968; Little et al, 1978; Somerville et al, 1979), and depressed serum Mg levels with the onset of spring grass tetany (Todd, 1971).

A further technique which is worthy of comment is the truest simple estimate of the flux (as distinct from the status) of a metabolite (in this case energy) through the ruminant animal. Stobbs and Brett (1974) have found that the ratio of short to long-chain fatty acids in milk is a measure of whether this energy is derived directly from dietary energy or from the catabolism of body reserves. When dairy cattle are fed restricted quantities of energy, the molar percentage of short-chain fatty acids (C4-C16) decreased from 72 to less than 60% while the quantity of oleic acid increased from 15 to 26%. The molar percentage of stearic acid also increases although there was no change in the blood glucose levels.

Laboratory Profiles

The use of various individual laboratory analyses to verify the existence of a nutrient deficiency has been discussed earlier. This section is concerned with techniques for monitoring the nutrient flux and/or status of apparently normal animals (i.e. those which are showing no signs of a nutrient deficiency).

The concept of "Metabolic Profiles" has been developed as a branch of preventative nutrition and represents the most comprehensive advance in this direction. The original aim of metabolic profile testing was to provide advice to dairy farmers on the need for nutrient supplementation to achieve optimal nutrient intake for milk production. Instead of waiting for signs of nutrient deficiency to occur in the most susceptible individual members of a herd, representative cattle from three production groups (non-lactating, medium and high producing cows) on a farm are monitored on a regular basis to attempt to detect the occurrence of sub-clinical nutrient deficiencies. Twelve blood metabolites or ions (blood packed cell volume, hemoglobin and glucose and serum urea, Ca, Mg, inorganic P, Na, K, total protein, albumin and globulin) were measured in the initial studies (Payne et al, 1970).

As Payne (1973) has said, theroretically there are no limits to the number of metabolites which can be included in the profile. However, in practice, the numbers of metabolites included will depend upon: (1) the convenience, reliability and cost of the analytical techniques; (2) the reliability of a result based upon single samples taken from a limited number of animals in a production group; and (3) the ability to interpret the results of a profile, with or without other production indices, and thus be able to provide corrective nutrient supplementation for the cattle.

A number of factors may interact which complicate both the diagnosis and resultant treatment of the animals from a metabolic profile. The concentration of a number of blood metabolites and ions are influenced by a number of non-nutritional factors which cause changes which may be confused with nutrient deficiencies. Lindsay (1978) has indicated that handling stress before sampling, the site of sampling and the time of sampling in relation to feeding time have effects upon the concentration of energy and protein metabolites. Season of the year may influence blood metabolite concentration (Rowlands et al, 1974, 1975). Chronic parasitism may cause a significant fall in serum albumin concentration (Sykes and Coop, 1976; Coop et al, 1976), and calving may cause a similar significant reduction in serum albumin (Rowlands, 1978). A recent discussion of blood profiles has been published (Anon., 1978) and provides a broader perspective of the use and function of profiles in the grazing animal industry and research.

The number of metabolites which may be usefully included in a profile as *indicators* of actual or impending limits to animal production will vary depending on the practical situation. The inclusion of other metabolites or ions (β-hydroxybutyric acid, ketone bodies, Fe, Cu, Zn, Mn) may be important in some situations; estimation of serum enzymes and hormones may be valuable in particular circumstances. The overall monetary and labor cost of profile testing has to be measured against the average improvement in production which can be expected from correcting the nutrient deficiencies revealed by the technique. Blowey (1975) has used a limited profile (involving measuring glucose, urea and albumin only) to provide information on the energy and N status of cattle. Blowey emphasizes that such blood analyses are *an aid* in the investigation of some aspects of nutrition and from frequently repeated analyses useful *normal* standards have been developed. The interpretation of abnormal profiles needs careful differential diagnosis and an interdisciplinary approach involving the veterinarian, nutritionist and farmer is desirable (Wilson and Meed, 1978).

PRINCIPLES OF PREVENTATIVE AND THERAPEUTIC NUTRITION

This section is intended to provide some information and suggestions on preventative therapeutic nutrition. It may be pointed out that information in many cases is quite sparse and some of the suggestions may or may not have a foundation of experimental or practical knowledge. Hopefully, this deficit will be remedied in the relatively near future.

A general recommendation for stressed and diseased animals that have been off feed for some days or for those that have had extensive treatments with oral antibiotics or sulfa drugs is that transplantation of fresh rumen fluid from a normal animal to the diseased animal is valuable (except for the milk-fed young). There are numerous examples where this type of treatment has been considered to hasten recovery. Jones (1965) points out that many European veterinary practitioners routinely carry fresh rumen ingesta which may be fortified with trace minerals in areas where deficiencies are common. It would seem likely that many large feedlots or dairies could very well afford to maintain a rumen-fistulated animal to serve as a donor.

Preventative Nutrition

The need for preventative nutrition will be diagnosed from the occurrence of a nutrient deficiency in one or more members of a group of animals, from the results of a clinical examination or from profile testing. The therapy will consist of adjusting the amount of the nutrient ingested for a primary deficiency or correcting the primary and secondary effects in the case of a secondary deficiency. For instance, Co deficiency may be corrected by the use of Co bullets (Vol. 2, Ch. 5), by drenching, by vitamin B_{12} injection, or less directly by distributing Co sulfate on the pasture in a fertilizer. When iodine deficiency (goiter) in sheep is due to the effects of goitrogens, a change of diet to reduce the intake of goitrogens is necessary as well as feeding extra iodine. Thiouracil, thiourea sulphonamides and L-5-vinyl-2-thioxazolidone, the latter found in a number of plants in the *Brassica* spp., are known to be goitrogenic (Kingsbury, 1964). The complex interrelationships between Cu, Mo and sulfur are well known. Although an observed Cu deficiency may be due to high levels of Mo and sulfur reducing the absorption of the Cu, the Cu deficiency may have to be treated as though it were a primary deficiency (Whitelaw et al, 1979).

Therapeutic Feeding

Nutrients may be supplied by voluntary or involuntary feeding or by various parenteral routes. Intravenous, intramuscular and subcutaneous injections of nutrients are the most regularly used parenteral routes.

When the ruminant animal will voluntarily ingest food, the therapeutic nutrients may be added to its diet. Extra protein supplement may be added to grain when the protein content of the pasture falls. Similarly, extra P may be added to the grain fed to cattle grazing on P deficient pastures. In some cases, an attractant may be needed to encourage the cattle to consume the supplementary nutrient. Molasses, molassine meal, grains and dried fodders are widely used attractants which may have the effect of disguising certain unpalatable nutrients.

Involuntary feeding involves the use of drenching guns or esophageal tubes. Virtually all supplementary nutrients fed by tube are given as liquids, the major exceptions being the Co bullet and the bloat capsule. Cattle and sheep are commonly drenched or injected with vitamin A-D emulsions in the middle of the dry season of the year.

Parenteral nutrition may involve complete, total or supplementary feeding (Shenkin and Wretlind, 1978). Complete parenteral nutrition involves feeding the animal its *entire* nutrient requirements by intravenous injection. There are no reports of this being attempted in ruminant animals although complete intravenous feeding of dogs throughout a pregnancy has been successfully completed. Complete intravenous feeding in preruminant calves has been described by Hoffsis et al (1977) and of dogs by Carter and Freedman (1977). The intravenous infusion technique has been described by Sherman et al (1976). Total parenteral feeding involves feeding the animals' total requirements of some nutrients by a parenteral route. Total parenteral feeding is a short-term alternative to oral feeding and the aim is to cover the basal requirements of the most immediately necessary nutrients and to compensate for increased losses or deficiencies due to disease or stress. For example, when a calf has severe diarrhea with dehydration and the veterinarian decides to use total parenteral feeding, intravenous infusions of isosmolar fluids containing Na, K, bicarbonate, P and glucose are used for 1-3 days. During this time no milk is fed and the animal is dependent upon the parenteral infusions for energy. No attempt is made to supply supplementary amino acids, other minerals or vitamins for the short period of the parenteral feeding. It is assumed that the calf has sufficient reserves of these other nutrients to last through the parenteral feeding period.

Supplementary parenteral feeding may involve injecting or infusing a single nutrient. These are most commonly single micronutrients or vitamins mixed in some slowly-released adjuvant. The route by which the additional nutrients are administered is mainly dependent upon the efficacy and ease of administration. When the animals are hand-fed or are fed a supplementary feed, extra therapeutics may be included in the diet. Thus correction of nutrient deficiencies may be made by injection of slowly-released formulations of the nutrient, e.g. supplying Cu as Cu glycinate injections or Co as Co bullets or by adding these minerals to the diet. On the other hand, many minerals may be spread on the pasture incorporated in a fertilizer, resulting in some cases in beneficial effects on pasture growth as well.

SPECIFIC NUTRITIONAL THERAPIES

Morris (1977) lists 23 general disease conditions in monogastric animals in which nutrition may be a beneficial component of the treatment of the animal. Many of these therapeutic nutritional regimens cannot be used in ruminants. This is because economic considerations are normally of more significance than for dogs and cats and the rumeno-reticulum complicates the relationship between ingested and absorbed nutrients. In addition, for adult cattle, the sheer volume of sterile fluids necessary to provide effective parenteral fluid therapy is large.

There are a number of distinct types of therapeutic nutrition which are important in ruminant animals. In decreasing order of importance in field situations, they are: fluid balance, blood electrolyte balance, energy supply, amino acid supply, vitamin supply and other (non-electrolyte) mineral supply. All are included in complete parenteral feeding regimens.

Fluid Therapy

Negative fluid balances result in the need for fluid therapy in many disease and stress conditions, including especially disorders resulting in diarrhea and fevers. Animals with diarrhea and high fevers may also have acidosis.

Estimations of the degree of dehydration are made on the basis of clinical signs (Ardington, 1977; Blood et al, 1979), although the latter have recommended also the use of packed cell volume and total serum solids. These two latter measurements will be satisfactory, provided the animal was not initially anemic or hypoproteinemic. The signs of dehydration will include muscular weakness, poor venous filling, dry mucous membranes, concentrated urine, reduced capillary refilling, reduced skin tissue turgour and elasticity, concentration of proteins, cells and ions in the blood and sunken eyes. As the degree of dehydration starts to exceed 5%, obvious depression is evident, the eyes become sunken in their sockets and, if the skin on the upper eyelid or the neck is folded, the fold remains folded for 2-4 secs. Higher degrees of dehydration may result in the eyes becoming more severely sunken and the folded skin remains 'tented' for longer—up to 45 secs.

Replacement fluids may be given orally by stomach tube in relatively mild dehydration ($<5\%$) and by the intravenous route if the degree of dehydration is more severe. One problem that has been recognized recently is that electrolyte solutions are poorly absorbed from the gastro-intestinal tract of calves with diarrhea. This occurs even though there is evidence showing that colonic absorption is increased. Recently, Bywater (1977) has reported that the addition of glycine to glucose-electrolyte solutions enhances the absorption of water and Na from the intestine of the diarrheic calf. Quickest and most effective recovery of calves from experimentally induced diarrhea occurred when a glucose-glycine-electrolyte solution containing 21.6 g glucose, 3.5 g glycine, 4.8 g NaCl, 2.3 g KH_2PO_4 and 0.07 g K citrate/ℓ were given orally with an antibiotic for 4 days.

Dalton's (1967) formula for oral replacement fluids for calves recognizes the need to replace Na and P losses, to combat the acidosis and to supply K and glucose which are depleted largely by the associated anorexia. Ardington (1977) stresses that *prolonged* anorexia results in hypokalaemia because cattle, in particular, are inefficient conservers of K and ruminant diets commonly have a high K content. However, some caution is necessary in the inclusion of K in parenteral infusions for fluid replacement resulting from diarrhea as the animal is initially hyperkalemic. The hypokalemia may be exacerbated by hemolysis, severe tissue destruction and diarrhea.

Parenteral (intravenous) fluids are given immediately when severe degrees of dehydration are evident and subsequently maintained orally. A technique for long-term parenteral feeding of calves has been described (Sherman et al, 1976).

Bicarbonate should be given parenterally, not orally, to animals with functional rumens. Ardington (1977) states that it is safe to give 4 mEq bicarbonate/kg liveweight by rapid intravenous infusion. In the case of grain engorgement, as well as intravenous bicarbonate infusions, Mg hydroxide is pumped into the rumen in mild or moderate cases but in severe cases a rumenotomy is performed, the ruminal contents washed out and 20-10 ℓ ruminal contents transferred from a donor animal.

Ruminant animals do have mechanisms for precisely regulating water and Na intake. As the animal starts to recover from the acute phases of the disease, provision of measured amounts of clean water and electrolyte solutions may be left in separate containers. Animals which are not drinking should be given further intraruminal infusions of isosmolar solutions at least once daily. The electrolyte solution should contain the equivalent of 80 g NaCl and 20 g KCl in 10 ℓ water (Ardington, 1977). The quantities of water and electrolyte solution consumed are monitored throughout the animals' recovery.

The energy intake of animals, especially young animals, is critical as liver glycogen reserves are depleted in less than 24 h of the onset of anorexia and β-oxidation of depot fats results in ketosis and acidosis. Energy requirements of animals are greatly increased during late pregnancy and early lactation, after shearing, during low ambient temperatures and in febrile conditions. Energy supplementation by the inclusion of glucose in parenteral and oral solutions is a valuable part of the therapy following anorexia associated with any of these conditions.

CORRECTION OF FEEDING AND NUTRITIONAL ABNORMALITIES

Anophagia, aphagia (anorexia), polyphagia and allotriophagia (pica) all occur in ruminant animals. In veterinary medicine it is important to overcome the reason for the abnormality in feeding behavior as well as to correct the nutrient deficiency or imbalance which results from the abnormality. When animals are aphagic for several days, fluid therapy may be used to supply essential nutrients (see above) and ruminal liquor from a normal ruminant infused into the rumen by stomach tube (Blood et al, 1979).

Making general recommendations on the most palatable type of food to offer to aphagic ruminants is not easy as the palatability of feeds is a sensation which is subjective to the animal. If the aphagia is diagnosed as being due to inflammation of the oral cavity, pharanyx or esophagus, the most acceptable feeds will be soft, low in fiber and free from spines and awns. Should the aphagia be due to grain engorgement, hay should be offered. Generally speaking, it takes at least 2 weeks for the digestive tract of ruminants to completely adapt to a substantial change in a feeding program. Conditions which involve diseases which compromise the digestion or absorption of the nutrients in the diets result in corrective nutrition being required for this time. Laxative foods (e.g. bran) would not be recommended for animals with diarrhea.

Rumen pH generally varies between 5.0 and 6.5 in healthy animals. Lower pH values are associated with high grain rations and the higher pH values are experienced with high roughage (especially poor quality) and high protein rations, particularly if urea is included. The carbohydrates are fermented with the production of excess amounts of lactic acid. The pH may drop to the range of 3.5 to 4.5. Ruminal stasis, acidosis and dehydreation then follow. On the other hand, fasting animals may have ruminal pH readings of 6.8 to 7.3. Despite this, anorectic animals, especially those with diarrheas, may have a metabolic acidosis. Because of this, concentrates should be limited in rations of stressed and diseased animals. Moreover, these animals nearly always prefer to eat forages or bulky rations.

The levels of the carbohydrate feed grains—such as corn, barley and sorghum—should be limited to 30% of dry weight or 40% (or less) of the caloric allowance of the total daily ration in diseased or stressed animals. Fine grinding of grains should be avoided and hays are best fed whole (not chopped). Whole rolled oats and wheat bran are useful in convalescent rations since they are palatable and do not depress rumen pH greatly. If corn silage is fed, it should be limited to 25% to 40% of the forage. Hay and grass pastures can and should be fed ad lib. Feeding a dry ration containing 55 to 70% TDN (2.0-2.5 Mcal ME/kg) at the daily rate of 1% of live body weight will usually maintain a morbid or convalescing animal. Force-feeding by intraruminal infusions is difficult, but the technique will often save anorectic animals.

Proteins are important in replacing normal N losses. The need for proteins undoubtedly increases when there is accelerated loss of tissue and fluid proteins. Diarrheas, especially hemorrhagic, deplete plasma proteins. The S-containing amino acids are important in healing processes. NPN sources probably should be avoided in convalescent rations. If they are fed, supplemental sulfur is indicated. Generally speaking, a protein level of 120 to

140 g/kg is probably adequate for diseased older animals.

Vitamins are necessary for ruminant metabolism, but supplementation with water-soluble vitamins is usually not considered important in adult ruminants. Vitamin A is associated with resistance to infection, especially of epithelial tissues. It is thus important in the prevention and treatment of diseases of the respiratory and gastro-intestinal tracts. If vitamin A stores are low or the nutritional status of vitamin A is unknown, each animal should receive daily 20,000 to 50,000 IU/head/day for 5-7 days to aid in the prevention of or recovery from the disease.

The water-soluble vitamins may benefit anorectic adult ruminants and are best administered in drinking water with electrolytes. It seems that in the future, certain vitamins may prove beneficial or even critical for ruminants. Thiamin given intramuscularly appears to aid in the treatment of polioencephalomalacia of cattle. Ascorbic acid is associated with the metabolism of adrenal steroids. Perhaps administration of the vitamin will prove beneficial during stress situations in cattle and other ruminants.

The requirements for non-electrolyte (elements other than Na and K) minerals are not generally believed to be critical in animals afflicted with non-nutritional diseases. Trace mineralized salt should be fed ad lib in most situations, as should Ca and P mineral sources. Of course, it is possible that a metabolic disease involving a mineral, such as Mg in grass tetany, may exist concurrently with an infectious disease. Then the mineral will be important in therapeutic nutrition.

Animals severely affected with internal parasites may suffer from a reduced food intake coupled with energy, protein, Ca, P, Mg, Fe, Co and vitamin A deficiencies. In addition, there may be some impairment of digestive and absorptive function. Therapy therefore should include supplementation with minerals and vitamin A, the feeding of readily digested carbohydrate and protein. Perhaps milk proteins should be considered even though their general use in ruminant animals is expensive and wasteful. In addition, appropriate anti-parasitic drugs will be administered.

The frequency of feeding of animals during disease and stress convalescence is important. All too frequently animals are allowed to eat large meals which expands the gastro-intestinal tract to an unnecessary degree and overtaxes the digestive and absorptive function of the tract, especially if the animal is hungry. Feeding little and often should be the preferred technique. When the animals are being fed 1% of liveweight daily (see above) this should be split into four meals spaced out as evenly as possible during the day.

In conclusion, it can be said that therapeutic nutrition regimens must be tailored to the individual disease or stress situation observed and diagnosed in the animal. The success of any therapy is dependent upon accurate diagnosis and a quantitative knowledge of the nutrient imbalances.

Literature Cited

Anderson, D.M. and R.W. Ask. 1971. Proc. Nutr. Soc. 30:34A.
Andrews, E.D., L.I. Hart and B.J. Stephenson. 1960. N. Z. J. Agr. Res. 3:364.
Anon. 1967. Merck Veterinary Manual. Merck & Co., Inc., Rahway, NJ.
Anon. 1978. The use of blood metabolites in animal production. Brit. Soc. Animal Prod. Occ. Publ. 1.
Ardington, P.C. 1977. J. S. Afr. Vet. Assoc. 48:215.
Baile, C.A. and J.M. Forbes. 1974. Physiol. Rev. 54:160.
Bawden, R.J. 1969. Aust. J. Agr. Res. 20:589.
Bawden, R.J. 1970. Brit. J. Nutr. 24:291.
Berry, C.I. and J.D. Dargie. 1978. Vet. Parasitol. 4:327.
Blaxter, K.L. 1979. Vet. Rec. 104:595.
Blaxter, K.L. and W.A. Wood. 1953. Vet. Rec. 65:889.
Blood, D.C., J.A. Henderson and O.M. Radostits. 1979. Veterinary Medicine. 5th ed. Bailliere Tindall, London.
Blowey, R.W. 1975. Vet. Rec. 97:324.
Butterfield, R.M. 1966. Res. Vet. Sci. 7:168.
Bywater, R.J. 1977. Amer. J. Vet. Res. 38:1983.

Carter, J.M. and A.B. Freedman. 1977. J. Amer. Vet. Med. Assoc. 171:71
Coop, R.L., C.J. Mapes and K.W. Angus. 1972. Res. Vet. Sci. 13:186.
Coop, R.L., A.R. Sykes and K.W. Angus. 1976. Res. Vet. Sci. 21:253.
Dalton, R.G. 1967. Brit. Vet. J. 123:237.
Fisher, E.W. and A.A. Martinez. 1975. Brit. Vet. J. 131:643.
Fisher, E.W. and A.A. Martinez. 1976. Vet. Ann. 16:22.
Garber, E.D. 1956. Amer. Naturalist 90:183.
Hoffsis, G.F., D.A. Gingerich, D.M. Sherman and R.R. Bruner. 1977. J. Amer. Vet. Med. Assoc. 171:67.
Houpt, T.R. 1970. In: Duk's Physiology of Domestic Animals. 8th ed. Cornell Univ. Press, Ithaca, NY.
Hudson, R.J., H.S. Saben and D. Emslie. 1974. Vet. Bull. 44:119.
Hungerford, T.G. 1975. Diseases of Livestock. Angus and Robertson, Sydney.
Jones, L.M. 1965. Veterinary Pharmacology and Therapeutics. 3rd ed. Iowa State Univ. Press, Ames.
Jordan, H.E., N.A. Cole, J.E. McCroskey and S.A. Ewing. 1977. Amer. J. Vet. Res. 38:1157.
Kingsbury, J.M. 1964. Poisonous Plants of the United States and Canada. Prentice-Hall, Inc.
Lebeda, M., J. Bouda and A. Kucera. 1970. Acta Vet. Brno. 39:415, 427.
Lindsay, D.B. 1978. In: The use of blood metabolites in animal production. Brit. Soc. Animal Prod. Occ. Pub. 1.
Little, D.A. 1975. Aust. Vet. J. 48:668.
Little, D.A. and N.P. McMeniman. 1973. Aust. J. Expt. Agr. An. Husb. 18:514.
Lloyd, W.E. 1970. Ph.D. Thesis, Iowa State Univ., Ames, Iowa.
Marston, H.R. 1970. Brit. J. Nutr. 24:615.
Mather, G.W. 1976. Southwestern Vet. 29:40.
McClure, T.J. 1968. Brit. Vet. J. 124:126.
Michell, A.R. 1970. J. Amer. Vet. Med. Assoc. 157:1540.
Michell, A.R. 1979. Vet. Rec. 104:542, 572.
Morris, M.L. 1977. In: Current Veterinary Therapy. W.B. Saunders Co., Philadelphia.
Murphy, G.M. and A.W. Plasto. 1973. Aust. J. Exp. Agr. An. Husb. 13:369.
Mylrea, P.J. 1968. Res. Vet. Sci. 9:5, 14.
Newberne, P.M. 1973. Adv. Vet. Sci. Comp. Med. 17:265.
Obara, Y. and K. Shimbayashi. 1979. Brit. J. Nutr. 42:497.
Payne, J.M. 1973. In: Production Disease in Farm Animals. Bailliere Tindall, London.
Payne, J.M., S.M. Dew, R. Manston and M. Faulks. 1970. Vet. Rec. 87:150.
Rader, W.A. 1969. Mod. Vet. Pract. 50(13):41.
Rai, G.S., O.N. Seth, M.D. Pandey and J.S. Rawat. 1973. Indian Vet. J. 50:512.
Rerat, A., T. Corring and J.P. Laplace. 1976. In: Protein Metabolism and Nutrition. European Assoc. Animal Prod. Publ. 16, Butterworth, London.
Reveron, A.E., J.H. Topps, D.C. MacDonald and G. Pratt. 1976. Res. Vet. Sci. 16:299.
Roseby, F.B. 1973. Aust. J. Agr. Res. 24:947.
Roseby, F.B. 1977. Aust. J. Agr. Res. 28:155
Rowlands, G.J. 1978. In: The use of blood metabolites in animal production. Brit. Soc. Anim. Prod. Occ. Pub. 1.
Rowlands, G.J., W. Little, R. Manston and S.M. Dew. 1974. J. Agr. Sci. 83:27.
Rowlands, G.J., R. Manston, R.M. Pocock and S.M. Dew. 1975. J. Dairy Res. 42:349.
Roy, J.H.B. 1969. Proc. Nutr. Soc. 28:160.
Roy, J.H.B. and J.H. Ternouth. 1972. Proc. Nutr. Soc. 31:53.
Shenkin, A. and A. Wretlind. 1978. World Rev. Nutr. Diet. 28:1.
Sherman, D.M., G.F. Hoffsis, D.A. Gingerich and R.R. Bruner. 1976. J. Amer. Vet. Med. Accos. 169:1310.
Siemon, N.J. and E.W. Moodie. 1973. Nature 243:541.
Siemon, N.J., E.W. Moodie and D.F. Robertson. 1974. Calc. Tissue Res. 15:189.
Skerman, K.D. et al. 1959. Amer. J. Vet. Res. 20:977.
Somerville, S.H., B.G. Lowmann and D.W. Deas. 1979. Vet. Rec. 104:95.
Sprunt, D.H. and C. Flanigan. 1960. Adv. Vet. Sci. 6:79.
Stobbs, T.H. and D.J. Brett. 1974. Aust. J. Agr. Res. 25:657.
Swenson, M.J. 1962. J. Amer. Vet. Med. Assoc. 141:1353.
Sykes, A.R. and R.L. Coop. 1976. J. Agr. Sci. 86:507.
Sykes, A.R., R.L. Coop and K.W. Angus. 1975. J. Comp. Path. 85:549.
Tagari, H. and J.H.B. Roy. 1969. Brit. J. Nutr. 23:763.

Ternouth, J.H. and J.H.B. Roy. 1973. Ann. Rech. Veter. 4:19.
Ternouth, J.H. and J.H.B. Roy. 1978. Brit. J. Nutr. 40:553.
Ternouth, J.H. et al. 1974b. Brit. J. Nutr. 32:37.
Ternouth, J.H., J.H.B. Roy and R.C. Siddons. 1974a. Brit. J. Nutr. 31:13.
Ternouth, J.H., I.J.F. Stobo and J.H.B. Roy. 1978. Proc. Nutr. Soc. 37:85A.
Todd, A.J. 1971. Vet. Rec. 88:645.
Vercoe, J.E. 1970. Brit. J. Nutr. 24:599.
Whitehair, C.K. 1969. In: Animal Growth and Nutrition. Lea & Febiger, Philadelphia.
Whitehair, C.K. 1970. In: Bovine Medicine and Surgery. Amer. Vet. Pub., Inc.
Whitelaw, A., R.H. Armstrong, C.C. Evans and A.R. Fawcett. 1979. Vet. Rec. 104:455.
Williams, V.J., J.H.B. Roy and C.M. Gillies. 1976. Brit. J. Nutr. 36:317.
Wilson, A.J. and K.F. Trueman. 1978. Aust. Vet. J. 54:121.
Wilson, P.N. and R.K. Medd. 1978. In: The use of blood metabolites in animal production. Brit. Soc. An. Prod. Occ. Pub. 1.
Whineray, S., B.J. Thomas, J.H. Ternouth and H.M.S. Davies. 1980. Inter. J. Appl. Rad. Isot. (in press).

Chapter 19—Nutrionally-Related Aspects of Pesticide Toxicity on Ruminants

by Lee R. Shull

According to the Federal Environmental Pesticide Control Act, a pesticide is "... (1) any substance or mixture of substances intended for preventing, destroying, repelling or mitigating any pest (insect, rodent, nematode, fungus, weed, other forms of terrestrial or acquatic plant or animal life or viruses, bacteria, or microorganisms on or in living man or other animal, which the Administrator declares to be a pest, and (2) any substance or mixture of substances intended for use as a plant regulator, defoliant or desiccant" (Hayes, 1975). Pesticides are an essential part of modern agriculture. Without them present world food demands would be difficult to meet. The prognosis is that use of pesticides must continue so that world food production and disease control programs can keep pace with world population growth. At the farm level the benefits from pesticides are a matter of simple economics. Headley (1968) calculated that for $1 spent on US farms for pesticides an additional $4 of product was produced. In England it is 5 units of production increase for every 1 unit expended for pesticides (Strickland, 1970). Although pesticides are intended to improve the environment that crops and livestock occupy, untoward effects cannot be totally avoided.

This chapter is intended to be a summarization of the nutritionally-related aspects of pesticide toxicity in ruminants. The relationship between pesticide exposure and animal performance is emphasized. The primary concern of the stockman is whether exposure will result in immediate or lasting adverse effects on performance. More specifically, pesticides that modify feeding behavior can change the profile of ingested nutrients; those that change the physiology or pathology of the GI track can modify the digestion and absorption of nutrients; finally, those that change the physiology or pathology of metabolic organs can alter the assimilation, utilization, storage, and excretion of nutrients. The inverse aspect is that nutritional status can be an important predisposing factor to pesticide toxicity.

During the course of the chapter reference to various pesticides will be made. See Table 19-12 for further information.

GENERAL TYPES OF PESTICIDES AND COMMON SOURCES OF EXPOSURE

Because they are the most widely dispersed, insecticides represent the greatest potential hazard to ruminants. Insecticides are composed primarily of organochlorine, organophosphate and carbamate compounds. They may be applied topically (dermally) to animals, introduced into the GI tract, or applied to feed crops. Whenever over-exposure occurs, it is generally the result of (1) the user not adhering to directions for safe dosage or usage, (2) accidental administration of the wrong chemical or (3) allowing livestock access to areas where insecticides are freshly applied or spilled. Aerial application is a common source of inadvertent contamination due to drift directly onto livestock areas or forage crops.

Herbicides are chemically either organic or inorganic. The organic types are used more extensively in agriculture and horticulture. Members of the herbicide family include the chlorophenoxy derivatives of fatty acids, amides, arsenicals, dipyridyl compounds, substituted urea compounds, thiocarbamates, triazines, benzoic acid derivatives, methyluracils, polychlorobicyclopentadiene isomers, pichloram, and dinitroaniline compounds. In general, exposure of ruminants to herbicides is usually by intentional or unintentional feeding or grazing of recently treated feed crops or by accidental administration. The potential for exposure of ruminants to other pesticide types (fungicides, acaricides, rodenticides, nematocides, and molluscacides) is quite remote but does occur periodically. In such cases exposure is usually accidental either by allowing animals uncontrolled access to the pesticides or by direct inadvertent administration.

TOXIC EFFECTS AND CLINICAL SIGNS

Organophosphates and Carbamates

Toxicity results from the inhibition of cholinesterase (CHE) enzymes and the consequential accumulation of acetylcholine, a neurosynaptic transmitter substance, in the region of the neuromuscular junction, autonomic ganglia and neuroeffector junction. The clinical signs are the same regardless whether the exposure is a single dose (acute) or multiple (chronic) doses over time. Toxic effects are the result of overstimulation of the nervous system.

Clinical signs are usually grouped into three different categories: muscarinic, nicotinic and central nervous system (CNS) signs (Radeleff, 1964, 1970; Buck et al, 1976). Muscarinic signs include anorexia, vomiting, abdominal pain, gastrointestinal hypermotility, sweating, excessive lacrimation and salivation, diarrhea, heat spasms, dyspnea, palor, miosis, lung edema, cyanosis and incontinence of feces and urine. Nicotinic signs which usually follow the muscarinic signs include shivering and tremors of stride and muscles followed by muscle weakness, ataxia and paralysis. CNS signs usually occur last and include restlessness, convulsions, and depression of respiratory and vasomotor centers. Death results from interference with respiratory or cardiac function. The toxicity of organophosphates and carbamates is quite variable but in general the minimum toxic dose to ruminants lies generally within the range of 1-100 mg/kg of body wt (Buck et al, 1976).

Organochlorine Pesticides

Toxic effects from organochlorines occur as a result of both stimulation and depression of the CNS. Although the mode of action is poorly understood, the site of reactivity is on the membrane of the nerve cell, particularly the sensory nerves. Apparently the threshhold for an action potential is lowered; DDT causes a reorientation of Na and K ions across the membrane resulting in partial depolarization (Buck et al, 1976).

Clinical signs associated with organochlorine-induced stimulation and/or depression of the CNS are generally of the neuromuscular type. These are described in much greater detail elsewhere (Radeleff et al, 1955; Radeleff and Bushland, 1960; Radeleff, 1964, 1970; Clarke and Clarke, 1975; Buck et al, 1976). Clinical signs resulting from CNS stimulation usually appear within 24 h after exposure to a toxic dose and may become progressively severe with time or may be explosive and fulminating. Stimulation usually precedes depression and is characterized by restlessness, hypersensitivity and beligerancy followed by intermittent or continuous muscle spasms of the eyelids and fore and hindquarters. CNS stimulation may progress into incoordination, frenzy or salivation. Single or several clonic-tonic seizure(s) accompanied by elevated body temperature may end in death. Some animals expire after a single seizure whereas others survive numerous seizures. Short to lengthy intervals of CNS depression often interrupt CNS stimulation. These signs are characterized by drowsiness, immobility and anorexia. If depression persists, emaciation and dehydration develop and death usually occurs. The minimum toxic dose of several organochlorine pesticides has been tabulated by Buck et al (1976). In general, the acute toxicity of aryl and cyclodiene insecticides such as aldrin, dieldrin, chlordane, endrin, heptachlor, lindane, mirex and toxaphene in ruminant species ranges from 10-100 mg/kg body wt/dose. Diphenyl aliphatic compounds such as DDT, methoxychlor and perthane are 10-20 times less toxic.

Chlorophenoxy Pesticides

The clinical signs associated with chlorophenoxy poisoning are generally the same regardless whether the mode of exposure involves single or multiple dosages. Anorexia, depression, rumen stasis, muscle weakness and sometimes diarrhea are common. Ulceration of oral mucosa and emaciation may occur after long exposure. Bloat may occur in come cases. Death is non-dramatic and does not include convulsions (Rowe and Hymas, 1955; Radeleff, 1964; Palmer, 1972b; Buck et al, 1976).

Other Pesticides

Readers are referred to Radeleff (1964, 1970), Clark and Clark (1975), and Buck et al (1976).

FEEDING BEHAVIOR AND PERFORMANCE

Feed consumption is perhaps the most

important determinant of performance in healthy ruminants. Pesticides can bring about either a beneficial or detrimental influence on performance. For example, enhanced weight gains in calves by controlling grubs with a systemic organophosphate exemplifies a beneficial impact (Campbell et al, 1973). Conversely, overexposure or exposure under the wrong conditions may end in impaired rather than enhanced performance.

Organophosphates and Carbamates

Members of this group are generally metabolized and excreted rapidly thereby preventing pesticide accumulation in the body. Similarly, they are not persistant in the environment. Thus, animal performance is more likely to be compromised due to a single (acute) exposure rather than from multiple (chronic) exposures.

Anorexia is listed as one of the muscarinic signs (see earlier section on clinical signs). Khan (1973) noted that cattle deaths due to organophosphate poisonings are frequently preceded by 1-4 weeks of progressive emaciation and dehydration attributed to anorexia. In general, the impact of anorexia on performance is directly dependent on the duration and reversibility of the effect.

Anorexia aresulting from a single dose generally appears within 1-3 days (Wilbur and Morrison, 1955; Radeleff and Woodward, 1957; Younger et al, 1963; Palmer, 1971; Khan, 1973). It is usually, but not always, accompanied by a reduction in blood CHE activity (Anderson and Mackin, 1969). For example, Brahman bulls treated with famfur were anorexic and losing weight within 2 days but blood CHE did not reach a low until 7 days post-treatment (Palmer, 1971; see Fig. 19-1). Usually, if antidotes such as atropine and certain oximes (2-PAM) are administered soon after poisoning, anorexia and other clinical signs are reversed rapidly (Buck et al, 1976) but blood CHE activity is not always restored (Anderson and Mackin, 1969). However, in situations where antidotes are not administered, recovery from anorexia appears to parallel the level of exposure and the rapidity of CHE inhibition (Wilbur and Morrison, 1955; Palmer, 1971). Carbamates tend to evoke a more short-lived anorexia. Animals that survive intoxication usually do not suffer lasting effects (Miles et al, 1971; Gardner, 1976).

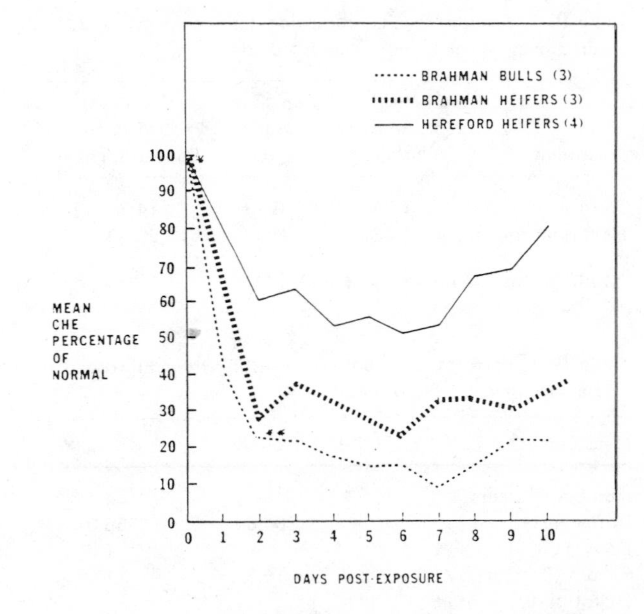

Figure 19-1. Mean whole blood cholinesterase (CHE) activity of Brahman bulls, Brahman heifers, and Hereford heifers after topical application of famfur as a pour-on at the rate of 40.5 mg/kg body weight. *Preexposure sample represents the normal CHE activity, with mean activity of the group at intervals after exposure as percentage. **Death of 1 calf. From Palmer (1971).

Chronic exposure can also affect body weight and feed intake adversely. Weight loss was reported in cattle administered Abate over a full year (Palmer, 1968) but not in sheep (McCarty et al, 1968). Sheep fed 1000 ppm ronnel continuously for 84 days consumed less feed, lost weight and utilized feed less efficiently (see Table 19-1, Crookshank et al, 1972). A chronic accumulation of toxic effects was observed in cattle that became anorexic, lost weight, and death to some after 2 weeks of continuous exposure to Thimet (Pugh, 1975). Similarly, in bulls fed Supracide continuously, blood CHE activity slowly declined and after 3 weeks the animals became anorexic and some died (Polan et al, 1969). However, chronic exposure is not necessarily always harmful. For example, ronnel fed to feedlot cattle enhanced daily gain and feed conversion (Rumsey et al, 1975; Riley and Ware, 1977; Wooden and Algeo, 1977; Rumsey, 1979). In one trial ronnel tended to increase intake (see Table 19-2; Rumsey et al, 1975). More recently, Rumsey (1979) reported improved growth rate (ca. 12%) regardless of intake. This effect has not

Table 19-1. Average feed consumption, daily gain and feed efficiency of sheep fed ronnel for 84 days.[a]

Treatment	Feed consumption kg/day	Gain, kg/day	Feed efficiency, kg feed/kg gain
Control	1.91	0.13	14.7
1000 ppm ronnel	1.43	0.07	20.4

[a]Modified from Crookshank et al (1972)

Table 19-2. Performance, apparent digestibility and nitrogen balance in steers fed ronnel.[a]

Measurement	-Ronnel	+Ronnel[b]
Number of steers	4	4
Initial weight, kg	364.5	355.8
Daily gain, kg	1.0	1.2
Feed intake, kg/day	10.3	11.0
Digestibility, % intake		
Dry matter	77.3	77.5
Gross energy	70.6	71.4
Acid detergent fiber	59.3	59.9
Crude protein	73.2	76.3
Ash	70.6	72.1
Nitrogen, % intake		
Fecal	26.8	23.7
Urinary	64.0	57.4
Retention	9.0	18.9

[a]Modified from Rumsey et al (1975)
[b]Ronnel was added with a 70% concentrate diet at a dosage of 4.5 mg/kg body wt/day and fed for 90 days.

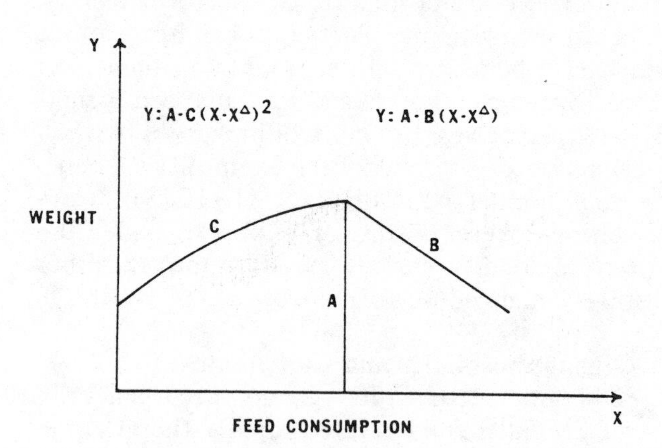

Figure 19-2. Long-term toxicologic effect of di-allate on total body weight of sheep as related to feed consumption. A = estimated body weight of sheep at time di-allate began to take effect; X^{Δ} = estimated week of experiment when di-allate began to take effect; B = decrease in body weight of sheep per week after X^{Δ}; C = increase in body weight of sheep as a function of the square $(X-X^{\Delta})$ or current time less X, at time X^{Δ} prior to X. From Palmer et al (1972).

been explained and is believed due to factors other than insecticidal or anthelmintic action of the chemical.

Anorexia and weight loss are common clinical signs of chronic carbamate poisoning in ruminants. Moderately reduced feed intake preceded body weight loss by 2½ weeks in sheep fed di-allate continuously (see Fig. 19-2; Palmer et al, 1972). It was suggested that the first signs of anorexia appear after a tolerance threshold is surpassed. However, ruminants are generally resistant to repeated exposures of carbamates for long periods without ill effects even though blood CHE may be 100% inhibited (Younger and Radeleff, 1963; Wiedenbach and Younger, 1962; Radeleff et al, 1963).

Members of the carbamate group are known to change the palatability of herbage. For example, pasture sprayed with demeton was grazed by sheep only when forced by starvation (McCarty et al, 1968). Similarly, reduced consumption of dasanit-containing silage by dairy cattle was attributed to palatability changes (Johnson et al, 1973). A different effect was noted in bulls administered Supracide by capsule for 5 weeks; the animals gradually shifted their preference from hay to grain (Polan et al, 1969).

Milk production, like growth, is generally decreased in anorexic cattle, and has been produced by organophosphates such as Dasanit (Johnson et al, 1973) and demeton (Dahm and Jacobson, 1956). However, cattle fed carbofuran-treated silage consumed less feed, namely 0.03 kg for every 1 ppm increase in residual pesticide up to the point of no observable clinical signs; moreover, milk production was not impaired (Miles et al, 1971). Conversely, dairy cattle fed Phosvel-containing silage over 8 weeks exhibited decreased milk production, increased body weight and unaltered feed intake or blood CHE (see Table 19-3; Johnson et al, 1971). However, chronic exposure without altered milk production was observed in cows fed silage containing 16.6 ppm fenthion for 8 weeks (Leuck et al, 1971) or 0.34 mg parathion/kg body wt for 61 days (Pankaskie et al, 1952).

Carbamates can impair wool growth. Gradual wool loss progressing to total baldness (alopecia) was observed in sheep fed di-allate for 10 days; rejuvenation of wool covering started ca. 40 days after first loss was

Table 19-3. Adjusted means for silage intake, whole blood cholinesterase activity, milk production, and means of body weight gains of cows fed silage containing residues of Phosvel.[a]

Item	Kilograms of Phosvel applied/ha of corn[b]			
	0.00	0.56	1.12	2.24
Silage intake,[c] kg/day	7.41	8.17	7.73	8.10
Cholinesterase activity[d]	1.29(1)[e]	1.23(1)	1.14(2)	1.05(3)
Milk production, kg/day	8.74(1)[e]	8.22(1)	7.70(1)	5.04(2)
Body weight gain,[f] kg/day	0.26(1)[e]	0.70(2)	0.54(1,2)	1.09(3)

[a]Modified from Johnson et al (1970)
[b]Means are adjusted for pretreatment differences in animal responses.
[c]Dry matter basis
[d]Micromoles of acetylcholine hydrolyzed
[e]Means within each row (across) not followed by the same number in parentheses are significantly different ($P < .05$).

noticed (Palmer, 1964). Similar effects were noted with dichlormate (Palmer, 1972b).

Organochlorines

In contrast to the organophosphates and carbamates, organochlorines tend to accumulate in body fat because of their metabolic stability and high lipid solubility. Also, persistency in the environment is characteristic of many organochlorines. Thus, animal performance is more likely to be compromised after a prolonged exposure during which the body burden may reach a critical threshhold resulting in toxic effects. Acute intoxication from a single exposure may also occur.

Almost without exception single exposures do not cause lasting effects. Anorexia, when observed, does not persist any longer than other clinical signs. For example, orally administered DDT in goats caused intermittant convulsive seizures during which feeding and drinking ceased; however, normal feeding behavior resumed immediately after seizures ceased. Goats and sheep drenched with toxaphene responded in a similar manner. A calf dermally administered toxaphene suckled normally between periods of convulsive seizure (Radeleff, 1949). In dairy cattle oversprayed with lindane, anorexia and decreased milk production were observed for only 1 day (Ely et al, 1953). Thus, single exposures may be acutely toxic and even induce anorexia but the effect does not persist long enough to have serious impact on performance.

Chronic exposures that are eventually lethal usually but not always involve feeding and body weight effects. For example, growth was normal in calves exhibiting tremors from DDT treatment (Bohman et al, 1952). Similarly, feed intake and growth rate was not impaired in steers exhibiting neurologic signs from toxaphene ingestion (Marsh, 1949). In cases where exposure to organochlorines cause loss of body weight, reduced feed intake is usually the cause. However, dieldrin may be an exception, since growing lambs administered 2 mg/kg body wt/day for 32 weeks gained less weight than controls but the effect was attributed to reduced deposition of body fat, not reduced feed intake (see Table 19-4; Davison, 1970). Similarly, slower rates of gain were observed in fawns suckling does fed dieldrin but neither fat deposition nor linear growth were reduced (Murphy and Korschgen, 1970). Conversely, kids born to goats fed 1 mg mirex/kg body wt/day for 61 weeks suckled milk containing ≤20 ppm (fat basis) mirex without ill effect (Smrek et al, 1977).

DDT is known to alter normal feeding behavior. For example, cattle dosed orally over 3 weeks selectively refused grain and continued normal hay intake but lost 13% of their body weight (Orr and Mott, 1945). Moreover, a palatability effect was indicated in cattle that refused DDT-containing bhoosa and the effect was dose related (Gupta et al, 1968).

Even though organochlorine residues accumulate in body fat, this does not imply toxicity. For example, milk production and body weight were normal in dairy cattle

Table 19-4. Effect of dieldrin on carcass composition and plasma vitamin A concentration of sheep fed dieldrin for 32 weeks.[a]

Dietary dieldrin mg/kg body wt	Carcass composition Dry matter, %	Fat,[b] %	Proteins,[c] %	Plasma vitamin A, mcg/dℓ
0	41	49	41	30.5
0.5	38	45	43	33.7
1	41	48	42	31.2
2	40	42	46	23.5

[a]Modified from Davison (1970)
[b]Averages of all data
[c]On freeze-dried carcass. This material contained 8-10% water, as determined by oven drying.

Table 19-5. Chronic exposure of various organochlorine pesticides without untoward affect on growth, body weight, milk production, or wool growth.

Organochlorine pesticide	Type of ruminant	Dose level	Duration of exposure	Reference
DDT	Dairy cows	50 mg/kg body wt	5 mo.	Wilson et al, 1946
	Steers	22.5 g weekly	3 yr	Roulston et al, 1954
	Calves	100 mg/kg DM*	8 mo.	Thomas et al, 1951
	Sheep	100 ppm, feed	16.5 wk	Wilson et al, 1946
Methoxychlor	Dairy cows	14 ppm, feed	16 wk	Biddulph et al, 1952
Lindane	Cattle	100 ppm, feed	12 wk	Radeleff, 1951
Toxaphene	Steers	320 ppm, feed	19 wk	Marsh, 1949
Dieldrin	Dairy cows	75 ppm, feed	16 wk	Gannon et al, 1959
	Calves	7 ppm, feed	25 wk	Link et al, 1951
	Sheep	25 ppm, feed	4 yr	Harris et al, 1966
	Deer	25 ppm, feed	3 yr	Murphy et al, 1970
Aldrin	Dairy cows	40 ppm, feed	16 wk	Gannon et al, 1959

*Dry matter

containing 380 ppm DDT in body fat (Wilson et al, 1946). Also, ADG was normal in calves with 340 ppm in body fat (Thomas et al, 1951), and body weight was maintained over a 3-year period in steers containing 425 ppm DDT in body fat (Roulston et al, 1953). The tolerance of various ruminant species to chronic exposure of various organochlorine insecticides is given in Table 19-5.

Chlorophenoxys

A single exposure may induce anorexia but the effect is usually shortlived. For example, anorexia, increased thirst and muscular weakness were observed within 24 h after exposure to 2,4-D salt in calves (Bjorkland and Erne, 1966) and cows (McLennan, 1974), but the signs disappeared within a few days.

Anorexia and weight loss are also consistent signs of chronic toxicity. Palmer (1972b) tested both the salts and esters of 2,4-D and 2,4,5-T, the salts of MCPB, MCPA and 2,4-DB in both sheep and cattle with dosages ranging from 50-250 mg/kg body wt over 10 days. A consistent observation was anorexia coupled with mild to severe weight loss. Recovery from both effects was rapid. Rations containing 1800 ppm 2,4,5-T or 2000 ppm silvex fed over a 2-3 week period induced mild anorexia in sheep and cattle together with weight loss, but the effect was rapidly reversed after exposure ended (Leng, 1972). A similar effect was observed in cattle dosed with Esteron (Rowe and Hymas, 1954). Because chlorophenoxy herbicides are excreted rapidly (Radeleff, 1964), cumulative effects do not occur. Thus, ruminants are generally able to tolerate long-term exposures without incident (Palmer, 1963; Palmer and Radeleff, 1964).

Chlorophenoxy herbicides can change the palatability characteristics of plants. When given a choice, cattle and sheep selected against pastures sprayed with 2,4-D and 2,4,5-T (Grigsby and Farwell, 1950). However, when forced to graze on pastures sprayed with 2,4-D, no adverse effects on feeding behavior or health were observed (Mitchell et al, 1946). Likewise, silvex-treated feed was refused by cattle and sheep but only when the levels of silvex were excessively high (Leng, 1972). There has been concern that spraying toxic plants which are normally unpalatable may improve palatability and lead to increased ingestion of plant toxicants. This subject has been discussed adequately elsewhere (Buck et al, 1976).

Dipyridyls

Loss of appetite and body weight are also commonly seen after exposure to dipyridyl herbicides. Doses of 10 mg paraquat or 5 mg diquat/kg body wt/day for 10 days caused anorexia and 5-10% weight loss (Palmer, 1972b). Recovery of lost body weight was very slow, requiring 6-8 weeks. Shultz et al (1976) noted that sheep and cattle grazing diquat-sprayed grass respond differently from one another. Sheep became anorexic whereas cattle exhibited increased thirst and a 40% drop in milk production. However, paraquat-killed herbage containing 400 ppm of the chemical was not toxic to heifers grazing it for 4 wk. Similarly, sheep and calves that consumed water containing 20 ppm paraquat for 1 mo. were not affected (Colderbank et al, 1968).

Other Pesticides

Technical pentachlorophenol fed at low levels (15 mg/kg body wt/day) to dairy heifers for 160 days affected growth and feed conversion adversely; the effect was due to toxic dioxin and furan impurities contained in industrial preparations and was dose related (McConnell et al, 1980; see Table 19-6). Smaller doses (0.2 and 2.0 mg/kg body wt) fed subchronically to lactating dairy cattle for 130-140 days did not affect performance (Kinzell et al, in press).

Organometallic pesticides include arsenicals, mercurials and cadmium-based compounds. Arsenic poisoning from MSMA, DSMA and cacodylic acid is characterized by anorexia and weight loss (Palmer, 1972b; Dickinson,

Table 19-6. Weekly feed consumption, weight gain, and feed efficiency in cattle fed pentachlorophenol (PCP).[a]

	Feed consumption,[b] kg	Weight gain, kg	Feed efficiency[c]
Control	61.1 ± 4.8[d]	154.3 ± 11.9	9.1 ± 0.8
Analytical PCP (aPCP)	61.9 ± 4.3	146.8 ± 12.0	9.8 ± 0.6
10% tPCP + 90% aPCP	63.7 ± 4.6	125.1 ± 1.7[e]	11.7 ± 0.7[e]
35% tPCP + 65% aPCP	56.5 ± 1.2	125.3 ± 7.6[e]	10.4 ± 0.7
Technical PCP (tPCP)	50.1 ± 3.2[f]	82.7 ± 8.7[fg]	14.1 ± 0.8[fg]
Dose response	$P < .01$	$P < .01$	$P < .01$

[a]Modified from McConnell et al (1980). All treated cattle received 20 mg of dose/kg body wt/day for 42 days followed by 15 mg/kg body wt/day for 118 days; technical (tPCP) and analytical pentachlorophenol (aPCP) were mixed in order to vary the exposure of dioxins and furans while maintaining constant the exposure to PCP.

[b]Feed consumption comparisons adjusted for pretreatment differences.

[c]Kg feed required to produce 1 kg weight gain.

[d]Values represent mean ± SE.

[e]$P < 0.05$, significantly different from control.

[f]$P < 0.01$, significantly different from control.

[g]$P < 0.01$, significantly different from aPCP.

1972). Sheep poisoned with DSMA that survived the treatment required a prolonged recovery period. Feeding behavior was impaired in calves fed methylmercury (Ceresan L) when the ataxia interfered with their ability to stand and feed (Herigstad et al, 1972). Feed consumption and dose were inversely related in sheep fed 50 to 500 ppm cadminate for 49 weeks and there was a trend toward improved efficiency with increased time of treatment (Wright et al, 1977).

In general, anorexia, weight loss or reduced weight gains are the most commonly observed effects resulting from chronic exposure to various herbicides at daily doses ranging from 5 to 500 mg/kg body wt for 10 days (Palmer, 1972b). The recovery period was sometimes quite extended. For example, one month was required after the last dose of trifluralin before sheep regained their original body wt. Also, delayed effects are sometimes observed. For example, sheep dosesd with naptalam lost weight during exposure but did not exhibit anorexia until 2 weeks after the last dose.

EFFECTS ON THE PHYSIOLOGY AND PATHOLOGY OF THE GASTROINTESTINAL TRACT

Pesticides of all types and uses are known to influence the physiology fo the GI tract. In many cases GI tract upsets such as diarrhea, vomiting, ruminal stasis or bloat are the only observable clinical signs. To the practicing veterinarian, these signs are not very useful for diagnostic purposes but nevertheless have important ramifications in terms of animal health. Effects of pesticides on the GI tract may be related to motility, pathology, or fermentation and digestibility of feedstuffs.

Motility

GI tract motility is under the control of the autonomic nervous system, and vagal nerve fibers of the cholinergic type innervate the reticulum and rumen causing rhythmatic contractions of the pillars and visceral walls. GI tract hypermotility accompanied by severe abdominal pain is generally listed as one muscarinic sign associated with organophosphate and carbamate poisoning. Hypermotility is believed to be the result of overstimulation of para-sympathetic nerves (Buck et al, 1976). It was shown, however, that stimulation of the dorsal vagal nucleus by implanted electrodes in conscious goats caused complete cessation of contractions in the rumen and reticulum (Anderson et al, 1959). Likewise, ruminal stasis was seen by Pugh (1975) in heifers that succumbed to chronic poisoning from Thimet. The autopsy report described the cattle as having rumens full of rancid-smelling contents, indicating lack of motility. Chemical rumenitis and reticulitis were also observed (Pugh, 1975), presumably caused by unabsorbed acids produced by fermentation. In other cases of organophosphate poisoning, the rumen contained semi-solid ingesta and the ruminal epithelia "peeled off" easily (Khan et al, 1961; McCarty et al, 1969). In 5-7 mo. old calves treated with crufomate for grubs, decreased average daily gain was thought to be due to rumen hypomotility which lasted for a 2-week period (Khan, 1973).

Various explanations as to how organophosphates and carbamates cause ruminal stasis have been suggested. First, the motor nerves that supply the rumen may be impaired due to the cholinergic action of the pesticides (Khan et al, 1972). Second, excessive amounts of adrenalin may be released from the adrenal medulla which is also under control of cholinergic nerves (Khan et al, 1972); antimotility properties of adrenalin have been described (Alexander and Moodie, 1960). Third, in acute organophosphate poisoning, a significant rise in blood glucose has been observed (Wilber and Morrison, 1955); elevated blood glucose is known to reduce rumen motility (Phillipson, 1977).

Ruminal stasis is also commonly observed in chlorophenoxy herbicide poisonings. Steers dosed with 500 mg Esteron/kg/body wt showed immediate ruminal stasis which persisted for 24 h (Rowe and Hymas, 1954). Similarly, dairy cattle accidentally exposed to 2,4-D lost rumen motility within 1 h after exposure but recovered within 24 h (McLennan, 1974). In each of these cases the dose was great enough to cause death but those that survived recovered rapidly, suggesting that loss of rumen function is a significant factor in fatalities. There are other known cases of chlorophenoxy poisonings where ruminal stasis was not reported (Palmer, 1972b). Chlorophenoxy-induced stasis differs from the organophosphate and carbamate-induced stasis in that the former is always accompanied by inflamation or congestion of the GI tract mucosa, particularly in the abomasum and small intestine (Rowe and Hymas, 1954; Palmer, 1963; Palmer, 1972b). However, the relationship between the pathologic events in the GI tract and ruminal stasis have not been established. Anorexia often accompanies stasis and may, in fact, be caused by a full rumen and one that cannot be emptied (Pugh, 1975). The relationship between gastric fill, rate of digesta passage and feeding behavior have been discussed in Ch. 11, Vol. 2.

Bloat

Ruminal bloat may or may not accompany stasis. In some poisoning cases involving organophosphates and carbamates, bloat is the most noticeable sign. Gardener (1976) reported a case where cows and heifers accidentally consumed the carbamate Furadan and the cause of death was bloat. Other carbamates such as dichlormate in cattle and pebulate in sheep were also found to cause bloat (Palmer, 1972b).

Several organophosphate preparations are used as systemic insecticides for the control of cattle grubs, the larval stage of *Hypoderma bovis* and *H. lineatum.* At certain points in the life cycle of the parasites, generally the fall of the year, the larvae are located in the esophagus or along the spinal cord. Treatment with therapeutic doses of organophosphates such as coumaphos or crufomate kill the larva but the resultant lesions in the esophagus frequently cause ruminal bloat (Scharf, 1962; Mullee et al, 1968; Khan, 1971; Khan et al, 1972). Khan et al (1972) explained that acute esophagitis is probably caused by release of proteolytic enzymes from dead larva which impairs the eructation process. Interference with nervous control of reticular contractions was also suggested by Khan et al (1972) as a cause of bloat. Organophosphate-induced bloat can happen regardless whether administration is dermal or oral.

Bloat is not generally seen in poisoning cases involving other pesticides. However, Palmer (1972b), observed it in sheep dosed with molinate, benefin and nitralin. Similarly, dairy cattle sprayed with lindane (Ely et al, 1953) and steers dosed repeatedly with 2,4-D amine salt (Bjorkland and Erne, 1966) developed mild to severe cases of bloat. Ruminal bloat and stasis frequently occur simultaneously. This would be expected since the same nerves that instigate the eructation process are also involved in other rumen contractions. However, in most cases, stasis is not reported unless death occurs at which time autopsy reveals the presence of undigested particles and foul smelling contents in the rumen.

Diarrhea

Diarrhea is a commonly observed clinical sign associated with many pesticide poisonings. Organophosphates, regardless whether administered orally, dermally or intramuscularly, can cause diarrhea. Onset and duration is not predictable and varies greatly among organophosphates. For example, cattle dosed with 37.5 mg coumaphos/kg body wt showed signs of toxicosis within 1 day but diarrhea did not set in until the 8th day and persisted for 8 days (Younger and Wright, 1971). On the other hand, dermal application of chlorpyrifos to calves resulted in clinical signs including diarrhea only from 5 to 90 min after treatment (Loomis et al, 1976). However, diarrhea generally develops along with and persists longer than other clinical signs. In sheep dosed with cythioate the severity of diarrhea correlated with blood CHE activity (Doval and Gupta, 1976). Sometimes it becomes most severe just before death (Wilber and Morrison, 1955; Pugh, 1975). In cases of marginal poisoning, diarrhea may be the only clinical sign observed, as has been shown in cattle administered coumaphos (Radeleff et al, 1963). On the other hand, all classical cholinergic effects except diarrhea were seen in calves treated with crotoxyphos (Greer et al, 1973). As a group, carbamate pesticides generally do not cause diarrhea, probably because cholinergic effects are very short-lived. However, long-term exposure may produce diarrhea as shown in sheep treated with zineb for 3 weeks (Palmer, 1963). Other compounds known to cause diarrhea in ruminant species are mostly herbicides such as chlorates, amides, substituted phenyl ureas, dinitroanilides, dipyridyls, phthalamic acids, arsenicals and mercurials.

Pathologic Changes

Organochlorine and chlorophenoxy pesticides rarely cause diarrhea, but members of both groups usually bring about significant pathologic changes in the GI tract. Necropsy reports on ruminant species that succumbed to organochlorine poisoning generally includes abomasal gastritis and/or enteritis, erosion and/or congestion of GI tract mucosa and cloudy swelling of the viscera and focal hemorrhage. The description will vary with pesticide and species. Radeleff et al (1955) described the intestines as being blanched or parboiled as a result of the extremely high body temperature (110-114°F) that accompanies convulsive seizures. Some heifers fed technical pentachlorophenol over 160 days developed abomasal lesions (McConnell et al, 1980).

Chlorphenoxy pesticides such as salts and esters of 2,4-D and 2,4,5-T (Rowe and Hymas, 1954; Palmer, 1972b) at toxic levels of oral exposure can cause inflamation, congestion or hemorrhage of the abomasum, and small or large intestine. Moreover, arsenicals (Palmer, 1972b; Dickinson, 1972; Selby et al, 1974), mercurials, e.g. Ceresan M, (Palmer, 1963) and various other herbicides (Palmer, 1972b) are known to cause pathologic changes in the GI tract. In general, pathologic changes of the GI tract are particularly apparent in animals that succumbed to pesticide poisoning, but changes have also been observed in sacrificed cases, indicating the effects are dose related.

Digestive Functions

The possibility that chronic exposure to pesticides may evoke subtle influences on GI tract function has prompted various investigations into effects of pesticides on rumen microbial ecology, digestibility and fermentation of feedstuffs, and nutrient absorption. In this regard, DDT has been the most studied. In lambs fed hay containing up to 44 ppm DDT for 112 days, digestibility of protein, fat, fiber and carbohydrates were not affected but nitrogen balance and DDT intake were inversely correlated (Harris et al, 1951). The same group conducted subsequent digestion and balance trials in 3-6 mo. old calves fed a diet containing up to 100 ppm DDT; as before dietary DDT concentrations were inversely correlated to nitrogen storage (Bohman et al, 1952). In rumen-fistulated cattle fed roughage containing 500-750 ppm DDT, feed consumption was reduced but no effect was observed on digestion of dry matter or cellulose as measured by the nylon bag technique (Gupta et al, 1968). Similarly, concentrations of DDT less than 1000 ppm had little effect on dry matter disappearance and VFA production when added to mixed cultures of rumen microorganisms from sheep (Kutches et al, 1970) or mule deer (Barber, 1972). However, Rumsey et al (1970) observed that even though DDT had little toxicologic effect on rumen microorganisms as measured by the gram reaction, total bacterial counts, proportions of protozoal species and proportions of VFA, the ecology of the rumen was modified in cattle fed 30 mg DDT/kg body wt/day over 6 weeks. They observed reductions in numbers of bacterial rods and protozoa and total VFA concentration, but these effects were more evident in cattle fed a 100% concentrate diet vs an 81% roughage diet (Rumsey et al, 1970; see Table 19-7). These workers theorized that lowered protozoal numbers may have caused the reduction of nitrogen retention observed by Harris et al (1951) in their DDT-fed lambs on the basis that lower nitrogen retention was observed in lambs free of ciliate protozoa.

Table 19-7. Effect of p, p' -DDT on microbial, protozoal, and VFA concentrations in ingesta from steers fed a concentrate or roughage diet.[a]

Ruminal ingesta measurements	Experimental diet: 100% Concentrate -DDT	100% Concentrate +DDT[b]	81% Roughage -DDT	81% Roughage +DDT[b]
Total bacteria per ml x 10^{-9}	13.0	13.8	12.0	11.3
Gram -, %	73.0	80.4	73.6	73.9
Gram +, %	27.0	19.6	26.4	26.1
Total protozoa per ml x 10^{-4}	28.1	14.0	14.2	13.2
Total VFA mmoles/liter	73.9	63.8	63.9	61.2

[a]Modified from Rumsey et al (1970)
[b]Cattle were fed 30 mg DDT/kg body wt/day for 6 weeks.

Other organochlorine compounds have not been studied as extensively as DDT. However, toxaphene added in vitro to mixed cultures of rumen microorganisms from sheep reduced dry matter disappearance and VFA production at a concentration of 500 ppm but reduced protozoa numbers at 100 ppm (Kutches et al, 1970). The greater sensitivity of protozoa than bacteria to DDT was noted previously (Rumsey et al, 1970). Moreover, Williams et al (1963) showed that oligotrich protozoa are more sensitive than isotrich protozoa to insecticides including lindane at concentrations of 500 ppm; rumen bacteria were not affected.

Organophosphates and carbamates appear to affect the ecology of rumen microorganisms as well. Dairy cattle fed Phosvel-containing silage for 8 weeks gained body weight but substantially reduced their milk production; the investigators hypothesized preferential formation of metabolic products as a possible explanation (Johnson et al, 1971; see Table 19-3). Similarly, 4.5 mg ronnel/kg body wt fed to steers for 90 days stimulated a higher proportion of ruminal acetate and a slightly higher ruminal pH; performance, ration digestibility and nitrogen balance were not affected (Rumsey et al, 1975; see Table 19-2). In a more recent study, feedlot steers fed ronnel tended to have lower ruminal butyric acid and ammonia levels and slightly elevated pH, but the changes were not great enough to explain the 12% greater rate of gain in the cattle (Rumsey, 1979).

In vitro studies with mixed rumen cultures demonstrated that rumen microorganisms are relatively resistant to organophosphates and carbamates. Malathion and Sevin at concentrations of ≤500 ppm have little effect on VFA production or dry matter disappearance (Kutches et al, 1970). Similar results were found with trifluralin (William and Feil, 1971). Barber (1972) concluded that toxicity of these pesticides to mixed cultures of rumen microorganisms is well above concentrations found in treated forages. However, Williams and Stolzenberg (1972) demonstrated that the carbamate Moban at 72 ppm while having no effect on mixed cultures of ruminal bacteria in vitro inhibited the growth of 2 of 10 pure culture strains isolated on selective media and also inhibited their ability to hydrolyze aromatic esters. Barber (1972) also reported that pure cultures are more susceptible than mixed cultures to organophosphates and carbamates. In general, organophosphates and carbamates, like organochlorines, have greater toxicity to protozoa than bacteria of the rumen (Williams et al, 1963; Kutches et al, 1970; Williams and Stolzenberg, 1972).

The toxicity of chlorophenoxy herbicides to rumen microorganisms parallels the toxicity to the whole animal; it is low in both cases. For example, 2,4-D and 2,4,5-T when incubated with sheep rumen fluid did not affect rate of fermentation or protozoa numbers until concentrations exceeded 500 ppm (Kutches et al, 1970). However, Barber (1972) noted that deer rumen microorganisms are more sensitive to 2,4-D than most other pesticides. Kutches et al (1970) tested a variety of other herbicides, including 2,3,6-TBA, Tordon, dicamba, diuron, simazine, and bromacil and observed no significant effects on mixed rumen cultures at concentrations ≤500 ppm. Similarly, no difference in molar percentages of VFA and protozoal numbers were noted when subtoxic doses of propazine were administered to sheep (Williams et al, 1968).

Rumen microorganisms appear to be relatively sensitive to the fungicide pentachlorophenol; rate of cellulose digestion and VFA production, particularly propionic acid, were impaired at concentrations ≤10 ppm with mixed cultures of bovine rumen microorganisms (Shull and McCarthy, 1978; see Table 19-8). However, heifers fed 15 mg/kg body wt purified pentachlorophenol for 160 days had normal growth rates, suggesting no apparent impairment of feed digestibility (McConnell et al, 1980). Commercial grade pentachlorophenol contains dioxin and furan impurities which impair growth when fed chronically at 15 mg/kg body wt/day (McConnell et a, 1980), but the impurities do not appear to interefere with VFA production or cellulose digestibility when tested with mixed cultures of bovine rumen microorganisms in vitro (Shull et al, 1977).

Table 19-8. Effect of pentachlorophenol (PCP) on the percentage of cellulose digestion and propionate production by rumen microorganisms in vitro after 24 and 48 hours incubation at 39 C.[a]

Treatment	% Cellulose digestion 24 h	% Cellulose digestion 48 h	Propionate production[b] 24 h	Propionate production[b] 48 h
Control − ethanol[c]	27.5[de]	46.8	19.9	40.6
Control + ethanol	24.1	43.9	16.3	34.9
10 ppm PCP	17.5[e]	38.1[e]	9.9[e]	29.1[e]
50 ppm PCP	5.2[e]	5.2[e]	0.2[e]	0.0[e]
100 ppm PCP	4.2[e]	5.4[e]	0.0	0.0[e]

[a]Modified from Shull and McCarthy (1978)

[b]Units are mmoles/liter of incubation medium per incubation period.

[c]PCP was solubilized in ethanol and added as such to buffered inoculum.

[d]Each value represents the mean of two replicates.

[e]$P < 0.05$, significantly different from control + ethanol.

The direct action of pesticides on the GI tract can result in numerous potential effects, most of which have never been investigated. Organophosphate and carbamate pesticides interfere with gut motility resulting in effects ranging from total stoppage of digesta flow as exemplified by the administration of Thimet to cattle (Pugh, 1975), to mild diarrhea produced by ronnel in cattle (Radeleff and Woodward, 1957). These influences are, of course, dose related.

To determine the effect of diarrhea on absorption of nutrients, Hunt and McCarty (1972) induced diarrhea in 3 to 10-day-old calves by administration of an organophosphate. They found that blood tocopherol, carotene and fatty acid concentrations were significantly reduced, indicating impairment of intestinal absorption of these nutrients. Feed consumption was not lowered, and no other clinical signs besides diarrhea were observed in these calves. In a similar study, tocopherol absorption was decreased in calves with diarrhea induced by treatment with tri-o-cresyl phosphate (Cowlishaw and Blaxter, 1955). A reduction in plasma Mg:Ca ratio was observed in calves dosed with 250 mg pebulate/kg body wt, but the effect was related to kidney damage, not malabsorption of the minerals (Hunt et al, 1971). Organophosphates and carbamates are known to stimulate the flow of gastric juices (Bush et al, 1965), but the consequences of this effect relative to feed digestibility and absorption of nutrients have not been studied. Little is known about such effects resulting from exposure to other pesticides. However, Chand (1974) suggested that hypoalbuminemia in sheep dosed with 100-150 mg DCPA/kg body wt could be due to imparied digestion and absorption of protein.

EFFECTS ON THE PHYSIOLOGY AND PATHOLOGY OF METABOLIC ORGANS

Generally, no specific or distinctive lesions are found in metabolic organs of ruminants that succumb to organophosphate or carbamate poisoning. However, congestion of the spleen, kidneys and liver including cloudy swelling and fatty degeneration of the liver has been observed in some cases (Khan, 1973). Whether modification of organ function accompanies pathologic changes is not clear. In lambs given an acute dose of Ruelene, clinical signs were observed but the bromosulfophthalein liver function test and blood creatinine (kidney function test) were not affected (Galvin et al, 1960). On the other hand, cattle poisoned with coumaphos had elevated serum indicator enzymes a few days after exposure, suggesting delayed hepatotoxicity and nephrotoxicity (Younger et al, 1971). Serum glutamic oxaloacetic transaminase (SGOT) and serum glutamic pyruvic transaminase (SGPT) activity increased in sheep administered a toxic dose of cythioate (Doval and Gupta, 1976). In cattle administered coumaphos, elevations of SGOT activity correlated with CHE depression (Wright et al, 1966); this was not the case in cattle acutely intoxicated with phosmet and ethion (Roe, 1969). Elevation of indicator enzymes in serum such as SGOT denote biological changes such as membrane permeability, cellular integrity, metabolism, stress or tissue damage (Younger and Wright, 1971).

Chronically poisoned animals often become emaciated and dehydrated (Khan, 1973), suggestin that some type of metabolic interference may be involved. For instance, when Phosvel-containing silage was fed to lactating dairy cattle, feed intake did not increase but the cows gained weight and simultaneously reduced milk production; an influence on disposition of nutrients was suggested as a possible explanation (Johnson et al, 1971; see Table 19-3). On the other hand, Osmanov (1976) showed that azidin administered to dairy cattle stimulated milk formation; enzymes involved in phosphorylation and glycogenolysis decreased in activity and hepatic proteosynthetic processes were stimulated. The metabolic implications of affects such as these are generally not explainable.

Concentrations of various nutrients in blood may fluctuate in organophosphate and carbamate poisoning. In chronic parathion poisoning of goats, hyperglycemia was observed in the terminal stages before death and was thought to be associated with cholinergic effects on the adrenal gland (Wilbur and Morrison, 1955). After long term chronic exposure of steers to ronnel, blood concentrations of total amino acids and ammonia were elevated, but Ca and Mg were not affected (Rumsey et al, 1975). In contrast, when an emulsifiable concentrate of pebulate was administered to cattle over a 5-day period, blood Ca and Mg concentrations were elevated because of impaired renal excretion due to tubular nephritis, but these effects were not observed in sheep (Hunt et al, 1971). Blood Cu was not altered in sheep dosed with di-allate for 29 weeks (Palmer et al, 1972). Ronnel fed at 1000 ppm to sheep for 84 days did not influence blood Ca, Cu, Fe, Mg, P, K, Na, creatinine, uric acid, urea N or total protein (Crookshank et al, 1972). Vitamin E concentrations in blood decreased after organophosphate exposure, but the effect was believed to be due to impaired absorption not metabolic interference (Cowlishaw and Blaxter, 1955; Hunt and McCarty, 1972).

Other pesticide-nutrient interactions were reported by Radeleff (1964). Animals fed fresh green forage were more sensitive to trichlorfon toxicity than those fed dry feeds. Similarly, feeding high levels of vitamin A appeared to increase the toxicity of coumaphos. In another case, cattle that had been sprayed with dioxathion responded to vitamin E/Se treatment as though they were deficient in these nutrients. However, for the most part very little is known about the impact of organophosphate and carbamate pesticides on biochemical processes related to nutrient utilization.

Only in chronic cases do organochlorine pesticides cause pathologic changes in organs of metabolic importance (Radeleff et al, 1955). Degenerative changes of various types are seen in the liver and kidneys (Buck et al, 1976). Enlargement of the liver and proliferation of the smooth endoplasmic reticulum is a characteristic response to most organochlorine pesticides, such as benzene hexachloride in cattle (Ray et al, 1975). Simultaneously, mixed function oxidase (MFO) enzymes are activated resulting in more rapid biotransformation of numerous types of compounds, possibly including some nutrients. Sheep liver MFO enzymes remained elevated in activity for several months after a single dose of dieldrin, DDT or lindane (Ford et al, 1976). This process, known as induction, has been observed in several ruminant species exposed to organochlorine insecticides such as DDT (Seawright et al, 1972; Seawright et al, 1973; Ford et al, 1976), and dieldrin and lindane (Ford et al, 1976). Even though these biological phenomena occur, organ functions are not necessarily altered. For example, deer fed diets containing 25 ppm dieldrin over a 3-year period had normal serum protein concentration and liver function (Tumbleson et al, 1968a), but SGOT and blood alkaline phosphatase were slightly elevated (Tumbleson et al, 1968b). Similarly, in sheep exposed for one year to 25 ppm dieldrin, both serum protein concentration and the albumin:globulin ratio were normal, indicating no liver damage; phenolsulfophthalein excretion rate was normal in these sheep, indicating no damage to the kidney (Harris et al, 1966). However, in sheep administered (acute or chronic) hexachlorophene, a liver lesion accompanied by glycogen depletion was described but was not correlative with clinical signs or death (Hall and Reid, 1974).

As discussed previously, organochlorine pesticides have been shown to interfere with N retention in lambs and calves fed DDT. These results could be explained by a modification of rumen ecology, but an influence of DDT on endogenous N metabolism has not been ruled out. Blood NPN concentrations were not changed in lambs and cows fed diets containing 7 ppm aldrin (Link et al, 1951).

Body concentrations of adipose tissue and viatmin A are affected by organochlorine pesticides. In wethers administered up to 2 mg dieldrin/kg body wt for 32 weeks, accumulation of body fat and plasma vitamin A concentrations were reduced (Davison, 1970; see Table 19-4). Liver vitamin A concentrations were not reduced in the wethers but in cattle fed forage containing 40-60 ppm DDT for 83 days, liver concentrations of vitamin A and carotenoids were reduced (Phillips and Hidiroglou, 1965; see Table 19-9). Conversely, serum levels of vitamin A in the steers

Table 19-9. Hepatic and serum vitamin A concentrations in yearling steers fed DDT-treated forages for 83 days.[a]

Treatment group	Number of steers	Days	Vitamin A: Hepatic, μg/g	Vitamin A: Serum, μg/dℓ
Control	20	0	48	34
		83	61	38
DDT[b]	20	0	48	21
		83	46	48

[a]Modified from Phillips and Hidiroglou (1965)

[b]Standing forage was sprayed with DDT at the rate of 1.5 lbs per acre, mechanically harvested 24 h after spraying and fed ad libitum. The chopped forage comprised the entire ration of the steers.

were elevated after 50 days of DDT treatment. Similarly, decreased concentrations of liver vitamin A in conjunction with increased serum vitamin A was observed in cattle fed chronically 40-60 ppm DDT (Suschetet and Causeret, 1974). However, unaltered blood vitamin A and carotene concentrations were measured in calves fed up to 2.9 mg DDT/kg body wt for as long as 230 days (Thomas et al, 1951) and dairy cattle fed 3.6 mg/kg body wt for up to 141 days (Wilson et al, 1946). However, an organochlorine-induced vitamin A deficiency has not been demonstrated. The interaction between organochlorine pesticides and vitamin A has not been resolved satisfactorily.

Other changes in blood chemistry have been observed in ruminants exposed to organochlorine pesticides. In yearling buffalo calves dosed with 25 mg dieldrin/kg body wt, blood glucose and lactic acid concentrations were increased 2.5 and 4-fold, respectively, when measured after clinical signs had subsided (Malik et al, 1973). In cattle administered 200 mg DDT or lindane/kg body wt, functions of the thyroid and adrenal cortex were impaired, resulting in reduced serum ascorbate concentrations (Evdokimov and Klochkova, 1974). Blood mineral concentrations are apparently not influenced by organochlorines (Thomas et al, 1951; Harris et al, 1951; Bohman et al, 1952; McParland eta al, 1973). In general, organochlorine pesticides appear to influence the metabolic profiles of N, lipid and vitamin A, but the mode of action has not been explained.

Chlorophenoxy compounds like the organophosphates, carbamates and organochlorines do not cause specific pathologic changes in metabolic organs. Congested kidneys and friable livers are commonly observed (Palmer, 1972b; Buck et al, 1976). It has been reported that 2,4-D and 2,4,5-T cause the thyroid to become swollen and congested (Hunt and McCarty, 1972; Palmer, 1972b). Hepatitis (Clark and Palmer, 1971), enlargement (Palmer, 1972b) and necrosis with focal hemorrage and fatty degeneration (Rowe and Hymas, 1954) are other liver changes that have been reported. A 4-fold elevation of SGOT was noted shortly after acute 2,4-D poisoning of dairy cattle, indicating cell injury (McLennon, 1974). However, the biological significance of elevated SGOT is difficult to interpret because it is not organ specific. Blood Mg and Ca ratio decreased and blood urea N increased in cattle treated with 250 mg of 2,4-D/kg body wt for 13-15 days; these blood responses reflected kidney damage (Hunt et al, 1970).

A chlorophenoxy-vitamin A interaction similar to that which has been noted with DDT has been observed. Calves raised on pastures treated with MCPA were unthrifty and showed several other clinical signs, and a response to vitamin A treatment was observed (Hidiroglou and Knutti, 1963). These workers also reported evidence that suggested MCPA or some constituents of MCPA-treated forage interfered with the animals' conversion of carotene to vitamin A, but no effect on vitamin A metabolism, storage, or transport in cattle fed MCPA-treated forage for 83 days was observed (Phillips and Hidiroglou, 1965).

The herbicide DCPA was reported to cause pathologic changes similar to those seen after chlorophenoxy poisoning (Agarwal et al, 1975). An acute dose of the herbicide to yearling cattle brought about a significant elevation of blood glucose and globulin simultaneous with a reduction of blood albumin and albumin:globulin ratio (Chand, 1974). The author suggested the elevated glucose was due to interference with carbohydrate metabolism and the albuminemia the result of impairment of liver protein synthesis.

The fungicide pentachloraphenol caused thyroid atrophy and reduced blood T_3 concentrations in heifers fed 15 mg/kg body wt for 160 days; however, growth rate and feed conversion were not impaired (McConnell et al, 1980). In the same study heifers fed a preparation of the chemical containing the toxic impurities (dioxins and furans) also developed anemia, numerous pathologic lesions, and performed poorly (see Table 19-6).

NUTRITIONAL FACTORS PREDISPOSING PESTICIDE TOXICITY

Whether exposure to a pesticide ends in a toxic response is dependent on numerous factors, some of which may be directly or indirectly related to the nutritional status of the animal. Factors that affect the extent and nature of toxicity create difficulties, especially for those conducting safety evaluation and efficacy studies, i.e., effects observed under one set of conditions (environment, diet, etc.) may differ substantially from effects observed under other conditions. Although this section will focus primarily on nutritionally related factors, it should be recognized there are numerous other factors that influence toxicity which can be grouped into two basic categories: (1) those related to the pesticide itself (physiochemical properties), and (2) those related to the host, such as age, sex, species, breed, pregnancy, lactation, etc., or environmental influences, such as nutrition, stress, disease, climate, etc.

Nutrition can affect the eventual toxicity of a pesticide as a result of certain preabsorption influences, such as rumen metabolism, or post-absorption influences, such as body metabolism, distribution into body compartments, and excretion.

Factors Associated with Rumen Metabolism and Absorption

The microflora of the reticulorumen includes bacteria and protozoa capable of metabolizing some pesticides but not others (Williams, 1977). Reactions known to occur include (1) reduction of organophosphates, such as parathion (Ahmed et al, 1958) and fenitrothion (Mihara et al, 1977); (2) hydrolysis of organophosphates, such as parathion (Ahmed et al, 1958), and carbamates, such as MCA-600 (Williams and Stolzenberg, 1972); (3) O-dealkylation, exemplified by disugran (Ivie et al, 1974), and N-dealkylation, exemplified by trifluralin (Williams and Feil, 1971) and benefin (Golab et al, 1970); (4) dechlorination of organochlorines such as DDT (McCully et al, 1966), methazole (Gutenmann, 1972), erbon (Wright et al, 1970) and dieldrin (Wedemeyer, 1968). Pesticides which are stable in the rumen include bromacil (Gutenmann and Lisk, 1970), triiodobenzoic acid (Gutenmann et al, 1967), and triazines and chlorophenoxy herbicides (James et al, 1975).

Ruminal metabolism generally is beneficial to the animal by degrading a pesticide to a less toxic form. This is true for most organophosphates and carbamates (James et al, 1975). Also, DDT is dechlorinated to DDD which is less toxic and less persistent (McCully et al, 1966). However, a ruminal metabolite formed from the herbicide dinobuton was found to have methemoglobin forming properties (Froslie, 1971). Thus, metabolites generated by microbial metabolism may possess greater or lesser toxicity than the parent compound. Metabolites are generally more water-soluble which in turn affects rate and extent of absorption through the gastric mucosa and subsequent excretion in urine. Nutrition is directly involved in the metabolic process in that the rumen population varies in nature and number with nutritional regimen. However, very few studies have investigated the effect of diet on pesticide metabolism.

The significance of diet was demonstrated in studies with DDT. Ruminal fluid from sheep fed an all-concentrate, high-energy diet converted DDT to DDD at a faster rate than ruminal fluid from sheep fed an all-roughage diet (Sink et al, 1972). Similar results were observed in cattle fed diets high in roughage vs concentrate (Rumsey and Bond, 1974). Conversely, rumen fluid from sheep fed a low-energy ration degraded disugran faster and more extensively than rumen fluid from sheep fed a high energy diet (Ivie et al, 1974).

Factors that influence rate and extent of pesticide absorption also affect resultant toxicity. Chemicals that are solubilized in tract fluids, non-ionized, and sufficiently lipid-soluble possess properties required for maximum absorption. Diet can influence the extent to which a pesticide is ionized by either increasing or decreasing the pH of the GI tract. This is particularly applicable to compounds having a dissociation constant (pKa) near neutrality. Rumen pH can decrease to 5.5 or less when animals are fed diets high in readily available carbohydrates; conversely, during fasting pH can rise to values above 7.0 (Phillipson, 1977). At pH 5.5 or less absorption of weak acids is increased whereas absorption of weak bases is decreased. The inverse is true at a more alkaline pH. Other parts of the GI tract (abomasum and intestine) maintain relatively constant pH regardless of diet (Phillipson, 1977). In a study conducted by Rumsey et al (1970), absorption of DDT and DDD was more rapid in cattle fed a high roughage diet compared to those fed high concentrate. It would seem that this effect would not be pH-related considering DDT has a pKa of 12. Other diet-related factors known to be associated with absorption of nutrients are rate of feed passage, GI tract motility, extent of fill (dilution), and binding of chemical substances to feed components such as lignin. The relationship of these factors to pesticide absorption has not been reported.

Factors Associated with Body Metabolism, Tissue Distribution and Excretion

Foreign compounds such as pesticides, if absorbed, may be excreted unchanged or metabolized to other compounds. Some tend to accumulate in various body compartments, such as adipose tissue. Metabolism generally accomplishes the conversion of lipophilic substances to more hydrophilic forms that are more compatible with renal excretion, thereby preventing accumulation. The enzymes responsible for this biotransformation activity are localized in several organs, primarily the liver, and are usually referred to as mixed function oxidases (MFO) because of their capacity to catalyze a variety of reactions (Parke, 1968). MFO enzymes via epoxidation convert aldrin to dieldrin (Bann et al, 1956) and heptachlor to heptachlor epoxide (Davidow and Redomski, 1952); the epoxides are both more toxic and resistant to renal excretion. This phenomenon is referred to as bioactivation. Conversely, many organophosphates and carbamates are bioinactivated by MFO enzymes and excreted rapidly. In general, chlorophenoxy herbicides do not require metabolic biotransformation in order to be excreted and organochlorine pesticides are generally resistant to metabolism. Nutritional status can have a marked influence on the activity of metabolic enzymes. This has been a subject of extensive research in nonruminant species (Campbell and Hayes, 1973), but not in ruminant species. However, in heifers fed chronically 1250 mg ethylenediamine dihydroiodide (EDDI), activities of certain liver MFO enzymes were significantly reduced (Shull et al, 1979). Thus, the implication is that rate of metabolism of foreign substances (pesticides) and possibly their toxicity and metabolic fate could vary markedly with nutritional regimen.

Distribution of pesticides into body compartments such as body fat has both residue and toxicity ramifications. From a toxicologic standpoint, the more that is taken up into body fat, the less there is available at other sites where potential toxicologic effects might be incurred. For example, cattle and sheep with less body fat reserves are not able to tolerate as large a dose of organochlorines as fatter animals. Emaciated sheep were more susceptible than well-fed sheep to chlordane, toxaphene, and lindane in dipping vats (Radeleff and Bushland, 1960; Radeleff et al, 1963). A similar situation exists for some organophosphates. Emaciated sheep and goats were more sensitive to dichlorofenthion than well-fed sheep; blood CHE activities were more suppressed in the emaciated sheep, presumably due to greater systemic concentrations (less storage in fat) of the organophosphate (Younger et al, 1962). On the other hand, cattle maintained on a minimum ration and then sprayed with coumaphos were not poisoned even though blood cholinesterase activities were depressed 89-95% (Radeleff et al, 1963).

Diet manipulation has been one approach used to reduce concentrations or organochlorine pesticides in body fat. However, mobilization of DDT and fat stores by a submaintenance diet and loss of body weight did not evoke toxicity (Roulston et al, 1954). Moreover, restricted intake did not enhance elimination of dieldrin from steers (Hironaka, 1968) as it did in sheep contaminated with DDT (Childers et al, 1972). Using another approach, after treatment with 3 to 4 times the normal dosages of vitamins A, D, and E, the half-life of dieldrin was shortened substantially in steers (Hironaka, 1968; see Table 19-10).

In lactating dairy cattle pretreated with DDT and then given thyroprotein, milk production increased and significantly more DDT was eliminated each day, thus shortening the half-life of the organochlorine (Stull et al, 1968). However, Fries et al (1970) found that thyroprotein is not effective unless additional energy is supplemented. Neither thyroprotein, milk-fat depressing rations such as alfalfa pellets, nor variable energy intake affected the rate of clearance of heptachlor epoxide (Bush et al, 1965; King, 1968) and dieldrin (Braund et al, 1969). However, the concentration of heptachlor epoxide in feces was increased (King, 1966). A low-energy ration increased the clearance of DDT from cows (Miller, 1967). Moreover, Fries et al (1969) found a difference in the DDT and DDD excretion-energy interaction; DDT excretion was more dependent on milk fat production whereas DDD excretion was more related to energy balance. Cows fed low-energy diets lost body fat reserves and DDT was concentrated in the fat whereas DDD was

Table 19-10. Effect of intramuscular injections of vitamins A, D, and E on the concentration of dieldrin in adipose tissue of cattle.[a]

	Steers		Heifers	
	+ Vitamins[b]	- Vitamins	+ Vitamins[b]	- Vitamins
No. of animals	7	8	4	6
Dieldrin concentration, ppm				
Initial	0.11 ± 0.09	0.12 ± 0.09	0.78 ± 0.73	0.84 ± 0.26
Final	0.06 ± 0.04	0.08 ± 0.05	0.08 ± 0.06	0.08 ± 0.06
Half concentration Time,[d] days	146.3 ± 73.5	245.6 ± 159.0	38.5 ± 9.1	85.7 ± 61.0

[a]Modified from Hironaka (1968)

[b]Three, 2 ml - intramuscular injections at 2 week intervals of a preparation containing 500,000 IU vitamin A, 50,000 IU vitamin D_2, and 50 IU vitamin E per ml.

[c]Values represent mean ± SD.

[d]Adjusted to 2 lb gain/day.

excreted more rapidly; a low-energy diet fed in early lactation accelerated the excretion of DDT. In general, a regimen that reduces milk production in dairy cattle prolongs the retention of organochlorine pesticides in body fat.

Accumulation of organochlorine pesticides in body fat may or may not be affected by nutritional status. For example, accumulation of dieldrin in sheep was not influenced by dietary energy (Davison, 1970). However, when beef heifers were fed low levels of dieldrin continuously from 6 weeks to 18 months of age together with DES implants, urea and concentrate or forage, more dieldrin was stored in body fat of cattle fed concentrate diets and less in cattle implanted with DES (Rumsey and Bond, 1974; see Table 19-11). The authors suggested the effect of DES may be due to its ability to accelerate metabolic rate. There tended to be slightly more residue in cattle fed urea. However, Clark et al (1974) have shown that urea feeding does not affect the toxicity of organochlorines. From Table 19-11 it also appears that there is more residue in concentrate-fed cattle. However, Link et al (1951) reported that calves and sheep fed 7 ppm aldrin over 145-173 days were more susceptible to aldrin poisoning when fed only hay rations. Similarly, in cattle fed an 81% roughage diet vs a 100% concentrate diet, more DDT was stored in body fat, less was metabolized to DDD and DDE, and ruminal protozoa were less sensitive to DDT (Rumsey et al, 1970). The authors suggested that absorption and metabolism of DDT are influenced by diet.

There are other diet-related strategies for preventing or reversing the accumulation pesticides in body fat. A dietary additive that binds to a pesticide in the gut and prevents its absorption is referred to as a mechanical antidote (Cook and Wilson, 1971). Activated charcoal (AC) is the most well-known of these substances. When fed free choice to sheep, goats, or cattle at the level of 2 to 4 g per kg of body wt, AC reduced the absorption of dieldrin from the rumen and brought about a 3-10 fold increase in excretion of dieldrin in feces (Wilson and Cook, 1970; see Fig. 19-3). A regimen of feeding AC in conjunction with administration of phenobarbital which increases the rate of recycling of pesticides back to the GI tract by way of the bile (Cook, 1970) has been applied effectively to dieldrin contamination cases (Cook and Wilson, 1971). A slightly different regimen was employed by Harr et al (1974) for purging dieldrin from steers. It consisted of an initial injection of testosterone propionate and vitamins A, D, and E and of daily oral doses of phenobarbital (first 20 days only) and AC in a finishing ration. The rate of dieldrin loss was increased six-fold and the residual half-life reduced from 150-250 days to about 25-80 days. The authors stated the rate of AC feeding appeared to be the most important component of the regimen.

AC is not as effective a mechanical antidote against other pesticides. Fries et al (1970) found it lowered the body load of DDT when

Table 19-11. Effect of type of diet, DES, and nitrogen supplements on the concentrations of dieldrin and aldrin residues in depot fat of beef heifers.[a]

Treatment[c]	Residue content of depot fat, ppm[b]		
	Dieldrin	Aldrin	Total
C - U	43.76[d]	0.26[d]	44.02[d]
C - S	35.13[de]	0.18[def]	35.31[de]
C + U	38.02[de]	0.20[de]	38.22[de]
C + S	30.04[e]	0.06[ef]	30.10[e]
F - U	31.81[e]	0.08[ef]	31.89[e]
F - S	29.09[ef]	0.05[f]	29.14[ef]
F + U	18.97[ef]	0.11[ef]	19.08[ef]
F + S	19.64[ef]	0.10[ef]	19.74[ef]

[a]Modified from Rumsey and Bond (1974). All cattle were fed from 42 days of age to 18 months 1.0 mg aldrin/kg body wt/day. DES (12 mg) was implanted at 168 days and again at 346 days.
[b]Each mean is an average of 16 samples and represents concentration in extracted fat; means within each column are different ($P < 0.05$) if they do not contain a common letter (d, e, f) in the superscript.
[c]Concentrate diet (C), forage diet (F), no DES (-), DES implanted (+) urea supplement (U), soybean meal supplement (S).

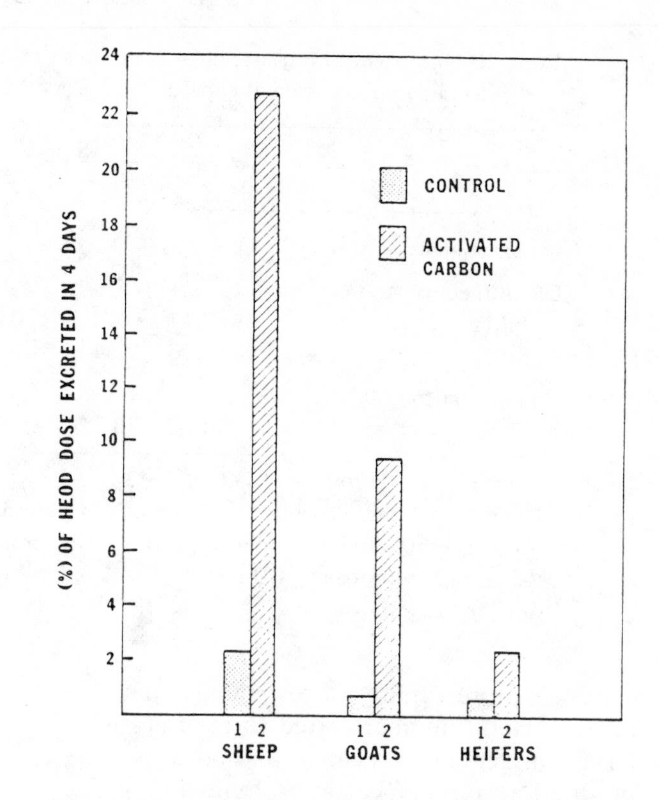

Figure 19-3. Percent of the total HEOD (dieldrin) added to the rumen that was excreted in 4 days. From Wilson and Cook (1970).

fed simultaneously with the pesticide but did not lower the residual concentration in milk when fed after DDT exposure. Similarly, Smalley et al (1971) showed that AC fed simultaneously with ronnel to sheep reduced the residue levels in omental fat by 10% compared to controls but did not affect the rate of excretion after ronnel feeding was terminated. In steers fed AC, 31% less ronnel accumulated in body fat (Rumsey et al, 1975). AC can be fed chronically at a level of 5% of the total ration without compromising health or performance in sheep (Crookshank et al, 1972) and steers (Rumsey et al, 1975).

Other mechanical antidotes that have been tested include mineral oil, animal and vegetable fat, calcium salts, bentonite, pectin, lignin and diatomaceous earth (Cook and Wilson, 1971). King (1968) found that mineral oil and fat supplementation slightly decreased heptachlor in milk fat and increased heptachlor in feces. Similarly, in lambs administered a dosage of ^{14}C-HCB sufficient to provide adipose tissue concentrations of 50-120 ppm, mineral oil given as 5% of the diet 2 weeks after the last dosage increased fecal excretion of HCB about 3-fold over levels of untreated controls (Smith et al, 1979). Moreover, mineral oil did not affect feed intake or the digestibility of dry matter and protein.

Table 19-12. General information on pesticides cited in Chapter 19.

Abbreviations used: I = insecticide; H = herbicide; F = fungicide; B = bacteriocide; G = grubicide; GR = growth regulator; N = nematocide; M = molluscacide; D = defoliant; A = Acaricide; IR = insect repellant

Common name	Trade name	Action	Principal Applications
Organochlorines			
Aldrin[ab]	Aldrex[cd]	I	Control of insects, on cotton, in soil, turf, and around buildings
BHC[b] or Benzene hexachloride[a]	Sopracide[cd]	I	Control of cotton insects
Chlordane[ab]	Octachlor[cd]	I	Control of institutional pests (ex. termites)
DDD[a] (TDE)[b]	Rhothane[c]	I	Control of insects on fruits and vegetables
DDT[ab]	Kopsol[cd]	I	Control of insects of human and animal disease, broad use on crops and forests
Dieldrin[ab]	Alvit[c]	I	Control of soil insects, public health insects, termites, etc. on crops
Endrin[b]	Endrex[c]	I	Control of insects on cotton and field crops, and pine mice in orchards
Ethylan[a]	Perthane[bc]	I	Control of pests on vegetables and pears, and moths and carpet beetles on textiles
Heptachlor[b]	Drinox[cd]	I	Control of soil and turf insects, ants, grasshoppers, and termites
Hexachlorophene[b]	Hexide[cd]	F,B	Control of fungal and bacterial diseases in vegetables, soil; seed fungicide
Lindane[ab] or Gamma BHC	Silvanol[cd]	I	Used as a seed treatment, control of grasshoppers, and other insects
Methoxychlor[ab]	Chemform[c]	I	Control of insects on fruit and shade trees, vegetables, garden, farm buildings, and livestock
Mirex[ab]	Dechlorane[c]	I	Control and irradication of many species of ants
Toxaphene[ab]	Camphochlor[cd]	I	Control of insects on cotton, field crops and animals
Organophosphates			
Azidin[e]		I	
Chlorpyrifos	Lorsban[bc]	I	Control of corn rootworms, peach tree borer; corn seed treatment against corn maggot
Coumphos	Co-Ral[cd]	G,I	Control of pests in livestock (flies, lice, and grubs)
Crotoxyphos	Ciodrin[cd]	I	Control of flies, lice, and ticks in livestock, contact stomach poison
Crufomate	Ruelene[bc]	G,I	Control of pests in livestock (flies, lice, and grubs)
Cyanofenphos[b]	Surecide[c]	I	Control of moth and butterfly larvae
Cythioate[b]	Proban[c]	I	Control of ectoparasites on domestic animals
Demeton[a]	Systox[bc]	I,Ac	Control of insects on forage crops
Dichlorofenthion	Mobilawn[bc]	I,N	Control of nematodes and insects on flowers and lawn
Dioxathion[a]	Delnav[bc]	I,Ac	Control of livestock pests and insects on fruits
EPN[b]		I,Ac	Control of insects on crops
Ethion[b]		I,Ac	Control of pests on food, fiber
Famfur[ab]	Warbex[c]	I,G	Control of cattle grubs and lice
Fensulfothion	Dasanit[bc]	I,N	Control of soil insects and nematodes
Fenthion[a]	Baytex[bcd]	I,G	Control of insects on ornamental plants, grubs, and lice in cattle
Fenitrothion[b]	Accothion[cd]	I,Ac	Control of insects on food and cotton crops; use in public health programs
Leptophos	Phosvel[bc]	I	Control of insects on food crops and cotton
Malathion[ab]	Cythion[cd]	I	Control of wide range of insects on crops, ornamentals, stored products, livestock (flies and lice)
Methidathion	Supracide[bc]	I,Ac	Control of alfalfa weevil, orchard, and vineyard pests
Parathion[ab]	Bladan[cd]	I	Control of many species of insects on crops
Phorate[a]	Thimet[bc]	I	Control of many species of insects on crops
Phosmet	Imidan[bc]	I	Control of many species of insects on crops
Ronnel[ab]	Korlan[cd]	I,G	Control of pests in cattle (flies, lice, grubs); residential uses
Temephos	Abate[bc]	I	Control of mosquito larvae
Trichlorfon[a]	Dylox[bc]	I,G	Control of wide variety of insects on crops and livestock (flies and lice)
Tri-o-cresyl phosphate[g]			
Carbamates			
Carbaryl[b]	Sevin[c]	I	Control of many species of insects on crops, citrus, forest, lawns, rangeland, and poultry and pets
Carbofuran	Furadan[bcd]	I,N	Control of soil pests, alfalfa weevil, and numerous other forages and food crop pests
Di-allate	Avadex[bc]	H	Control of wild oats in most leguminous crops, barley, corn, potatoes, and sugar beets
Dichlormate[g]			

Table 19-12. Continued

Common name	Trade name	Action	Principal Applications
MCA - 600	Mobam[bc]	I	Control of cockroaches, flies, grasshoppers and a variety of crop insects
Pebulate	Tellam[bc]	H	Pre-emergence control of grasses and broad leaf weeds on sugarbeets, tomatoes, and tobacco
Zineb[b]	Zinosan[cd]	F	Control of seed- and soil-born organisms
Chlorophenoxys			
2,4-D[bc]		H	Control of weeds on cereals, turf, pastures and non-cropland
2,4-DB[b]	Butoxone[cd]	H	Control of broad leaf weeds in forage legumes
Erbon[b]	Novege[c]	H	Control of weeds in non-crop areas; residual sterilant
MCPA[b]	Agroxone[cd]	H,GR	Control of weeds in various crops such as food crops, turf, legumes
MCPB[b]	Can-Trol[cd]	H	Control of weeds in peas
Mecoprop[b] (MCPP)	Compitox[cd]	H	Control of boradleaf weeds in cereals and grasses
Silvex[ab]	Kuron[cd]	H,GR	Controls turf weeds, aquatic plant pests, weeds in food crops
2,4,5,-T[b]	Esteron[cd]	H	Control of brush on rangelands, rights of way, forests
Dipyridyls			
Diquat[ab]	Aquacide[cd]	H,D	Control of non-crop weeds; seed crop dessicant, preharvest top killer and general aquatic herbicide
Paraquat[ab]	Dextrone X[cd]	H,D	Weed control during seed crop establishment, weed control in orchards and non-crop areas
Organometallics			
Cadminate[bc]		F	Plant fungicide primarily for turf
DSMA[a]	Arsinyl[cd]	H	Postemergent control of grassy weeds in non-crop areas and turf
Methylmercury	Ceresan L[bc]	F	Seed-treating agent
	Ceresan M[c]		Seed-treating agent
MSMA[b]	Ansar[cd]	H	Postemergent control of grassy weeds in non-crop areas, turf, citrus-bearing and non-bearing orchards
Cacodylic acid[b]	Dilic[cd]	H,D	Non-selective herbicide, cotton defoliant, and tree killer
Other			
Benefin	Balan[bcd]	H	Preemergent control of grassy and broadleaf weeds in forage crops
Bromacil[ab]	Cynogan[cd]	H	Control of weeds in non-cropland areas
Magnesium chlorate	Magron[cd]	H,D	Control of grasses, broadleaf weeds (annuals and perennials), kills trees, and stumps
Sodium chlorate	Klorex[cd]	H,D	Control of grasses, broadleaf weeds (annuals and perennials), kills trees, and stumps
DCPA	Dacthal[bc]	H	Control of grassy and broadleaf weeds
Dicamba	Banvel[bcd]	H	Control of brush and broadleaf weeds on both non-cropland and crops
Dinobuton[b]	Acrex[cd]	Ac,F	Control of mites on fruits, vegetables and some crops
Disugran	Racuza[c]	GR	Increases recoverable sugar in sugarcane and sugar beets, affects ripening of grapefruit, increases yields of grapes
Diuron[ab]	Diurex[cd]	H	Control of broadleaf and grassy weeds in numerous crops, weed killer, soil sterilant
Methazole	Probe[bc]	H	Pre- and post emergence control of both grassy and broadleaf weeds
Molinate	Ordram[bc]	H	Control of watergrass on rice, other weeds
Naptalam	Alanap[bc]	H	Control of numerous leaf weeds in soybeans, peanuts, and vine crops
Nitralin	Planavin[bc]	H	Pre-emergence control of weeds in various fruit and vegetable crops
Pentachlorophenol (PCP)[b]	Pentacon[cd]	F,B,M,D	Wood preservative, preharvest dessicant, snail control
Dibutyl- or dimethyl phthalate[b]		IR	Insect repellent for mosquitoes, fleas, chiggers
Pichloram	Tordon[bc]	H	Control of woody and broadleaf weeds in pastures, rangeland, and small grains
Propazine	Milogard[bcd]	H	Preemergence control of annual broadleaf weeds and grasses in milo and sorghum
Simazine[ab]	Aquazine[cd]	H	Control of most annual grassy and broadleaf weeds in many forage and orchard crops; algae control in ponds
2,3,6-tri-chlorobenzoic acid (TBA)	Benazc[cd]	H	Control of many species of broadleaf weeds in non-cropland areas.
TIBA[b]	Regim - 8[c]	GR	Initiates flower budding on apple trees, promotes earlier maturity and improves harvestability of soybeans
Trifluralin[ab]	Treflan[cd]	H	Pre-emergent control of grassy and broadleaf weeds on numerous crops

[a]Common name officially designated by USDA; [b]Listed under this name in the Farm Chemicals Handbook (1980); [c]Registered trade name; [d]Several trade names exist; this is one randomly chosen; [e]Not registered in the USA; [f]For export use only; [g]Not a registered pesticide.

Literature Cited

Agarwal, N.C., H.S. Bahga, U.K. Sharma and B.K. Soni. 1974. Indian J. Animal Sci. 44:101.
Ahmed, M.K. and J.E. Casida. 1958. J. Econ. Ent. 51:59.
Alexander, F. and E.W. Moodie. 1960. Res. Vet. Sci. 1:248.
Anderson, B., R.L. Kitchell and N. Persson. 1959. Acta Physiol. Scand. 46:319.
Anderson, P.H. and A.F. Machin. 1969. Vet. Rec. 85:484.
Bann, J.M., T.J. DeCinto, N.W. Earle and Y.P. Sun. 1956. J. Agr. Food Chem. 4:937.
Barber, T.A. 1972. Diss. Abstr. Int. 31:475/B.
Biddulph, C. et al. 1952. J. Dairy Sci. 35:445.
Bjorklund, N.E. and K. Erne. 1966. Acta Vet. Scand. 7:364.
Bohman, V.R. et al. 1952. J. Dairy Sci. 35:6.
Braund, D.G. et al. 1969. J. Dairy Sci. 52:172.
Buck, W.B., G.D. Osweiller and G.A. VanGelder. 1976. Clinical and Diagnostic Veterinary Toxicology. Kendall/Hunt Publ. Co., Dubuque, Iowa.
Bush, M.B., R.L. King and R.W. Hemken. 1965. J. Dairy Sci. 48:794.
Calderbank, A., M.A. Stevens and J.K. Walley. 1968. J. Sci. Food Agr. 19:246.
Campbell, J.B., W. Woods, A.F. Hagen and E.C. Howe. 1973. J. Econ. Ent. 66:429.
Campbell, T.C. and J. Hayes. 1974. Pharmacol. Rev. 26:171.
Chand, N. 1974. Indian J. Med. Sci. 28:76.
Childers, A.B., C.A. Carson and W.A. Landmann. 1972. J. Agr. Food Chem. 20:878.
Clark, D.E. and J.S. Palmer. 1971. J. Agr. Food Chem. 19:761.
Clark, D.E., H.E. Smalley, H.R. Crookshank and F.M. Farr. 1974. Pest. Monit. J. 8:180.
Clarke, E.G.C. and M.L. Clarke. 1975. Veterinary Toxicology. The Williams and Wilkins Co., Baltimore, MD.
Cook, R.M. 1970. J. Agr. Food Chem. 18:434.
Cook, R.M. and K.A. Wilson. 1971. J. Dairy Sci. 54:712.
Cowlishaw, B. and K.L. Blaxter. 1955. Proc. Nutr. Soc. 14:iii.
Crookshank, H.R., H.E. Smalley and R.D. Radeleff. 1972. J. Animal Sci. 34:322.
Dahm, P.A. and N.L. Jacobson. 1956. J. Agr. Food Chem. 4:150.
Davidow, B. and J. Radomski. 1952. Fed. Proc. 11:336.
Davison, K.L. 1970. J. Anima. Sci. 31:567; J. Agr. Food Chem. 18:1156.
Dickinson, J.O. 1972. Am. J. Vet. Res. 33:1889.
Doval, C.P. and I. Gupta. 1976. Indian J. Exp. Biol. 14:193.
Eknathrao, D.S. 1966. Indian Vet. J. 43:85.
Ely, R.E. et al. 1953. J. Amer. Vet. Med. Assoc. 123:448.
Evdokimov, E.S. and G.M. Klochkova. 1974. Izv. Akad. Nauk. Turkm. 6:25.
Ford, E.J.H., O. Araya, R. Rivera and J. Evans. 1976. Toxicol. Appl. Pharmacol. 35:475.
Fries, G.F., W.F. Flatt and L.A. Moore. 1969. J. Dairy Sci. 52:684.
Fries, G.F. et al. 1970. J. Dairy Sci. 53:1632.
Froslie, A. 1971. Acta Vet. Scand. 12:300.
Galvin, T.J., R.R. Bell and R.D. Turk. 1960. Amer. J. Vet. Res. 21:1058.
Gannon, N., R.P. Link and G.C. Decker. 1959. J. Agr. Food Chem. 7:829.
Gardener, D.L. 1976. Vet. Med. Small Anim. Clin. 71:1589.
Golab, T. et al. 1970. J. Agr. Food Chem. 18:838.
Greer, N.I., M.J. Janes and D.W. Beardsley. 1973. Fla. Entomol. 56:243.
Grigsby, B.H. and E.D. Farwell. 1950. Mich. Agr. Exp. Sta. Quart. Bul. 32:378.
Gupta, B.N., V. Mahadeven and H.P. Singh. 1968. Indian Vet. J. 45:1037.
Gutenmann, W.H., C.A. Bache and D.J. Lisk. 1967. J. Agr. Food Chem. 15:812.
Gutenmann, W.H. and D.J. Lisk. 1970. J. Agr. Food Chem. 18:128.
Gutenmann, W.H., J.W. Serum and D.J. Lisk. 1972. J. Agr. Food Chem. 20:991.
Hall, G.A. and I.M. Reid. 1974. J. Path. 114:241.
Harr, J.R., J.W. Gillett, J.H. Exon and D.E. Clark. 1974. Bull. Env. Contam. Tox. 12:433.
Harris, L.E. et al. 1951. J. Animal Sci. 10:581.
Harris, L.E. et al. 1966. (quoted by: Hodge, H.C. et al., Toxicol. Appl. Pharmacol. 10:613, 1967).
Hayes, W.J. 1975. Toxicology of Pesticides. The Williams and Wilkins Co., Baltimore, MD.

Headley, J.C. 1968. Amer. J. Agr. Econ. 50:13.
Herigstad, R.R. et al. 1972. J. Amer. Vet. Med. Assoc. 160:173.
Hidiroglou, M. and H.J. Knutti. 1963. Can. J. Animal Sci. 43:113.
Hironaka, R. 1968. Can. Vet. J. 9:167.
Hunt, L.M., B.N. Gilbert and J.S. Palmer. 1970. Bull. Environ. Contam. Toxicol. 5:54.
Hunt, L.M., B.N. Gilbert, J.S. Palmer and R.L. Younger. 1971. Bull. Environ. Contam. Toxicol. 6:263.
Hunt, L.M. and R.T. McCarty. 1972. Bull. Environ. Contam. Toxicol. 8:297.
Ivie, G.W., D.E. Clark and D.D Rushing. 1974. J. Agr. Food Chem. 22:632.
James, L.T., M.J. Allison and E.T. Littedike. 1975. In: Digestion and Metabolism in the Rumen. U. of New England Publishing Unit, Armidale, Aust.
Johnson, J.C., M.C. Bowman, D.B. Leuck and F.E. Knox. 1971. J. Dairy Sci. 54:1840.
Johnson, J.C., M.C. Bowman, D.B. Leuck and F.E. Knox. 1973. J. Dairy Sci. 56:775.
Khan, M.A. 1961. Amer. J. Vet. Res. 23:516.
Khan, M.A. 1971. Can. J. Animal Sci. 51:411.
Khan, M.A. 1973. Vet. Res. 92:411; Res. Vet. Sci. 15:180.
Khan, M.A., W.N. Harries and W.J. Dorward. 1972. Can. Vet. J. 13:129.
Khan, M.A., T. Kramer and R.J. Avery. 1961. Can. Vet. J. 2:207.
King, R.L. et al. 1966. Proc. Maryland Nutr. Conf. p. 71.
King, R.L. 1968. Report to ARS - USDA.
Kinzell, J.H. et al. (in press). J. Dairy Sci.
Kutches, A.J., D.C. Church and F.L. Duryee. 1970. J. Agr. Food Chem. 18:430.
Leng, M.L. 1972. Down to Earth 28:12.
Leuck, D.B. et al. 1971. J. Econ. Ent. 64:1394.
Link, R.P., G.C. Decker, and C.C. Morrill. 1951. (quoted by: Hodge, H.C. et al. 1967. Tox. Appl. Pharm. 10:613).
Loomis, E.C., A.N. Webster and P.G. Lobb. 1976. Vet. Rec. 91:168.
Malik, J.K., H.S. Bahaz and S.C. Cud. 1973. Indian J. Animal Sci. 43:711.
Marsh, H., L.H. Johnson, R.S. Clark and J.H. Pepper. 1949. Montana Agr. Exp. Sta. Rept. 477:2.
McCarty, R.T., M. Haufler and C.A. McBeth. 1968. J. Amer. Vet. Med. Assoc. 152:279; Amer. J. Vet. Res. 29:911.
McCarty, R.T., M. Haufler, M.G. Osborn and C.A. McBeth. 1969. Amer. J. Vet. Res. 30:1149.
McConnell, E.E. et al. 1980. Toxicol. Appl. Pharmacol. 53:(in press)
McCully, K.A. et al. 1966. J.A.O.A.C. 49:966.
McLennan, M.W. 1974. Aust. Vet. J. 50:578.
McParland, P.J., R.M. McCracken, M.B. O'Hare and A.M. Raven. 1973. Vet. Rec. 93:369.
Mihara, K., Y. Okuno, Y. Misaki and J. Miyamoto. 1978. J. Pestic. Sci. 2:233.
Miles, J.T. et al. 1971. J. Dairy Sci. 54:478.
Miller, D.D. 1967. J. Dairy Sci. 50:1444.
Mitchell, J.W., R.E. Hodgson and C.F. Gaetjen. 1946. J. Animal Sci. 5:226.
Mullee, M.T., D.L. Nelson and D.D. Cox. 1968. Vet. Med. Small Animal Clin. 63:876.
Murphy, D.A. and L.J. Korschgen. 1970. J. Wildl. Mgmt. 34:887.
Orr, L.W. and L.O. Mott. 1945. J. Econ. Ent. 38:428.
Osmanov, A.R. 1976. Dokl. Vses. Akad. Skh. Nauk. 12:24.
Palmer, J.S. 1963. J. Amer. Vet. Med. Assoc. 143:398, 994, 1385.
Palmer, J.S. 1964. J. Amer. Vet. Med. Assoc. 145:917.
Palmer, J.S. 1968. J. Amer. Vet. Med. Assoc. 152:282.
Palmer, J.S. 1971. J. Amer. Vet. Med. Assoc. 159:1263.
Palmer, J.S. 1972a. J. Amer. Vet. Med. Assoc. 160:338.
Palmer, J.S. 1972b. U.S. Dept. Agr. Prod. Res. Rpt. 137:1.
Palmer, J.S. and R.D. Radeleff. 1964. Ann. N.Y. Acad. Sci. 111:729.
Palmer, J.S. et al. 1972. Amer. J. Vet. Res. 33:543.
Pankaskie, J.E., F.C. Fountaine and P.S. Dahm. 1952. J. Econ. Ent. 45:51.
Parke, D.V. 1968. The Biochemistry of Foreign Compounds. Pergamon Press, N.Y.
Phillips, W.E.J. and M. Hidiroglou. 1965. J. Agr. Food Chem. 13:254.
Phillipson, A.T. 1977. In: Duke's Physiology of Domestic Animals, 9th edition. Comstock Publishing Assoc.
Polan, C.E., J.T. Huber, R.W. Young and J.C. Osborne. 1969. J. Agr. Food Chem. 17:857.

Pugh, W.S. 1975. Can. Vet. J. 16:56.
Radeleff, R.D. 1949. Vet. Med. 44:436.
Radeleff, R.D. 1951. Vet. Med. 46:105.
Radeleff, R.D. 1964. Veterinary Toxicology. Lea and Febiger, Philadelphia, PA.
Radeleff, R.D. 1970. Veterinary Toxicology. 2nd ed. Lea and Febiger, Philadelphia, PA.
Radeleff, R.D. and R.C. Bushland. 1960. In: The Nature and Fate of Chemicals Applied to Soils, Plants, and Animals. USDA ARS20-9, pp. 134-159.
Radeleff, R.D. et al. 1963. J. Econ. Ent. 56:272; J. Amer. Vet. Med. Assoc. 142:624.
Radeleff, R.D. and G.T. Woodard. 1957. J. Econ. Ent. 50:249.
Radeleff, R.D., G.T. Woodard, W.J. Nickerson and R.C. Bushland. 1955. U.S. Dept. Agr. Tech. Bull. 1122.
Ray, A.G., J.D. Morris and J.C. Reagor. 1975. J. Amer. Vet. Med. Assoc. 166:1180.
Riley, J.G. and D.R. Ware. 1977. J. Animal Sci. 45(suppl. 1):253.
Roe, R.T. 1969. Aust. Vet. J. 45:332.
Roulston, W.J., L.F. Hitchcock, A.W. Turner and A.D. Campbell. 1953. Aust. J. Agri. Res. 4:469.
Rowe, V.K. and T.A. Hymas. 1954. Amer. J. Vet. Res. 15:622.
Rumsey, T.S. 1979. J. Animal Sci. 49:1059.
Rumsey, T.S. and J. Bond. 1972. J. Animal Sci. 35:978.
Rumsey, T.S. and J. Bond. 1974. J. Agr. Food Chem. 22:664.
Rumsey, T.S., L.L. Slyter, S.M. Shepherd and D.L. Kern. 1970. J. Agr. Food Chem. 18:485.
Rumsey, T.S., E.E. Williams and A.D. Evans. 1975. J. Animal Sci. 40:743.
Scharf, D.K. 1962. J. Econ. Ent. 55:191.
Seawright, A.A., D.P. Steele, A.W. Mudie and R. Bishop. 1972. Res. Vet. Sci. 13:245.
Seawright, A.A., L.J. Filippich and D.P. Steele. 1973. Res. Vet. Sci. 15:158.
Selby, L.A., A.A. Case, C.R. Dorn and D.J. Wagstaff. 1974. J. Amer. Vet. Med. Assoc. 165:1010.
Shull, L.R., S.K. McCarthy and D.J. Crosswhite. 1977. Proc. Conf. on Rumen Function, p. 16-17.
Shull, L.R. and S.K. McCarthy. 1978. J. Dairy Sci. 61:260.
Shull, L.R., R. Gartner and S.K. McCarthy. 1979. Proc. Soc. Tox., A172.
Shulz, O., K. Kirchner, P. Meuller and R. Rothe. 1976. Monatsch. Veterinaermed. 31:647.
Sink, J.D., H. Valera-Alvarez and C. Hess. 1972. J. Agr. Food Chem. 20:7.
Smalley, H.E., H.R. Crookshank and R.D. Radeleff. 1971. J. Agr. Food Chem. 19:331.
Smith, G.S. et al. 1979. Proc. of the 15th National Rumen Function Conference.
Smrek, A.L., S.R. Adams, J.A. Liddle and R.D. Kimbrough. 1977. J. Agr. Food Chem. 25:1321.
Strickland, A.H. 1970. In: Concepts of Pest Management. Proc. of Conference at No. Caro. St. Univ., Raleigh.
Stull, J.W. et al. 1968. J. Dairy Sci. 51:56.
Suschetet M. and J. Causeret. 1974. Ann. Nutr. Aliment. 28:307.
Thomas, J.W., P.E. Hubankds, R.H. Carter and L.A. Moore. 1951. J. Dairy Sci. 34:203.
Tumbleson, M.E. et al. 1968. Am. J. Vet. Clin. Pathol. 2:121, 127.
Wedemeyer, G. 1968. Appl. Microbiol. 16:661.
Wiedenbach, C.P. and R.L. Younger. 1962. J. Econ. Ent. 55:793.
Wilber, C.G. and R.A. Morrison. 1955. Amer. J. Vet. Res. 16:308.
Williams, P.P. 1977. Residue Rev. 66:63.
Williams, P.P., K.L. Davison and E.J. Thacker. 1968. J. Animal Sci. 27:1472.
Williams, P.P. and V.J. Feil. 1971. J. Agr. Food Chem. 19:1198.
Williams, P.P., J.D. Robbins, J. Gutierrez and R.E. Davis. 1963. Appl. Microbiol. 11:517.
Williams, P.P. and R.L. Stolzenberg. 1972. Appl. Microbiol. 23:745.
Wilson, H.E. et al. 1946. J. Econ. Ent. 39:801.
Wilson, K.A. and R.M. Cook. 1970. J. Agr. Food Chem. 18:437.
Wooden, G.R. and J.W. Algeo. 1977. J. Animal Sci. 45 (Suppl. 1):269.
Wrich, M.J. 1971. Diss. Abst. Int. 32:2779.
Wright, F.C., L.M. Hunt and J.S. Palmer. 1966. Amer. J. Vet. Res. 27:177.
Wright, F.C., J.S. Palmer and J.C. Schlinke. 1970. J. Agr. Food Chem. 18:845.
Wright, F.C. et al. 1977. J. Agr. Food Chem. 25:293.
Younger, R.L., R.D. Radeleff and D.P. Weidenbach. 1962. J. Econ. Ent. 55:249.
Younger, R.L. and R.D. Radeleff. 1963. 1963. J. Econ. Ent. 56:592.
Younger, R.L., R.D. Radeleff and J.B. Jackson. 1963. J. Econ. Ent. 56:757.
Younger, R.L. and F.C. Wright. 1971. Amer. J. Vet. Res. 32:1053.

Appendix

Composition of Feeds

International feed names are used in most instances in this book. The nomenclature of the feeds under which the analytical data are shown primarily follows the "International Feed Vocabulary" as outlined by Harris et al (Utah Agr. Expt. Sta. Bul. 501; 1978). Many feeds in the United States and Canada have official names and definitions as designated by the Association of American Feed Control Officials (AAFCO) and the Canada Feed Act (CFA). Frequently, these names are common or trade names and the origin of the feed name does not always follow a standardized naming system. A name should clearly state the source of the material and describe any process, alteration or special circumstance which affects the nutritional value of that feed.

The International Feed Vocabulary is designed to give a comprehensive name to each feed as concisely as possible. Each feed name was coined by using descriptors taken from one or more of 6 facets: (a) origin consisting of scientific name (genus, species, variety); common name (generic and kind, and chemical formula as appropriate); (2) part fed to animals as affected by process(es); (3) process(es) and treatment(s) to which the part has been subjected; (4) stage of maturity or development (applicable to forages and animals); (5) cutting (primarily applicable to forages); (6) grade (official grades with guarantees).

Feeds are grouped into classes on the basis of their composition and in the way they are used for feeding animals or in formulating diets. These classes, by necessity, are arbitrary and, in borderline cases, a feed is assigned to a class according to the most common use made of it in usual feeding practice.

Each feed is assigned a unique 5-digit number for positive identification. Numbers are assigned consecutively as new feed names are created. In addition, the feed class number as identified above is put in front of the international feed number (IFN). Therefore, the full IFN is a 6-digit number.

The data were compiled by the International Feedstuffs Institute at Utah State University. To assist in making the tables more useful, source data values were generated for missing data by using regression equations. In some cases data were estimated from similar feeds, such as for stage of maturity of forages.

Tables on Animal Requirements

The appendix tables listing nutrient requirements for the various species and classes of animals within species have been taken from the latest National Research Council (NRC) publications on beef and dairy cattle, sheep and a forthcoming publication on goats. These various publications have been authored by committees considered to have expertise on that particular topic. These tables are reproduced with permission of the National Academy of Science.

Appendix Table 1. Composition of feeds consumed by ruminants.

Name	Ident. #	Stage of maturity, etc.	Typical DM, %	Composition, % dry basis			
				CP	C. fiber	ADF	NDF
				....	%	....	
ROUGHAGES							
Pasture-Range Grasses							
Alkali Scaton *(Sporobolus airoides)*	2-08-691	mid bloom		7.5	30		
Bermudagrass *(Cynodon dactylon)*	2-00-706	ear. veg.	32	9.6	30		
	2-08-354	mid bloom	35	10.3	28		
Bluegrass, Kent. *(Poa pratensis)*	2-00-777	ear. veg.	31	17.4	25		
	2-00-779	ear. bloom	35	16.6	27		
Bluegrass, Sandberg *(P. secunda)*	2-00-804	ear. bloom		11.0	33		
Bluestem *(Andropogon* spp.)	2-00-821	ear. veg.	27	12.8	25		
	2-00-825	mature	59	5.8	34		
Bluestem, Little *(A. scoparius)*	2-00-849	fresh	58	5.6	33		80
Brome, Cheatgrass *(Bromus tectorum)*	2-00-908	ear. veg.	22	15.8	23		
	2-00-909	full bloom		10.0	32		
	2-00-911	mature	55	5.3	35		
Brome, Ripgut *(B. rigidus)*	2-00-938	fresh		8.8	27		
Brome, Smooth *(B. inermis)*	2-00-956	ear. veg.	29	22.7	23	27	48
	2-00-364	mature	55	6.0	35		
Buffalograss *(Buchloe dactyloides)*	2-01-010	fresh	46	10.3	27	36	74
Buffelgrass *(Cenchrus ciliaris)*	2-10-360	fresh	32	5.5	34		
Canarygrass, Reed *(Phalaris arundinacea)*	2-01-113	fresh	23	18.9	22	28	46
Cordgrass, Marshhay *(Spartina patens)*	2-01-591	ear. veg.		11.1	32		
	2-01-592	mature		5.0	33		
Cordgrass, Prairie *(S. pectinata)*	2-01-593	mature		7.1	36		
Dropseed, Sand *(Sporobolus cryptandrus)*	2-01-821	ear. bloom	36	12.2	32		
Fescue, Alta *(Festuca arundinacea)*	2-01-882	ear. veg.	25	20.3	22	32	48
	2-01-883	ear. bloom	21	13.6	21	38	58
Fescue, Foxtail *(F. megalura)*	2-01-891	ear. veg.		15.5	24		
	2-01-892	full bloom		7.6	33		
Fescue, Idaho *(F. idahoensis)*	2-01-898	fresh		5.6	38		
Fescue, Meadow *(F. elatior)*	2-08-420	ear. veg.	25	17.6	24		
Fescue, Sixweeks *(F. octoflora)*	2-01-935	ear. veg.		19.7	24		
Foxtail, Meadow *(Alopecurus pratensis)*	2-02-073	ear. veg.	26	17.2	22		
Galleta *(Hilaria jamesii)*	2-05-594	stem cured	86	5.5	33		
Grama, Black *(Bouteloua eriopoda)*	2-02-169	ear. veg.	50	11.5	30		
	2-02-170	ear. bloom		9.3	31		
Grama, Blue *(B. gracilis)*	2-02-180	ear. veg.	48	11.8	29	33	
	2-08-426	dough stage	65	6.3	29	41	
Guineagrass *(Panicum maximum)*	2-02-338	late veg.	24	9.2	33	45	69
Lovegrass *(Eragrostis* spp.)	2-02-647	ear. veg.	43	12.6	31		
	2-02-648	full bloom	45	8.9	33		
Millet, Foxtail *(Setaria italica)*	2-03-101	fresh	29	9.6	32		
	2-03-100	late bloom	26	9.2	31		
Mixed Grasses (English data)							
Very leafy			18	22.2	20		
Leafy			19	17.4	24		
Early Flowering			21	14.3	26		
Flowering			23	10.4	27		
Mature			25	8.4	30		
Napiergrass *(Pennisetum purpureum)*	2-03-158	late veg.	20	8.7	33	45	70
	2-03-162	late bloom	23	7.8	39	47	75
Needle and Thread *(Stipa comata)*	2-07-989	stem cured	92	4.1		43	83
Oatgrass, Tall *(Arrhenatherum elatius)*	2-03-261	ear. veg.	22	19.7	19		
	2-03-262	ear. bloom		12.2	31		
Oats *(Avena sativa)*	2-03-286	ear. veg.	16	15.6	25		
Oats, Wild *(Avena fatua)*	2-03-393	fresh	36	7.1	24		
Orchardgrass *(Dactylis glomerata)*	2-03-439	ear. veg.	22	30.6	27	31	55
	2-03-442	ear. bloom	24	12.8	32	31	54
Prairie Grasses, Midwest USA	2-07-853	stem cured	90	5.1	33		
Redtop *(Agrostis alba)*	2-03-891	full bloom	26	8.1	25		
Rhodesgrass *(Chloris gayana)*	2-03-916	fresh	26	7.6	37		

						Composition, % dry basis									
TDN,	DE	ME	NEm	NEg	NEl	Ca	Mg	P	K	S	Co	Cu	Fe	Mn	Zn
%	. . . .	Mcal/kg	. . . .			. . . .		%			. . . .		ppm		
57*	2.51*	2.09*	1.23*	0.53*	1.27*	0.38		0.15							
57*	2.53*	2.10*	1.22*	0.52*	1.26*										
59	2.61*	2.19*	1.31*	0.65*	1.35*	0.40	0.17	0.20	1.57						
72	3.02*	2.60*	1.64*	1.03*	1.64*	0.50	0.18	0.44	2.27	0.17			300		
69	3.06*	2.64*	1.57	0.95*	1.58*	0.46	0.11	0.39	2.01	0.17			300		
58*	2.56*	2.14*	1.23*	0.53*	1.27*	0.45		0.30							
68	2.99*	2.57*	1.52*	0.91*	1.55*	0.63		0.20	1.72			47	890	106	
53	2.34*	1.92*	1.14*	0.40*	1.18*	0.40	0.06	0.12	0.51			26	1,075	61	
55	2.44*	2.02*	1.14*	0.40*	1.18*	0.32		0.07							
67*	2.95*	2.53*	1.54*	0.93*	1.56*	0.64		0.28							
58*	2.57*	2.15*	1.24*	0.56*	1.28*	0.41		0.25							
54*	2.37*	1.95*	1.12*	0.37*	1.16*	0.38		0.27							
62*	2.76*	2.33*	1.37*	0.73*	1.41*	0.48		0.39	2.92						
73	3.23*	2.81*				0.55	0.32	0.45	3.16						
53	2.30*	1.87*	1.11*	0.34*	1.14*	0.26		0.16				2			
56	2.54*	2.11*	1.29*	0.62*	1.33*	0.57	0.14	0.21	0.71						
40	1.76	1.44	0.81	0.31*	1.13*	0.06		0.22					329		
65	2.70*	2.28*				0.36		0.33	3.64						
57*	2.50*	2.08*	1.19*	0.48*	1.24*										
55*	2.41*	1.98*	1.12*	0.36*	1.16*	0.37		0.09							
55*	2.42*	1.99*	1.11*	0.35*	1.15*	0.25		0.11							
58*	2.55*	2.13*	1.25*	0.57*	1.29*	0.45	0.24	0.14	1.53						
						0.48		0.36	2.22						
						0.29		0.30	1.09						
51*	2.27*	1.84*	1.01*	0.15*	1.03*	0.38		0.09							
71	3.43*	3.01*	2.01*	1.35*	1.94*	0.76		0.48							
68	2.98*	2.56*	1.56*	0.94*	1.57*	0.45	0.21	0.37	2.35						
73*	3.20*	2.78*	1.83*	1.20*	1.80*	0.57		0.46							
51	1.83	1.50	1.31*	0.65*	1.35*	0.70	0.08	0.07	0.48	0.10	0.7	19	511	79	23
57	2.53*	2.11*				0.32	0.05	0.12	0.55		0.5	9	303	48	20
58*	2.56*	2.14*	1.25*	0.58	1.30*	0.42		0.16							
56	2.46*	2.03*				0.46	0.08	0.14	0.61		0.06	14	425	45	30
43	2.19*	1.77*	1.21*	0.51*	1.25*	0.22		0.12							
58	2.51*	2.09*	1.21*	0.51*	1.26*	0.51	0.29	0.24	2.10			31	1465	184	43
63*	2.77*	2.35*	1.40*	0.77*	1.44*	0.47		0.24							
60*	2.64*	2.22*	1.30*	0.64*	1.34*	0.33		0.18							
63	2.66*	2.24*	1.26*	0.58*	1.30*	0.32		0.19	1.94						
64	2.73*	2.31	1.29*	0.63*	1.34*										
70		2.58													
69		2.56													
68		2.51													
64		2.36													
62		2.29													
55	2.44*	2.01*	1.23*	0.55*	1.28*	0.60	0.26	0.41	1.31	0.10	0.04	27	900	33	49
53	2.34*	1.91*	1.23*	0.55*	1.28*	0.35	0.26	0.30	1.31	0.10					
48	2.09	1.81				1.08		0.06							
77*	3.41*	2.99*	2.02*	1.36*	1.95*										
62*	2.73*	2.31*	1.37*	0.74*	1.41*										
64*	2.80*	2.38*	1.42*	0.79*	1.45*					0.08					
64	3.32*	2.91*	2.07*	1.40*	1.99*	0.25	0.23	0.27							
72						0.60	0.27	0.58	3.58			7		31	
66	2.77*	2.35*	1.47*	0.85*	1.50*	0.25	0.31	0.39	3.38	0.3		33	785	104	
						0.46		0.10							
62	2.77*	2.35*	1.36*	0.72*	1.40*	0.62	0.25	0.37	2.35	0.16			200		
58	2.56	2.17*	1.26*	0.59*	1.24*	0.51	0.21	0.39	2.29	0.08			65		

Name	Ident. #	Stage of maturity, etc.	Typical DM, %	CP	C. fiber	ADF	NDF
				Composition, % dry basis			
				....	%	....	
Rye *(Secale cereale)*	2-04-018	fresh	20	17.6	29		
Ryegrass, Italian *(Lolium multiflorum)*	2-04073	fresh	23	17.9	21		
	2-04-071	ear. bloom	35	5.8	30		
Ryegrass, Perennial *(L. perenne)*	2-04-086	fresh	27	11.3	25		
Saltgrass *(Distichlis* spp.)	2-04-170	fresh	74	6.5	30		
	2-04-169	post ripe	74	4.2	35		
Squirreltail *(Sitanion* spp.)	2-95-566	stem cured	50	3.1			
Sudangrass *(Sorghum vulgare sudanense)*	2-04-484	ear. veg.	18	16.8	31		
	2-04-485	mid bloom	23	8.8	36		
Switchgrass *(Panicum virgatum)*	2-04-800	fresh	55	6.4	35		75
Threeawn, Red *(Aristida longiseta)*	2-04-852	fresh		8.3	36		74
Timothy *(Phleum pratense)*	2-04-903	late veg.	26	9.9	32		
	2-04-905	late bloom	29	9.1	34	37	64
Tobosa *(Hilaria mutica)*	2-08-578	ear. veg.	40	11.7	31		70
Vine Mesquite *(Panicum obtusum)*	2-05-138	ear. veg.		12.6	30		
Wheat *(Triticum aestivum)*	2-05-176	ear. veg.	22	28.6	17	30	52
Wheatgrass, Bluebunch *(Agropyron spicatum)*	2-05-394	fresh		13.8		35	
Wheatgrass, Crested *(A. desertorum)*	2-05-420	ear. veg.	28	21.5	22		
	2-05-422	ear. bloom	37	7.3	31		
	2-05-424	full bloom	45	9.8	30		
	2-05-427	mature	60	5.5	39		
	2-05-428	post ripe	80	3.1	40		
Wheatgrass, Slender *(A. trachycaulum)*	2-05-439	fresh	35	7.8	32		
Wild Rye, Canada *(Elymus canadensis)*	2-05-459	fresh		8.4	31		
Wild Rye, Fuzzyspike *(E. caput-medusae)*	1-08-285	ear. bloom		6.2	31		
Pasture-Range Legumes							
Acacia Pods *(Acacia* spp.)	1-00-007		91	10.7	22		
Acacia Twigs	1-00-007		88	11.2	36		
Alfalfa *(Medicago sativa)*	2-00-181	late veg.	20	20.0	23	29	38
	2-00-185	mid bloom	25	17.0	30	35	46
	2-00-187	full bloom	27	14.7	34		
Alfalfa-Brome (Smooth)	2-00-261	ear. bloom	21	19.6	25		
Alfileria, Redstem *(Erodium cicutarium)*	2-00-352	ear. veg.		26.3	12		
Burclover, Toothed *(Medicago hispida)*	2-01-035	fresh	27	23.0			
Clover, Alsike *(Trifolium hybridum)*	2-01-314	ear. veg.	19	24.1	18		
	2-08-378	mid bloom	24	15.7	27		
	2-01-315	full bloom	24	15.7	27		
Clover, Crimson *(T. incarnatum)*	2-01-336	fresh	18	17.0	28		
Clover, Hop *(T. agrarium)*	2-01-361	fresh	25	17.1	20		
Clover, Ladino *(T. repens)*	2-01-380	fresh	20	17.5	23		
Clover, Ladino - Timothy	2-01-521	fresh	38	10.7	19		
Clover, Red *(T. pratense)*	2-01-428	ear. bloom	20	19.4	23	33	44
	2-01-429	full bloom	26	14.6	26		
Clover, Sweet *(Melilotus officinalis)*	2-04-766	fresh	23	18.7	30		
Clover, White *(T. repens)*	2-01-468	fresh	18	28.2	16		
	2-01-466	full bloom	18	28.3	14		
Lespedeza, Korean *(Lespedeza stipulacea)*	2-02-596	mature	35	12.7	45		
Mesquite, Common *(Prosopis chilensis)*	2-03-081	fresh	35	21.1	27		
Mesquite, Seeds	4-03-084		89	9.5	23		
Trefoil, Birdsfoot *(Lotus corniculatus)*	2-20-786	fresh	22	24.8	18		
	2-01-766	ear. veg.		26.7	17		
	2-01-767	full bloom		19.9	22		
Vetch *(Vicia* spp.)	2-05-111	fresh	22	20.9	28		
Miscellaneous Pasture-Range Forage							
Bitterbrush, Antelope *(Purshia tridentata)*	2-00-732	fresh	33	11.8	21		
Chamise *(Adenostoma fasciculatum)*	1-07-860	ear. veg.		7.1	21		
Kale *(Brassica oleracea medullosa)*	2-02-456	fresh	14	14.2	16		
Kudzu *(Pueraria* spp.)	2-02-482	fresh	28	17.6	30		
Oak, Live, Leaves *(Quercus virginiana)*	2-29-859	fresh	50	10.2		45	

						Composition, % dry basis									
TDN,	DE	ME	NEm	NEg	NEl	Ca	Mg	P	K	S	Co	Cu	Fe	Mn	Zn
%		. . . . Mcal/kg					. . . . %					. . . . ppm			
67	2.96*	2.54*	1.33*	0.69*	1.37*	0.45	0.31	0.38	3.40						
61	2.70*	2.28*	1.36*	0.72*	1.40*	0.65	0.35	0.41	2.00	0.1			650		
57	2.53*	2.10*	1.25*	0.58*	1.30*										
68	3.37	2.95	2.03	1.37*	1.96*	0.55	0.19	0.27	1.91	0.3	0.06	13		571	9
53*	2.35*	1.93*	1.08*	0.29*	1.11*	0.22	0.30	0.08	0.24				190	155	
53*	2.35*	1.93*	1.08*	0.29*	1.11*	0.23	0.30	0.07							
50	2.02	1.70				0.37		0.06							
70	2.94*	2.52*	1.58*	0.97*	1.60*	0.43	0.35	0.41	2.14	0.11			200		
63	2.79	2.20*	1.39*	0.76*	1.38*	0.43	0.35	0.36	2.14	0.11			200		
55*	2.43*	2.00*	1.12*	0.36*	1.16*	0.29		0.10							
55*	2.41*	1.99*	1.19*	0.47*	1.23*	0.76		0.13							
68	2.83*	2.40*	1.50*	0.88*	1.52*	0.39	0.15	0.32	2.40	0.13			200		
63	2.81*	2.39*	1.47*	0.85*	1.50*	0.38	0.14	0.30	2.06	0.13		11	179	193	
54	2.41*	1.98*				0.47	0.09	0.14	0.54		0.7	14	580	82	27
58*	2.56*	2.13*	1.23*	0.55*	1.28*	0.56	0.26	0.25	2.48						
73	3.22*	2.80*	1.67*	1.06*	1.67*	0.42	0.21	0.40	3.50	0.22			100		
								0.28							
75	3.10*	2.68*	1.45*	0.83*	1.48*	0.46	0.28	0.34							
57	2.51*	2.08*	1.22*	0.53*	1.26*	0.46		0.24							
61*	2.69*	2.27*	1.36*	0.72*	1.40*	0.39		0.28							
54	2.21*	1.78*	1.00*	0.13*	1.02*	0.27		0.15							
49*	2.15*	1.72*	0.95*	-0.01*	0.95*	0.27		0.07			0.25	8		53	
58*	2.56*	2.13*	1.28*	0.61*	1.32*	0.47	0.36	0.14							
56*	2.49*	2.07*	1.19*	0.48*	1.24*										
54*	2.38*	1.96*	1.18*	0.47*	1.23*										
60	2.66*	2.24*	1.23*	0.55*	1.28*										
37	2.06*	1.63*	1.23*	0.54*	1.27*										
63	2.65*	2.17	1.31*	0.66*	1.38*	2.19	0.27	0.33	2.14	0.48	0.17	11	111	41	41
58	2.65	2.23*	1.31*	0.66*	1.29*	2.01	0.26	0.28	2.06	0.29	0.25	19		38	38
54	2.47*	2.05*	1.25*	0.58*	1.30*										
62	2.73*	2.31*	1.36*	0.72*	1.40*	1.52	0.35	0.37	3.87	0.23			130		
						2.70		0.58	3.95						
66*	2.91*	2.49*	1.44*	0.82*	1.47*	1.19	0.34	0.42	2.31						
67	2.89*	2.48*	1.40*	0.77*	1.44*	1.27	0.30	0.25	2.59	0.13		6	466	117	
63*	2.79*	2.37*	1.41*	0.78*	1.44*	1.30		0.29	2.58				470	117	
65	2.84*	2.42*	1.41*	0.78*	1.44*	1.38	0.29	0.29	3.10	0.28					
72	3.53*	3.12*	2.12*	1.45*	2.04*	1.09	0.19	0.33	1.94		0.04		250	99	
64	2.83*	2.41*	1.41*	0.78*	1.45*	1.50		0.40							
77*	3.38*	2.96*	1.96*	1.32*	1.91*										
69	3.02*	2.60*	1.56*	0.95*	1.58*	2.26	0.51	0.38	2.49	0.17			300		
64	2.83*	2.41*	1.41*	0.78*	1.45*	1.01	0.51	0.27	1.96	0.17			300		
64	2.82*	2.40*	1.41*	0.78*	1.44*	1.32	0.33	0.27	1.65	0.49		10	140	125	
73	3.06*	2.64*	1.47*	0.86*	1.50*	1.40	0.45	0.51	2.14	0.33		28	337	307	26
						1.70	0.32	0.44							
						1.00		0.20							
69*	3.03*	2.61*	1.59*	0.97*	1.60*	1.94	0.23	0.19	1.41						
65	3.03*	2.62*	1.44*	0.81*	1.47*	0.11		0.18							
75	3.24*	2.82*	1.73*	1.11*	1.72*	1.98	0.28	0.23	2.07	0.25	0.21		400		
69*	3.03*	2.61*													
68*	3.01*	2.59*	1.55*	0.94*	1.57*										
59	2.71*	2.29*	1.40*	0.76*	1.43*					0.15	0.30				
89*	3.96*	3.55*	2.86*	2.00*	2.62*					0.13					
48	2.67*	2.25*	1.67*	1.06*	1.67*										
72	3.16*	2.74*	1.59*	0.97*	1.60*										
65	2.81*	2.39*	1.38*	0.74*	1.41*	3.14		0.23							
								0.11							

Name	Ident. #	Stage of maturity, etc.	Typical DM, %	Composition, % dry basis			
				CP	C. fiber	ADF	NDF
				. . . . %			
Prickly Pear *(Opuntia* spp.)	2-01-061	fresh	15	5.3	13	23	30
Rabbitbrush, Small *(Chrysothamnus stenophyllus)*	2-07-997	stem cured	85	5.7			
Rape *(Brassica* spp.)	2-03-867	fresh	17	17.6	15		
Russian Thistle *(Salsola kali)*	2-08-000	stem cured	88	11.2		44	64
Sage, Black *(Salvia mellifera)*	2-05-564	stem cured	65	8.5			
Sagebrush, Bud *(Artemisia spinescens)*	2-07-991	ear. veg.	23	17.3			
	2-04-124	late veg.	32	17.5	23		
Sagebrush, Fringed *(A. frigida)*	2-04-129	mid bloom	43	9.4	33		
	2-04-130	mature	60	7.1	32	35	46
Saltbush, Fourwing *(Atriplex canescens)*	2-04-160	fresh	41	19.4	14		
Saltbush, Nuttall *(A. nuttallii)*	2-07-993	stem cured	55	7.2			
Saltbush, Shadscale *(A. confertifolia)*	2-05-565	stem cured	80	7.7			
Sedge *(Carex* spp.)	2-04-194	ear. veg.		17.2	26		
Sorrell, Sheep *(Rumex acetosella)*	2-13-333	fresh		9.7	33		
Turnip Tops *(Brasiea rapa rapa)*	2-05-063	fresh	14	20.3	10		
Willow Browse *(Salix* spp.)	2-05-472	fresh	41	9.8	27		
Winterfat, Common *(Eurotia lanata)*	2-26-142	stem cured	80	10.8		44	72
Grass Hays							
Barley Hay *(Hordeum vulgare)*	1-00-495	sc	88	8.8	27		
Bermudagrass *(Cynodon dactylon)*	1-00-703	sc	91	9.1	30	33	75
	1-09-213	dehy, pel	90	15.2	26	24	40
Bluegrass, Kent. *(Poa pratensis)*	1-00-776	sc	89	10.2	30		
Brome *(Bromus* spp.)	1-00-890	sc	91	9.5	33	39	69
Fescue, Meadow *(Festuca elatoir)*	1-00-912	sc	87	9.5	32	43	65
Grass Hay	1-02-212	sc, ear. veg.	89	16.4	24		
	1-02-244	sc, full bloom	89	9.5	32		
	1-02-246	sc, mature	91	3.5	34		
	1-02-247	sc, post ripe		4.2	37		
	1-08-595	weathered, mature	90	3.7	38		
Grass Meal, Dehy.	1-02-209	ear. veg.	88	26.1	15		
Meadow Hay, Intermountain	1-03-181	sc	95	8.7	32		
Millet, Foxtail *(Setaria italica)*	1-03-009	sc	86	9.9	29		
Oat Hay *(Avena sativa)*	1-03-280	sc	91	9.5	32	36	66
Orchardgrass *(Dactylis glomerata)*	1-03-438	sc	89	10.5	35	40	72
Prairie Hay, Midwest USA	1-07-956	sc, mid bloom	95	5.4	35		
	1-03-187	sc, mature	90	6.5	32		
	1-03-188	sc, post ripe	92	4.5	34		
Ryegrass, Italian *(Lolium multiflorum)*	1-04-064	sc, ear, veg.	89	15.2	20		
	1-04-066	sc, ear. bloom	88	5.5	36		
	1-04-067	sc, full bloom	86	6.6	31	45	57
Ryegrass, Perennial *(L. perenne)*	1-04-074	sc, late veg.	90	11.5	29	30	41
	1-04-075	sc, ear. bloom	88	6.0	35		
	1-04-076	sc, full bloom	89	6.2	30		
Sudangrass *(Sorghum vulgare sudanense)*	1-04-480	sc	90	12.2	30	43	68
Timothy *(Phleum pratense)*	1-04-883	sc, mid bloom	89	8.8	34	43	68
Wheatgrass, Intermediate *(Agropyron intermedium)*	1-05-431	sc	90	8.6	34	40	
Wheatgrass, Slender *(A. trachycaulum)*	1-05-436	sc	91	8.4	37		
Legume Hays							
Alfalfa *(Medicago sativa)*	1-00-050	sc, ear. veg.	90	27.0	24	26	35
	1-00-059	sc, ear. bloom	91	23.0	26	30	41
	1-00-063	sc, med bloom	91	19.0	27	32	44
	1-00-071	sc, mature	91	13.3	38	44	58
	1-08-851	sc, weathered	91	16.1	36		
Alfalfa Leaves	1-00-146	sc	89	25.1	15	24	34
Alfalfa Stems	1-00-164	sc	89	10.9	44		
Alfalfa Hay, Pelleted	1-00-124	sc	92	17.1	27	33	
Alfalfa Meal, Dehy.	1-00-023		92	18.9	26	35	45
Alfalfa - Smooth Brome *(Bromus inermis)*	1-00-251	sc, ear. veg.	93	27.1	30		
	1-00-252	sc, full bloom	92	14.3	34		

Composition, % dry basis															
TDN, %	DE	ME	NEm	NEg	NEl	Ca	Mg	P	K	S	Co	Cu	Fe	Mn	Zn
	.... Mcal/kg					.... %					.... ppm				
57	2.60	2.13*	1.29*	0.62*	1.26*	9.61	1.38	0.12	2.21	0.23					
44	2.12	1.50				1.86		0.08							
79	3.55*	3.13*	1.93*	1.29*	1.88*	1.33	0.07	0.39	2.98	0.58		8	182	46	
50	2.14	1.78				3.31		0.12							
50	2.12	1.05				0.81		0.17							
51	2.56	2.01				0.97		0.33							
52*	2.30*	1.87*	1.04*	0.20*	1.06*	0.60	0.49	0.42							
58*	2.54*	2.12*	1.21*	0.51*	1.25*										
50*	2.23*	1.80*	1.01*	0.13*	1.02*										
35	2.17*	1.74*	0.83*	-0.46*	0.73*	1.19	0.58	0.15	0.81	0.41			100	155	
36	1.49	1.32				2.21		0.12							
31	1.29	0.88				2.23		0.08							
71*	3.14*	2.72*	1.75*	1.13*	1.74*										
57*	2.51*	2.09*	1.20*	0.49*	1.24*										
71	3.12*	2.70*				3.04	0.80	0.44	3.00	0.27		18	403	408	
55	3.11*	2.70*	2.06*	1.39*	1.98*										
35	1.66	1.31				1.98		0.12							
56	2.46*	1.96	1.17*	0.44*	1.25*	0.24	0.16	0.28	1.47	0.17	0.07	4	300	39	
47	2.12	1.55	1.05	0.29	0.96*	0.47	0.17	0.17	1.53	0.21	0.12		290		
61	2.66	2.24*	1.32*	0.66*	1.30*	0.34		0.25							
61	2.64*	2.21*	1.28*	0.62*	1.32*	0.45	0.21	0.30	1.87	0.13		10	280	86	
53	2.73	2.31*	1.24*	0.56*	1.22*	0.36	0.10	0.16	1.64	0.20			200		
61	2.68*	2.26*	1.36*	0.72*	1.40*	0.37	0.50	0.29	1.84		0.13	19	300	24	17
60	2.69*	2.27*	1.35*	0.71*	1.39*	0.46	0.11	0.13	0.82						
55*	2.41*	1.98*	1.20*	0.49*	1.24*	0.57	0.18	0.24	1.34	0.12					
53*	2.35*	1.92*	1.17*	0.45*	1.22*										
47	2.11*	1.68*	1.01*	0.14*	1.02*										
46	2.15*	1.73*	1.11*	0.35*	1.15*	0.37		0.10							
74	3.25*	2.84*													
58	2.54*	2.12*	1.25*	0.59		0.61	0.17	0.18	1.58						
60	2.61*	2.18*	1.25*	0.58*	1.30*	0.33	0.23	0.18	1.94	0.16			138		
61	2.48	2.14	1.15	0.39	1.35*	0.29	0.24	0.25	1.31	0.25	0.07	4	520	83	17
55	3.03	2.36*	1.24*	0.56*	1.29*	0.38	0.18	0.26	3.00	0.26	0.38	14	150	120	36
51	2.26*	1.83*	1.09*	0.31*	1.12*						0.07		100		29
48	2.13*	1.70*				0.41	0.29	0.13				23	111	49	
54	2.30*	1.88*	1.16*	0.42*	1.20*	0.38	0.28	0.12	0.79			23	110	48	
68	2.74*	2.31*	1.41*	0.78*	1.44*										
54	2.28*	1.85*	1.05*	0.23*	1.08*										
55	2.39*	1.96*	1.14*	0.39*	1.18*										
68	2.74*	2.32*	1.21*	0.51*	1.25*										
53	2.25*	1.82*	1.05*	0.23*	1.08*	0.65		0.37	1.92						
56	2.39*	1.96*	1.12*	0.36*	1.15*										
56	2.46*	2.04*	1.18*	0.47	1.25*	0.51	0.37	0.31	1.87	0.06	0.13	31	164	76	38
59	2.62*	2.19*	1.28*	0.62*	1.33*	0.48	0.13	0.22	1.59			16	149		43
53	2.54	1.96*	1.09*	0.31*	1.12*	0.37	0.06	0.19	1.65						
57	2.43*	2.00*	1.14*	0.39*	1.18*	0.35	0.24	0.25	2.68	0.12					
66	2.78*	2.36*	1.37	0.78	1.42*	1.61	0.20	0.37	2.84	0.63	0.25	12	359	58	36
60	2.76	2.35	1.13	0.39	1.34*	1.64	0.33	0.21	2.51	0.30	0.31	12	220	32	30
58	2.53	2.11*	1.03	0.59*	.124*	1.41	0.31	0.23	1.71	0.28	0.36	13	134	28	23
50	2.43*	2.01*	1.19*	0.47*	1.23*	1.11	0.20	0.21	1.91	0.25	0.40	14	142	44	25
57	2.47*	2.04*	1.19*	0.48*	1.23*										
72	3.11*	2.69*	1.77*	1.15*	1.75*	2.48	0.45	0.27	2.30				380	81	
46	2.17*	1.74*	.112*	0.36*	1.16*	0.92	0.29	0.19	2.48				168	15	
58	2.40	2.06*	1.27*	0.60*	1.31*	1.61		0.22		0.26					
60	2.69*	2.32	1.30	0.71	1.37*	1.52	0.32	0.25	2.60	0.24	0.33	10	441	34	21
64*	2.81*	2.39*	1.42*	0.79*	1.46*										
56*	2.47*	2.05*	1.22*	0.52*	1.26*	1.14	0.78	0.26	1.40		0.09	18	120	37	

Name	Ident. #	Stage of maturity, etc.	Typical DM, %	Composition, % dry basis			
				CP	C. fiber	ADF	NDF
				........%			
Alfalfa - Grass	1-08-331	sc	92	16.3	34	39	
Alfalfa - Orchardgrass	1-00-321	sc, cut 1	92	14.4	30		
Alfalfa - Timothy	1-00-336	sc	90	13.8	35		
Clover, Alsike *(Trifolium hybridum)*	1-01-313	sc	88	14.2	30		
Clover, Crimson *(T. incarnatum)*	1-01-328	sc	88	16.8	32		
Clover, Ladino *(T. repens)*	1-08-381	sc, late veg.	88	24.2	21		
	1-01-369	sc, ear. bloom	94	20.4	21		
	1-01-372	sc, mature	91	15.8	28		
Clover, Red *(T. pratense)*	1-01-415	sc	87	14.9	32	41	56
Lespedeza, Korean *(Lespedeza stipulacea)*	1-02-584	sc, ear. bloom	94	15.3	30		
Lespedeza, Sericea *(L. cuneata)*	1-02-607	sc	90	12.3	33		
Peavine *(Lathyrus* spp.)	1-03-666	sc	91	21.8	28		
Trefoil, Birdsfoot *(Lotus corniculatus)*	1-05-044	sc	91	15.3	32		
Vetch *(Vicia* spp.)	1-05-106	sc	89	20.7	28		
Silages, Corn and Sorghum							
Corn, Dent *(Zea mays indentata)*	3-28-239		43	9.3			
	3-02-817	ear. veg.	24	9.7	25		
	3-28-246	dough stage	27	7.7	24	31	
	3-02-821	post ripe	30	7.9	24		
Corn Stover Silage	3-28-251	no ears	30	6.1	32	55	68
Corn (Sweet) Cannery Residue	3-07-955		32	7.7	35		
Sorghum Heads	3-17-403	post ripe	69	9.6	5		
Sorghum	3-04-323		32	9.2	26	38	
	3-04-321	dough stage	29	6.2	28		
	3-04-322	mature	32	7.5	25	21	43
Legume, Grass, Miscellaneous Silages							
Alfalfa	3-00-204	late veg.	20	21.2	31	50	
	3-00-205	ear. bloom	28	17.6	33		
	3-00-206	mid bloom	31	17.3	33		
Alfalfa Silage, Wilted	3-00-215	late veg.	35	20.7	28		
	3-00-217	mid bloom	38	16.8	30	35	45
	3-00-218	full bloom	36	14.9	33	38	51
Alfalfa + Formic Acid	3-08-850		26	19.0	30		
Alfalfa + Molasses	3-00-238		30	18.7	26		
Barley *(Horedum vulgare)*	3-00-512		32	10.3	30		
Beet Tops, Sugar *(Beta vulgaris)*	3-00-660		27	12.7	12		
Clover, Ladino *(Trifolium repens)*	3-01-384		25	23.5	21		
Clover, Red *(T. pratense)* + Molasses	3-07-900	cut 2		15.4	23		
Clover, Red, Wilted	3-01-443	mid bloom	31	12.9	33		
Citrus Pulp	3-01-234		20	7.1	16		
Grass -Legume + Molasses	3-02-309		28	11.8	32		
Grass - Legume	3-02-303		29	11.3	32		
Grass	3-02-217	ear. veg.	28	13.2	32		
	3-02-218	ear. bloom	23	12.1	31	38	66
Mint Residue *(Mentha* spp.)	3-20-697		28	16.4	37		
Oats *(Avena sativa)*	3-03-298		30	9.5	31		
Peavine *(Pisum* spp.)	3-03-596	no seeds	25	13.1	30	49	59
Soybean *(Glycine max)*	3-04-581		27	17.0	29		
Sudangrass *(Sorghum vulgare)*	3-04-499		23	11.2	34		
Sugarcane *(Saccharum officinarum)*	3-04-693		22	5.1	34		
Sunflower *(Helianthus* spp.)	3-04-728	late veg.	26	10.2	33		
	3-04-729	ear. bloom	21	9.7	33		
	3-04-730	mid bloom	23	9.3	33		
	3-04-732	late bloom	26	8.2	38		
Straws							
Barley *(Hordeum vulgare)*	1-00-498		91	4.3	41	59	80
Bean *(Phaseolus* spp.)	1-00-585		90	9.2	38		
Flax, Common *(Linum usitatissimum)*	1-02-038		93	5.3	52		
Grass	1-08-430		89	5.3	41		

Composition, % dry basis															
TDN,	DE	ME	NEm	NEg	NEl	Ca	Mg	P	K	S	Co	Cu	Fe	Mn	Zn
%	. . . .	Mcal/kg	. . . .			. . . .		%			. . . .		ppm	. . . .	
55	2.44*	2.01*	1.23*	0.53*	1.27*	1.65	0.34	0.23	2.60					34	
58	2.50*	2.07*	1.25*	0.58*	1.30*	1.42	0.32	0.20	1.49						
54	2.41*	1.98*	1.20*	0.49*	1.24*	1.20	0.18	0.21	2.16		0.08			36	
58	2.55*	2.12*	1.25*	0.56*	1.29*	1.30	0.45	0.25	2.22	0.19		6	260	69	
57	2.52*	2.10*	1.34*	0.69*	1.38*	1.40	0.28	0.22	2.40	0.28			700	171	
66	2.90*	2.48*									0.14	8	340	48	25
65	2.85*	2.42*				1.31		0.31			0.14	8	283	51	29
						1.40	0.37	0.25	2.16		0.14	8	201	54	31
55	3.02	2.38*	1.27*	0.60*	1.31*	1.28	0.33	0.29	2.36	0.16	0.16	21	224	108	37
58*	2.55*	2.12*	1.25*	0.57*	1.30*										
46	2.28*	1.85*	1.24*	0.55*	1.28*	1.03	0.22	0.25	1.10				291	101	
61	2.74*	2.32*	1.40*	0.77*	1.43*										
59	2.22	2.03*	1.33*	0.69*	1.38*	1.70	0.51	0.23	1.92	0.25	0.11	9	228	29	
62	2.69*	2.27*	1.36*	0.72*	1.40*	1.36	0.27	0.34	2.12	0.15	0.36	10	490	61	
72	3.40*	2.99*	1.64*	1.03*	1.64*	0.10	0.12	0.29	0.49	0.13			80		
65	2.85*	2.43*	1.37*	0.74*	1.41*	0.52	0.31	0.31	1.64				490		
69	2.95	2.50	1.57	0.95*	1.50*	0.27	0.18	0.19	0.95	0.14			180		
67*	2.97*	2.55*	1.46*	0.84*	1.49*	0.67		0.13							
55	2.38	1.93*	1.16*	0.44*	1.13*	0.38	0.31	0.31	1.54	0.11					
72	2.98*	2.56*	1.64*	1.03*	1.64*	0.30	0.24	0.90	1.15	0.11			200		
84*	3.72*	3.31*	1.92*	1.28*	1.87*										
63	2.65	1.98	1.47	0.66	1.29*	0.39	0.28	0.23	1.44	0.09	1.14	15	332	106	27
55	2.49	2.07*	1.22*	0.53*	1.19*										
54	2.12	2.08*	1.03*	0.19*	0.96*	0.81									
63	2.13	1.89*	1.23*	0.54*	1.27*										
60	2.54*	2.11*	1.19*	0.48	1.23*	1.49	0.30	0.27	2.08	0.30	0.09	13	200	31	17
58	2.65*	2.23*	1.31*	0.66*	1.35*	1.28	0.35	0.20	2.00	0.25	0.09	13	200	33	17
63	2.78*	2.36*	1.38*	0.75*	1.42*										
58	2.70*	2.28*	1.33*	0.68*	1.37*										
55	2.58*	2.16*	1.24*	0.55*	1.28*										
65	2.92*	2.49*	1.49*	0.87*	1.52*										
61	2.94	2.41	.149*	0.87*	1.47*	1.64	0.34	0.30	2.61	0.36			300		
51	2.22*	1.76	1.05*	0.22*	1.13*	0.34	0.13	0.28	2.10	0.11	0.67	5	200	86	23
49	2.18*	1.75*	1.16*	0.43*	1.20*	1.65	1.07	0.21	5.74	0.57			200		
69	3.06*	2.65*													
65	2.66	2.24	1.38*	0.75*	1.42*										
57	2.58*	2.15*	1.23*	0.55*	1.28*										
88	3.21*	2.80*	2.12*	1.45*	2.04*	2.04	0.16	0.15	0.62	0.02			160		16
58	2.55*	2.12*	1.23*	0.55*	1.28*	1.07	0.32	0.34	1.92	0.24			520		
56	2.71	2.30*	1.35*	0.71*	1.33	0.85	0.32	0.27	1.80	0.54	0.13		520	55	
60*	2.64*	2.21*	1.28*	0.61*	1.32*										
58*	2.58*	2.16*	1.24*	0.56*	1.28*										
54	2.38*	1.95*	1.19*	0.49*	1.24*										
56	2.39*	1.95	1.10	0.57*	1.29*	0.34	0.30	0.24	2.66	0.19	0.06	5	282	91	24
57	2.48*	2.07*	1.21*	0.51*	1.25*	1.31	0.39	0.24	1.40	0.25			100		
54	2.38*	1.96*	1.16*	0.43*	1.20*	1.36	0.38	0.47	0.93	0.30		9	400	113	
56	2.48*	2.05*	1.20*	0.50*	1.25*	0.50	0.42	0.21	2.61	0.06	0.27	37	120	99	
61	2.58*	2.16*	1.22*	0.52*	1.26*	0.35	0.22	0.18							
50	2.33*	1.90*	1.21*	0.51*	1.25*										
53	2.35*	1.92*	1.24*	0.56*	1.29*										
54	2.39*	1.96*	1.02*	0.17*	1.04*										
43	1.94*	1.51*	0.94*	-0.03*	0.94*										
46	2.03*	1.54	1.00	0.15	1.00*	0.30	0.23	0.07	2.37	0.17	0.07	5	201	17	7
46	2.15*	1.72*	1.04*	0.20*	1.06*										
40	1.91	1.59*	1.04*	0.21*	1.07*	0.55	0.31	0.06	1.74					8	
44	2.18*	1.75*	1.11*	0.34*	1.15*										

Name	Ident. #	Stage of maturity, etc.	Typical DM, %	Composition, % dry basis			
				CP	C. fiber	ADF	NDF
				....	%	....	
Oats *(Avena sativa)*	1-03-283		92	4.4	40	47	70
Pea *(Pisum* spp.)	1-03-577		89	8.9	40		
Rice *(Oryza sativa)*	1-03-925		91	4.4	35	57	78
Rice Straw, Ammoniated			87	9.0	39	53	68
Rye *(Secale cereale)*	1-04-007		90	2.9	43		
Ryegrass *(Lolium* spp.)			89	4.0	40		
Soybean *(Glycme max)*	1-04-567		88	5.2	44	54	70
Wheat *(Triticum avestivum)*	1-05-175		89	3.6	42	54	85
Miscellaneous Roughages							
Almond Hulls *(Prunus amygdalus)*	4-00-359		90	2.1	13	27	31
Coffee Grounds	1-01-576	wet	74	11.0	29	68	80
Corn Plant, Whole	1-02-768	dehy	90	8.2	24		
Corn Stover	1-02-776	no ears	86	6.5	35	39	67
Corn Cobs	1-02-782	ground	90	3.1	36	43	89
Cotton Bolls	1-01-596	sc	92	11.0	32		
Cotton Gin Trash (By-product)	1-08-413		90	7.4	34		
Cottonseed Hulls	1-01-599		91	4.1	48	67	86
Oat Hulls	1-03-281		92	3.9	33	44	81
Peanut Pods	1-08-028		91	8.2	65	65	74
Rice Hulls	1-08-075		92	3.3	43	70	80
Seaweed, Blade Kelp *(Laminaria* spp.)	4-08-507	dehy	87	10.5	10		
Sorghum Stover	1-04-302	no heads	93	4.4		46	
Soybean Hulls (Seed Coats)	1-04-560		91	10.0	40	50	67
Sugarcane Bagasse	1-04-686	dried	91	1.7	46	55	86

*Estimated values for energy.

Name	Ident. #	Typical DM, %	Density lb/ft^3 as fed	Composition, % DM basis			
				CP	C. fiber	ADF	NDF
				....	%	....	
ENERGY FEED							
Cereal Grains and By-products w <20% CP							
Barley grain	4-00-549	90	38-43	13.2	5.8	7	19
Barley grain, ground	4-00-526	88	24-26	13.6	5.9	8	37
Barley grain, Pac. Coast	4-07-939	89		10.8	7.3		
Barley grain, hull-less	4-00-552	92		18.1	2.4		
Barley grain, pearled	4-00-527	89		9.9	<1		
Barley flour by-product	4-00-547	89		16.3	10.4		
Barley groats	4-00-543	90		13.7	2.4		
Barley malt, dehy	4-00-578	93	30-31	14.1	3.6		
Barley screenings	4-00-542	89		13.1	8.3		
Corn grain, #2 grade	4-02-931	89	45	10.1	2.2	3	10
Corn grain, Opaque 2	4-28-253	90		11.2	3.3		
Corn grain, flaked		89		11.0	1.7		
Corn grain, ground	4-26-023	90	38-40	10.8	2.5	3	9
Corn grain, high moisture	4-20-770	78		10.7	2.6	5	
Corn and cob meal	4-28-238	87	36	9.0	9.4	10	25
Corn ears w husks, ground	4-02-850	88		8.3	10.9		
Corn bran	4-02-991	89	13	9.4	10.8		
Corn grits (hominy)	4-03-010	92		10.8	8.3		
Corn grits by-product (hominy feed)	4-03-011	90		11.7	6.0	13	55
Grain screenings	4-02-156	90	22-24	13.4	13.4		
Oats grain	4-03-309	90	25-35	13.4	12.0	15	
Oats grain, ground	4-08-471	90	20-25	13.7	11.3	17	32
Oats grain, Pac. Coast	4-07-999	91		10.0	12.3		
Oats grain, white	4-03-377	89		13.6	12.3		
Oats groats	4-03-331	90	46-47	17.7	2.8		
Oats middlings	4-03-303	91	38	16.2	3.9		

						Composition, % dry basis									
TDN, %	DE	ME	NEm	NEg	NEl	Ca	Mg	P	K	S	Co	Cu	Fe	Mn	Zn
	. . . . Mcal/kg					. . . . %					. . . . ppm				
46	1.86		1.11*	0.34*	1.15*	0.24	0.18	0.06	2.57	0.23		10	175	37	6
46	2.16*	1.73*	1.25*	0.56*	1.29*										
39	1.94*	1.51*	0.70	0.07	1.01*	0.21	0.11	0.08	1.32					345	
45	0.90		0.44	0.03		0.30		0.10	1.00	0.11					
40	1.89*	1.46*	0.89*	-0.18*	0.86	0.24	0.08	0.09	0.97	0.11		4		7	
43				0.32*	1.13*										
41	1.79*	1.36*	0.87*	-0.28*	0.81*	1.59	0.92	0.06	0.56	0.26			300	51	
40	2.12	1.72	1.02*	0.17*	0.96*	0.18	0.12	0.05	1.42	0.19	0.04	3	157	41	6
60	3.27*	2.82	1.59	0.98	1.61*	0.23		0.11	0.53	0.11					
52						0.12		0.08							
60	2.69*	2.26*	1.34*	0.70*	1.38*	1.96		0.14					11		
56	2.63*	2.21*	1.22	0.59	1.35*	0.54	0.38	0.10	1.28	0.17		5	200		
49	2.23*	1.80	1.05	0.32	1.07*	0.12	0.07	0.04	0.87	0.47	0.13	7	230	6	
44	1.93*	1.50*	0.92*	-0.08*	0.91*	0.90	0.28	0.12	2.73						
46	2.37	1.78*	1.16*	0.43*	1.12*	0.92		0.20							
50	2.10	1.45	1.03	0.22	0.95*	0.15	0.14	0.19	0.87	0.09	0.02	13	131	119	22
34	1.52*	1.09*	0.82*	-0.48*	0.72*	0.15	0.09	0.15	0.62	0.15		4	111	20	
22	0.79*	0.35*		-1.79*	0.26*	0.26	0.17	0.07	0.95	0.10	0.12	18	310	69	24
12	0.65*	0.38*		-2.23*	0.19*	0.10	0.83	0.08	0.57	0.09			166	333	
40	1.77*	1.34*				1.50	0.71	0.27	4.40			9	10	7	
51	2.21	1.78*	1.08*	0.28*	1.02*				1.20						
76	2.88*	2.46*	1.77*	1.15*	1.75*	0.49		0.21	1.27	0.09	0.12	18	324	11	24
48	2.12	1.68*	1.03*	0.19*	0.96*	0.51	0.08	0.29	0.37	0.10		15	206	24	

						Composition, % DM basis										
TDN, %	DE	ME	NEm	NEg	NEl	Ca	Mg	P	K	S	Co	Cu	Fe	Mn	Se	Zn
	. . . . Mcal/kg					. . . . %					. . . . ppm					
85	3.56	2.91	1.95	1.30	2.05*	0.05	0.15	0.39	0.51	0.17	0.10	9	78	18	0.2	19
78	3.24	3.04*	1.68*	1.07*	1.66*	0.08	0.15	0.39	0.55						0.4	
85	3.75*	3.34*	1.94*	1.30*	1.89*	0.06	0.14	0.39	0.58	0.16	0.10	9	97	18	0.1	17
84*	3.72*	3.31*	1.95*	1.30*	1.90*	0.04		0.50								
						0.03	0.07	0.26	0.27		1.07	5	20	8		24
81	3.42*	3.00	1.70*	1.27*	1.86*	0.08	0.13	0.40	0.60							
89	3.75*	3.34*	1.91*	1.24*	1.83*	0.07	0.20	0.51	0.47		0.06	6	66	20		
83	3.64*	3.23*	1.87*	0.82*	1.47*	0.17		0.46								
80	3.53*	3.12*	1.88*	1.24*	1.84*	0.32	0.14	0.33	0.75	0.15			60			
90	3.96	3.55*	2.18	1.49*	2.11*	0.02	0.13	0.35	0.37	0.14	0.04	4	26	6		16
89	3.77*	3.35*	1.93*	1.29*	1.88	0.03	0.14	0.22	0.39	0.11						
93	4.10*	3.39*	2.24*	1.53*	2.71*											
85	3.53	3.04	2.00*	1.35*	1.84*	0.03	0.14	0.26	0.36	0.11	1.02	4	50	8		23
93	4.09*	3.69*	2.24*	1.53*	2.12*	0.02	0.14	0.31	0.34	0.14		4	30	6		18
83	3.70	3.21	1.98	1.32	1.95*	0.07	0.14	0.27	0.53	0.16	0.31	8	91	14	0.1	14
76	3.36*	2.94*	1.72*	1.11*	1.72*	0.07	0.16	0.29	0.80	0.13				12		
76	3.36*	2.94*	1.77*	1.15*	1.75*	0.04	0.29	0.22	0.72	0.08		108	115	18		
82	3.61*	3.19*	1.88*	1.24*	1.84*	0.03		0.18					44			
94	4.15*	3.74*	2.36*	1.63*	2.22*	0.07	0.18	0.57	0.69	0.03	0.07	15	86	17	0.1	3
69	3.02*	2.60*	1.54	0.98	1.53*	0.37	0.14	0.39	0.34					49		
75	2.96	2.51	1.50	1.25	1.74*	0.09	0.16	0.38	0.43	0.23	0.06	6	76	42	0.2	39
76	3.33*	2.91*	1.72*	1.10*	1.71*	0.10	0.18	0.48	0.48		0.07	7	89	43		30
77	3.42*	3.00*	1.79*	1.17*	1.77*	0.11	0.19	0.34	0.42	0.22			80	42	0.1	
76	3.35*	2.93	1.75*	1.13*	1.73*	0.06	0.14	0.36	0.46			7	86	41	0.3	39
97	4.24*	3.59	2.47*	1.71*	2.17*	0.08	0.13	0.48	0.39	0.22		7	82	31		
95	4.19*	3.78*	2.27*	1.56*	2.15*	0.08	0.16	0.49	0.55	0.24	0.05	6	421	48		154

Name	Ident. #	Typical DM, %	Density lb/ft^3 as fed	Composition, % DM basis			
				CP	C. fiber	ADF	NDF
				. . . .	. %	. . . .	
Oat shorts	4-03-304	91		14.1	14.8		
Rice grain, ground	4-03-938	89		8.4	9.7		
Rice bran w germs	4-03-928	91		14.2	12.8	18	33
Rice feed	4-03-937	90	42-45	16.2	6.8		
Rice groats (Brown rice)	4-03-936	88		9.2	0.9		
Rice groats, polished	4-03-942	89		8.1	0.4		
Rice polishing, dehy	4-03-943	90		13.4	3.6	72	82
Rye grain	4-04-047	89	43-45	13.1	2.4		
Rye bran	4-04-022	91	21	17.5	7.6		
Rye middlings	4-04-031	89	42	18.2	5.4		
Rye millrun	4-04-034	90	32-36	18.5	5.1		
Sorghum grain, kafir	4-04-428	89	40-46	12.3	2.3		
Sorghum grain, milo	4-04-444	89	40-45	11.4	2.5	8	17
Sorghum grain, milo, ground	4-05-643	89	32-36	11.0	2.9		
Sorghum grain heads	4-04-446	90		10.0	8.9		
Wheat, hard red spring	4-05-258	89		17.2	2.8		
Wheat, hard red winter	4-05-268	89		14.3	2.8	4	12
Wheat, soft red winter	4-05-294	88		13.1	2.6		
Wheat, soft white winter	4-08-112	87		9.6	2.1		
Wheat, Pac. Coast	4-08-142	89	45-52	11.1	3.0		
Wheat bran	4-05-190	89	11-16	17.4	11.2	15	44
Wheat flour by-product	4-05-203	88		17.6	3.2		
Wheat mill run	4-05-206	90	20-21	17.1	9.1		
Wheat shorts	4-05-201	89	18-25	18.8	7.7		
Wheat screenings	4-05-216	89		15.8	7.7		
Miscellaneous Energy Feeds							
Apple pomace, dehy	4-00-423	89		4.9	20.0	45	
Bakery waste, dehy	4-00-466	92		11.2	1.4		
Bean cannery residue	5-00-587	9		23.5	13.5		
Beet pulp (sugar), wet	4-00-671	11		11.2	21.3	34	59
Beet pulp w molasses, dehy	4-00-672	92	11-16	10.1	16.5	25	44
Beet pulp w steffen filt., dehy	4-00-675	92		12.2	16.6	21	42
Buckwheat grain	4-00-994	88		12.5	11.8		
Cassava meal, dehy	4-01-152	90		2.6	5.0		
Citrup pulp, dehy	4-01-237	91	20-21	6.7	12.7	22	23
Corn cannery residue	2-02-810	28		7.9	30.0	36	59
Garbage, institutional, boiled	4-07-867	18		15.5	2.7		
Garbage, municipal, boiled	4-07-864	22		16.8	11.8	50	59
Grape pomace	1-02-208	91		13.0	31.9	54	56
Grass seed screenings	4-26-071	88		8.0	20.0		
Molasses dist. sol., condensed	4-04-697	96		12.6	1.4		
Molasses dist. sol., dehy	4-04-698	88		8.8	0.7		
Molasses, sugarcane, dehy	4-04-695	94		10.3	6.7		
Orange pulp, dehy	4-01-254	88		8.5	9.6		21
Pineapple bran	4-03-722	87		4.6	20.9	37	73
Tomato pulp, dehy	5-05-041	92		23.5	26.4	50	55
Whey, condensed	4-01-180	65		13.7			
Whey, dehy	4-01-182	93	40	14.1			
Whey, fresh	4-08-134	7		13.0			
Whey, low lactose, cond.	4-01-185	55		15.9			
Roots and Tubers							
Beet, mangel, fresh	4-00-637	11		11.8	7.4		
Beet, sugar, fresh	4-00-677	20		6.8	5.9		
Potato, white, cull	4-03-787	23		9.1	2.4		
Potato peelings	4-03-774	23		8.9	3.1		
Potato waste, dehy	4-03-775	89		8.4	7.3		
Potato waste, fresh	4-03-777	12		9.9	3.6		
Potato silage	4-03-768	25		7.6	4.0		
Turnip, fresh	4-05-067	9		11.9	11.6	34	44

	Composition, % DM basis															
TDN, %	DE	ME	NEm	NEg	NEl	Ca	Mg	P	K	S	Co	Cu	Fe	Mn	Se	Zn
	.... Mcal/kg					.... %					.... ppm					
63	2.80*	2.37*	1.37*	0.73*	1.41*											
71	3.13*	2.72*	1.39*	0.76*	1.43*	0.07	0.14	0.36	0.36	0.05				20		
71	2.67	2.40	1.48*	0.86*	1.30*	0.08	1.04	1.70	1.92	0.20		14	210	415	0.4	32
78	3.46*	3.04*	1.89*	1.25*	1.85*	0.44	0.83	2.07	1.17		1.33	8	150	294		65
88	3.89*	3.47*	2.00*	1.35*	1.94*	0.04	0.07	0.25	0.21	0.05		4	32	14		6
88	3.96	3.58*	2.18*	1.49*	2.11*	0.03	0.02	0.13	0.12	0.09		3	16	12		2
89	3.82	3.50*	2.18	1.45	2.02*	0.05	0.87	1.48	1.27	0.19		4	178	14		29
84	3.57	2.97	1.94*	1.30*	1.86*	0.08	0.13	0.38	0.52	0.17		9	81	82		34
61	2.94*	2.52*	1.64*	1.03*	1.65*											
83	3.68*	3.27*	2.08*	1.41*	2.00*	0.07		0.70	0.70					49		
75	3.41	2.99	1.86	1.23*	1.82*	0.08	0.26	0.71	0.92	0.04						
63	2.77*	2.35*			0.88*	0.04	0.17	0.35	0.38	0.18	0.43	8	71	18	0.9	15
86	3.52	2.81	1.81*	1.18*	1.83*	0.04	0.18	0.32	0.38	0.13	0.53	12	56	19	0.1	24
84	3.68*	3.27*	1.90*	1.26*	1.86*	0.03	0.17	0.34	0.39		0.12	16		14		
81	3.58*	3.16*	1.81*	1.19*	1.79*	0.13	0.17	0.25	0.56	0.13						
88	3.88*	3.47*	2.12*	1.45*	2.04*	0.03	0.17	0.44	0.43	0.17	0.14	7	66	42	0.3	46
88	3.88*	3.47*	2.14*	1.46*	2.05*	0.05	0.13	0.42	0.50	0.15	0.16	6	38	33	0.3	40
89	3.91*	3.50*	2.15*	1.47*	2.06*	0.06	0.11	0.40	0.46	0.12	0.12	8	33	38		48
89						0.03	0.02	0.09	0.15		0.06	1	11	4		4
89	3.92	3.21	2.18	1.43	2.52	0.14		0.34								
70	3.12	2.68*	1.69	1.13	1.58*	0.15	0.62	1.29	1.38	0.25	0.10	14	174	129	0.7	132
82	3.67	3.47*	1.97*	1.33*	1.92*	0.05	0.21	0.57	0.59	0.27	0.13	7	56	59	0.4	74
79	3.58	2.81	1.91	1.08	1.87*	0.11	0.52	1.13	1.33	0.34	0.23	21	105	116		
73	3.66*	3.11	2.06*	1.40*	1.98*	0.11	0.27	0.91	1.04	0.22	0.12	12	84	131	0.5	128
71	3.14*	2.73*	1.52*	0.91*	1.54*	0.15	0.18	0.39	0.58	0.22			60	32	0.7	
71	3.13*	2.71*	1.58*	0.97*	1.59*	0.13	0.07	0.12	0.49	0.02			299	8		
93	4.10	3.46	2.98	1.63*	2.06*	0.16	0.18	0.25	0.39	0.02	1.40	13	205	59		19
78	3.18*	2.76*	1.52*	0.91*	1.55*	0.87	0.22	0.10	0.19	0.22			330			1
75	3.33*	2.91*	2.03*	1.35	1.69*	0.61	0.16	0.10	1.78	0.42	0.23	16	207	26		2
66*	2.91*	2.49*	1.21*	0.51*	1.25*											
71	3.16*	2.74*	1.58*	0.97*	1.60*	0.11	0.12	0.37	0.51	0.16	0.06	11	50	38		10
79	3.49*	3.08*	1.84*	1.21*	1.81*	0.15		0.10	0.26							
83	3.41*	2.71	1.72*	1.11*	1.87*	1.84	0.17	0.12	0.79	0.08	0.16	6	378	7		15
70																
105*	4.62*	4.22*	2.65*	1.84*	2.45*											
78	3.43*	3.01*														
33	1.17*	0.73*				0.61		0.06	0.62					41		24
65	2.86	2.35*														
69*	3.06*	2.64*														
75	2.71*	2.29*	1.02*	0.17*	1.05*	1.20		0.07								
70	3.21*	2.80*	1.58*	0.97*	1.60*	1.10	0.47	0.15	3.60	0.46	1.21	79	250	57		33
82	3.49*	3.07*	1.55*	0.94*	1.57*	0.71		0.11								
68	3.23*	2.81*	1.72*	1.10*	1.71*	0.23		0.13					560			
58	2.89*	2.47*	1.44*	0.81*	1.47*	0.43	0.20	0.60	3.63			33	4600	51		
82	3.63*	3.22*				0.60		0.91								
82	3.57*	3.03	2.06	1.35	1.83*	0.92	0.14	0.82	1.23	1.12	0.12	50	181	6		3
94	4.15*	3.75*				0.73		0.65	2.75				290	3		
						0.95		1.44								
80	3.51*	3.09*	1.81*	1.19*	1.79*	0.18	0.20	0.22	2.30	0.20		5	154			
84	3.69*	3.27*	1.87*	1.23*	1.83*	0.24	0.18	0.24	1.52	0.06		8	61	206		
81	3.58*	3.16*	1.99	1.33	1.82*	0.03	0.14	0.23	2.22	0.09		28	92	42		
79	3.36*	2.95*	1.68*	1.06*	1.68*	0.14		0.19								
89	3.64*	3.23*	1.75*	1.13*	1.74*	0.16		0.25								
74	3.24*	2.82*				0.33		0.67	1.33							
82	3.62*	3.21*	1.94*	1.30*	1.89*	0.04	0.14	0.23	2.13	0.23			90			
83	3.68*	3.27*	2.00*	1.35*	1.94*	0.63	0.22	0.31	2.99	0.43		21-	112	43		

Name	Ident. #	Typical DM, %	Density lb/ft^3 as fed	Composition, % DM basis CP	C. fiber	ADF	NDF
				. . . . %	. . . .		
Fats and Oils							
Animal fat, feed grade							
for preruminants		98	54				
for ruminants		98	54				
Vegetable oil	4-05-076	99					
Other Liquid Feeds (<20%, CP, DM basis)							
Beet molasses	4-00-668	78		8.5			
Citrus syrup	4-01-241	68		8.2			
Hemicellulose extract	4-08-030	76		0.7	1.0		
Lignin sulfonate, dehy	8-02-627	53		10.3	1.0		
Propylene glycol	8-03-809	100					
Soybean solubles, cond.		60		13.0			
Sugarcane molasses	4-04-696	74	88	5.9	9.9		
Protein Supplements w >20% CP							
Oil Seed Meals							
Coconut, solv-extd	5-01-573	91		23.4	15.4		
Cottonseed, solv-extd, 41% CP	5-01-621	91	37-40	45.3	13.3	22	30
Cottonseed, low gossypol, solv-extd	5-01-633	93		44.8	13.7		
Cottonseed, prepress-solv-extd, 50% CP	5-07-874	93		54.0	8.8		
Crambe, wo hulls, solv-extd	5-08-167	92		49.8	4.6		
Linseed, solv-extd	5-02-048	90	25-33	38.4	10.1	19	25
Peanut, solv-extd	5-03-650	92		52.3	10.8		
Mustard, solv-extd	5-12-149						
Rape, solv-extd	5-03-871	91		40.6	13.2		
Safflower, solv-extd	5-04-110	92		26.9	31.3	41	58
Safflower, wo hulls, solv-extd	5-07-959	91		47.1	14.0		
Sesame, mech-extd	5-04-220	93		49.6	6.0	17	
Soybean, solv-extd, toasted	5-04-607	91	35-38	50.5	6.0	10	12
Soybean, wo hulls, solv-extd	5-04-612	90	41-43	55.1	3.7	6	8
Soybean grits, solv-extd	5-04-592	92		54.0	2.6		
Soybean protein conc. >70% CP	5-08-038	92		91.9	0.1		
Sunflower, solv-extd	5-09-340	90		25.9	35.1	33	40
Sunflower, wo hulls, solv-extd	5-04-739	93		51.5	12.2		
Other Plant Protein Sources							
Alfalfa seed screenings	5-08-326	90		34.3	12.5		
Barley brewers grains, dehy	5-00-516	91	14-15	22.0	19.9		
Barley distillers grains, dehy	5-00-518	92		30.1	11.0		
Barley malt sprouts, dehy	5-00-545	94		28.1	16.0	18	46
Bean seed, Navy	5-00-623	89		25.6	5.0		
Clover, red, seed screenings	5-08-005	90		31.6	14.1		
Corn distillers grains, dehy	5-28-235	94	19	29.7	12.2		
Corn distillers grains w sol., dehy	5-28-236	92		29.5	9.9	18	44
Corn distillers solubles, dehy	5-28-237	93	26	29.5	4.9	7	23
Corn germ meal	5-08-083	91	35	21.8	12.0		
Corn gluten feed	5-28-243	90	30	25.6	9.7		
Corn gluten meal	5-28-241	91	38	47.3	4.9	9	37
Cottonseed, whole	5-13-749	91		24.0	21.4		
Flax seed screenings, mech extd	5-02-054	91	26-28	21.9	12.9		
Hops, spent, dehy	5-02-396	93		24.8	24.3		
Pea seeds, field, cull	5-08-481	91	48-50	26.0	6.5		
Potato distillers residue, dehy	5-03-773	96		23.9	21.5		
Rye distillers grains, dehy	5-04-023	92		23.5	13.4		
Rye distillers sol., dehy	5-04-026	93		37.2	3.6		
Safflower seeds	4-07-958	94		17.4	28.6		
Sorghum distillers grains, dehy	5-04-374	94		33.2	13.0		
Sorghum distillers sol., dehy	5-04-376	93		28.5	4.2		
Sorghum gluten, wet milled, dehy	5-04-388	91		50.0	11.1		
Sorghum gluten feed	5-04-389	92		29.7	8.3		

						Composition, % DM basis										
TDN, %	DE	ME	NEm	NEg	NEl	Ca	Mg	P	K	S	Co	Cu	Fe	Mn	Se	Zn
	. . . . Mcal/kg					. . . . %					. . . . ppm					
199	8.75	7.18*														
174	7.65	6.30*	4.53	2.59	5.25											
195	8.60															
78	3.38*	2.94	1.92*	1.28*	1.83*	0.17	0.29	0.03	6.07	0.60	0.46	22	87	6		18
75	3.29*	2.81	1.81*	1.18*	1.71*	1.72	0.21	0.13	0.14	0.23	0.16	108	508	38		137
58	1.77*	1.34*	0.85*		1.35*	1.03		0.09								
						3.75				4.50						
158*	6.10*	5.71	4.16*	3.16*	3.75*											
						2.0	0.6	0.8	4.3						0.1	
72	3.60*	3.31	2.02	1.27	1.89*	1.00	0.43	0.11	4.01	0.47	1.59	79	249	56		23
75	3.31	2.90*	1.67	1.10	1.70*	0.19	0.36	0.66	1.63	0.37	0.14	10	750	72		
75	3.56	2.82	1.57	1.60	1.71*	0.18	0.60	1.21	1.55	0.28	0.16	22	228	23		68
71*	3.12*	2.70*	1.50*	0.88*	1.52*											
75	3.31*	2.89*	1.73*	1.11*	1.72*	0.19	0.50	1.24	1.56	0.56	0.04	16	120	25		79
86*	3.77*	3.36*	2.03*	1.37*	1.96*											
80	3.33*	2.74*	1.79	1.20	1.76*	0.43	0.66	0.89	1.53	0.43	0.21	28	354	42	0.9	
77	3.40	2.74*	1.79*	1.17*	1.76*	0.29	0.17	0.68	1.23	0.33	0.12	17	154	19		22
69	3.04*	2.62*	1.55*	0.94*	1.57*	0.67	0.60	1.04	1.36	1.25					1.1	
58	2.32*	1.99	1.18*	0.47*	1.29*	0.36	0.38	0.75	0.83	0.14		11	520	20		44
76	3.24*	2.83*	1.76*	1.14*	1.74*	0.37	1.01	1.58	1.15	0.06	2.22	9	525	44		36
77	3.38*	2.97*	1.73*	1.11*	1.72*	2.17	0.50	1.44	1.33	0.35			100	51		107
82	3.63*	3.21*	2.04*	1.38*	1.97*	0.37		0.64								
87	3.82*	3.41*				0.29	0.33	0.71	2.36	0.48	0.07	22	370	41	0.1	63
88	3.90*	3.49*	2.14*	1.46*	2.05*											
76*	3.16*	2.74	1.75*	1.14*	1.74*	0.12	0.02	0.74	0.19	0.76	0.42	15	149	6	0.1	36
44	1.94*	1.51*	0.93*		0.92*	0.23	0.75	1.03	1.06	0.33						
65	3.10*	2.68*	1.44*	0.82*	1.47*	0.44	0.77	0.98	1.14			4	33	20		
87	3.63*	3.21*	1.80*	1.18*	1.78*	0.17		0.55								
67	3.03*	2.61*	1.60	0.99*	1.61*	0.21		0.50								
86*	3.79*	3.38*	1.75*	1.13*	1.74*	0.20		0.45								
70	3.11*	2.69*	1.58*	0.97*	1.60*	0.23	0.20	0.75	0.23	0.85				34	0.5	
84	3.77*	3.36*	1.98*	1.33*	1.93*	0.19	0.15	0.61	1.47	0.26		11	110	24		
78	3.32*	2.90*	1.67*	1.05*	1.67*	0.46		0.74								
87	3.64*	3.02	2.22	1.46	1.99*	0.10	0.07	0.42	0.18	0.46	0.09	48	223	23	0.4	35
88	3.73*	3.22	2.20	1.47	2.06*	0.15	0.18	0.73	0.44	0.33	0.18	58	260	25	0.4	
87	3.68*	3.19	2.13	1.40	2.03*	0.35	0.65	1.39	1.80	0.40	0.21	89	610	80	0.4	91
77	3.38*	2.96*				0.04		0.55	0.22					18		
83	3.52*	2.96	1.94*	1.30*	1.91*	0.36	0.36	0.82	0.64	0.23	0.10	52	470	26	0.3	72
85	3.62*	2.99	1.96*	1.32*	1.92*	0.16	0.06	0.51	0.03	0.22	0.08	30	420	8	1.1	190
95	3.81*	3.38*	2.01	1.20	2.21*	0.16	0.35	0.76	1.22	0.26		54	152	12		
58	2.56*	2.13*	1.26*	0.58*	1.30*	0.45	0.43	0.58	0.84	0.25			100			
34	2.16*	1.73*	1.42*	0.79*	1.45*											
83	3.66*	3.24*	1.88*	1.24*	1.84*	0.18	0.10	0.42	1.14		0.41	6	55	11	0.4	29
62*	2.74*	2.32*	1.31*	0.65*	1.35*											
59	2.59*	2.17*	1.15*	0.41*	1.19*	0.16	0.18	0.52	0.08	0.48				20		
87*	3.83*	3.42*	2.25*	1.55*	2.14*	0.37		1.26								
89	3.91*	3.50*	1.57	1.00	2.06*	0.26	0.36	0.67	0.79			11	500	20		
82	3.72*	3.31*	1.94*	1.30*	1.89*	0.15	0.19	0.63	0.38	0.18			50			
85*	3.74*	3.33*	2.02*	1.36*	1.95*	0.73		1.48								
85*	3.74*	3.32*	1.92*	1.28*	1.87*											
74*	3.27*	2.86*	1.68*	1.07*	1.68*											

Name	Ident. #	Typical DM, %	Density lb/ft³ as fed	Composition, % DM basis			
				CP	C. fiber	ADF	NDF
				%			
Soybean seeds, gr	5-04-610	92	25-34	41.8	5.8	10	
Sunflower seeds	5-08-530	94		18.4	31.8		
Wheat germ meal	5-05-218	88	28-32	27.6	3.6		
Animal and Fish By-Products†							
Cattle manure, dried		92		15.0	34	37	57
Crab process residue, dehy	5-01-663	92		32.3	11.6		
Feather meal, hydro.	5-03-795	91	34	89.5		20	20
Fish meal, Herring, mech extd	5-02-000	92	30-34	78.3	0.7		
Fish solubles, condensed	5-01-969	50		62.1	0.9		
Fish waste, liquified		24		16-18			
Hair meal, hydrol.	5-08-997	92		96.0			
Meat meal	5-00-385	94	37	54.7	2.8		
Meat and bone meal	5-00-388	93		54.1	2.4		
Poultry waste (manure), dehy	5-14-015	91		28.2	13.2	15	38
Poultry litter, dehy	5-05-587	86		27.0	17.8		
Shrimp process residue, dehy	5-04-226	90	25	44.2	15.6	17	
Milk and Milk By-Products							
Buttermilk, dehy	5-01-160	92		34.4			
Milk, skim	5-01-170	10		31.2			
Milk, skim, dehy	5-01-175	94		35.8			
Milk, whole	5-01-168	12		26.7			
Milk, whole, dehy	5-01-167	96		26.5			
Whey (listed as energy source)							
Non-Protein-Nitrogen Sources							
Biuret	5-09-824	98		230.0			
Dehy-100 (alfalfa-urea)		94		106.0	17.9		
Diammonium phosphate	6-09-370	99		112.5			
Monoammonium phosphate	6-09-338	98		69.9			
Grain-urea, extruded (starea)	5-14-506	92		25.0			
Urea, 45% N	5-05-070	99	38	285.0			
Single Cell Protein Sources							
Pulpmill SCP, dried w fat		92		40		35	
Yeast, active, dehy	7-05-524	90		15.7	3.8		
Yeast, Brewers, dehy	7-05-527	94	41	46.5	3.1		
Yeast, Grain distillers, dehy	7-05-519	94		51.0			
Yeast, irradiated, dehy	7-05-529	94		51.2	6.6		
Yeast, molasses distillers, dehy	7-05-532	82		25.4	5.9		
Yeast, torula, dehy	7-05-534	93		51.9	2.2		

*Estimated values for energy.

†Note. Many others are available, but they are not commonly used for ruminant animals.

						Composition, % DM basis										
TDN, %	DE	ME	NEm	NEg	NEl	Ca	Mg	P	K	S	Co	Cu	Fe	Mn	Se	Zn
			Mcal/kg					%					ppm			
92	4.04*	3.63*	2.16	1.48*	2.07*	0.27	0.29	0.65	1.80	0.24		20	90	40	0.1	62
83	3.66*	3.25*	2.01*	1.36*	1.95*	0.27		0.59	0.71				32	23		
94	4.14*	3.73*	2.34*	1.61*	2.21*	0.07	0.27	1.06	1.09	0.31	0.13	10	62	140	0.5	134
41	0.82		0.41	0.0*		1.30		1.00	0.50							240
29	1.29*	0.85*	0.78*			15.72	1.02	1.71	0.49	0.27		35	4700	144		
72	2.87*		2.03	1.35	1.54*	0.28	0.22	0.71	0.31	1.61	0.05	7	81	14	0.9	74
83	3.67*	3.26*	1.85*	1.22*	1.82*	2.39	0.16	1.82	1.18	0.50	0.06	6	5340	6	2.1	142
84	3.77*	3.42	2.33*	1.60*	1.92*	0.37	0.06	1.17	3.22	0.25	0.13	92	445	27	3.9	87
71	2.81*	2.08*	1.25*	0.56*	1.63*	9.44	0.29	4.74	0.61	0.50	0.14	10	470	10	0.5	85
71	3.11*	2.70*	1.63*	1.01*	1.63*	11.06	1.09	5.48	1.43	0.27	0.19	2	735	14	0.3	96
58	1.89	1.46*	0.93*			9.31	0.64	2.52	2.25	0.18		88	2000	406		434
66	2.44	2.18				2.87	0.46	1.75	1.68	1.26		255	778	288	0.7	450
46	2.02*	1.59*	0.94*			10.80	0.60	2.05	0.92				116	33		33
89	3.73*	3.08	2.04*	1.38*	2.03*	1.44	0.52	1.01	0.90	0.09		1	9	4		44
92	4.11	3.67*	2.29*	1.57*	2.20*	1.31	0.12	1.04	1.90	0.32	0.11	12	10	2		51
85	3.35*	2.51	1.57*	0.95*	1.94*	1.36	0.13	1.09	1.70	0.34	0.12	12	10	2	0.1	41
128	5.75	5.29*	3.41*	2.45*	3.22*	0.95	0.10	0.76	1.12	0.32		1	10			23
119	5.24*	4.84*	2.96*	2.08*	2.70*	0.95	0.10	0.74	1.08	0.32		1	10			23
						0.51	0.46	20.25		2.53		81	5152	505		303
						0.42	0.46	25.26	0.14	0.73		85	1940	535		320
61	2.68															
						0.16		0.98								
78	3.34*	2.85	1.81	1.22	1.82*	0.13	0.27	1.49	1.79	0.45	0.20	35	117	6	1.0	41
76	3.37*	2.95*				0.83		1.51	2.28							
57	2.51*	2.17*														
80	3.53*	2.98*	1.86*	1.24*	2.05*	0.61	0.14	1.81								

No Table

Appendix Table 2. Mineral composition of selected salts and mineral supplements used in livestock rations, as fed basis, feed grade purity. +

Name	Basic chemical formula	Elemental composition, % feed grade products										Acid-base reaction
		Ca	Cl	F	K	Mg	Mn	N	Na	P	S	
Ammonium chloride	NH_4Cl		66.28					26.18				Acid pH 4.5-5.5
Ammonium phosphate												
monobasic	$(NH_4)H_2PO_4$			.0025*				12.18		26.93		Acid pH 4.5
dibasic	$(NH_4)_2HPO_4$			.0025*				21.21		23.48		Basic pH 8.0
Animal bone charcoal		27.1				.5		1.4		12.7		
Animal bone meal, steamed		29.0				.6	trace	1.9	.5	13.6		
Calcium												
Clacium carbonate	$CaCO_3$	40.04										Basic
Limestone	$CaCO_3$	35.8				2.0	trace			trace		Basic
Limestone, dolomitic	$CaCO_3$	22.3			.4	10.0	trace					Basic
Oyster shell	$CaCO_3$	38.0			.1	.3	trace		.21	.07		Basic
Calcium phosphates												
Monocalcium phosphate	$CaH_4(PO_4)_2$	15.9		.0025*						24.6		Acid pH 3.9
Dicalcium phosphate	$CaHPO_4$	23.35		.02						18.21		Acid pH 5.5
Triclacium phosphate	$Ca_3(PO_4)_2$, CaO	38.76		.18*						19.97		sl. Acid pH 6.9
Defluorinated rock phosphates		32.0		.18*	.09				4.0	18.0		
Gypsum	$CaSO_4$	29.44									23.55	
Magnesium carbonate	$MgCO_3$	.02				25.2						sl. Basic
Magnesium oxide	MgO					60.3			.5			sl. Basic
Manganese sulfate	$MnSO_4{+}H_2O$ MnO						28.7				14.7	
Potassium chloride	KCl		47.6		52.4							Neutral pH 7.00
Sodium chloride	NaCl		60.66						39.34			Neutral
Sodium phosphate												
Monobasic	$NaH_2PO_4 \cdot H_2O$			.0025*					16.6	22.40		Acid pH 4.4
Dibasic	Na_2HPO_4			.0025*					32.39	21.80		Basic pH 9.0
Tripoly	$Na_5P_3O_{10}$			.0025*					24.94	30.85		Basic pH 9.9

+Elemental content varies according to source; use guaranteed analysis where available.

*Maximum.

Appendix Table 3. Nutrient requirements for growing-finishing steer calves and yearlings (daily nutrients per animal). NRC 1976.

Weight[a]		Daily gain		Minimum dry matter consumption[b]		Rough-age,[b]	Protein		Energy					Ca,	P,	Vitamin A,
kg	lb	kg	lb	kg	lb	%	Total, kg	Digestible, kg	NE_m, Mcal	NE_g, Mcal	ME,[b] Mcal	TDN,[b,c] kg	TDN,[b,c] lb	g	g	(Thousands IU)
100	220	0	0	2.1	4.6	100	0.18	0.10	2.43	0	4.2	1.2	2.6	4	4	5
		0.5	1.1	2.9	6.4	70-80	0.36	0.24	2.43	0.89	6.6	1.8	4.0	14	11	6
		0.7	1.5	2.7	6.0	50-60	0.40	0.28	2.43	1.27	7.1	2.0	4.4	19	13	6
		0.9	2.0	2.8	6.2	25-30	0.46	0.33	2.43	1.68	7.7	2.1	4.6	24	16	7
		1.1	2.4	2.7	6.0	15	0.49	0.36	2.43	2.10	8.4	2.3	5.1	28	19	7
150	331	0	0	2.8	6.2	100	0.23	0.13	3.30	0	5.6	1.6	3.5	5	5	6
		0.5	1.1	4.0	8.8	70-80	0.44	0.28	3.30	1.20	9.0	2.5	5.5	14	12	9
		0.7	1.5	3.9	8.6	50-60	0.49	0.33	3.30	1.73	9.6	2.7	6.0	18	14	9
		0.9	2.0	3.8	8.4	25-30	0.54	0.37	3.30	2.27	10.7	3.0	6.6	23	17	9
		1.1	2.4	3.7	8.2	15	0.58	0.41	3.30	2.84	11.3	3.1	6.8	28	20	9
200	441	0	0	3.5	7.7	100	0.30	0.17	4.10	0	7.0	1.9	4.2	6	6	8
		0.5	1.1	5.8	12.8	80-90	0.57	0.35	4.10	1.49	12.1	3.4	7.5	14	13	12
		0.7	1.5	5.7	12.6	70-80	0.61	0.39	4.10	2.14	13.0	3.6	7.9	18	16	13
		0.9	2.0	4.9	10.8	35-45	0.61	0.40	4.10	2.82	13.3	3.7	8.2	23	18	13
		1.1	2.4	4.6	10.1	15	0.63	0.43	4.10	3.52	14.1	3.9	8.6	27	20	13
250	551	0	0	4.4	9.7	100	0.35	0.20	4.84	0	8.2	2.3	5.1	8	8	9
		0.7	1.5	5.8	12.8	55-65	0.62	0.39	4.84	2.53	14.4	4.0	8.8	18	16	14
		0.9	2.0	6.2	13.7	45-50	0.69	0.44	4.84	3.33	16.2	4.5	9.9	22	19	14
		1.1	2.4	6.0	13.2	20-25	0.73	0.48	4.84	4.17	17.0	4.7	10.4	26	21	14
		1.3	2.9	6.0	13.2	15	0.76	0.51	4.84	5.04	18.6	5.2	11.5	30	23	14
300	661	0	0	4.7	10.4	100	0.40	0.23	5.55	0	9.4	2.6	5.7	9	9	10
		0.9	2.0	8.1	17.9	55-65	0.81	0.50	5.55	3.82	19.5	5.4	11.9	22	19	16
		1.1	2.4	7.6	16.8	20-25	0.82	0.52	5.55	4.78	20.4	5.6	12.3	25	22	16
		1.3	2.9	7.1	15.6	15	0.83	0.54	5.55	5.77	21.6	6.0	13.2	29	23	16
		1.4[d]	3.1	7.3	16.1	15	0.87	0.57	5.55	6.29	22.5	6.2	13.7	31	25	16
350	772	0	0	5.3	11.7	100	0.46	0.26	6.24	0	10.6	2.9	6.4	10	10	12
		0.9	2.0	8.0	17.6	45-55	0.80	0.49	6.24	4.29	20.8	5.8	12.8	20	18	18
		1.1	2.4	8.0	17.6	20-25	0.83	0.52	6.24	5.36	22.4	6.2	13.7	23	20	18
		1.3	2.9	8.0	17.6	15	0.87	0.55	6.24	6.48	24.2	6.8	15.0	26	22	18
		1.4[d]	3.1	8.2	18.1	15	0.90	0.57	6.24	7.06	25.3	7.0	15.4	28	24	18

Appendix Table 3. Continued.

Weight[a]		Daily gain		Minimum dry matter consumption[b]		Rough-age,[b]	Protein		Energy					Ca,	P,	Vitamin A,
							Total	Digestible,	NE_m,	NE_g,	ME,[b]	TDN,[b,c]				
kg	lb	kg	lb	kg	lb	%	kg	kg	Mcal	Mcal	Mcal	kg	lb	g	g	(Thousands IU)
400	882	0	0	5.9	13.0	100	0.51	0.29	6.89	0	11.8	3.3	7.3	11	11	13
		1.0	2.2	9.4	20.7	45-55	0.87	0.54	6.89	5.33	24.5	6.8	15.0	21	20	19
		1.2	2.6	8.5	18.7	20-25	0.87	0.54	6.89	6.54	25.4	7.0	15.4	23	21	19
		1.3	2.9	8.6	19.0	15	0.90	0.56	6.89	7.16	26.5	7.3	16.1	25	22	19
		1.4[d]	3.1	9.0	19.8	15	0.94	0.59	6.89	7.80	28.0	7.7	17.0	26	23	19
450	992	0	0	6.4	14.1	100	0.54	0.31	7.52	0	12.8	3.6	7.9	12	12	14
		1.0	2.2	10.3	22.7	45-55	0.96	0.57	7.52	5.82	26.7	7.4	16.3	20	20	20
		1.2	2.6	10.2	22.5	20-25	0.97	0.58	7.52	7.14	28.6	7.9	17.4	23	22	20
		1.3	2.9	9.3	20.5	15	0.97	0.59	7.52	7.83	29.0	8.0	17.6	24	23	20
		1.4[d]	3.1	9.8	21.6	15	0.98	0.60	7.52	8.52	30.5	8.4	18.5	25	23	20
500	1,102	0	0	7.0	15.4	100	0.60	0.34	8.14	0	13.9	3.8	8.4	13	13	15
		0.9	2.0	10.5	23.1	45-55	0.95	0.56	8.14	5.60	27.1	7.5	16.5	19	19	23
		1.1	2.4	10.4	22.9	20-25	0.96	0.57	8.14	7.01	29.2	8.1	17.8	20	20	23
		1.2	2.6	9.6	21.2	15	0.96	0.58	8.14	7.73	29.7	8.2	18.1	21	21	23
		1.3[d]	2.9	10.0	22.0	15	0.97	0.60	8.14	8.47	31.4	8.7	19.2	22	22	23

[a] Average weight for a feeding period.

[b] Dry matter consumption, ME and TDN allowances are based on NE requirements and the general types of diet indicated in the roughage column. Most roughages will contain 1.9-2.2 Mcal of ME/kg dry matter and 90-100% concentrate diets are expected to contain 3.1-3.3 Mcal of ME/kg.

[c] TDN was calculated by assuming 3.6155 Mcal of ME per kg of TDN.

[d] Most steers of the weight indicates, and not exhibiting compensatory growth, will fail to sustain the energy intake necessary to maintain this rate of gain for an extended period.

Appendix Table 4. Nutrient requirements for growing-finishing heifer calves and yearlings (daily nutrients per animal). NRC 1976.

Weight[a]		Daily gain		Minimum dry matter consumption[b]		Roughage,[b]	Protein		Energy					Ca,	P,	Vitamin A,
							Total,	Digestible	NE_m,	NE_g,	ME,[b]	TDN,[b,c]				
kg	lb	kg	lb	kg	lb	%	kg	kg	Mcal	Mcal	Mcal	kg	lb	g	g	(Thousands IU)
100	220	0	0	2.1	4.6	100	0.18	0.10	2.43	0	4.2	1.2	2.6	4	4	5
		0.5	1.1	3.0	6.6	70-80	0.37	0.25	2.43	0.99	6.9	1.9	4.2	14	11	6
		0.7	1.5	2.9	6.4	50-60	0.42	0.29	2.43	1.44	7.5	2.1	4.6	19	14	6
		0.9	2.0	3.0	6.6	25-30	0.48	0.34	2.43	1.92	8.3	2.3	5.1	24	17	7
		1.1	2.4	3.0	6.6	<15	0.53	0.39	2.43	2.43	9.2	2.5	5.5	29	19	7
150	331	0	0	2.8	6.2	100	0.24	0.14	3.30	0	5.6	1.6	3.5	5	5	6
		0.5	1.1	4.1	9.0	70-80	0.45	0.29	3.30	1.34	9.4	2.6	5.7	14	12	9
		0.7	1.5	4.0	8.8	50-60	0.50	0.33	3.30	1.95	10.4	2.8	6.2	18	14	9
		0.9	2.0	4.0	8.8	25-30	0.54	0.37	3.30	2.60	11.3	3.1	6.8	23	17	9
		1.1	2.4	4.0	8.8	<15	0.60	0.42	3.30	3.30	12.4	3.4	7.5	28	20	9
200	441	0	0	3.5	7.7	100	0.30	0.17	4.10	0	7.0	1.9	4.2	6	6	8
		0.3	0.7	5.4	11.9	100	0.49	0.29	4.10	0.95	10.8	3.0	6.6	10	10	12
		0.5	1.1	6.0	13.2	80-90	0.58	0.35	4.10	1.66	12.7	3.5	7.7	14	13	13
		0.7	1.5	6.0	13.2	70-80	0.61	0.39	4.10	2.42	13.8	3.8	8.4	18	16	13
		0.9	2.0	5.3	11.7	35-45	0.62	0.40	4.10	3.23	14.3	4.0	8.8	22	17	13
		1.1	2.4	5.0	11.0	<15	0.64	0.43	4.10	4.09	15.4	4.3	9.5	25	19	13
250	551	0	0	4.1	9.0	100	0.35	0.20	4.84	0	8.3	2.3	5.1	7	7	9
		0.3	0.7	6.4	14.1	100	0.57	0.33	4.84	1.13	12.8	3.5	7.8	12	12	14
		0.5	1.1	6.5	14.3	80-90	0.62	0.37	4.84	1.96	14.2	3.9	8.6	13	13	14
		0.7	1.5	5.8	12.8	55-65	0.62	0.38	4.84	2.86	15.0	4.1	9.1	17	15	14
		0.9	2.0	5.9	13.0	35-45	0.65	0.42	4.84	3.81	16.5	4.6	10.1	21	17	14
		1.1	2.4	6.5	14.3	20-25	0.74	0.48	4.84	4.84	18.7	5.2	11.5	25	20	14
		1.2	2.6	6.3	13.9	<15	0.75	0.49	4.84	5.37	19.4	5.4	11.9	27	21	14
300	661	0	0	4.7	10.4	100	0.40	0.23	5.55	0	9.5	2.6	5.7	9	9	10
		0.3	0.7	7.4	16.3	100	0.63	0.36	5.55	1.29	14.5	4.0	8.4	13	13	16
		0.5	1.1	7.4	16.3	80-90	0.67	0.40	5.55	2.25	16.3	4.5	9.9	14	14	16
		0.7	1.5	6.6	14.6	55-65	0.67	0.40	5.55	3.37	17.1	4.7	10.4	16	15	16
		0.9	2.0	6.8	15.0	35-45	0.70	0.44	5.55	4.37	19.0	5.2	11.5	19	17	16
		1.1	2.4	7.5	16.5	20-25	0.78	0.49	5.55	5.55	21.5	6.0	13.2	23	20	16
		1.2	2.6	7.2	15.9	<15	0.79	0.50	5.55	6.16	22.3	6.2	13.7	24	20	16

Appendix Table 4. Continued.

Weight[a]		Daily gain		Minimum dry matter consumption[b]		Rough-age,[b]	Protein		Energy					Ca,	P,	Vitamin A,
							Total,	Digestible,	NE_m,	NE_g,	ME,[b]	TDN,[b,c]				
kg	lb	kg	lb	kg	lb	%	kg	kg	Mcal	Mcal	Mcal	kg	lb	g	g	(Thousands IU)
350	772	0	0	5.3	11.7	100	0.46	0.26	6.24	0	10.6	2.9	6.4	10	10	12
		0.3	0.7	8.2	18.1	100	0.69	0.39	6.24	1.45	16.5	4.6	10.0	15	15	18
		0.5	1.1	8.3	18.3	80-90	0.73	0.42	6.24	2.52	18.3	5.1	11.2	15	15	18
		0.7	1.5	7.9	17.4	55-65	0.73	0.43	6.24	3.68	19.7	5.4	11.9	15	15	18
		0.9	2.0	8.1	17.9	35-45	0.77	0.46	6.24	4.91	21.8	6.0	13.2	17	17	18
		1.1	2.4	8.3	18.3	20-25	0.81	0.50	6.24	6.23	24.0	6.6	14.5	20	19	18
		1.2[d]	2.6	8.1	17.9	<15	0.81	0.50	6.24	6.91	25.0	6.9	15.2	21	20	18
400	882	0	0	5.9	13.0	100	0.51	0.29	6.89	0	11.8	3.3	7.3	11	11	13
		0.3	0.7	9.1	20.0	100	0.76	0.43	6.89	1.61	18.2	5.0	11.1	16	16	19
		0.5	1.1	8.5	18.7	70-80	0.78	0.43	6.89	2.79	19.5	5.4	11.9	15	15	19
		0.7	1.5	8.7	19.2	55-65	0.79	0.46	6.89	4.06	21.7	6.0	13.2	16	16	19
		0.9	2.0	8.4	18.5	20-25	0.79	0.47	6.89	5.43	23.5	6.5	14.3	17	17	19
		1.1[d]	2.4	8.3	18.3	<15	0.81	0.49	6.89	6.88	25.9	7.2	15.9	19	18	19
450	992	0	0	6.4	14.1	100	0.55	0.31	7.52	0	12.9	3.6	7.9	12	12	14
		0.2	0.4	8.7	19.2	100	0.74	0.41	7.52	1.14	17.4	4.8	10.6	16	16	19
		0.5	1.1	9.3	20.5	70-80	0.80	0.46	7.52	3.05	21.3	5.9	13.0	17	17	20
		0.8	1.8	9.1	20.1	35-45	0.82	0.48	7.52	5.17	24.5	6.8	15.0	16	16	20
		1.0[d]	2.2	8.5	18.7	<15	0.83	0.48	7.52	6.71	26.8	7.4	16.3	19	19	20

[a] Average weight for a feeding period.

[b] Dry matter consumption, ME and TDN allowances are based on NE requirements and the general type of diet indicated in the roughage column. Most roughages will contain 1.9-2.2 Mcal of ME/kg dry matter and 90-100% concentrate diets are expected to have 3.1 to 3.3 Mcal of ME/kg.

[c] TDN was calculated by assuming 3.6155 kcal of ME per g of TDN.

[d] Most heifers of the weight indicated, and not exhibiting compensatory growth, will fail to sustain the energy intake necessary to maintain this rate of gain for an extended period.

Appendix Table 5. Nutrient requirements for beef cattle breeding herd (daily nutrients per animal). NRC 1976.

Weight[a]		Daily gain		Minimum dry matter consumption[b]		Rough-age,[b]	Protein		Energy					Ca,	P,	Vitamin A,
							Total,	Digestible,	NE_m,	ME_g,	ME,[b]	TDN,[b,c]				
kg	lb	kg	lb	kg	lb	%	kg	kg	Mcal	Mcal	Mcal	kg	lb	g	g	(Thousands IU)
Pregnant yearling heifers—Last 3-4 months of pregnancy																
325	716	0.4[c]	0.9	6.6	14.5	100[d]	0.58	0.34	5.89	0.62	12.6	3.5	7.7	15	15	19
		0.6	1.3	8.5	18.7	100	0.75	0.42	5.89	1.52	16.2	4.5	9.9	18	18	23
		0.8	1.8	9.4	20.7	85-100	0.85	0.50	5.89	2.49	20.1	5.6	12.3	22	20	26
350	772	0.4[c]	0.9	6.9	15.2	100	0.61	0.35	6.23	0.65	13.2	3.7	8.1	15	15	19
		0.6	1.3	8.9	19.6	100	0.78	0.45	6.23	1.60	16.9	4.7	10.3	19	19	25
		0.8	1.8	10.0	22.0	85-100	0.88	0.51	6.24	2.63	21.1	5.8	12.9	22	21	28
375	827	0.4[c]	0.9	7.2	15.9	100	0.63	0.36	6.56	0.68	13.7	3.8	8.4	15	15	20
		0.6	1.3	9.3	20.5	100	0.81	0.46	6.56	1.68	17.7	4.9	10.8	19	19	26
		0.8	1.8	11.0	24.2	85-100	0.96	0.55	6.56	2.76	22.1	6.1	13.5	22	22	31
400	882	0.4[c]	0.9	7.5	16.5	100	0.65	0.38	6.89	0.71	14.2	3.9	8.7	16	16	21
		0.6	1.3	9.7	21.4	100	0.84	0.48	6.89	1.76	18.5	5.1	11.3	19	19	27
		0.8	1.8	11.6	25.6	85-100	1.01	0.57	6.89	2.90	23.0	6.4	14.0	22	22	33
425	937	0.4[c]	0.9	7.8	17.2	100	0.69	0.40	7.21	0.74	14.8	4.1	9.0	16	16	22
		0.6	1.3	10.1	22.3	100	0.88	0.50	7.21	1.84	19.2	5.3	11.7	19	19	28
		0.8	1.8	12.1	26.7	85-100	1.05	0.60	7.21	3.03	24.0	6.6	14.6	22	22	34
Dry pregnant mature cows—Middle third of pregnancy																
350	772			5.5	12.2	100[d]	0.32	0.15	6.23		10.8	3.0	6.6	10	10	15
400	882			6.1	13.4	100	0.36	0.17	6.89		11.9	3.3	7.3	11	11	17
450	992			6.7	14.8	100	0.39	0.19	7.52		13.0	3.6	7.9	12	12	19
500	1,102			7.2	15.9	100	0.42	0.20	8.14		14.1	3.9	8.6	13	13	20
550	1,213			7.7	17.0	100	0.45	0.22	8.75		15.1	4.2	9.2	14	14	22
600	1,323			8.3	18.3	100	0.49	0.23	9.33		16.1	4.4	9.8	15	15	23
650	1,433			8.8	19.4	100	0.52	0.25	9.91		17.1	4.7	10.4	16	16	25
Dry pregnant mature cows—Last third of pregnancy																
350	772	0.4[c]	0.9	6.9	13.9	100[d]	0.41	0.19	7.8		13.2	3.6	8.0	12	12	19
400	882	0.4	0.9	7.5	15.4	100	0.44	0.21	8.4		14.3	4.0	8.7	14	14	21
450	992	0.4	0.9	8.1	16.5	100	0.48	0.23	9.1		15.4	4.2	9.4	15	15	23
500	1,102	0.4	0.9	8.6	17.9	100	0.51	0.24	9.7		16.4	4.5	10.0	15	15	24
550	1,213	0.4	0.9	9.1	19.0	100	0.54	0.25	10.3		17.5	4.8	10.7	16	16	26
600	1,323	0.4	0.9	9.7	20.3	100	0.57	0.27	10.9		18.5	5.1	11.2	17	17	27
650	1,433	0.4	0.9	10.2	22.4	100	0.60	0.29	11.5		19.6	5.4	11.9	18	18	29

Appendix Table 5. Continued.

Weight[a]		Daily gain		Minimum dry matter consumption[b]		Roughage,[b]	Protein		Energy					Ca,	P,	Vitamin A,
							Total	Digestible,	NE_m,	NE_g,	ME,[b]	TDN,[b,c]				
kg	lb	kg	lb	kg	lb	%	kg	kg	Mcal	Mcal	Mcal	kg	lb	g	g	(Thousands IU)
Cows nursing calves—Average milking ability[e]—First 3-4 months postpartum																
350	772			8.2	18.1	100[d]	0.75	0.44	9.2		15.9	4.4	9.7	24	24	19
400	882			8.8	19.4	100	0.81	0.48	9.9		17.0	4.7	10.4	25	25	21
450	992			9.3	20.5	100	0.86	0.50	10.5		18.1	5.0	11.0	26	26	23
500	1,102			9.8	21.6	100	0.90	0.53	11.1		19.2	5.3	11.7	27	27	24
550	1,213			10.5	23.1	100	0.97	0.57	11.9		20.3	5.6	12.3	28	28	26
600	1,323			11.0	24.2	100	1.01	0.59	12.3		21.3	5.9	13.0	28	28	27
650	1,433			11.4	25.1	100	1.05	0.62	12.9		22.3	6.2	13.7	29	29	29
Cows nursing calves—Superior milking ability[f]—First 3-4 months postpartum																
350	772			10.2	22.4	100[g]	1.11	0.65	12.3		21.0	5.8	12.8	45	40	32
400	882			10.8	23.8	100	1.17	0.69	13.0		22.1	6.1	13.5	45	41	34
450	992			11.3	24.9	100	1.23	0.72	13.6		23.2	6.4	14.1	45	42	36
500	1,102			11.8	26.0	100	1.29	0.76	14.2		24.3	6.7	14.8	46	43	38
550	1,213			12.4	27.3	100	1.35	0.79	14.9		25.3	7.0	15.4	46	44	41
600	1,323			12.9	28.4	100	1.41	0.83	15.5		26.4	7.3	16.1	46	44	43
650	1,433			13.4	29.5	100	1.46	0.86	16.2		27.5	7.6	16.8	47	45	45
Bulls, growth and maintenance (moderate activity)																
300	661	1.00	2.2	8.8	19.4	70-75	0.90	0.55	5.6	3.8	20.4	5.6	12.3	27	23	34
400	882	0.90	2.0	11.0	24.2	70-75	1.03	0.62	6.9	4.1	25.2	7.0	15.4	23	23	43
500	1,102	0.70	1.5	12.2	26.9	80-85	1.07	0.62	8.5	3.7	27.0	7.5	16.5	22	22	48
600	1,323	0.50	1.1	12.0	26.4	80-85	1.02	0.60	9.8	3.0	26.4	7.3	16.1	22	22	48
700	1,543	0.30	0.7	12.9	28.4	90-100[g]	1.08	0.60	11.0	2.0	27.7	7.7	17.0	23	23	50
800	1,764	0	0	10.5	23.1	100[g]	0.89	0.50	12.2	0	21.0	5.8	12.8	19	19	41
900	1,984	0	0	11.4	25.1	100[g]	0.99	0.55	13.3	0	22.8	6.3	13.9	21	21	44
1000	2,205	0	0	12.4	27.3	100[g]	1.05	0.60	14.4	0	24.8	6.9	15.2	22	22	48

[a] Average weight for a feeding period.
[b] Dry matter consumption, ME and TDN requirements are based on the general type of diet indicated in the roughage column.
[c] Approximately 0.4 ± 0.1 kg of weight gain/day over the last third of pregnancy is accounted for by the products of conception. These nutrients and energy requirements include the quantities estimated as necessary for conceptus development.
[d] Average quality roughage containing about 1.9-2.0 Mcal ME/kg dry matter.
[e] 5.0 ± 0.5 kg of milk/day. Nutrients and energy for maintenance of the cow and for milk production are included in these requirements.
[f] 10 ± 0.5 kg of milk/day. Nutrients and energy for maintenance of the cow and for milk production are included in these requirements.
[g] Good quality roughage containing at least 2.0 Mcal ME/kg dry matter.

Appendix Table 6. Nutrient requirements for growing-finishing steer calves and yearlings (nutrient concentration in diet dry matter).[a] NRC 1976.

Weight[b]		Daily gain[c]		Minimum dry matter consumption[c]		Rough-age,[c]	Protein		Energy								
							Total,	Digestible,	NE_m,[f]		NE_g,[f]		ME,[f]		TDN,[d,f]	Ca,	P,
kg	lb	kg	lb	kg	lb	%	%	%	Mcal/kg	Mcal/lb	Mcal/kg	Mcal/lb	Mcal/kg	Mcal/lb	%	%	%
100	220	0	0	2.1	4.6	100	8.7	5.0	1.17	0.53	–	–	2.0	0.91	55	0.18	0.18
		0.5	1.1	2.9	6.4	70-80	12.4	8.3	1.35	0.60	0.75	0.23	2.2	1.00	62	0.48	0.38
		0.7	1.5	2.7	6.0	50-60	14.8	10.7	1.60	0.71	1.00	0.43	2.5	1.13	70	0.70	0.48
		0.9	2.0	2.8	6.2	25-30	16.4	11.8	1.81	0.82	1.18	0.54	2.8	1.27	77	0.86	0.57
		1.1	2.4	2.7	6.0	<15	18.2	13.3	2.07	0.94	1.37	0.62	3.1	1.41	86	1.04	0.70
150	331	0	0	2.8	6.2	100	8.7	5.0	1.17	0.53	–	–	2.0	0.91	55	0.18	0.18
		0.5	1.1	4.0	8.8	70-80	11.0	7.0	1.35	0.60	0.75	0.23	2.2	1.00	62	0.35	0.32
		0.7	1.5	3.9	8.6	50-60	12.6	8.5	1.60	0.71	1.00	0.43	2.5	1.13	70	0.46	0.36
		0.9	2.0	3.8	8.4	25-30	14.1	9.7	1.81	0.82	1.18	0.54	2.8	1.27	77	0.61	0.45
		1.1	2.4	3.7	8.2	<15	15.6	11.1	2.07	0.94	1.37	0.62	3.1	1.41	86	0.76	0.54
200	441	0	0	3.5	7.7	100	8.5	4.8	1.17	0.53	–	–	2.0	0.91	55	0.18	0.18
		0.5	1.1	5.8	12.8	80-90	9.9	6.0	1.25	0.56	0.60	0.27	2.1	0.95	58	0.24	0.22
		0.7	1.5	5.7	12.6	70-80	10.8	6.8	1.40	0.64	0.78	0.35	2.3	1.04	64	0.32	0.28
		0.9	2.0	4.9	10.8	35-45	12.3	8.2	1.70	0.78	1.10	0.50	2.7	1.22	75	0.47	0.37
		1.1	2.4	4.6	10.1	<15	13.6	9.3	2.07	0.94	1.37	0.62	3.1	1.41	86	0.59	0.43
250	551	0	0	4.1	9.7	100	8.5	4.8	1.17	0.53	–	–	2.0	0.91	55	0.18	0.18
		0.7	1.5	5.8	12.8	55-65	10.7	6.7	1.56	0.71	0.95	0.43	2.5	1.13	70	0.31	0.28
		0.9	2.0	6.2	13.7	45-50	11.1	7.1	1.64	0.74	1.02	0.46	2.6	1.18	72	0.35	0.31
		1.1	2.4	6.0	13.2	20-25	12.1	8.0	1.81	0.82	1.18	0.54	2.8	1.27	77	0.43	0.35
		1.3	2.9	6.0	13.2	<15	12.7	8.5	2.07	0.94	1.37	0.62	3.1	1.41	86	0.50	0.38
300	661	0	0	4.7	10.4	100	8.6	4.8	1.17	0.53	–	–	2.0	0.91	55	0.18	0.18
		0.9	2.0	8.1	17.9	55-65	10.0	6.2	1.56	0.71	0.95	0.43	2.5	1.18	70	0.27	0.23
		1.1	2.4	7.6	16.8	20-25	10.8	6.8	1.81	0.82	1.18	0.54	2.8	1.27	77	0.33	0.29
		1.3	2.9	7.1	15.6	<15	11.7	7.6	1.98	0.90	1.31	0.59	3.0	1.36	83	0.41	0.32
		1.4	3.1	7.3	16.1	<15	11.9	7.8	2.07	0.94	1.37	0.62	3.1	1.41	86	0.42	0.34
350	772	0	0	5.3	11.7	100	8.5	4.8	1.17	0.53	–	–	2.0	0.91	55	0.18	0.18
		0.9	2.0	8.0	17.6	45-55	10.0	6.1	1.64	0.74	1.02	0.46	2.6	1.18	72	0.25	0.22
		1.1	2.4	8.0	17.6	20-25	10.4	6.5	1.81	0.82	1.18	0.54	2.8	1.27	80	0.29	0.25
		1.3	2.9	8.0	17.6	<15	10.8	6.9	1.98	0.90	1.31	0.59	3.0	1.36	83	0.32	0.28
		1.4[e]	3.1	8.2	18.1	<15	10.9	7.0	2.07	0.98	1.37	0.62	3.1	1.41	86	0.34	0.29

Appendix Table 6. Continued.

Weight[b]		Daily gain[c]		Minimum dry matter consumption[c]		Roughage,[c]	Protein		Energy								
							Total,	Digestible,	NE_m,[f]		NE_g,[f]		ME,[f]		TDN,[d,f]	Ca,	P,
kg	lb	kg	lb	kg	lb	%	%	%	Mcal/kg	Mcal/lb	Mcal/kg	Mcal/lb	Mcal/kg	Mcal/lb	%	%	%
400	882	0	0	5.9	13.0	100	8.5	4.8	1.17	0.53	–	–	2.0	0.91	55	0.18	0.18
		1.0	2.2	9.4	20.7	45-55	9.4	5.7	1.64	0.74	1.02	0.46	2.6	1.18	72	0.22	0.21
		1.2	2.6	8.5	18.7	20-25	10.2	6.3	1.81	0.82	1.18	0.54	2.8	1.27	80	0.27	0.25
		1.3	2.9	8.6	19.0	<15	10.4	6.5	2.07	0.98	1.37	0.62	3.1	1.41	86	0.29	0.26
		1.4[e]	3.1	9.0	19.8	<15	10.5	6.6	2.07	0.98	1.37	0.62	3.1	1.41	86	0.29	0.26
450	992	0	0	6.4	14.1	100	8.5	4.8	1.17	0.53	–	–	2.0	0.91	55	0.18	0.18
		1.0	2.2	10.3	22.7	45-55	9.3	5.5	1.64	0.75	1.02	0.46	2.6	1.18	72	0.19	0.19
		1.2	2.6	10.2	22.5	20-25	9.5	5.7	1.81	0.82	1.18	0.54	2.8	1.27	80	0.23	0.22
		1.3	2.9	9.3	20.5	<15	10.4	6.3	2.07	0.98	1.31	0.62	3.1	1.41	86	0.26	0.25
		1.4[e]	3.1	9.8	21.6	<15	10.0	6.1	2.07	0.98	1.37	0.62	3.1	1.41	86	0.26	0.23
500	1,102	0	0	7.0	15.4	100	8.5	4.8	1.17	0.53	–	–	2.0	0.91	55	0.18	0.18
		0.9	2.0	10.5	23.1	45-55	9.1	5.3	1.64	0.74	1.02	0.46	2.6	1.18	72	0.18	0.18
		1.1	2.4	10.4	22.9	20-25	9.2	5.5	1.81	0.82	1.18	0.54	2.8	1.27	80	0.19	0.19
		1.2	2.6	9.6	21.2	<15	10.0	6.0	2.07	0.98	1.31	0.62	3.1	1.41	86	0.22	0.22
		1.3[d]	2,9	10.0	22.0	<15	9.7	6.0	2.07	0.98	1.37	0.62	3.1	1.41	86	0.22	0.22

[a] The concentration of vitamin A in all diets for finishing steers is 2,200 IU/kg of dry diet.
[b] Average weight for a feeding period.
[c] Dry matter consumption, ME and TDN allowances are based on NE requirements and the general types of diet indicated in the roughage column. Most roughages will contain 1.9-2.2 Mcal of ME/kg dry matter and 90-100 percent concentrate diets are expected to contain 3.1-3.3 Mcal of ME/kg.
[d] TDN was calculated by assuming 3.6155 Mcal of ME per kg of TDN.
[e] Most steers of the weight indicated, and not exhibiting compensatory growth, will fail to sustain an energy intake necessary to maintain this rate of gain for an extended period.
[f] Due to conversion and rounding variation, the figures in these columns may not be in exact agreement with a similar energy concentration figure calculated from the data of Table 7.

Appendix Table 7. Nutrient requirements for growing-finishing heiver calves and yearlings (nutrient concentration in diet dry matter).[a] NRC 1976.

Weight[b]		Daily gain		Minimum dry matter consumption[c]		Roughage,[c]	Protein		Energy								
							Total,	Digestible,	NE_m,[f]		NE_g,[f]		ME,[f]		TDN,[d,f]	Ca,	P,
kg	lb	kg	lb	kg	lb	%	%	%	Mcal/kg	Mcal/lb	Mcal/kg	Mcal/lb	Mcal/kg	Mcal/lb	%	%	%
100	220	0	0	2.1	4.6	100	8.7	5.0	1.17	0.53	—	—	2.0	0.91	55	0.18	0.18
		0.5	1.1	3.0	6.6	70-80	12.4	8.3	1.32	0.60	0.70	0.32	2.2	1.00	61	0.47	0.37
		0.7	1.5	2.9	6.4	50-60	14.4	10.0	1.56	0.71	0.95	0.43	2.5	1.13	69	0.66	0.48
		0.9	2.0	3.0	6.6	25-30	15.9	11.3	1.81	0.82	1.18	0.54	2.8	1.27	77	0.80	0.57
		1.1	2.4	3.0	6.6	<15	17.8	13.0	2.07	0.94	1.37	0.62	3.1	1.41	86	0.97	0.62
150	331	0	0	2.8	6.2	100	8.7	5.0	1.17	0.53	—	—	2.0	0.91	55	0.18	0.18
		0.5	1.1	4.1	9.0	70-80	11.0	7.1	1.32	0.60	0.70	0.32	2.2	1.00	61	0.34	0.29
		0.7	1.5	4.0	8.8	50-60	12.4	8.2	1.56	0.71	0.95	0.43	2.5	1.13	69	0.45	0.35
		0.9	2.0	4.0	8.8	25-30	13.5	9.2	1.81	0.82	1.18	0.54	2.8	1.27	77	0.57	0.42
		1.1	2.4	4.0	8.8	<15	15.0	10.5	2.07	0.94	1.37	0.62	3.1	1.41	86	0.70	0.50
200	441	0	0	3.5	7.7	100	8.5	4.9	1.17	0.53	—	—	2.0	0.91	55	0.18	0.18
		0.3	0.7	5.4	11.9	100	9.1	5.4	1.17	0.53	0.50	0.23	2.0	0.91	55	0.18	0.18
		0.5	1.1	6.0	13.2	80-90	9.6	5.8	1.24	0.56	0.60	0.27	2.1	0.95	58	0.23	0.22
		0.7	1.5	6.0	13.2	70-80	10.2	6.5	1.40	0.64	0.87	0.39	2.3	1.04	64	0.30	0.27
		0.9	2.0	5.3	11.7	35-45	11.7	7.5	1.72	0.78	1.10	0.50	2.7	1.22	75	0.41	0.32
		1.1	2.4	5.0	11.0	<15	12.8	8.6	2.07	0.94	1.37	0.62	3.1	1.41	86	0.50	0.38
250	551	0	0	4.1	9.0	100	8.5	4.9	1.17	0.53	—	—	2.0	0.91	55	0.18	0.18
		0.3	0.7	6.4	14.1	100	8.9	5.2	1.17	0.53	0.50	0.23	2.0	0.91	55	0.18	0.18
		0.5	1.1	6.5	14.3	80-90	9.5	5.7	1.24	0.56	0.60	0.27	2.1	0.95	58	0.20	0.20
		0.7	1.5	5.8	12.8	55-65	10.5	6.5	1.64	0.74	1.02	0.46	2.6	1.18	72	0.29	0.26
		0.9	2.0	5.9	13.0	35-45	11.1	7.1	1.81	0.82	1.18	0.54	2.8	1.27	77	0.36	0.29
		1.1	2.4	6.5	14.3	20-25	11.4	7.4	1.89	0.86	1.25	0.57	2.9	1.31	80	0.38	0.31
		1.2	2.6	6.3	13.9	<15	11.9	7.8	2.07	0.94	1.37	0.62	3.1	1.41	86	0.43	0.33
300	661	0	0	4.7	10.4	100	8.6	4.9	1.17	0.53	—	—	2.0	0.91	55	0.18	0.18
		0.3	0.7	7.4	16.3	100	8.5	4.9	1.17	0.53	0.50	0.23	2.0	0.91	55	0.18	0.18
		0.5	1.1	7.4	16.3	80-90	9.2	5.4	1.32	0.60	0.70	0.32	2.2	1.00	61	0.19	0.19
		0.7	1.5	6.6	14.6	55-65	10.1	6.1	1.64	0.74	1.02	0.46	2.6	1.18	72	0.24	0.23
		0.9	2.0	6.8	15.0	35-45	10.4	6.5	1.81	0.82	1.18	0.54	2.8	1.27	77	0.28	0.25
		1.1	2.4	7.5	16.5	20-25	10.4	6.5	1.89	0.86	1.25	0.57	2.9	1.31	80	0.31	0.27
		1.2	2.6	7.2	15.9	<15	10.9	6.9	2.07	0.94	1.37	0.62	3.1	1.41	86	0.33	0.28

Appendix Table 7. Continued.

Weight[b]		Daily gain		Minimum dry matter consumption[c]		Roughage,[c]	Protein		Energy							Ca,	P,
							Total,	Digestible,	NE_m,[f]		NE_g,[f]		ME,[f]		TDN,[d,f]		
kg	lb	kg	lb	kg	lb	%	%	%	Mcal/kg	Mcal/lb	Mcal/kg	Mcal/lb	Mcal/kg	Mcal/lb	%	%	%
350	772	0	0	5.3	11.7	100	8.5	4.8	1.17	0.53	—	—	2.0	0.91	55	0.18	0.18
		0.3	0.7	8.2	18.1	100	8.5	4.8	1.17	0.53	0.50	0.23	2.0	0.91	55	0.18	0.18
		0.5	1.1	8.3	18.3	80-90	8.7	5.1	1.32	0.60	0.70	0.32	2.2	1.00	61	0.18	0.18
		0.7	1.5	7.9	17.4	55-65	9.2	5.4	1.56	0.71	0.95	0.43	2.5	1.13	69	0.19	0.19
		0.9	2.0	8.1	17.9	35-45	9.5	5.7	1.72	0.78	1.10	0.50	2.7	1.22	75	0.21	0.21
		1.1	2.4	8.3	18.3	20-25	9.9	6.0	1.89	0.86	1.25	0.57	2.9	1.31	80	0.24	0.23
		1.2[e]	2.6	8.1	17.9	<15	10.0	6.2	2.07	0.94	1.37	0.62	3.1	1.41	86	0.26	0.25
400	882	0	0	5.9	13.0	100	8.5	4.8	1.17	0.53	—	—	2.0	0.91	55	0.18	0.18
		0.3	0.7	9.1	20.0	100	8.5	4.8	1.17	0.53	0.50	0.23	2.0	0.91	55	0.18	0.18
		0.5	1.1	8.5	18.7	70-80	8.8	5.1	1.40	0.64	0.78	0.35	2.3	1.04	64	0.18	0.18
		0.7	1.5	8.7	19.2	55-65	9.0	5.3	1.56	0.71	0.95	0.43	2.5	1.09	66	0.18	0.18
		0.9	2.0	8.4	18.5	20-25	9.4	5.6	1.81	0.82	1.18	0.54	2.8	1.27	77	0.20	0.20
		1.1[e]	2.4	8.3	18.3	<15	9.7	5.9	2.07	0.94	1.37	0.62	3.1	1.41	86	0.23	0.22
450	992	0	0	6.4	14.1	100	8.5	4.8	1.17	0.53	—	—	2.0	0.91	55	0.18	0.18
		0.2	0.4	8.7	19.2	100	8.5	4.7	1.17	0.53	0.50	0.23	2.0	0.91	55	0.18	0.18
		0.5	1.1	9.3	20.5	70-80	8.6	4.9	1.40	0.64	0.78	0.35	2.3	1.04	64	0.18	0.18
		0.8	1.8	9.1	20.1	35-45	9.0	5.3	1.72	0.78	1.10	0.50	2.7	1.22	75	0.18	0.18
		1.0[e]	2.2	8.5	18.7	<15	9.5	5.6	2.07	0.94	1.37	0.62	3.1	1.41	86	0.22	0.22

[a] The concentration of vitamin A in all diets for finishing heifers is 2,200 IU/kg of dry diet.

[b] Average weight for a feeding period.

[c] Dry matter consumption, ME and TDN allowances are based on NE requirements and the general type of diet indicated in the roughage column. Most roughages will contain 1.9-2.2 Mcal of ME/kg dry matter and 90-100% concentrate diets are expected to have 3.1 to 3.3 Mcal of ME/kg.

[d] TDN was calculated by assuming 3.6155 kcal of ME per g of TDN.

[e] Most heifers of the weight indicated, and not exhibiting compensatory growth, will fail to sustain the energy intake necessary to maintain this rate of gain for an extended period.

[f] Due to conversion and rounding variation, the figures in these columns may not be in exact agreement with a similar energy concentration figure calculated from the data of Table 7A.

Appendix Table 8. Nutrient requirements for beef cattle breeding herd (nutrient concentration in diet dry matter).[a] NRC 1976.

Weight[b]		Daily gain		Minimum dry matter consumption		Rough-age,[c]	Protein		Energy							Ca,	P,
							Total,	Digestible,	NE_m,[d]		NE_g,[d]		ME,[d]		TDN,[d]		
kg	lb	kg	lb	kg	lb	%	%	%	Mcal/kg	Mcal/lb	Mcal/kg	Mcal/lb	Mcal/kg	Mcal/lb	%	%	%
Pregnant yearling heifers—Last third of pregnancy																	
325	716	0.4[e]	0.9	6.6	14.5	100[f]	8.8	5.1	1.09	0.49	0.38	0.17	1.9	0.86	52	0.23	0.23
		0.6	1.3	8.5	18.7	100	8.8	5.1	1.09	0.49	0.38	0.17	1.9	0.86	52	0.21	0.21
		0.8	1.8	9.4	20.7	85-100	9.0	5.3	1.24	0.56	0.60	0.27	2.1	0.95	58	0.23	0.21
350	772	0.4[e]	0.9	6.9	15.2	100	8.8	5.1	1.09	0.49	0.38	0.17	1.9	0.86	52	0.22	0.22
		0.6	1.3	8.9	19.6	100	8.8	5.1	1.09	0.49	0.38	0.17	1.9	0.86	52	0.21	0.21
		0.8	1.8	10.0	22.0	85-100	8.8	5.1	1.24	0.56	0.60	0.27	2.1	0.95	58	0.22	0.21
375	827	0.4[e]	0.9	7.2	15.9	100	8.7	5.0	1.09	0.49	0.38	0.17	1.9	0.86	52	0.21	0.21
		0.6	1.3	9.3	20.5	100	8.7	5.0	1.09	0.49	0.38	0.17	1.9	0.86	52	0.20	0.20
		0.8	1.8	11.0	24.2	85-100	8.7	5.0	1.17	0.53	0.50	0.23	2.0	0.91	55	0.20	0.20
400	882	0.4[e]	0.9	7.5	16.5	100	8.7	5.0	1.09	0.49	0.38	0.17	1.9	0.86	52	0.21	0.21
		0.6	1.3	9.7	21.4	100	8.7	5.0	1.09	0.49	0.38	0.17	1.9	0.86	52	0.20	0.20
		0.8	1.8	11.6	25.6	85-100	8.7	5.0	1.17	0.53	0.50	0.23	2.0	0.91	55	0.19	0.19
425	937	0.4[e]	0.9	7.8	17.2	100	8.8	5.1	1.09	0.49	0.38	0.17	1.9	0.86	52	0.20	0.20
		0.6	1.3	10.1	22.3	100	8.7	5.0	1.09	0.49	0.38	0.17	1.9	0.86	52	0.19	0.19
		0.8	1.8	12.1	26.7	85-100	8.7	5.0	1.17	0.53	0.50	0.23	2.0	0.91	55	0.18	0.18
Dry pregnant mature cows—Middle third of pregnancy																	
350	772	—	—	5.5	12.2	100[f]	5.9	2.8	1.09	0.49	—	—	1.9	0.86	52	0.18	0.18
400	882	—	—	6.1	13.4	100	5.9	2.8	1.09	0.49	—	—	1.9	0.86	52	0.18	0.18
450	992	—	—	6.7	14.8	100	5.9	2.8	1.09	0.49	—	—	1.9	0.86	52	0.18	0.18
500	1,102	—	—	7.2	15.9	100	5.9	2.8	1.09	0.49	—	—	1.9	0.86	52	0.18	0.18
550	1,213	—	—	7.7	17.0	100	5.9	2.8	1.09	0.49	—	—	1.9	0.86	52	0.18	0.18
600	1,323	—	—	8.3	18.3	100	5.9	2.8	1.09	0.49	—	—	1.9	0.86	52	0.18	0.18
650	1,433	—	—	8.8	19.4	100	5.9	2.8	1.09	0.49	—	—	1.9	0.86	52	0.18	0.18
Dry pregnant mature cows—Last third of pregnancy																	
350	772	0.4[e]	0.9	6.9	13.9	100[f]	5.9	2.8	1.09	0.49	—	—	1.9	0.86	52	0.18	0.18
400	882	0.4	0.9	7.5	15.4	100	5.9	2.8	1.09	0.49	—	—	1.9	0.86	52	0.18	0.18
450	992	0.4	0.9	8.1	16.5	100	5.9	2.8	1.09	0.49	—	—	1.9	0.86	52	0.18	0.18
500	1,102	0.4	0.9	8.6	17.9	100	5.9	2.8	1.09	0.49	—	—	1.9	0.86	52	0.18	0.18
550	1,213	0.4	0.9	9.1	19.0	100	5.9	2.8	1.09	0.49	—	—	1.9	0.86	52	0.18	0.18
600	1,323	0.4	0.9	9.7	20.3	100	5.9	2.8	1.09	0.49	—	—	1.9	0.86	52	0.18	0.18
650	1,433	0.4	0.9	10.2	22.4	100	5.9	2.8	1.09	0.49	—	—	1.9	0.86	52	0.18	0.81

Appendix Table 8. Continued.

Weight[b]		Daily gain		Minimum dry matter consumption		Rough-age,[c]	Protein		Energy							Ca,	P,
							Total,	Digestible,	NE_m,[d]		NE_g,[d]		ME,[d]		TND,[d]		
kg	lb	kg	lb	kg	lb	%	%	%	Mcal/kg	Mcal/lb	Mcal/kg	Mcal/lb	Mcal/kg	Mcal/lb	%	%	%
Cows nursing calves—Average milking ability[g]—First 3-4 months postpartum																	
350	772	—	—	8.2	18.1	100[f]	9.2	5.4	1.09	0.49	—	—	1.9	0.86	52	0.29	0.29
400	882	—	—	8.8	19.4	100	9.2	5.4	1.09	0.49	—	—	1.9	0.86	52	0.28	0.28
450	992	—	—	9.3	20.5	100	9.2	5.4	1.09	0.49	—	—	1.9	0.86	52	0.28	0.28
500	1,102	—	—	9.8	21.6	100	9.2	5.4	109	0.49	—	—	1.9	0.86	52	0.28	0.28
550	1,213	—	—	10.5	23.1	100	9.2	5.4	1.09	0.49	—	—	1.9	0.86	52	0.27	0.27
600	1.323	—	—	11.0	24.2	100	9.2	5.4	1.09	0.49	—	—	1.9	0.86	52	0.25	0.25
650	1,433	—	—	11.4	25.1	100	9.2	5.4	1.09	0.49	—	—	1.9	0.86	52	0.25	0.25
Cows nursing calves—Superior milking ability[h]—First 3-4 months postpartum																	
350	772	—	—	10.2	22.4	100[i]	10.9	6.4	1.17	0.53	—	—	2.0	0.91	55	0.44	0.39
400	882	—	—	10.8	23.8	100	10.9	6.4	1.17	0.53	—	—	2.0	0.91	55	0.42	0.38
450	992	—	—	11.3	24.9	100	10.9	6.4	1.17	0.53	—	—	2.0	0.91	55	0.40	0.37
500	1,102	—	—	11.8	26.0	100	10.9	6.4	1.17	0.53	—	—	2.0	0.91	55	0.39	0.36
550	1,213	—	—	12.4	27.3	100	10.9	6.4	1.17	0.53	—	—	2.0	0.91	55	0.37	0.35
600	1,323	—	—	12.9	28.4	100	10.9	6.4	1.17	0.53	—	—	2.0	0.91	55	0.36	0.34
650	1,433	—	—	13.4	29.5	100	10.9	6.4	1.17	0.53	—	—	2.0	0.91	55	0.35	0.33
Bulls, growth and maintenance (moderate activity)																	
300	661	1.00	2.2	8.8	19.4	70-75	10.2	6.3	1.40	0.64	0.78	0.35	2.3	1.04	64	0.31	0.26
400	882	0.90	2.0	11.0	24.2	70-75	9.4	5.6	1.40	0.64	0.78	0.35	2.3	1.04	64	0.21	0.21
500	1,102	0.70	1.5	12.2	26.9	80-85	8.8	5.1	1.32	0.60	0.70	0.32	2.2	1.00	61	0.18	0.18
600	1,323	0.50	1.1	12.0	26.4	80-85	8.8	5.0	1.32	0.60	0.70	0.32	2.2	1.00	61	0.18	0.18
700	1,543	0.30	0.7	12.9	28.4	90-100[i]	8.5	4.8	1.17	0.53	0.50	0.17	2.0	0.91	55	0.18	0.18
800	1,764	0.00	0.0	10.5	23.1	100[i]	8.5	4.8	1.17	0.53	—	—	2.0	0.91	55	0.18	0.18
900	1,984	0.00	0.0	11.4	25.1	100[i]	8.5	4.8	1.17	0.53	—	—	2.0	0.91	55	0.18	0.18
1,000	2,205	0.00	0.0	12.4	27.3	100[i]	8.5	4.8	1.17	0.53	—	—	2.0	0.91	55	0.18	0.18

[a] The concentration of vitamin A in all diets for pregnant heifers and cows is 2,800 IU/kg dry diet; for lactating cows and breeding bulls, 3,900 IU/kg.
[b] Average weight for a feeding period.
[c] Dry matter consumption, ME and TDN requirements are based on the general type of diet indicated in the roughage column.
[d] Due to conversion and rounding variation, the figures in these columns may not be in exact agreement with a similar figure calculated from the data in Table 7B.
[e] Approximately 0.4 ± 0.1 kg of weight gain/day over the last third of pregnancy is accounted for by the products of conception.
[f] Average quality roughage containing about 1.9-2.0 Mcal ME/kg dry matter.
[g] 5.0 ± 0.5 kg of milk/day.
[h] 10 ± 1 kg of milk/day.
[i] Good quality roughage containing 2.0 Mcal ME/kg dry matter.

Appendix Table 9. Net energy requirements of growing and finishing steers (after NRC, 1976).

NEm, Mcal/day	Bodyweight, kg																			
	75	100	125	150	175	200	225	250	275	300	325	350	375	400	425	450	475	500	525	550
	1.96	2.43	2.89	3.30	3.70	4.10	4.47	4.84	5.20	5.55	5.89	6.23	6.56	6.89	7.21	7.52	7.83	8.14	8.45	8.74
Daily gain, kg	NEg, Mcal/day																			
0.1	0.14	0.17	0.20	0.23	0.26	0.28	0.31	0.34	0.36	0.39	0.41	0.43	0.46	0.48	0.50	0.52	0.54	0.56	0.59	0.61
0.2	0.28	0.34	0.40	0.46	0.52	0.57	0.63	0.68	0.73	0.78	0.83	0.88	0.92	0.97	1.01	1.06	1.10	1.14	1.19	1.23
0.3	0.42	0.52	0.61	0.70	0.79	0.87	0.95	1.03	1.11	1.18	1.26	1.33	1.40	1.47	1.54	1.61	1.67	1.74	1.80	1.87
0.4	0.57	0.70	0.82	0.95	1.07	1.18	1.29	1.40	1.50	1.60	1.70	1.80	1.89	1.99	2.08	2.17	2.26	2.34	2.43	2.52
0.5	0.72	0.89	1.05	1.20	1.35	1.49	1.63	1.77	1.90	2.02	2.15	2.27	2.39	2.51	2.63	2.74	2.86	2.97	3.08	3.19
0.6	0.87	1.08	1.27	1.46	1.64	1.81	1.98	2.15	2.30	2.46	2.61	2.76	2.90	3.05	3.19	3.33	3.47	3.60	3.74	3.87
0.7	1.03	1.27	1.50	1.73	1.94	2.14	2.34	2.53	2.72	2.90	3.08	3.26	3.43	3.60	3.77	3.93	4.10	4.25	4.41	4.57
0.8	1.19	1.47	1.74	2.00	2.24	2.47	2.70	2.93	3.14	3.36	3.56	3.77	3.97	4.17	4.36	4.55	4.74	4.92	5.11	5.29
0.9	1.35	1.67	1.98	2.27	2.55	2.81	3.08	3.34	3.58	3.82	4.06	4.29	4.52	4.74	4.96	5.18	5.39	5.60	5.81	6.02
1.0	1.52	1.88	2.23	2.55	2.87	3.16	3.46	3.75	4.02	4.29	4.56	4.82	5.08	5.33	5.57	5.82	6.06	6.29	6.53	6.76
1.1	1.69	2.09	2.48	2.84	3.19	3.52	3.85	4.17	4.48	4.78	5.07	5.37	5.65	5.93	6.20	6.47	6.74	7.00	7.27	7.53
1.2	1.86	2.31	2.73	3.13	3.52	3.88	4.25	4.60	4.94	5.27	5.60	5.92	6.23	6.55	6.84	7.14	7.44	7.73	8.02	8.30
1.3	2.04	2.53	2.99	3.43	3.85	4.26	4.65	5.04	5.41	5.77	6.13	6.49	6.82	7.17	7.50	7.82	8.15	8.46	8.78	9.10
1.4	2.22	2.76	3.26	3.74	4.20	4.63	5.07	5.49	5.89	6.29	6.68	7.06	7.43	7.81	8.16	8.52	8.87	9.22	9.57	9.90
1.5		2.99	3.53	4.05	4.55	5.02	5.49	5.95	6.38	6.81	7.23	7.65	8.05	8.46	8.84	9.23	9.61	9.98	10.36	10.73
1.6			3.81	4.37	4.90	5.42	5.92	6.40	6.88	7.34	7.80	8.24	8.68	9.11	9.53	9.95	10.36	10.77	11.17	11.57
1.7					5.26	5.82	6.36	6.88	7.39	7.89	8.37	8.85	9.32	9.78	10.24	10.69	11.13	11.57	12.00	
1.8							6.80	7.36	7.48	8.44	8.96	9.47	9.98	10.47	10.96	11.44	11.91			
1.0									8.43	9.00	9.56	10.10	10.64	11.17	11.69					
2.0											10.17	10.75	11.32	11.88						

Appendix Table 10. Net energy requirements for growing and finishing heifers (NRC, 1976).

Daily gain, kg	Bodyweight, kg								
	100	150	200	250	300	350	400	450	500
	NE_m required, Mcal/day								
	2.43	3.30	4.10	4.84	5.55	6.24	6.89	7.52	8.14
Heifers									
0.1	0.18	0.25	0.30	0.36	0.41	0.46	0.51	0.56	0.61
0.2	0.37	0.50	0.62	0.74	0.84	0.95	1.05	1.14	1.24
0.3	0.57	0.77	0.95	1.13	1.29	1.45	1.60	1.75	1.90
0.4	0.77	1.05	1.30	1.54	1.76	1.98	2.18	2.39	2.58
0.5	0.99	1.34	1.66	1.96	2.25	2.53	2.79	3.05	3.30
0.6	1.21	1.64	2.03	2.40	2.75	3.09	3.41	3.73	4.03
0.7	1.44	1.95	2.42	2.85	3.27	3.68	4.06	4.44	4.80
0.8	1.67	2.28	2.81	3.33	3.82	4.28	4.73	5.17	5.59
0.9	1.92	2.60	3.23	3.81	4.37	4.91	5.42	5.93	6.41
1.0	2.17	2.94	3.65	4.32	4.95	5.56	6.14	6.71	7.26
1.1	2.43	3.30	4.09	4.84	5.55	6.23	6.88	7.52	8.13
1.2	2.70	3.66	4.55	5.37	6.16	6.92	7.64	8.35	9.03
1.3	2.98	4.04	5.01	5.92	6.79	7.63	8.42	9.21	9.96
1.4	3.26	4.42	5.49	6.49	7.44	8.36	9.23	10.09	10.91
1.5	3.56	4.82	5.98	7.07	8.11	9.11	10.06	11.00	11.90

Appendix Table 11. Estimated total digestible nutrient (TDN) requirements for beef cattle growing at varying rates.[a]

Ave. wt	Expected average daily gain																			
	.2	.4	.6	.8	1.0	1.2	1.4	1.6	1.8	2.0	2.2	2.4	2.6	2.8	3.0	3.2	3.4	3.6	3.8	4.0
300	2.88	3.28	3.68	4.08	4.48	4.88	5.28	5.68	6.08	6.48	6.88	7.28	7.68							
325	3.04	3.46	3.88	4.30	4.73	5.15	5.57	5.99	6.41	6.84	7.26	7.68	8.10							
350	3.20	3.64	4.08	4.52	4.97	5.41	5.85	6.30	6.74	7.18	7.63	8.07	8.50	8.96						
375	3.34	3.81	4.27	4.74	5.20	5.66	6.13	6.59	7.06	7.52	7.98	8.45	8.91	9.37						
400	3.49	3.98	4.46	4.94	5.43	5.91	6.40	6.88	7.37	7.85	8.34	8.82	9.30	9.79	10.27					
425	3.64	4.14	4.64	5.15	5.65	6.16	6.66	7.17	7.67	8.17	8.68	9.18	9.69	10.19	10.70					
450	3.78	4.30	4.83	5.35	5.87	6.40	6.92	7.45	7.97	8.49	9.02	9.54	10.06	10.59	11.11	11.64				
475	3.92	4.46	5.00	5.55	6.09	6.63	7.17	7.72	8.26	8.80	9.35	9.89	10.43	10.98	11.52	12.06				
500	4.05	4.61	5.17	5.74	6.30	6.86	7.42	7.98	8.54	9.11	9.67	10.23	10.79	11.35	11.91	12.48				
525	4.19	4.77	5.35	5.93	6.51	7.09	7.67	8.25	8.83	9.41	9.99	10.58	11.16	11.74	12.32	12.90	15.48			
550	4.33	4.93	5.52	6.13	6.73	7.33	7.92	8.53	9.13	9.73	10.33	10.93	11.53	12.13	12.73	13.33	13.93	14.53		
575	4.45	5.07	5.68	6.30	6.92	7.53	8.15	8.77	9.38	10.00	10.62	11.23	11.85	12.45	13.09	13.70	14.32	14.94		
600	4.58	5.21	5.85	6.48	7.12	7.75	8.38	9.02	9.65	10.29	10.92	11.56	12.19	12.83	13.46	14.10	14.73	15.37	16.00	
625	4.70	5.36	6.01	6.66	7.31	7.96	8.62	9.27	9.92	10.57	11.23	11.88	12.53	13.18	13.83	14.49	15.14	15.79	16.44	
650	4.83	5.50	6.17	6.84	7.51	8.18	8.85	9.52	10.19	10.85	11.52	12.19	12.86	13.53	14.20	14.87	15.54	16.21	16.88	17.55
675	4.95	5.64	6.32	7.01	7.70	8.38	9.07	9.76	10.44	11.13	11.82	12.50	13.19	13.88	14.56	15.25	15.94	16.62	17.31	18.00
700	5.07	5.78	6.48	7.18	7.89	8.59	9.29	10.00	10.70	11.40	12.11	12.81	13.51	14.22	14.92	15.62	16.33	17.03	17.73	18.44
725	5.19	5.91	6.63	7.35	8.07	8.79	9.51	10.23	10.95	11.67	12.39	13.11	13.83	14.56	15.28	16.00	16.72	17.44	18.16	18.88
750	5.31	6.05	6.78	7.52	8.26	8.99	9.73	10.47	11.20	11.94	12.68	13.41	14.15	14.89	15.62	16.36	17.10	17.83	18.57	19.31
775	5.43	6.18	6.93	7.69	8.44	9.19	9.95	10.70	11.45	12.20	12.96	13.71	14.46	15.22	15.97	16.72	17.47	18.23	18.98	19.73
800	5.54	6.31	7.08	7.85	8.62	9.39	10.16	10.93	11.70	12.47	13.23	14.00	14.77	15.54	16.31	17.08	17.85	18.62	19.39	20.15
825	5.66	6.44	7.23	8.01	8.80	9.58	10.37	11.15	11.94	12.72	13.51	14.29	15.08	15.86	16.65	17.43	18.22	19.00	19.79	20.57
850	5.77	6.58	7.38	8.18	8.98	9.78	10.58	11.38	12.18	12.98	13.78	14.58	15.38	16.19	16.97	17.79	18.59	19.39	20.19	20.99
875	5.89	6.70	7.52	8.34	9.15	9.97	10.78	11.60	12.42	13.23	14.05	14.87	15.68	16.50	17.31	18.18	18.95	19.76	20.58	21.40
900	6.00	6.83	7.66	8.49	9.33	10.16	10.99	11.82	12.65	13.49	14.32	15.15	15.98	16.81	17.65	18.48	19.31	20.14	20.97	21.80
925	6.11	6.96	7.80	8.65	9.50	10.35	11.19	12.04	12.89	13.73	14.58	15.43	16.28	17.12	17.97	18.92	19.66	20.51	21.36	22.21
950	6.22	7.08	7.94	8.81	9.67	10.53	11.39	12.25	13.12	13.98	14.84	15.70	16.57	17.43	18.29	19.15	20.01	20.88	21.74	22.60
975	6.33	7.20	8.08	8.96	9.84	10.71	11.59	12.47	13.35	14.22	15.10	15.98	16.86	17.73	18.61	19.50	20.38	21.23	22.12	23.00
1000	6.42	7.31	8.20	9.09	9.98	10.87	11.76	12.65	13.54	14.43	15.32	16.21	17.10	17.99	18.88	19.78	10.67	21.56	22.45	23.34
1025	6.55	7.46	8.36	9.27	10.18	11.09	12.00	12.90	13.81	14.72	15.63	16.54	17.44	18.35	19.26	20.17	21.08	21.98	22.89	23.80
1050	6.65	7.57	8.50	9.42	10.34	11.26	12.19	13.11	14.03	14.95	15.87	16.80	17.71	18.64	19.56	20.49	21.41	22.33	23.25	
1075	6.75	7.69	8.63	9.56	10.50	11.44	12.37	13.31	14.25	15.18	16.12	17.06	18.00	18.93	19.87	20.91	21.74	22.67	23.62	
1100	6.86	7.81	8.76	9.71	10.66	11.61	12.56	13.51	14.46	15.41	16.37	17.32	18.27	19.22	20.17	21.12	22.07	23.02		
1125	6.96	7.93	8.89	9.86	10.82	11.79	12.75	13.72	14.68	15.65	16.61	17.58	18.54	19.51	20.47	21.44	22.40	23.37		
1150	7.06	8.04	9.02	10.00	10.98	11.96	12.94	13.92	14.90	15.88	16.86	17.84	18.81	19.79	20.77	21.75	22.73			
1175	7.16	8.15	9.14	10.14	11.13	12.12	13.12	14.08	15.10	16.09	17.09	18.08	19.07	20.06	21.06	22.05	23.04			
1200	7.27	8.28	9.29	10.30	11.31	12.32	13.33	14.33	15.34	16.35	17.36	18.37	19.38	20.39	21.39	22.40				

[a]From Winchester (1953). TDN = $0.0553\ BW^{2/3}(1 + 0.805\ gain)$, where gain is in pounds.

Appendix Table 12. Equivalent concentrations of energy in feedstuffs.[a]

TDN, %	DE	ME	NE	NE	DE	ME	NE	NE
	Mcal/lb				Mcal/kg			
45	.90	.74	.44	.03	1.98	1.63	.97	.06
46	.92	.76	.45	.05	2.03	1.66	.99	.10
47	.94	.77	.46	.07	2.07	1.70	1.01	.15
48	.96	.79	.47	.09	2.12	1.74	1.03	.19
49	.98	.80	.48	.11	2.16	1.77	1.05	.24
50	1.00	.82	.49	.13	2.20	1.81	1.07	.28
51	1.02	.84	.49	.14	2.25	1.84	1.09	.32
52	1.04	.85	.50	.16	2.29	1.88	1.11	.36
53	1.06	.87	.51	.18	2.34	1.92	1.13	.40
54	1.08	.89	.52	.20	2.38	1.95	1.15	.44
55	1.10	.90	.53	.22	2.43	1.99	1.17	.48
56	1.12	.92	.54	.23	2.47	2.02	1.19	.51
57	1.14	.94	.55	.25	2.51	2.06	1.22	.55
58	1.16	.95	.56	.27	2.56	2.10	1.24	.59
59	1.18	.97	.57	.28	2.60	2.13	1.26	.62
60	1.20	.98	.58	.30	2.65	2.17	1.28	.66
61	1.22	1.00	.59	.31	2.69	2.21	1.31	.69
62	1.24	1.02	.61	.33	2.73	2.24	1.33	.73
63	1.26	1.03	.62	.34	2.78	2.28	1.36	.76
64	1.28	1.05	.63	.36	2.82	2.31	1.38	.79
65	1.30	1.07	.64	.37	2.87	2.35	1.41	.82
66	1.32	1.08	.65	.39	2.91	2.39	1.44	.86
67	1.34	1.10	.66	.40	2.95	2.42	1.46	.89
68	1.36	1.12	.68	.42	3.00	2.46	1.49	.92
69	1.38	1.13	.69	.43	3.04	2.49	1.52	.95
70	1.40	1.15	.70	.44	3.09	2.53	1.54	.97
71	1.42	1.17	.71	.46	3.13	2.57	1.57	1.00
72	1.44	1.18	.73	.47	3.17	2.60	1.60	1.03
73	1.46	1.20	.74	.48	3.22	2.64	1.63	1.06
74	1.48	1.21	.75	.49	3.26	2.68	1.66	1.09
75	1.50	1.23	.77	.51	3.31	2.71	1.69	1.11
76	1.52	1.25	.78	.52	3.35	2.75	1.73	1.14
77	1.54	1.26	.80	.53	3.40	2.78	1.76	1.16
78	1.56	1.28	.81	.54	3.44	2.82	1.79	1.19
79	1.58	1.30	.83	.55	3.48	2.86	1.82	1.21
80	1.60	1.31	.84	.56	3.53	2.89	1.86	1.24
81	1.62	1.33	.86	.57	3.57	2.93	1.89	1.26
82	1.64	1.35	.87	.58	3.62	2.96	1.93	1.29
83	1.66	1.36	.89	.59	3.66	3.00	1.96	1.31
84	1.68	1.38	.91	.60	3.70	3.04	2.00	1.33
85	1.70	1.40	.92	.61	3.75	3.07	2.04	1.35
86	1.72	1.41	.94	.62	3.79	3.11	2.07	1.37
87	1.74	1.43	.96	.63	3.84	3.15	2.11	1.40
88	1.76	1.44	.98	.64	3.88	3.18	2.15	1.42
89	1.78	1.46	1.00	.65	3.92	3.22	2.19	1.44
90	1.80	1.48	1.01	.66	3.97	3.25	2.23	1.46

[a] Data calculated from NRC publications by Speth (22).

Appendix Table 13. Mineral and vitamin requirements of beef cattle (in % of DM or amount/kg of dry diet). NRC (1976).

Nutrient		Growing and Finishing Steers and Heifers	Dry Pregnant Cows	Breeding Bulls and Lactating Cows	Possible Toxic Levels (mg/kg diet)[e]
Vitamin A activity	IU[a,b]	2,200	2,800	3,900	
Vitamin D	IU	275	275	275	
Vitamin E	IU	15–60	—[d]	15–60	
Minerals					
Sodium	%	0.06	0.06	0.06	
Calcium[b]	%	0.18–1.04	0.18	0.18–0.44	
Phosphorus[b]	%	0.18–0.70	0.18	0.18–0.39	
Magnesium	%	0.04–0.10	—[d]	0.18	
Potassium	%	0.6–0.8	—[d]	—[d]	
Sulfur	%	0.1	—[d]	—[d]	
Iodine	μg	[c]	50–100	50–100	100
Iron	mg	10	—[d]	—[d]	400
Copper	mg	4	—[d]	—[d]	115
Cobalt	mg	0.05–0.10	0.05–0.10	0.05–0.10	10–15
Manganese	mg	1.0–10.0	20.0	—[d]	150
Zinc	mg	20–30	—[d]	—[d]	900
Selenium	mg	0.10	0.05–0.10	0.05–0.10	5

[a] May be vitamin A or provitamin A equivalence.
[b] See Table 1 for more detailed data on requirements.
[c] Very small, but unknown.
[d] Unknown. It is suggested that the level for the growing and finishing animal be used.
[e] The level of mineral that is toxic is at best an estimate and is dependent upon such factors as length of intake, availability of the mineral in the feedstuff or compound, and other mineral levels.

Appendix Table 14. Daily nutrient requirements of dairy cattle (NRC, 1978).

Body Weight (kg)	Breed Size, Age (wk)	Daily Gain (g)	Feed DM (kg)	Feed Energy: NE_m (Mcal)	Feed Energy: NE_g (Mcal)	Feed Energy: ME (Mcal)	Feed Energy: DE (Mcal)	Feed Energy: TDN (kg)	Total Crude Protein (g)	Minerals: Ca (g)	Minerals: P (g)	Vitamins: A (1,000 IU)	Vitamins: D (IU)
Growing Dairy Heifer and Bull Calves Fed Only Milk													
25	S-1[a,b]	300	0.45	0.85	0.53	2.14	2.38	0.54	111	6	4	1.1	165
30	S-3	350	0.52	0.95	0.63	2.49	2.77	0.63	128	7	4	1.3	200
42	L-1	400	0.63	1.25	0.70	2.98	3.31	0.75	148	8	5	1.8	280
50	L-3	500	0.76	1.40	0.90	3.61	4.01	0.91	180	9	6	2.1	330
Growing Dairy Heifer and Bull Calves Fed Mixed Diets													
50		300	1.31	1.45	0.57	3.91	4.45	1.01	150	9	6	2.1	330
50	S-10	400	1.40	1.45	0.76	4.36	4.94	1.12	176	9	6	2.1	330
50	L-3	500	1.45	1.45	0.96	4.82	5.42	1.23	198	10	6	2.1	330
50		600	1.45	1.45	1.16	5.01	5.69	1.29	221	11	7	2.1	330
50		700	1.45	1.45	1.35	5.36	5.95	1.35	243	12	7	2.1	330
75		300	2.10	1.96	0.58	5.17	6.05	1.37	232	11	7	3.2	495
75		400	2.10	1.96	0.77	5.56	6.53	1.46	254	12	7	3.2	495
75	S-19	500	2.10	1.96	0.98	5.96	6.94	1.55	275	13	7	3.2	495
75		600	2.10	1.96	1.17	6.36	7.31	1.64	296	14	8	3.2	495
75	L-10	700	2.10	1.96	1.37	6.71	7.67	1.72	318	15	8	3.2	495
75		800	2.10	1.96	1.56	7.08	7.94	1.80	341	16	8	3.2	495
Growing Dairy Heifers													
100		300	2.80	2.43	0.60	6.27	7.45	1.69	317	14	7	4.2	660
100		400	2.80	2.43	0.84	6.78	7.96	1.81	336	15	8	4.2	660
100	S-26	500	2.80	2.43	1.05	7.17	8.35	1.89	360	16	8	4.2	660
100		600	2.80	2.43	1.26	7.64	8.81	2.00	380	17	9	4.2	660
100	L-16	700	2.80	2.43	1.47	8.09	9.26	2.10	402	18	9	4.2	660
100		800	2.80	2.43	1.68	8.47	9.63	2.18	426	19	10	4.2	660
150		300	4.00	3.30	0.72	8.44	10.14	2.30	433	16	10	6.4	990
150		400	4.00	3.30	0.96	8.90	10.59	2.40	455	17	11	6.4	990
150	S-40	500	4.00	3.30	1.20	9.42	11.11	2.52	474	17	11	6.4	990
150		600	4.00	3.30	1.44	9.97	11.65	2.64	491	18	11	6.4	990
150	L-26	700	4.00	3.30	1.68	10.49	12.17	2.76	510	19	12	6.4	990
150		800	4.00	3.30	1.92	11.03	12.70	2.88	528	20	12	6.4	990
200		300	5.00	4.10	0.84	10.44	12.57	2.85	533	18	12	8.5	1320
200		400	5.20	4.10	1.12	11.20	13.41	3.04	571	19	13	8.5	1320
200	S-54	500	5.20	4.10	1.40	11.86	14.06	3.19	586	20	13	8.5	1320
200		600	5.20	4.10	1.68	12.39	14.59	3.31	604	21	14	8.5	1320
200	L-36	700	5.20	4.10	1.96	13.01	15.20	3.45	620	21	14	8.5	1320
200		800	5.20	4.10	2.24	13.52	15.70	3.56	640	22	15	8.5	1320
250		300	5.89	4.84	0.93	12.05	14.55	3.30	610	20	15	10.6	1650
250		400	6.30	4.84	1.24	13.15	15.83	3.59	665	21	15	10.6	1650
250	S-69	500	6.30	4.84	1.55	13.81	16.49	3.74	678	22	16	10.6	1650
250		600	6.30	4.84	1.86	14.57	17.24	3.91	689	22	16	10.6	1650
250	L-47	700	6.30	4.84	2.17	15.20	17.86	4.05	704	23	17	10.6	1650
250		800	6.30	4.84	2.48	15.82	18.47	4.19	719	23	17	10.6	1650
300		300	6.67	5.55	1.02	13.64	16.47	3.74	671	20	15	12.7	1980
300		400	7.00	5.55	1.36	14.80	17.77	4.03	713	22	17	12.7	1980
300	S-83	500	7.20	5.55	1.70	15.69	18.74	4.25	746	23	17	12.7	1980
300		600	7.20	5.55	2.04	16.49	19.53	4.43	755	23	17	12.7	1980
300	L-57	700	7.20	5.55	2.38	17.07	20.11	4.56	771	24	18	12.7	1980
300		800	7.20	5.55	2.72	17.83	20.86	4.73	782	24	18	12.7	1980
350		300	7.23	6.24	1.08	15.27	18.34	4.16	701	22	16	14.8	2310
350	S-97	400	7.42	6.24	1.44	15.99	19.14	4.34	738	23	17	14.8	2310
350		500	8.00	6.24	1.80	17.42	20.81	4.72	804	25	18	14.8	2310
350		600	8.00	6.24	2.16	18.21	21.60	4.90	812	25	19	14.8	2310
350	L-67	700	8.00	6.24	2.52	18.88	22.26	5.05	826	25	19	14.8	2310
350		800	8.00	6.24	2.88	19.56	22.93	5.20	841	26	19	14.8	2310

Appendix Table 14. Continued.

Body Weight (kg)	Breed Size, Age (wk)	Daily Gain (g)	Feed DM (kg)	Feed Energy NE_m (Mcal)	NE_g (Mcal)	ME (Mcal)	DE (Mcal)	TDN (kg)	Total Crude Protein (g)	Minerals Ca (g)	P (g)	Vitamins A (1,000 IU)	D (IU)
400	S-115	200	7.26	6.89	0.76	14.85	17.94	4.07	692	21	16	17.0	2640
400		400	8.50	6.89	1.52	17.76	21.38	4.85	833	24	19	17.0	2640
400		600	8.60	6.89	2.28	19.61	23.24	5.27	856	25	20	17.0	2640
400	L-77	700	8.60	6.89	2.66	20.40	24.03	5.45	864	25	20	17.0	2640
400		800	8.60	6.89	3.04	21.11	24.73	5.61	876	26	21	17.0	2640
450		200	7.87	7.52	0.80	16.09	19.44	4.41	749	23	18	19.1	2970
450		400	9.00	7.52	1.60	19.02	22.84	5.18	867	26	20	19.1	2970
450		600	9.10	7.52	2.40	21.03	24.87	5.64	883	27	21	19.1	2970
450	L-87	700	9.10	7.52	2.80	21.82	25.66	5.82	892	27	21	19.1	2970
450		800	9.10	7.52	3.20	22.67	26.50	6.01	898	28	21	19.1	2970
500		200	8.46	8.14	0.84	17.30	20.90	4.74	788	24	19	21.2	3300
500		400	9.50	8.14	1.68	20.26	24.29	5.51	900	27	21	21.2	3300
500	L-98	600	9.50	8.14	2.52	22.26	26.28	5.96	903	27	21	21.2	3300
500		800	9.50	8.14	3.36	24.00	28.00	6.35	916	28	21	21.2	3300
550		200	9.05	8.75	0.88	18.50	22.34	5.07	835	25	19	23.3	3630
550	L-109	400	9.80	8.75	1.76	21.33	25.48	5.78	913	27	20	23.3	3630
550		600	9.80	8.75	2.64	23.38	27.51	6.24	914	27	20	23.3	3630
550		800	9.80	8.75	3.52	25.08	29.19	6.62	928	28	21	23.3	3630
600	L-127	200	9.58	9.33	0.90	19.60	23.68	5.37	879	25	18	25.4	3960
600		300	9.72	9.33	1.35	20.78	24.87	5.64	895	25	18	25.4	3960
600		400	10.00	9.33	1.80	22.22	26.45	6.00	918	26	19	25.4	3960
600		500	10.00	9.33	2.25	23.34	27.56	6.25	916	26	19	25.4	3960
Growing Dairy Bulls													
100		500	2.80	2.43	1.05	7.17	8.35	1.89	361	16	8	4.2	660
100	S-26	600	2.80	2.43	1.26	7.64	8.81	2.00	381	17	9	4.2	660
100		700	2.80	2.43	1.47	8.09	9.26	2.10	403	18	9	4.2	660
100	L-15	800	2.80	2.43	1.68	8.47	9.63	2.18	427	19	10	4.2	660
100		900	2.80	2.43	1.89	8.84	10.00	2.27	450	20	10	4.2	660
150		500	4.00	3.30	1.15	9.42	11.11	2.52	476	18	11	6.4	990
150		600	4.00	3.30	1.38	9.91	11.59	2.63	497	19	11	6.4	990
150	S-38	700	4.00	3.30	1.61	10.30	11.98	2.72	520	20	12	6.4	990
150		800	4.00	3.30	1.84	10.84	12.52	2.84	539	21	12	6.4	990
150		900	4.00	3.30	2.07	11.47	13.14	2.98	555	21	13	6.4	990
150	L-24	1000	4.00	3.30	2.30	11.73	13.40	3.04	583	22	13	6.4	990
200		500	5.20	4.10	1.25	11.46	13.66	3.10	602	20	13	8.5	1320
200		600	5.20	4.10	1.50	12.01	14.21	3.22	622	21	14	8.5	1320
200	S-48	700	5.20	4.10	1.75	12.59	14.78	3.35	640	21	14	8.5	1320
200		800	5.20	4.10	2.00	13.07	15.26	3.46	660	22	15	8.5	1320
200		900	5.20	4.10	2.25	13.52	15.70	3.56	688	23	16	8.5	1320
200	L-31	1000	5.20	4.10	2.50	14.05	16.23	3.68	702	23	16	8.5	1320
250		500	6.30	4.84	1.35	13.44	16.11	3.65	684	22	16	10.6	1650
250		600	6.30	4.84	1.62	14.00	16.67	3.78	702	23	16	10.6	1650
250	S-58	700	6.30	4.84	1.89	14.62	17.28	3.92	718	23	17	10.6	1650
250		800	6.30	4.84	2.16	15.20	17.86	4.05	736	24	17	10.6	1650
250		900	6.30	4.84	2.43	15.78	18.43	4.18	753	25	17	10.6	1650
250	L-38	1000	6.30	4.84	2.70	16.13	18.78	4.26	778	25	18	10.6	1650
300		500	7.33	5.69	1.48	15.45	18.56	4.21	777	24	18	12.7	1980
300		600	7.40	5.69	1.77	16.13	19.27	4.37	800	25	19	12.7	1980
300	S-68	700	7.40	5.69	2.07	16.89	20.02	4.54	811	26	19	12.7	1980
300		800	7.40	5.69	2.36	17.51	20.63	4.68	827	26	19	12.7	1980
300		900	7.40	5.69	2.66	18.09	21.21	4.81	845	27	19	12.7	1980
300	L-45	1000	7.40	5.69	2.95	18.67	21.78	4.94	862	27	20	12.7	1980
350		500	8.10	6.54	1.60	17.27	20.71	4.70	828	25	19	14.8	2310
350		600	8.30	6.54	1.92	18.13	21.65	4.91	863	26	20	14.8	2310
350	S-79	700	8.30	6.54	2.24	18.93	22.44	5.09	873	27	20	14.8	2310
350		800	8.30	6.54	2.56	19.60	23.10	5.24	887	27	20	14.8	2310

Appendix Table 14. Continued.

Body Weight (kg)	Breed Size, Age (wk)	Daily Gain (g)	Feed DM (kg)	Feed Energy NE$_m$ (Mcal)	NE$_g$ (Mcal)	ME (Mcal)	DE (Mcal)	TDN (kg)	Total Crude Protein (g)	Minerals Ca (g)	P (g)	Vitamins A (1,000 IU)	D (IU)
350		900	8.30	6.54	2.88	20.22	23.72	5.38	903	28	20	14.8	2310
350	L-52	1000	8.30	6.54	3.20	20.89	24.38	5.53	917	28	21	14.8	2310
400		500	9.00	7.41	1.75	19.24	23.06	5.23	891	27	21	17.0	2640
400		600	9.00	7.41	2.10	20.00	23.81	5.40	902	27	21	17.0	2640
400	S-89	700	9.00	7.41	2.45	20.84	24.64	5.59	910	28	22	17.0	2640
400		800	9.00	7.41	2.80	21.60	25.40	5.76	921	28	22	17.0	2640
400		900	9.00	7.41	3.15	22.36	26.15	5.93	932	28	22	17.0	2640
400	L-60	1000	9.00	7.41	3.50	22.93	26.72	6.06	947	29	23	17.0	2640
450		200	8.41	8.27	0.76	17.20	20.77	4.71	762	23	19	19.1	2970
450		400	9.33	8.27	1.52	19.90	23.86	5.41	868	27	21	19.1	2970
450	S-90	600	9.50	8.27	2.28	21.83	25.84	5.86	898	28	22	19.1	2970
450		800	9.50	8.27	3.04	23.52	27.52	6.24	914	28	22	19.1	2970
450	L-67	1000	9.50	8.27	3.80	25.08	29.07	6.59	934	29	23	19.1	2970
500		100	8.26	8.95	0.40	16.90	20.41	4.63	740	22	18	21.2	3300
500		300	9.30	8.95	1.20	19.83	23.77	5.39	855	25	21	21.2	3300
500	S-111	500	10.00	8.95	2.00	22.22	26.45	6.00	941	28	23	21.2	3300
500		700	10.00	8.95	2.80	23.60	27.82	6.31	967	29	23	21.2	3300
500	L-74	900	10.00	8.95	3.60	25.56	29.76	6.75	973	29	23	21.2	3300
550		100	8.86	9.62	0.42	18.11	21.87	4.96	789	24	18	23.3	3630
550	S-125	300	10.20	9.62	1.25	21.29	25.62	5.81	935	28	22	23.3	3630
550		500	10.50	9.62	2.08	23.56	28.00	6.35	967	29	22	23.3	3630
550	L-82	700	10.50	9.62	2.91	25.51	29.94	6.79	976	29	22	23.3	3630
550		900	10.50	9.62	3.74	27.16	31.57	7.16	994	30	23	23.3	3630
600	S-149	100	9.42	10.27	0.43	19.27	23.28	5.28	833	25	19	25.4	3960
600		300	10.52	10.27	1.29	22.44	26.90	6.10	947	28	22	25.4	3960
600		500	10.80	10.27	2.15	24.72	29.28	6.64	980	29	23	25.4	3960
600	L-92	700	10.80	10.27	3.01	26.58	31.13	7.06	988	29	23	25.4	3960
650		100	9.96	10.90	0.44	20.37	24.60	5.58	875	26	20	27.6	4290
650		300	10.69	10.90	1.32	23.29	27.82	6.31	947	28	22	27.6	4290
650	L-102	500	11.10	10.90	2.20	25.75	30.44	6.90	992	29	23	27.6	4290
650		700	11.10	10.90	3.08	27.78	32.45	7.36	995	29	23	27.6	4290
700		100	10.51	11.53	0.45	21.50	25.97	5.89	918	27	21	29.7	4620
700		300	11.40	11.53	1.35	24.61	29.45	6.68	1005	29	23	29.7	4620
700	L-117	500	11.40	11.53	2.25	26.94	31.75	7.20	998	30	23	29.7	4620
700		700	11.40	11.53	3.15	28.99	33.78	7.66	1001	30	23	29.7	4620
750		100	11.02	12.14	0.45	22.53	27.21	6.17	960	28	22	31.8	4950
750	L-131	300	11.70	12.14	1.35	25.48	30.44	6.90	1024	30	23	31.8	4950
750		500	11.70	12.14	2.25	27.86	32.80	7.44	1014	30	23	31.8	4950
800		100	11.52	12.74	0.45	23.55	28.44	6.45	999	29	23	33.9	5280
800		300	12.00	12.74	1.35	26.35	31.44	7.13	1040	30	23	33.9	5280
800		500	12.00	12.74	2.25	28.62	33.68	7.64	1035	30	23	33.9	5280
Growing Veal Calves Fed Only Milk													
35	—	500	0.67	0.98	0.90	3.17	3.52	0.80	173	7	4	1.5	231
45	L-1.0	800	1.06	1.36	1.52	5.04	5.60	1.27	259	8	5	1.9	297
55	L-2.8	900	1.20	1.55	1.73	5.74	6.38	1.45	292	11	7	2.3	363
65	L-4.4	1000	1.36	1.76	1.95	6.48	7.20	1.63	324	13	8	2.8	429
75	L-5.8	1050	1.48	1.96	2.10	7.05	7.83	1.78	334	15	9	3.2	495
100	L-9.2	1100	1.69	2.43	2.31	8.05	8.94	2.03	357	17	10	4.2	660
125	L-12.4	1200	1.95	2.88	2.64	9.30	10.33	2.34	392	19	11	5.3	825
150	L-15.4	1300	2.22	3.30	2.99	10.58	11.75	2.66	428	20	12	6.4	990
Maintenance of Mature Breeding Bulls													
500	—	—	7.80	9.36	—	15.95	19.27	4.37	673	20	15	21	—
600	—	—	8.95	10.74	—	18.29	22.09	5.01	766	23	17	25	—
700	—	—	10.04	12.05	—	20.52	24.78	5.62	852	26	19	30	—
800	—	—	11.10	13.32	—	22.52	27.20	6.17	942	29	21	34	—
900	—	—	12.13	14.55	—	24.79	29.94	6.79	1017	31	23	38	—
1000	—	—	13.12	15.75	—	26.83	32.41	7.35	1093	34	25	42	—

Appendix Table 14. Continued.

Body Weight (kg)	Breed Size, Age (wk)	Daily Gain (g)	Feed DM (kg)	Feed Energy NE$_m$ (Mcal)	NE$_g$ (Mcal)	ME (Mcal)	DE (Mcal)	TDN (kg)	Total Crude Protein (g)	Minerals Ca (g)	P (g)	Vitamins A (1,000 IU)	D (IU)
1100	—	—	14.10	16.91	—	28.84	34.83	7.90	1169	36	27	47	—
1200	—	—	15.05	18.05	—	30.77	37.17	8.43	1244	39	29	51	—
1300	—	—	15.98	19.17	—	32.67	39.46	8.95	1316	41	31	55	—
1400	—	—	16.88	20.27	—	34.49	41.66	9.45	1386	43	33	59	—

[a] Breed size: S for small breeds (e.g., Jersey); L is for large breeds (e.g., Holstein).
[b] Age in weeks indicates probable age of S or L animals when they reach the weight indicated.

Appendix Table 15. Daily nutrient requirements of lactating and pregnant cows (NRC, 1978).

Body Weight (kg)	Feed Energy NE$_l$ (Mcal)	ME (Mcal)	DE (Mcal)	TDN (kg)	Total Crude Protein (g)	Calcium (g)	Phosphorus (g)	Vitamin A (1,000 IU)
Maintenance of Mature Lactating Cows[a]								
350	6.47	10.76	12.54	2.85	341	14	11	27
400	7.16	11.90	13.86	3.15	373	15	13	30
450	7.82	12.99	15.14	3.44	403	17	14	34
500	8.46	14.06	16.39	3.72	432	18	15	38
550	9.09	15.11	17.60	4.00	461	20	16	42
600	9.70	16.12	18.79	4.27	489	21	17	46
650	10.30	17.12	19.95	4.53	515	22	18	50
700	10.89	18.10	21.09	4.79	542	24	19	53
750	11.47	19.06	22.21	5.04	567	25	20	57
800	12.03	20.01	23.32	5.29	592	27	21	61
Maintenance Plus Last 2 Months of Gestation of Mature Dry Cows								
350	8.42	14.00	16.26	3.71	642	23	16	27
400	9.30	15.47	17.98	4.10	702	26	18	30
450	10.16	16.90	19.64	4.47	763	29	20	34
500	11.00	18.29	21.25	4.84	821	31	22	38
550	11.81	19.65	22.83	5.20	877	34	24	42
600	12.61	20.97	24.37	5.55	931	37	26	46
650	13.39	22.27	25.87	5.90	984	39	28	50
700	14.15	23.54	27.35	6.23	1035	42	30	53
750	14.90	24.79	28.81	6.56	1086	45	32	57
800	15.64	26.02	30.24	6.89	1136	47	34	61
Milk Production—Nutrients Per Kg Milk of Different Fat Percentages								
(% Fat)								
2.5	0.59	0.99	1.15	0.260	72	2.40	1.65	
3.0	0.64	1.07	1.24	0.282	77	2.50	1.70	
3.5	0.69	1.16	1.34	0.304	82	2.60	1.75	
4.0	0.74	1.24	1.44	0.326	87	2.70	1.80	
4.5	0.78	1.31	1.52	0.344	92	2.80	1.85	
5.0	0.83	1.39	1.61	0.365	98	2.90	1.90	
5.5	0.88	1.48	1.71	0.387	103	3.00	2.00	
6.0	0.93	1.56	1.81	0.410	108	3.10	2.05	
Body Weight Change During Lactation—Nutrients Per Kg Weight Change								
Weight loss	−4.92	−8.25	−9.55	−2.17	−320			
Weight gain	5.12	8.55	9.96	2.26	500			

[a] To allow for growth of young lactating cows, increase the maintenance allowances for all nutrients except vitamin A by 20 percent during the first lactation and 10 percent during the second lactation.

Appendix Table 16. Recommended nutrient content of rations for dairy cattle (NRC, 1978).

	Lactating Cow Rations					Nonlactating Cattle Rations					
	Cow Wt (kg)	Daily Milk Yields (kg)									Maximum Concentrations (All Classes)
Nutrients (Concentration in the Feed Dry Matter)	≤400 500 600 ≥700	< 8 <11 <14 <18	8–13 11–17 14–21 18–26	13–18 17–23 21–29 26–35	>18 >23 >29 >35	Dry Pregnant Cows	Mature Bulls	Growing Heifers and Bulls	Calf Starter Concentrate Mix	Calf Milk Replacer	
Ration No.		I	II	III	IV	V	VI	VII	VIII	IX	Max.
Crude Protein, %		13.0	14.0	15.0	16.0	11.0	8.5	12.0	16.0	22.0	—
Energy											
NE_l, Mcal/kg		1.42	1.52	1.62	1.72	1.35	—	—	—	—	—
NE_m, Mcal/kg		—	—	—	—	—	1.20	1.26	1.90	2.40	—
NE_g, Mcal/kg		—	—	—	—	—	—	0.60	1.20	1.55	—
ME, Mcal/kg		2.36	2.53	2.71	2.89	2.23	2.04	2.23	3.12	3.78	—
DE, Mcal/kg		2.78	2.95	3.13	3.31	2.65	2.47	2.65	3.53	4.19	—
TDN, %		63	67	71	75	60	56	60	80	95	—
Crude Fiber, %		17	17	17	17[a]	17	15	15	—	—	—
Acid Detergent Fiber, %		21	21	21	21	21	19	19	—	—	—
Ether Extract, %		2	2	2	2	2	2	2	2	10	—
Minerals[b]											
Calcium, %		0.43	0.48	0.54	0.60	0.37	0.24	0.40	0.60	0.70	—
Phosphorus, %		0.31	0.34	0.38	0.40	0.26	0.18	0.26	0.42	0.50	—
Magnesium, %[c]		0.20	0.20	0.20	0.20	0.16	0.16	0.16	0.07	0.07	—
Potassium, %		0.80	0.80	0.80	0.80	0.80	0.80	0.80	0.80	0.80	—
Sodium, %		0.18	0.18	0.18	0.18	0.10	0.10	0.10	0.10	0.10	—
Sodium chloride, %[d]		0.46	0.46	0.46	0.46	0.25	0.25	0.25	0.25	0.25	5
Sulfur, %[d]		0.20	0.20	0.20	0.20	0.17	0.11	0.16	0.21	0.29	0.35
Iron, ppm[d,e]		50	50	50	50	50	50	50	100	100	1,000
Cobalt, ppm		0.10	0.10	0.10	0.10	0.10	0.10	0.10	0.10	0.10	10
Copper, ppm[d,f]		10	10	10	10	10	10	10	10	10	80
Manganese, ppm[d]		40	40	40	40	40	40	40	40	40	1,000
Zinc, ppm[d,g]		40	40	40	40	40	40	40	40	40	500
Iodine, ppm[h]		0.50	0.50	0.50	0.50	0.50	0.25	0.25	0.25	0.25	50
Molybdenum, ppm[i,j]		—	—	—	—	—	—	—	—	—	6
Selenium, ppm		0.10	0.10	0.10	0.10	0.10	0.10	0.10	0.10	0.10	5
Fluorine, ppm[j]		—	—	—	—	—	—	—	—	—	30
Vitamins[k]											
Vit A, IU/kg		3,200	3,200	3,200	3,200	3,200	3,200	2,200	2,200	3,800	—
Vit D, IU/kg		300	300	300	300	300	300	300	300	600	—
Vit E, ppm		—	—	—	—	—	—	—	—	300	—

[a] It is difficult to formulate high-energy rations with a minimum of 17 percent crude fiber. However, fat percentage depression may occur when rations with less than 17 percent crude fiber or 21 percent ADF are fed to lactating cows.

[b] The mineral values presented in this table are intended as guidelines for use of professionals in ration formulation. Because of many factors affecting such values, they are not intended and should not be used as a legal or regulatory base.

[c] Under conditions conducive to grass tetany (see text), should be increased to 0.25 or higher.

[d] The maximum safe levels for many of the mineral elements are not well defined; estimates given here, especially for sulfur, sodium chloride, iron, copper, zinc, and manganese, are based on very limited data; safe levels may be substantially affected by specific feeding conditions.

[e] The maximum safe level of supplemental iron in some forms is materially lower than 1,000 ppm. As little as 400 ppm added iron as ferrous sulfate has reduced weight gains (Standish *et al.*, 1969).

[f] High copper may increase the susceptibility of milk to oxidized flavor (see text).

[g] Maximum safe level of zinc for mature dairy cattle is 1,000 ppm.

[h] If diet contains as much as 25 percent strongly goitrogenic feed on dry basis, iodine provided should be increased two times or more.

[i] If diet contains sufficient copper, dairy cattle tolerate substantially more than 6 ppm molybdenum (see text).

[j] Maximum safe level of fluorine for growing heifers and bulls is lower than for other dairy cattle. Somewhat higher levels are tolerated when the fluorine is from less-available sources such as phosphates (see text). Minimum requirement for molybdenum and fluorine not yet established.

[k] The following minimum quantities of B-complex vitamins are suggested per unit of milk replacer: niacin, 2.6 ppm; pantothenic acid, 13 ppm; riboflavin, 6.5 ppm; pyridoxine, 6.5 ppm; thiamine, 6.5 ppm; folic acid, 0.5 ppm; biotin, 0.1 ppm; vitamin B_{12}, 0.07 ppm; choline, 0.26 percent. It appears that adequate amounts of these vitamins are furnished when calves have functional rumens (usually at 6 weeks of age) by a combination of rumen synthesis and natural feedstuffs.

Appendix Table 17. Daily nutrient requirements of sheep (100% dry matter basis).*

Body weight		Gain or loss		Dry matter[a] per animal			Nutrients per animal											
							Energy											
kg	lb	g	lb	kg	lb	% of live wt	TDN, kg	DE,[b] Mcal	ME, Mcal	Total protein, g	DP,[c] g	Grams DP per Mcal DE	Ca, g	P, g	Carotene, mg	Vitamin A, IU	Vitamin D, IU
EWES[d]																	
Maintenance																	
50	110	10	.02	1.0	2.2	2.0	0.55	2.42	1.98	89	48	20	3.0	2.8	1.9	1275	278
60	132	10	.02	1.1	2.4	1.8	0.61	2.68	2.20	98	53	20	3.1	2.9	2.2	1530	333
70	154	10	.02	1.2	2.6	1.7	0.66	2.90	2.38	107	58	20	3.2	3.0	2.6	1785	388
80	176	10	.02	1.3	2.9	1.6	0.72	3.17	2.60	116	63	20	3.3	3.1	3.0	2040	444
Nonlactating and first 15 weeks of gestation																	
50	110	30	.07	1.1	2.4	2.2	0.60	2.64	2.16	99	54	20	3.0	2.8	1.9	1275	278
60	132	30	.07	1.3	2.9	2.1	0.72	3.17	2.60	117	64	20	3.1	2.9	2.2	1530	333
70	154	30	.07	1.4	3.1	2.0	0.77	3.39	2.78	126	69	20	3.2	3.0	2.6	1785	388
80	176	30	.07	1.5	3.3	1.9	0.82	3.61	2.96	135	74	20	3.3	3.1	3.0	2040	444
Last 6 weeks of gestation or last 8 weeks of lactation suckling singles[e]																	
50	110	175(+45)	.39	1.7	3.7	3.3	0.99	4.36	3.58	1.58	88	20	4.1	3.9	6.2	4250	278
60	132	180(+45)	.40	1.9	4.2	3.2	1.10	4.84	3.97	177	99	20	4.4	4.1	7.5	5100	333
70	154	185(+45)	.41	2.1	4.6	3.0	1.22	5.37	4.40	195	109	20	4.5	4.3	8.8	5950	388
80	176	190(+45)	.42	2.2	4.8	2.8	1.28	5.63	4.62	205	114	20	4.8	4.5	10.0	6800	444
Last 8 weeks of lactation suckling singles or last 8 weeks of lactation suckling twins[f]																	
50	110	-25(+80)	-.06	2.1	4.6	4.2	1.36	5.98	4.90	218	130	22	10.9	7.8	6.2	4250	278
60	132	-25(+80)	-.06	2.3	5.1	3.9	1.50	6.60	5.41	239	143	22	11.5	8.2	7.5	5100	333
70	154	-25(+80)	-.06	2.5	5.5	3.6	1.63	7.17	5.88	260	155	22	12.0	8.6	8.8	5950	388
80	176	-25(+80)	-.06	2.6	5.7	3.2	1.69	7.44	6.10	270	161	22	12.6	9.0	10.0	6800	444
First 8 weeks of lactation suckling twins																	
50	110	-60	-.13	2.4	5.3	4.8	1.56	6.86	5.63	276	173	25	12.5	8.9	6.2	4250	278
60	132	-60	-.13	2.6	5.7	4.3	1.69	7.44	6.10	299	187	25	13.0	9.4	7.5	5100	333
70	154	-60	-.13	2.8	6.2	4.0	1.82	8.01	6.57	322	202	25	13.4	9.5	8.8	5950	388
80	176	-60	-.13	3.0	6.6	3.7	1.95	8.58	7.04	345	216	25	14.4	10.2	10.0	6800	444
Replacement lambs and yearlings[g]																	
30	66	180	.40	1.3	2.9	4.3	0.81	3.56	2.92	130	75	21	5.9	3.3	1.9	1275	166
40	88	120	.26	1.4	3.1	3.5	0.82	3.61	2.96	133	74	20	6.1	3.4	2.5	1700	222
50	110	80	.18	1.5	3.3	3.0	0.83	3.65	2.99	133	73	20	6.3	3.5	3.1	2125	278
60	132	40	.09	1.5	3.3	2.5	0.82	3.61	2.96	133	72	20	6.5	3.6	3.8	2550	333

Appendix Table 17. Continued.

RAMS																	
Replacement lambs and yearlings[g]																	
40	88	250	.55	1.8	4.0	4.5	1.17	5.15	4.22	184	108	21	6.3	3.5	2.5	1700	222
60	132	200	.44	2.3	5.1	3.8	1.38	6.07	4.98	219	122	20	7.2	4.0	3.8	2550	333
80	176	150	.33	2.8	6.2	3.5	1.54	6.78	5.56	249	134	20	7.9	4.4	5.0	3400	444
100	220	100	.22	2.8	6.2	2.8	1.54	6.78	5.56	249	134	20	8.3	4.6	6.2	4250	555
120	265	50	.11	2.6	5.7	2.2	1.43	6.29	5.16	231	125	20	8.5	4.7	7.5	5100	666
LAMBS																	
Finishing[h]																	
30	66	220	.44	1.3	2.9	4.3	0.83	3.65	2.99	143	87	24	4.8	3.0	1.1	765	166
35	77	220	.48	1.4	3.1	4.0	0.94	4.14	3.39	154	94	23	4.8	3.0	1.3	892	194
40	88	250	.55	1.6	3.5	4.0	1.12	4.93	4.04	176	107	22	5.0	3.1	1.5	1020	222
45	99	250	.55	1.7	3.7	3.8	1.19	5.24	4.30	187	114	22	5.0	3.1	1.7	1148	250
50	110	220	.48	1.8	4.0	3.6	1.26	5.54	4.54	198	121	22	5.0	3.1	1.9	1275	278
55	121	200	.44	1.9	4.2	3.5	1.33	5.85	4.80	209	127	22	5.0	3.1	2.1	1402	305
Early-weaned[i]																	
10	22	250	.55	0.6	1.3	6.0	0.44	1.94	1.59	96	69	36	2.4	1.6	1.2	850	67
20	44	275	.60	1.0	2.2	5.0	0.73	3.21	2.63	160	115	36	3.6	2.4	2.5	1700	133
30	66	300	.66	1.4	3.1	4.7	1.02	4.49	3.68	196	133	30	5.0	3.3	3.8	2550	200

[*] From NRC (1975).

[a] To convert dry matter to an as-fed basis, divide dry matter by percentage of dry matter.

[b] 1 kg TDN = 4.4 Mcal DE (digestible energy). DE may be converted to ME (metabolizable energy) by multiplying by 82%.

[c] DP = digestible protein.

[d] Values are for ewes in moderate condition, not excessively fat or thin. Fat ewes should be fed at the next lower weight, thin ewes at the next higher weight. Once maintenance weight is established, such weight would follow through all production phases.

[e] Values in parentheses are for ewes suckling singles last 8 weeks of lactation.

[f] Values in parentheses are for ewes suckling twins last 8 weeks of lactation.

[g] Requirements for replacement lambs (ewe and ram) start when the lambs are weaned.

[h] Maximum gains expected. If lambs are held for later market, they should be fed as replacement ewe lambs are fed. Lambs capable of gaining faster than indicated should be fed at a higher level. Lambs finish at the maximum rate if they are self-fed.

[i] A 40-kg early-weaned lamb should be fed the same as a finishing lamb of the same weight.

Appendix Table 18. Daily nutrient requirements of sheep.

Body Weight		Daily Gain or Loss		Daily Dry Matter[a] Per Animal		Daily Dry Matter[a] % Live Wt	Energy TDN	Energy DE[b]	Energy ME	Total Protein	DP[c]	Ca	P	Carotene	Vitamin A	Vitamin D
(kg)	(lb)	(g)	(lb)	(kg)	(lb)		(%)	(Mcal/kg)	(Mcal/kg)	(%)	(%)	(%)	(%)	(mg/kg)	(IU/kg)	(IU/kg)
EWES[d]																
Maintenance																
50	110	10	.02	1.0	2.2	2.0	55	2.4	2.0	8.9	4.8	.30	.28	1.9	1275	278
60	132	10	.02	1.1	2.4	1.8	55	2.4	2.0	8.9	4.8	.28	.26	2.0	1391	303
70	154	10	.02	1.2	2.6	1.7	55	2.4	2.0	8.9	4.8	.27	.25	2.2	1488	323
80	176	10	.02	1.3	2.9	1.6	55	2.4	2.0	8.9	4.8	.25	.24	2.3	1569	342
Nonlactating and first 15 weeks of gestation																
50	110	30	.07	1.1	2.4	2.2	55	2.4	2.0	9.0	4.9	.27	.25	1.7	1159	253
60	132	30	.07	1.3	2.9	2.1	55	2.4	2.0	9.0	4.9	.24	.22	1.7	1177	256
70	154	30	.07	1.4	3.1	2.0	55	2.4	2.0	9.0	4.9	.23	.21	1.9	1275	277
80	176	30	.07	1.5	3.3	1.9	55	2.4	2.0	9.0	4.9	.22	.21	2.0	1360	296
Last 6 weeks of gestation or last 8 weeks of lactation suckling singles[e]																
50	110	175(+45)	.39	1.7	3.7	3.3	58	2.6	2.1	9.3	5.2	.24	.23	3.6	2500	164
60	132	180(+45)	.40	1.9	4.2	3.2	58	2.6	2.1	9.3	5.2	.23	.22	3.9	2684	175
70	154	185(+45)	.41	2.1	4.6	3.0	58	2.6	2.1	9.3	5.2	.21	.20	4.2	2833	185
80	176	190(+45)	.42	2.2	4.8	2.8	58	2.6	2.1	9.3	5.2	.21	.20	4.5	3091	202
First 8 weeks of lactation suckling singles or last 8 weeks of lactation suckling twins[f]																
50	110	−25(+80)	−.06	2.1	4.6	4.2	65	2.9	2.4	10.4	6.2	.52	.37	3.0	2024	132
60	132	−25(+80)	−.06	2.3	5.1	3.9	65	2.9	2.4	10.4	6.2	.50	.36	3.3	2217	145
70	154	−25(+80)	−.06	2.5	5.5	3.6	65	2.9	2.4	10.4	6.2	.48	.34	3.5	2380	155
80	176	−25(+80)	−.06	2.6	5.7	3.2	65	2.9	2.4	10.4	6.2	.48	.34	3.8	2615	171
First 8 weeks of lactation suckling twins																
50	110	−60	−.13	2.4	5.3	4.8	65	2.9	2.4	11.5	7.2	.52	.37	2.6	1771	116
60	132	−60	−.13	2.6	5.7	4.3	65	2.9	2.4	11.5	7.2	.50	.36	2.9	1962	128
70	154	−60	−.13	2.8	6.2	4.0	65	2.9	2.4	11.5	7.2	.48	.34	3.1	2125	139
80	176	−60	−.13	3.0	6.6	3.7	65	2.9	2.4	11.5	7.2	.48	.34	3.3	2267	148

Appendix Table 18. Continued.

Replacement lambs and yearlings[g]															
30	66	180	.40	1.3	2.9	4.3	62	2.7	2.2	10.0	5.8	.45	.25	1.5	981	128
40	88	120	.26	1.4	3.1	3.5	60	2.6	2.1	9.5	5.3	.44	.24	1.8	1214	159
50	110	80	.18	1.5	3.3	3.0	55	2.4	2.0	8.9	4.8	.42	.23	2.1	1417	185
60	132	40	.09	1.5	3.3	2.5	55	2.4	2.0	8.9	4.8	.43	.24	2.5	1700	222
RAMS																
Replacement lambs and yearlings[g]																
40	88	250	.55	1.8	4.0	4.5	65	2.9	2.4	10.2	6.0	.35	.19	1.4	944	123
60	132	200	.44	2.3	5.1	3.8	60	2.6	2.1	9.5	5.3	.31	.17	1.7	1109	145
80	176	150	.33	2.8	6.2	3.5	55	2.4	2.0	8.9	4.8	.28	.16	1.8	1214	159
100	220	100	.22	2.8	6.2	2.8	55	2.4	2.0	8.9	4.8	.30	.17	2.2	1518	198
120	265	50	.11	2.6	5.7	2.2	55	2.4	2.0	8.9	4.8	.33	.18	2.9	1962	256
LAMBS																
Finishing[h]																
30	66	200	.44	1.3	2.9	4.3	64	2.8	2.3	11.0	6.7	.37	.23	0.8	588	128
35	77	220	.48	1.4	3.1	4.0	67	3.0	2.4	11.0	6.7	.34	.21	0.9	637	139
40	88	250	.55	1.6	3.5	4.0	70	3.1	2.5	11.0	6.7	.31	.19	0.9	638	139
45	99	250	.55	1.7	3.7	3.8	70	3.1	2.5	11.0	6.7	.29	.18	1.0	675	147
50	110	220	.48	1.8	4.0	3.6	70	3.1	2.5	11.0	6.7	.28	.17	1.1	708	154
55	121	200	.44	1.9	4.2	3.5	70	3.1	2.5	11.0	6.7	.26	.16	1.1	738	161
Early-weaned[i]																
10	22	250	.55	0.6	1.3	6.0	73	3.2	2.6	16.0	11.5	.40	.27	2.0	1417	112
20	44	275	.60	1.0	2.2	5.0	73	3.2	2.6	16.0	11.5	.36	.24	2.5	1700	133
30	66	300	.66	1.4	3.1	4.7	73	3.2	2.6	14.0	9.5	.36	.24	2.7	1821	143

[a] To convert dry matter to an as-fed basis, divide dry matter by percentage of dry matter.
[b] 1 kg TDN = 4.4 Mcal DE (digestible energy). DE may be converted to ME (metabolizable energy) by multiplying by 82%. Because of rounding errors, calculations between Table 1 and Table 2 may not give the same values.
[c] DP = digestible protein.
[d] Values are for ewes in moderate condition, not excessively fat or thin. Fat ewes should be fed at the next lower weight, thin ewes at the next higher weight. Once maintenance weight is established, such weight would follow through all production phases.
[e] Values in parentheses are for ewes suckling singles last 8 weeks of lactation.
[f] Values in parentheses are for ewes suckling twins last 8 weeks of lactation.
[g] Requirements for replacement lambs (ewe and ram) start when the lambs are weaned.
[h] Maximum gains expected. If lambs are held for later market, they should be fed as replacement ewe lambs are fed. Lambs capable of gaining faster than indicated should be fed at a higher level. Lambs finish at the maximum rate if they are self-fed.
[i] A 40-kg early-weaned lamb should be fed the same as a finishing lamb of the same weight.

Appendix Table 19. Mineral requirements of sheep.[a]

Nutrient	Requirement	Toxic level
Macrominerals, % of DM		
Sodium	0.04-0.10	
Chlorine	—	
Calcium	0.21-0.52	
Phosphorus	0.16-0.37	
Magnesium	0.04-0.08	
Potassium	0.50	
Sulfur	0.14-0.26	
Trace minerals, ppm		
Iodine	0.10-0.80[b]	8+
Iron	30-50	—
Copper	5	8-25
Molybdenum	>0.5	5-20
Cobalt	0.1	100-200
Manganese	20-40	—
Zinc	35-50	1,000
Selenium	0.1	>2
Fluorine	—	60-200

[a]Values are estimates based on experimental data. From NRC (1975).
[b]High level for pregnancy and lactation in diets not containing goitrogens; should be increased if diets contain goitrogens.

Appendix Table 20. Maintenance requirements of goats (per animal/day).

Body weight,		Degree of exercise											
		Confined				Semi-intensive				Intensive			
kg	lb	TDN,[a] kg	SE,[b] kg	DE, Mcal	ME,[c] Mcal	TDN, kg	SE, kg	DE, Mcal	ME, Mcal	TDN, kg	SE, kg	DE, Mcal	ME, Mcal
10	22	0.15	0.13	0.65	0.53	0.18	0.15	0.78	0.64	0.21	0.19	0.94	0.77
15	33	0.20	0.17	0.87	0.72	0.24	0.21	1.05	0.86	0.28	0.26	1.26	1.03
20	44	0.25	0.22	1.08	0.89	0.30	0.26	1.30	1.07	0.35	0.30	1.52	1.25
25	55	0.29	0.25	1.28	1.05	0.35	0.30	1.54	1.26	0.41	0.36	1.79	1.47
30	66	0.34	0.30	1.48	1.21	0.40	0.35	1.77	1.45	0.47	0.41	2.06	1.69
35	77	0.38	0.33	1.50	1.23	0.43	0.37	1.80	1.49	0.50	0.46	2.32	1.90
40	88	0.42	0.37	1.66	1.36	0.45	0.39	1.99	1.63	0.53	0.57	2.56	2.10
45	99	0.45	0.39	2.00	1.64	0.54	0.47	2.39	1.97	0.64	0.56	2.80	2.30
50	110	0.49	0.43	2.16	1.77	0.59	0.51	2.59	2.12	0.69	0.60	3.02	2.48
55	121	0.53	0.46	2.32	1.90	0.63	0.55	2.78	2.28	0.74	0.64	3.24	2.66
60	132	0.56	0.49	2.48	2.03	0.68	0.59	2.98	2.44	0.79	0.69	3.46	2.84

[a]Total digestible nutrients; 1 kg TDN = 4.4 Mcal DE; [b]starch equivalent; [c]ME = DE x .82.

Appendix Table 21. Digestible protein requirements for maintenance of goats (g/animal/day).

Weight (kg)	Dig. protein (DP)
10	11.2
15	15.2
20	18.8
25	22.2
30	25.5
35	28.6
40	31.7
45	34.6
50	37.4
55	40.2
60	42.9

Appendix Table 22. Nutrient requirements for lactation of goats (per kg of milk).

Fat content of milk, %	SE, g	TDN, g	ME, Mcal	DP, g	Ca, g	P, g
3.5	262	301	1.09	47	0.8	0.7
4.0	280	322	1.16	52	0.9	0.7
4.5	296	340	1.23	59	0.9	0.7
5.0	314	361	1.30	66	1.0	0.7
5.5	331	381	1.37	73	1.1	0.7

[a]From Devendra (1970)

Appendix Table 23. Energy and protein requirements for growth of goats (per animal/day).

Body weight		Daily live weight gain,	ME			Dry matter intake[a]		DP[b] requirements,
kg	lb	g	Maint., Kcal	Gain, Kcal	Total, Kcal	g	% BW	g
10	22	50	530	422	952	414	4.1	23.2
		100		844	1374	597	6.0	33.5
		150		1266	1796	781	7.8	43.8
20	44	50	892	422	1314	571	2.9	32.0
		100		844	1736	755	3.8	42.3
		150		1266	2158	938	4.7	52.6
30	66	50	1208	422	1630	709	2.4	39.8
		100		844	2052	893	3.0	50.1
		150		1266	2474	1076	3.6	60.3
40	88	50	1500	422	1922	836	2.1	46.9
		100		844	2344	1019	2.6	57.2
		200		1266	2766	1203	3.0	67.5
50	110	50	1773	422	2195	954	1.9	53.5
		150		844	2617	1138	2.3	68.3
		200		1266	3039	1321	2.6	74.1
60	132	50	2034	422	2456	1068	1.8	59.9
		100		844	2878	1251	2.1	70.2
		200		1266	3300	1435	2.4	80.5

[a] Dry matter, based on an ME concentration of 2.3 Mcal/kg DM.
[b] Digestible protein, calculated according to 1 Mcal DE = 20 g DP.

Appendix Table 24. Energy and protein requirements for goats during pregnancy (per animal/day).

Body weight		DM intake		Total	DP[b]
kg	lb	g	% of BW	Mcal	g
10	22	484	4.8	1.21	29.6
15	33	656	4.4	1.64	40.0
20	44	816	4.1	2.04	49.8
25	55	960	3.8	2.40	58.6
30	66	1104	3.7	2.76	67.4
35	77	1240	3.5	3.10	75.6
40	88	1368	3.4	3.42	83.4
45	99	1496	3.3	3.74	91.2
50	110	1620	3.2	4.05	98.8
55	121	1736	3.2	4.34	105.9
60	132	1856	3.1	4.64	113.2

[a] Dry matter, based on an ME concentration of 2.5 Mcal/kg DM.
[b] Digestible protein, calculated according to 1 Mcal DE = 20 g DP.

GENERAL TIPS ON FEEDING, FACILITIES AND MANAGEMENT OF SOME RUMINANT ANIMALS*

BEEF CATTLE

Beef Calves in the Dry Lot

Feed and Water

Have top of feed bunk 24" high for calves (animals up to 600 lb) and 30" high for older cattle.

Provide feed bunks 8" deep for calves; 8-10" deep for older cattle.

Allow the following amount of feeder space/head for grains and other concentrates:

Hand feeding	Self feeding
Calves–18-24"	Calves–6-8"
Older cattle–24-30"	Older cattle–8-12"

Provide following roughage rack space for free-choice feeding with liberal grain or other concentrates:

Dry roughage	6-9"/head
Silage	12-18"/head

When little or no grain or other concentrate is fed, rack space should be 12-18"/head.

Make feed bunks 24-30" wide when feeding from 1 side and 36" wide when feeding from both sides.

Provide plenty of clean, fresh water at all times. Allow 1' of linear open water tank space for each 10 cattle; or one automatic watering bowl for each 25 cattle.

Water temperature may range from a low of 35-40°F in winter to a high of 80°F in summer. Warming water to a minimum of 50°F is desirable.

Feedlot Facilities

Provide the following lot space:

A. Paved lots–50 to 100 square feet/head.

B. Dirt lots–150 to 200 square feet/head. More space may be desired under some soil and weather conditions.

Provide a minimum paved area of 10' around waterers, feed bunks, and roughage racks.

Allow slope of ½ to ¼"/foot in paved lots and ½" or more in dirt lots, depending on soil and climate.

Provide housing as economically as possible. Open sheds or wind breaks are usually adequate. In open sheds, allow 20-30 square feet of space/head for calves; allow 40-50 square feet/head for older cattle.

Provide artificial shade in hot climates when cattle have no access to natural shade. For calves, allow 15-25 square feet/head and 25-35 square feet for older cattle. Build shade 8-10' high.

Provide bedding, except in mild climates, to keep cattle dry, comfortable and healthy. Straw, corn cobs, sawdust, shavings, and similar materials may be used. Use that readily available at lowest cost. Mounds covered with cobs or other bedding are desirable in open lots.

Build fences at least 60" in height, using lowest-cost, readily available material.

Provide facilities for restraining and handling individual animals to minimize production losses and prevent injury to personnel.

Replacement and Breeding Beef Cattle

Feed, Water and Lot Facilities

Allow following amount of feeder space (linear inches)/head for grains and/or other concentrates:

	Hand Feeding	Self Feeding
Calves	18-24	6-8
Yearlings	24-30	8-12
Cows	24-30	8-12
Mature bulls	24-30	12-16

Provide this rack or bunk space (linear inches) for free choice feeding of roughage:

	Dry Roughage	Silage
Calves	3-6	8-12
Yearlings	6-8	8-12
Cows	8-12	12-18
Bulls	8-12	12-18

Allow the following rack or bunk space (linear inches) for hand feeding of roughage:

	Dry Roughage	Silage
Calves	14-16	14-16
Yearlings	16-18	16-18
Cows	18-24	18-24
Bulls	24-30	24-30

*Excerpted from Management Guide, published in 1980 by the American Feed Manufacturers Association, Inc.

Provide feed bunks with the following specifications (inches):

	Calves	Yearlings	Cows and Bulls
Height of top of bunk	20-24	24-30	24-30
Depth of bunk	6-8	8-10	10-12
Width of trough with one side used	24	24-30	24-30
Width of trough with 2 sides used	36	36-40	40-48

If high levels of salt are used to control feed intake, place feeders as near water as practical.

Provide plenty of clean, fresh water at all times. Allow 1 linear foot of open water tank space or equivalent for each 10 head; or one automatic watering bowl for each 25 animals.

Water temperature may range from a low of 35-40°F in winter to a high of 80°F in summer. Warming water to a minimum of 50°F is desirable.

If shelter (shade in summer) is needed, provide for 20-50 square feet/head, depending on size of animals.

Provide a minimum paved area of 10 feet around waterers, feed bunks and roughage racks.

When breeding cattle are confined to a lot, provide the following space:

A. Paved lots—50 to 100 square feet/head
B. Dirt lots—150 to 200 square feet/head; more may be desirable under some soil and weather conditions.

DAIRY CATTLE

Dairy Calves

Housing

Keep dairy calves in individual pens or tie stalls until at least 1 week after milk or milk replacer is discontinued. If necessary to pen in groups, tie calves, particularly during milk feeding and for 30 min. thereafter to reduce sucking of navels, ears, etc.

Calves may be raised in groups beginning 1 week after milk or milk replacer is discontinued.

10 calves should be a maximum for any one group, provided floor and feeding space is adequate and calves are liberally fed. Smaller groups (6-8 calves) are preferable.

Maximum age difference between calves in a group should not exceed 2 months so calves are of similar size. Make sure all calves are actually eating their fair share.

Space Needs

Minimum pen size for individual calves is 24 square feet. Minimum tie stall size is 2' wide by 4' long.

Pens for older calves in groups, with no outside run, should provide 30 square feet/calf.

Feeding Equipment

Calf ration feed box for individual pens or tie stalls should be 8" x 10" x 6" deep. Make boxes removable for easy cleaning.

When calves are raised in groups, feed boxes should be 10" wide x 6" deep, allowing up to 2 linear feet/calf. Two troughs/pen are preferred. Provide stanchions or feeding gates to insure desired consumption of feed by each calf.

Top of feed boxes should be 20" from the floor.

Locate feed boxes where convenient for calves and away from waterers.

Water Devices

Automatic drinking cups are preferred for calves in individual pens. Where pails are used, keep clean and well filled with fresh water.

Automatic drinking cups are also preferred for calves penned in groups. Where watering tanks are used for calves in outside runs, keep water fresh and clean.

Tops of drinking cups or tanks should not be more than 20" from the floor or ground, depending on calf breed.

Provide 1 automatic drinking cup for each 5 to 8 calves/pen.

Keep watering equipment away from feed boxes.

Temperature and Ventilation

Calves can thrive in a wide range of temperatures if pens are dry and draft-free. Dryness is particularly important.

Solid partitions between pens or tie stalls aid in dreducing drafts. Pen fronts may be slatted to permit feeding.

General Management

Leave newborn calf with cow as long as necessary. It is very important that the calf receives colostrum the first 2-3 days, either by nursing or drinking.

Teach young calves to drink from a pail or nipple feeding device. Keep pails or other feeding equipment scrupulously clean at all times to avoid digestive disturbances.

Turn calves out to pasture as soon as practicable after 4 mo. of age. Calves require grain and hay with pasture to make normal growth. If necessary, limit pasture so other roughage will be consumed.

Feed high-quality hay to calves from the start.

Ensilage may be fed after calves are 6 mo. old. Amount fed will depend on moisture content. Feed some hay; do not replace completely with ensilage.

A safe age for weaning to a total ration of calf starter, grain mixture and roughage will depend on calf vigor. About 6 weeks is safe age for this change, though healthy, vigorous calves may be switched up to 2 weeks earlier with good results.

Feed calves a minimum of twice a day.

Dehorn after 10 days of age when horn buttons can be felt. Use an electric dehorner or caustic potash.

Growing Dairy Cattle

Feed and Water

Top of feed bunks should be 24" above ground for small cattle (up to 600 lb) and 30" above ground for large cattle.

Depth of feed bunks should be 8" for grains and other concentrates and 8-12" for silage or cut green forage.

Bunks permitting feeding from one side only should be 18" wide for small cattle and 24" for large cattle. If feeding is permitted from both sides, bunk should be 36-48" wide.

Allow the following amount of feeder space (linear inches)/head:

	Grain	Roughage
Small cattle	12	18
Large cattle	18	24

Provide plenty of clean, fresh water at all times. Cattle will consume 10-15 gallons of water/day depending on size, feed and climate. Allow 1 linear foot of open water tank space for each 10 animals; or automatic watering bowl for each 25 animals.

Water temperature may range from a low of 35 to 40°F in winter to a high of 80°F in summer. Warming water to a minimum of 50°F is desirable.

Growing Facilities

Depending upon climate and type of soil, provide the following lot space:

A. Paved lots—50-70 square feet/head
B. Paved and dirt lots—75-100 square feet/head
C. Dirt lots—100-150 square feet/head

Provide a scarified paved area, 15-20' wide if possible, around waterers, feed bunks, roughage racks and entrances to sheds.

Allow a slope of ¼ to ½"/foot in paved lots and ½" or more in dirt lots, depending on soil and weather conditions.

Except in mild climates, provide open sheds for shelter. Allow 20-30 square feet/head for small cattle; 30-40 square feet/head for large cattle.

Provide economical bedding in sheds as needed to keep cattle dry and comfortable.

In hot weather provide artificial shade, if necessary. Build shade 8-10' high. Allow 20 square feet/head for small cattle; 30 square feet for large cattle.

Dairy Cows

Feed and Water

Provide 24-30 linear inches/head of manger space for roughage feeding.

Make feed bunks and roughage racks 24-30" wide when cattle feed from one side and 36-48" when feeding from both sides.

Provide plenty of clean, fresh water at all times. Allow 1 linear foot of open tank per 8-10 head; or 1 automatic bowl/15 head.

Water temperature may range from a low of 35-40°F in winter to a high of 80°F in summer. Warming water to a minimum of 50°F is desirable.

Lot Facilities

Build hard-surface lots with a slope of ¼ to ½"/foot. In dirt lots provide ½" or more/foot depending upon soil and weather conditions.

Provide a scarified paved area, 15-20' wide if possible, around waterers, feed bunks, roughage racks and shed entrances.

With a loose housing system use open sheds. Free-stall housing saves bedding, keeps cows cleaner and saves labor. Recommended stall sizes are: 7' x 4' for small breeds and 7½' x 4' for large breeds.

Provide economical bedding as needed to keep cows dry and comfortable. Bedding materials should not contain any pesticide residue which may appear in milk if some bedding is eaten.

Supply up to 40 gallons of water/day/head.

Depending on climate and soil, provide minimum lot space/head as follows:

A. Paved lots–100 square feet
B. Paved and dirt lots–150 square feet
C. Dirt lots–200 square feet

General Management

Desirable gain in body weight/cow during dry period depends upon animal condition at beginning of dry period. A guide to desirable gains is as follows: 800-1000 pound cows–75 to 150 lb gain; 1000 to 1200 lb cows–100 to 200 lb gain; 1200-1400 lb cows–125 to 250 lb gain. Feed for optimum condition at calving.

Unless otherwise provided, feed salt and other needed minerals free-choice.

Worm dairy cows when necessary, preferably during dry period. Worm during lactation only under veterinarian's direction.

Temperatures in stall barns may range from 40°F upwards but should be kept as cool as practical. Temperatures in milking parlors can be higher for operator comfort.

Provide dairy cows with a dry period of 6 weeks if possible. Stop milking to dry off cows producing less than 20-25 lb/day. To dry up cows milking over 25 lb daily, cut milking schedule to once a day until they drop to 20 lb. Then stop milking. Use the same procedure with problem cows.

Breed healthy cows during first heat period occurring 60 days after calving.

SHEEP

Growing and Fattening Lambs

Feed and Water

Provide a minimum of 8 linear inches of feeder space/head for hand-feeding roughage or concentrate.

When grain or concentrates are self-fed, allow 3" of linear feeder space/head.

When feeding a complete ration, allow 4" of linear feeder space.

When roughage racks are needed, provide 4" of linear rack space/head.

Provide at least 1 gallon of water/head/day.

Provide watering space as follows: open tank or trough–1 linear foot/20 head; automatic bowl–1/30 head.

General Management

Provide economical housing for lambs. Open sheds or wind breaks are usually adequate. In open sheds, provide at least 6 square feet/head.

Provide minimum lot space as follows: all dirt–20 square feet/head; paved and dirt–16 square feet/head; paved–12 square feet/head.

Build hard-surfaced lots with a slope of ¼ to ½"/foot. In dirt lots, provide ½" or more slope/foot depending on soil and weather conditions.

Provide a minimum paved area of at least 5' around waterers, feed bunks, roughage racks and shed entrances.

Provide artificial shade, unless sheep have access to natural shade. Allow 6-8 square feet/head. Build shade 8-10' in height.

Breeding Sheep

Feeding Management

Provide a minimum of 1 linear foot of feeder space/head for hand-feeding roughage or concentrate.

When grain or concentrates are self-fed, allow 6 linear inches feeder space/ewe. If salt or other feed intake inhibitors are used, less space may be needed. Provide a minimum of 6 linear inches of roughage rack space/ewe for self-feeding hay or silage.

Provide at least 3 gallons of water/ewe/day.

Provide watering space as follows: 1 linear foot of open tank/each 20 head or 1 automatic bowl/30 head.

Provide a minimum of 16 square feet of space for lamb and ewe during lambing.

INDEX